QMC 336270 7
a30213 003362707b

D1420572

Coal Conversion Technology

ENERGY SCIENCE AND TECHNOLOGY

A Series of Graduate Textbooks, Monographs, Research Papers

Series Editors: C. Y. WEN and E. STANLEY LEE

Consulting Editor: Richard Bellman

No. 1 RICHARD C. BAILIE
Energy Conversion Engineering, 1978

No. 2 C. Y. WEN and E. STANLEY LEE (Eds.)
Coal Conversion Technology, 1979

Other Numbers in preparation

Contributors

S. DUTTA — pp. 57–170
Department of Chemical Engineering
West Virginia University
Morgantown, West Virginia*

SABRI ERGUN — pp. 1–56
Lawrence Berkeley Laboratory
University of California
Berkeley, California

ROBERT H. ESSENHIGH — pp. 171–312
Department of Mechanical Engineering
The Ohio State University
Columbus, Ohio

E. STANLEY LEE — pp. 428–545
Department of Industrial Engineering
Kansas State University
Manhattan, Kansas

LESTER G. MASSEY — pp. 313–427
Consolidated Natural Gas Service Company, Inc.
Cleveland, Ohio

C. Y. WEN — pp. 57–170
Department of Chemical Engineering
West Virginia University
Morgantown, West Virginia

*Current affiliation is Department of Chemical Engineering and Chemistry
New Jersey Institute of Technology
Newark, New Jersey

Coal Conversion Technology

Edited by

C. Y. Wen

West Virginia Unviersity
Morgantown, West Virginia

E. Stanley Lee

Kansas State University
Manhattan, Kansas

1979

Addison-Wesley Publishing Company
Advanced Book Program
Reading, Massachusetts

London · Amsterdam · Don Mills, Ontario · Sydney · Tokyo

Library of Congress Cataloging in Publication Data

Main entry under title:

Coal conversion technology.

(Energy science and technology ; no. 2)
Includes bibliographical references and index.
1. Coal–Combustion. 2. Coal gasification. 3. Coal liquefaction. 4. Pyrolysis. I. Wen, Chin-Yung, 1928– II. Lee, Eugene Stanley, 1930– III. Series.
TP325.C5149 662'.625 79-12975
ISBN 0-201-08300-0

Manufactured in the United States of America

Contents

ISBN 0-201-08300-0

ISBN 0-201-08300-0

ISBN 0-201-08300-0

ISBN 0-201-08300-0

ISBN 0-201-08300-0

Series Editors' Foreword

Accelerated research and development for alternate energy resources to augment the dwindling supply of petroleum and natural gas is currently under way throughout the world. The energy problem has suddenly created new demands on the effective dissemination of large volumes of information as well as on science and engineering education. A serious gap already exists between research publications and the more comprehensive and systematic treatment of subjects found in a monographic text. One of the purposes of this Series is to fill this gap. A second purpose is to provide a vehicle for use as a teaching text. In fact, the magnitude of the energy problem and its solution through the creative application of scientific and engineering principles requires that essentially all scientists and engineers possess some background in the basic concepts of energy production, distribution, utilization, and related economics and social implications. However, most of the currently available books and monographs are neither adequate nor suitable for use as textbooks. Much of the material published in the last several years addresses itself either to the general public or to a group of specialists involved in research and development.

Needed are books that are logically developed to emphasize underlying principles and the applications of these principles to energy, complete with sample problems suitable for use by instructors and students in classes. To these ends, this Series attempts to deal with subjects such as:

Energy Conversion Engineering
Coal Conversion Technology
Geothermal Modeling
Trace Elements and Pollution in Energy Production
Water Resources in Energy Production
Solar Energy: Materials and Models.

It is the sincere wish of the Editors that this Series will be able to play a key role in filling the existing gaps in information and in education.

Coal, representing our most abundant fossil fuel, now plays an ever-increasing, important role in fulfilling the world's energy needs. To fully utilize coal as an energy source and to understand the rapid development in coal conversion technology, a more systematic approach is required for interpreting and organizing the

ISBN 0-201-08300-0

voluminous amount of published literature. This, the second volume of the series, addresses this need. Special emphasis is placed on processing details, the fundamental theories, and the principles of coal conversion. This book will serve both as a convenient guide to the researchers working in the field and as a useful introduction to coal conversion technology for scientists and engineers either entering into or redirecting their activities to this area of research.

C. Y. WEN
E. STANLEY LEE

ISBN 0-201-08300-0

Preface

Coal is clearly our most abundant fossil resource and must play a key role in supplying energy and chemicals for the remainder of this and all of the next century. There is an urgent need for a strong, balanced energy program involving the direct combustion of coal and the conversion of coal to gas and liquid fuels. In addition, new and improved technology in these areas must assure environmental protection.

Coal combustion and conversion processes are presently in various stages of development; they range from commercially available or currently being tested at pilot-plant scale to those that are formulated conceptually and still to be tested. The combustion and gasification of coal as it is carried out today in furnaces and reactors is still more a practical art relying on past experience rather than science.

In recent years, considerable advances have been made in the understanding of the physical and chemical properties of coal and coal conversion reactions. In large part, these advances are due to intensive research and development effort; aimed at improving coal combustion and conversion technology to meet societal, economic, and environmental requirements. However, relatively few attempts have been made to systematically organize the subject matter and to critically evaluate the vast amount of available information based mainly on technological and engineering points of view. A major difficulty in accomplishing this task is due to the complexity and inhomogeneity of the structure of coal and its behavior under different environments, which precludes any attempt to draw generalizations. Additional difficulty stems from the fact that coal conversion technology is in a state of flux as numerous alternative processes are still being developed. It is therefore impossible and indeed undesirable to cover all aspects of coal conversion technology in one volume.

In this monograph an attempt is made to critically review the available literature and to present materials that each contributor considers as the most reliable information in his subject area.

The material presented is intended primarily for researchers, practicing engineers, and scientists who are engaged in research, development, and design of coal combustion and conversion systems. The monograph can also serve as a text or reference book for graduates and advanced undergraduates on subjects related to coal combustion, gasification, and liquefaction.

The initial discussion concentrates on the physical and chemical characterization of coals needed for the study of coal combustion and conversion systems.

ISBN 0-201-08300-0

Since a complete and detailed exposition of this subject is not practical because of space limitation, such basic essentials as classification of coal by rank and by grade, proximate and ultimate analyses, chemical and mechanical properties, are emphasized.

The next chapter deals with the reaction rates of coal pyrolysis and gasification. Discussion is based on the fundamentals of a single particle behavior and emphasizes how complex physical processes, such as diffusion and heating rates, affect the heterogeneous coal pyrolysis and gasification reactions. This treatment paves the way for the discussions of the subsequent chapters dealing with processing of coal in combustors, gasifiers, and other reactors.

Coal combustion systems, the essentials of the equipment used, the understanding of the behavior of experimental coal combustion systems, and our current knowledge of the underlying fundamentals are described in the succeeding chapter.

Coal gasification for the production of high and low Btu fuels is then treated in detail. This chapter emphasizes the principles and unit operations of coal gasification. The thermodynamics and the reaction kinetics of gasification are discussed initially. The principles developed are then extended to the discussion of gasifier operations. Rather than attempting to cover the numerous individual gasifiers developed or being developed, the coal gasifiers are classified, based on their unique gas-solids contacting modes: fixed bed, fluidized bed, entrained bed, and molten bed in order to stress the similarity as well as the dissimilarity of the gasifiers. In addition the downstream processes, such as gas purification, shift conversion, methanation, are also discussed.

The last chapter deals with the topic of coal liquefaction with emphasis on principles rather than the individual or commercial processes. Since one of the most important aspects of coal processing is the removal of pollutants such as sulfur, the behavior and reactions of sulfur in liquefaction are emphasized. The use of catalysts to promote the removal of sulfur is another aspect of importance in liquefaction.

Although some attempts were made to achieve basic uniformity in writing style for each chapter, the involvement of multiple authors in the monograph has resulted in a display of individualities and differences. In fact, it is most refreshing and informative to observe that a few differences in opinions and emphasis exist on technical subjects among the individual contributors.

It is hoped that this monograph will serve as a catalyst to encourage the development of a more coherent and systematic approach to this complex but important area of coal conversion technology.

C. Y. WEN
E. STANLEY LEE

ISBN 0-201-08300-0

1. Coal Classification and Characterization

S. Ergun

1. THE ORIGIN OF COAL

There is ample evidence that coal is formed from plant substances preserved from complete decay in a favorable environment and later altered chemically and physically by various environmental effects. Coal geologists consider there are two broad stages in the process of coal formation: (1) the biochemical and peat stage, and (2) the dynamochemical or metamorphic stage.

According to Thiessen [1], the origin of different types of coal depends upon the environment. He cited factors such as (1) the nature of the topographic depression and the consequent size and form of the peat swamp it produced; (2) the nature of the underlying rock and soil; (3) the prevailing climate; (4) the physiography of the surrounding area, including inlets and outlets to the swamp; and (5) the nature of the swamp water under both surface and subsurface conditions, that is, its pH, oxygen, and mineral matter contents both in solution and in suspension. Additional factors that may be cited are (a) the type of plant associations that thrived in the peat swamp environment, and (b) the biochemical process in the swamp.

The accumulation of plant remains as peat deposits is considered as a biochemical process. Plants are heterogeneous in composition and contain cellulose, hemicellulose, lignin, resins, waxes, fats, and so on. Cellulose is easily oxidized into carbon dioxide and water through simple sugars by aerobic organisms which acquire atmospheric oxygen directly. In anaerobic fermentation, cellulose is a source of oxygen to the organisms and the products are no longer simply CO_2 and H_2O. Lignin is more resistant to bacterial action and is slowly oxidized by aerobic organisms to complex humic acids that form a large part of the peat. Fats, resins, and waxes, on the other hand, are quite resistant to the attack of anaerobic bacteria

ISBN 0-201-08300-0

C. Y. Wen and E. Stanley Lee, Editors, Coal Conversion Technology, ISBN 0-201-08300-0 and are oxidized more readily by aerobic organisms. It is clear that aerobic activity must be severely curtailed for peat formation.

The cell wall material of plants is largely cellulose. However, lignin is the "glue" that holds the cells of the wood in place. Since cellulose is the major constituent of plants, it was assumed by many workers that cellulose was the principal plant compound that contributed to coal formation. However, on the basis of chemical studies of coals, cellulose, and lignin, Fischer and Schrader [2] concluded that in peat formation cellulose is easily decomposed to CO_2 and H_2O by the action of microorganisms and consequently plays no important part in coal formation. Although their contention was supported by some later workers [cf. Ref. 3], it was also recognized that humic acids and phenolic compounds formed by the decomposition of lignin can render the swamp unsuitable for the growth of bacteria that attack cellulose and that intermediate compounds formed from cellulose may be a part of the coal substance.

On the basis of experiments involving hydrothermal reactions of cellulose and lignin at high pressures, Bergins [4], on the other hand, maintained that coal can be formed directly from both lignin and cellulose by chemical reaction at high temperatures over a short time or at low temperatures under reasonable pressure over a long period of time. He did not consider bacterial action essential in the hydrothermal process. Bergins' results also found support from many researchers.

From the data obtained by various lines of research one may conclude the following:

> Depending upon the conditions of the swamp formations and subsequent geological upheavals (many coal seams are buried very deep), both cellulose and lignin may have contributed to coal formation to different degrees.
>
> Since, by bacterial action, lignins appear to yield products significantly different from those produced by cellulose, the composition of the coals formed will reflect the selective contributions by lignins and cellulose, that is, we may expect the absence of a norm in coal composition.

The concern about the metamorphic stage stems from the fact that coals differ considerably in composition and many other properties. As one may expect, coals have been classified to reflect the degree of metamorphosis the peats have undergone, usually referred to as the rank of coal. Coal classification by rank has been the subject of numerous studies and has evoked much controversy. This subject will be dealt with in Section 3; however, the rank concept will be introduced here, for it is not possible to ignore it. Designations of solid fossil fuels as peat, lignite, subbituminous, bituminous, and anthracite undoubtedly resulted from the obvious differences in their burning characteristics. Classification attempts were simply intended to place the fuels in a logical order by some useful and easily determinable parameters. Peat is not regarded as coal, although it is generally

ISBN 0-201-08300-0

recognized to be the precursor of most coals. Lignites (or brown coals) are nearest in composition to peats and are believed to have undergone least metamorphosis. In increasing order of rank after lignites, coals are designated as subbituminous, bituminous, semianthracite, anthracite, and meta-anthracite. Subbituminous and bituminous coals are further subclassified. It is sufficient here to state that anthracites are less abundant than the younger coals. In the past, interest has been largely confined to bituminous coals because of their abundance and greater utility. Subbituminous coals and lignites have been regarded as inferior coals because of their lower calorific values. However, they are more abundant than bituminous coals.

Coal beds usually contain various minerals, such as clays, shales, and pyrites. Depending upon their rank, coals also contain moisture at ambient temperatures, up to 45% by weight. The major elements commonly found in organic coal are carbon, hydrogen, oxygen, nitrogen, and sulfur. The sulfur contents of coals range from a fraction of 1 wt % to several weight percent. Nitrogen, likewise, is a minor element in coals, ranging from 0.5 to slightly over 2 wt %. Carbon, hydrogen, and oxygen are considered to be the key elements in the structure of coals; undoubtedly the compositions of cellulose and lignins have been kept in mind in such a judgment.

In Figure 1.1 are shown the atomic percentages of carbon, hydrogen, and oxygen (on a mineral matter-, moisture-, sulfur-, and nitrogen-free basis) of 75 coal samples ranging in rank from lignite to meta-anthracite. From the figure it is seen that coals containing more than 55 atomic % carbon contain very little oxygen on an atomic basis, that is, about 1.5 atomic %. It is also evident that further increase in carbon content with increase in rank occurs at the expense of hydrogen being removed. The four points at the extreme right of Figure 1.1 belong to meta-anthracites, and they suggest that coals ultimately tend to become carbon; X-ray diffraction patterns point to graphite formation [5–7].

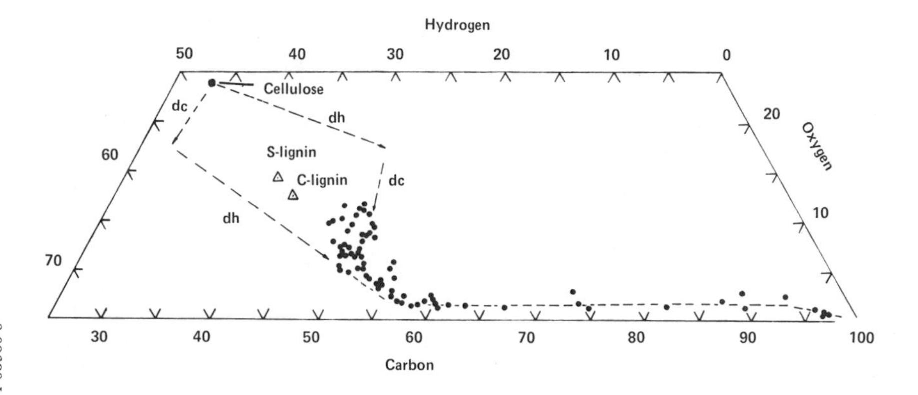

Figure 1.1. Carbon, hydrogen, and oxygen contents, atomic %, of some American coals.

ISBN 0-201-08300-0

The hydrogen and oxygen contents of coals containing less than 55% carbon fluctuate widely. Figure 1.1 does not represent the results of a sufficient number of analyses to demonstrate the true population density of the spectrum of coal composition. However, the extremes in composition variations are probably indicated because at least one sample from every coal producing region in the United States is included in the plot.

The triangle diagram shown in Figure 1.1 definitely shows a break in the composition trend. The oxygen contents of coals containing less than 39 atomic % hydrogen are low and more or less invariant until they reach the meta-anthracite stage. The change in composition in this range cannot be explained by dehydration or decarboxylation reactions because there is not enough oxygen in the coal to account for the hydrogen lost in reaching the meta-anthracite stage. Obviously external oxidizing agents were responsible for the removal of hydrogen.

Of particular interest in exploring coal metamorphism is the wide fluctuation in the atomic oxygen/carbon ratios of coals containing more than 39 atomic % hydrogen. The dot shown in the upper left corner of Figure 1.1 corresponds to the composition of cellulose. The compositions of coals obviously do not fall on a line drawn from this point to the break point in the composition diagram (e.g., C = 59, H = 39%, O = 2%). Such a change could be explained stoichiometrically by

$$C_6H_{10}O_5 = C_{5.225}H_{3.454}O_{0.177} + 0.775CO_2 + 3.273H_2O \qquad (1.1)$$

The stoichiometry shown in Equation 1.1 requires that twice as much oxygen be removed by dehydration reactions as by decarboxylation reactions. The composition of every coal included in Figure 1.1 obviously can be expalined by dehydration and decarboxylation reactions, which are indicated in both biochemical and thermochemical reactions of cellulose. Also shown in the figure are the compositions resulting from stepwise dehydration (dh) and decarboxylation (dc) of cellulose. Although the stepwise procedure is hypothetical, the dotted lines in Figure 1.1 show that coal formation from cellulose is explainable, provided, of course, that cellulose undergoes decomposition yielding widely different amounts of CO_2 and H_2O. But why the decomposition leads to such a wide spectrum in the composition of solid products is subject to much speculation.

Although the structure of cellulose has been well established, the structure of lignin remains unknown. The uncertainty about even the composition of lignin is one of the disappointments of organic chemistry [8]. Softwood lignins are believed to be composed of units built up from corniferyl alcohols by linkage of hydroxyl groups. This postulation leads to repeating units having the composition $C_{10}H_{11}O_3$, shown in Figure 1.1 as C-lignin. On the other hand, hardwood lignins have been found to contain sinapyl alcohol, which, if linked by hydroxyl groups, would lead to repeating units having the composition $C_{11}H_{13}O_4$ shown in Figure 1.1 as S-lignin. A smooth line with a slight curvature can be drawn through the points representing the composition of cellulose, the compositions of two lignins,

ISBN 0-201-08300-0

and the breakpoint in the coal composition spectrum. The compositions of many coals shown in Figure 1.1 would not be far away from the line drawn. However, a cluster of coal samples have high oxygen contents and would be distant from the line.

Figure 1.1 also shows that lignins are closer to coals in composition than is cellulose. Since lignins are more resistant to bacterial action than cellulose, are phenolic in nature, and are easily converted into humins, humic acids, they may have played a great role in preventing the complete decay of cellulose by providing conditions detrimental to the growth of decaying bacteria. Thus the differences in the relative amounts of compounds formed from cellulose and lignins involved in the formation of peats could be responsible for the wide fluctuation in the oxygen/carbon ratios of coals observed in Figure 1.1.

The metamorphosis of peat, that is, its physical and chemical alteration to form coals ranging in rank from lignite to anthracite, has received considerable attention. The biochemical degradation of various plant components, although highly complex, can be and has been studied in laboratories. On the other hand, peat metamorphosis took millions of years. It is logical to expect that heat, pressure, and other environmental factors played a great role in the process of coal formation. To attempt to produce a coal-like substance from peat in a laboratory is tantamount to squeezing millions of years into days. Nevertheless, this consideration proved to be no deterrent to many researchers. Indeed, most of the investigative efforts were diverted to the synthesis of a coal-like substance from lignin, cellulose, resins, glucose, and so on, thus incorporating the biochemical stage as well in the coal formation process. In addition to the academic interest in the coal formation process, the chemistry of the process was explored by thermochemical treatment of substances of known structure at elevated pressures.

2. PETROGRAPHIC CONSTITUTION AND CLASSIFICATION

It has long been recognized that coals are heterogeneous in appearance as well as in composition. Most American bituminous coals are made up of strata easily distinguished by the eye and are called banded coals. The bands are classified as bright or splint on the basis of appearance alone. In some cases the distinction is not clear-cut; hence some bands are termed semisplint. Often, coal seams are also called bright, splint, or semisplint according to the predominant band. Some coals are not made up of easily distinguishable strata and are termed nonbanded coals; these are grouped into subtypes as cannel and boghead coals. In some publications cannel and boghead coals are referred to as sapropelic coals, meaning that they are rich in organic matter.

Although curiosity about the origin and formation of coal was the main reason for interest in coal types, the importance of these types to coal utilization (e.g., combustion and carbonization) did not remain unrecognized. The appearance of coals served as a guide in their selection based on past experience. From a

ISBN 0-201-08300-0

scientific and engineering viewpoint, however, coal types did not make much sense until coals were examined under the microscope in greater detail.

Because of the intimate connections between coal conversion and the properties of coal material, detailed discussions of the influences of coal properties on conversion are given in other chapters. For example, liquefaction and petrographic constitution are discussed in Chapter 5, and pyrolysis behavior in Chapter 2.

2.1. Coal Petrography

Coal petrography deals with the description of what is seen in coals by microscopic observation and relates the different constituents observed to plants and plant parts.

The usual and simpler procedure for microscopic observation is to polish the coal samples either in block forms or as granules embedded in a resin. Differences in the reflectances of the various coal components permit their identification from the polished surface. A more elaborate procedure is to prepare thin sections, 5–15 μ in thickness, and examine the sections in transmitted light. The former technique was first proposed by Stopes in England [9]; the latter procedure was developed by Thiessen et al. [10, 11] in the United States.

The transmitted light method of petrographic examination is frequently referred to as the American system of coal petrography, in contrast to the more frequently used reflected light method, known as the European system of coal petrography. As one might expect, a different petrographic terminology has emerged in Europe based on the reflected light examination.

2.2. Petrographic Constituents and Classification

Thiessen's examinations under the microscope of thin sections prepared from coals of various coal seams and from various banded layers of the coal revealed that the various coal types are composed of more or less the same petrographically distinct entities and that the differences in their appearances and properties stem from the differences in the relative amounts of the different entities.

The different petrographic entities present in coals are often referred to as macerals or petrographic components. Thiessen named the coal components to be seen in thin sections as anthraxylon, attritus, and fusain. Depending upon the translucency revealed by the thin sections, he subclassified the attritus as translucent and opaque. Anthraxylon occurs in bands up to several millimeters in thickness. In thin sections its color ranges from orange to dark red, depending upon the rank of the coal, provided that the thickness of the section is controlled (e.g., about 8 μ). Anthraxylon is derived from woody tissues and often shows the original plant cell structure, as shown in Figure 2.1. Attritus may include finely divided anthraxylon; yellow bodies such as resins, pollen, cuticles, and algae; semitranslucent brown matter; and opaque matter. In Figures 2.2 and 2.3 are shown microphotographs of spores and resins which are present in some coals. The size and

ISBN 0-201-08300-0

shape of the metamorphosed spores usually reflect the types of plants which existed in the swamps. Fusain is usually identified as particles or lenses, is generally opaque, and shows a fibrous cellular structure. Figure 2.4 shows a band of fusain as found in some American coals. Thiessen's classification of coal components is shown in Table 2.1. (Figures 2.1–2.4, see color plate following p. 16.)

Table 2.1

Microscopic Composition of Coal According to Thiessen[a]

Component		Constituents of Attritus
Anthraxylon (bands $>14\ \mu$ thick)		
Attritus	Translucent	Humic matter (anthraxylon-like shreds less than 14 μ thick)
		Brown matter
		Spores, pollen, cuticles
		Resinous bodies
		Algal remains
	Opaque	Granular and amorphous opaque matter
		Fusain splinters ($<37\ \mu$)
Fusain (bands $>37\ \mu$)		

[a] Modified from B. C. Parks and H. V. O'Donnell, *U.S. Bur. Mines Bull.* 530, 1956.

That coal types could be quantified in terms of the petrographically distinct entities present in them was also recognized in Europe. These entities were called macerals, and the coal types were often termed rock types. In a pioneering study Stopes [9] classified coal types as vitrain, clarain, durain, and fusain and was responsible for European terminology. This terminology was adapted from the French words *vitro, clara, dur,* and *fusain,* meaning "glass," "bright," "hard," and "charcoal," respectively. Since vitrain was the most homogeneous rock type, its major compound was called vitrinite, which, in fact, is analogous to Thiessen's anthraxylon. Clarain and durain were found to contain, in addition to fusain, weakly reflecting exinitic material (spores, cutines, resins, etc.) and strongly reflecting granular opaque matter. These components were termed exinite and micrinite, respectively.

ISBN 0-201-08300-0

The European or Stopes-Heerlen system of petrographic components classification is shown in Table 2.2. Column 3 of this table describes the appearance of the macerals in reflected light, column 4 lists the terminology used for macerals in transmitted light, and column 2 lists the terminology used by Thiessen. As seen from Table 2.2., the two systems of maceral classification seem, for the most part, to be compatible. Many petrographers, nevertheless, disagreed with Thiessen in making a distinction between anthraxylon and translucent humic matter based on size. But Thiessen had a point in that, in particles of less than 14 μ, the cell structure could not be observed. From a practical point of view, however, they can be

Table 2.2

Description of Macerals in Thiessen-Bureau of Mines System and Comparison with Stopes-Heerlen System

Appearance in Transmitted Light[a]	Thiessen System	Appearance in Reflected Light[a]	Maceral	
Yellowish orange to dark red; material defined as collinite in RL may show structure in TL	Anthraxylon and humic matter	Groundmass for other macerals; often cell filling of telinite; no visible structure; gray to yellowish white	Vitrinite	Colinite
		Occurs with collinite in discrete bands; shows cell structure; gray to yellowish white		Telinite
Golden-yellow to reddish brown	Spores and pollen	Elongate, discrete bodies; dark gray to light gray	Exinite	Sporinite
Orange-yellow to rust	Cuticles	Narrow bands one edge of which is often serrate; dark gray to light gray		Cutinite
May sometimes show cell structure, yellow to orange	Algal remains	Derived from algae; weaker reflectivity than associated vitrinite and sporinite		Alginite
Yellow to reddish orange	Resinous and waxy substances	Discrete, small bodies; round, oval, or rod-shaped; black to gray		Resinite
Generally opaque; in very thin sections dark brown	Opaque matter, brown matter (in part)	Variable form; finely to coarsely granular; light to white	Inertinite	Micrinite
Opaque	Fusain	Discrete lenses, bands, and fragments, good cellular structure; yellowish white		Fusinite
Orange-red to opaque	Fusain (in part), brown matter (in part)	Intermediate between vitrinite and fusinite; cell structure not as well defined as fusinite; light gray to white		Semifusinite
Dark red-brown to opaque	Fusain (in part), brown matter (in part)	Round or oval bodies or interlaced fibrous masses; light gray to yellowish white		Sclerotinite

[a] The color ranges given represent changes with rank. In reflected light constituents become lighter in color with increase in rank; in transmitted light they become darker.

ISBN 0-201-08300-0

combined and called anthraxylon or vitrinite. Thiessen's translucent attritus (see Table 2.1) is analogous to exinite in the Stopes-Heerlen nomenclature. Generally, no difficulty is encountered in correlating the two systems of maceral description in regard to vitrinite and exinite because there has been no difficulty in identifying the origin of these macerals, and because they usually have distinct physicochemical properties. In addition, there is no difficulty in reconciling Thiessen's fusain with the fusinite of the European system of classification. However, it is difficult to correlate the brown matter components which are named granular opaque and amorphous opaque matters by Thiessen (Table 2.1) with those named micrinite, semifusinite, and sclerotinite in the Stopes-Heerlen system.

Such difficulty stems from the controversy regarding the origin of these macerals. It has been suggested that micrinite was formed by the coalification of humic acid flocculations. Perhaps Thiessen was more accurate in his classification of these macerals because his thin-section samples undoubtedly revealed more details.

Table 2.3

Bureau of Mines Type Classification[a]

Transmitted Light		Reflected Light	
Types of Coal	Quantitative Statements	Microlithotypes	
Banded coals			
Bright coal	More than 5% anthraxylon	Vitrite	
	Less than 20% opaque attritus	Clarite	
Semisplint coal	More than 5% anthraxylon	Duroclarite	
	20–30% opaque attritus	Vitrinertite	
Splint coal	More than 5% anthraxylon More than 30% opaque attritus	Clarodurite	
Nonbanded coals			
Cannel coal	Attritus with little or no algae	Durite	Cannel
Boghead coal	Attritus with predominant oil algae		Boghead

[a] Parks and O'Donnell [12].

ISBN 0-201-08300-0

Nevertheless, it is probably more practical to lump this group of macerals with fusinite and term them all as inertinite. It is advisable, however, to draw a distinction between micrinite and fusinite (fusain) in banded American coals containing very distinct fusain bands. On the basis of the above petrographic components classifications, U.S. Bureau of Mines researchers were able to quantify the petrographic compositions of the various coal types encountered in the United States. This petrographic quantification of types is shown in Table 2.3, which gives the Bureau of Mines petrographic description (Thiessen's terminology), as well as the European petrographic terminology of types.

Coal beds are generally heterogenous, mainly in the direction perpendicular to the bed, as illustrated in Figure 2.5, which shows the petrographic makeup of a typical banded, high volatile bituminous coal bed. The first column of Figure 2.5 shows the distribution of the different coal type layers throughout the bed, the second column shows the measured thicknesses of the various layers, and the fifth through the eighth columns show the percentages of the various petrographic entities. These analyses were obtained from microscopic examinations of 56 thin sections prepared from a columnar sample taken from the coal mine.

A more extensive study of the petrographic makeup of 160 coal beds is given in a Bureau of Mines bulletin [12] which provides detailed information collected mostly from columnar-profile studies. Although the thin-section, transmitted light procedure used by the Bureau of Mines in these studies is not generally adopted by all other American coal petrography laboratories, the method is familiar to most American coal petrographers, and the bulletin cited above can be considered as the standard reference for characterizing the petrographic compositions of American coals.

In Figure 2.6 is shown the distribution of coal components in 217 coals of varied rank, type, and source studied at the Bureau of Mines. The distribution shown does not necessarily represent the coal resources of the United States as a whole because the numbers of samples examined from the various locations are not in proportion to the reserves of the coals. Nevertheless, the figure gives an idea of the relative proportions of the various petrographic components in American coals because they were taken from 90 counties in 27 states.

Figure 2.6 shows that fusain is the least prevalent component in coals, although fusinite concentration as high as 20% has been reported. The average fusain contained in 217 coal samples was 2.6%. The opaque matter (total inerts less fusain) evidently occurs in higher concentration than does fusain; for example, 80% of the samples examined contained less than 3.5% fusain, as compared to less than 14.0% opaque attritus. The average for the opaque matter was 9.5%.

Anthraxylon and translucent attritus constitute the major portion of American coals, the average of the former in 217 samples being about 50.5% and that of the latter, 37.4%; however, they exhibit considerably different distributions. Inasmuch as about one half of the translucent attritus is brown matter (in Thiessen's terminology), which would be classified as vitrinite in European terminology, it is apparent that the average vitrinite contents of American coals are in

ISBN 0-201-08300-0

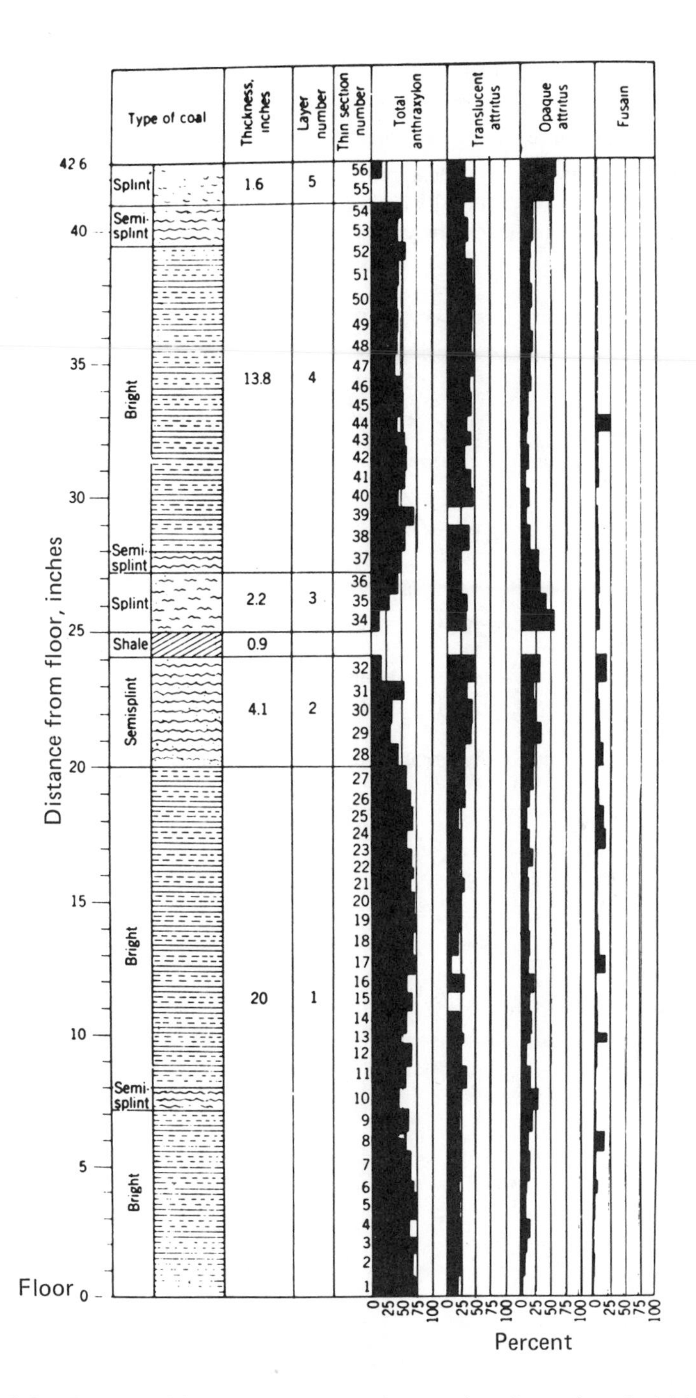

Figure 2.5. Petrographic components and types in a bed of typical high volatile bituminous coal.

ISBN 0-201-08300-0

excess of 70%, their inertinite fraction being on the order of 10–15%. Figure 2.7 shows the frequency of occurrences of fusinite plus opaque attritus at various concentrations. The figure indicates that the distribution of inert coal components (inertinite) is bimodal, if not multimodal, the average being 13%. A statistical analysis of the data is unwarranted because sampling involved no statistical procedure; it was simply a pioneering survey.

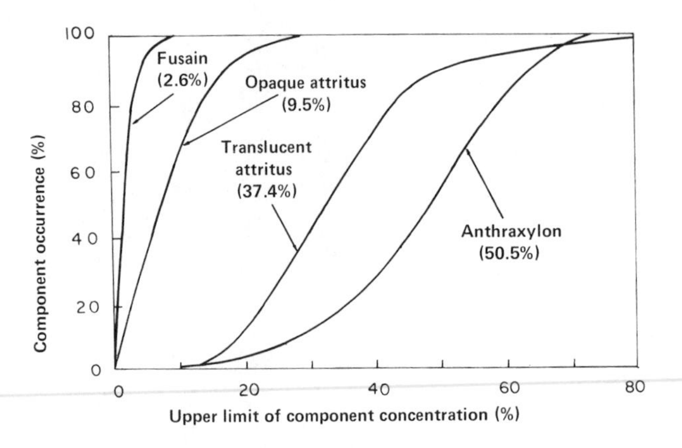

Figure 2.6. Distribution of petrographic components in American coals.

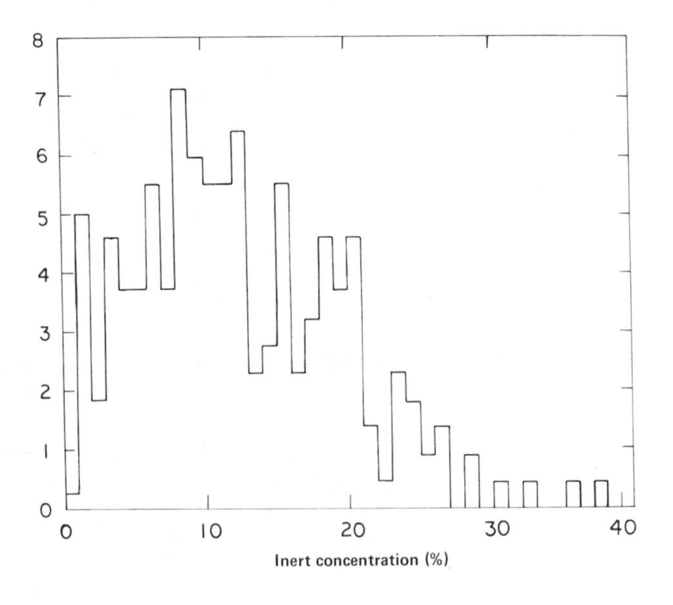

Figure 2.7. Frequency of occurrence of inerts in American coals.

ISBN 0-201-08300-0

2.3. Some Selected Properties of Coal Components

Petrographic analysis of coal (using reflected or transmitted light) is made on the basis of shape, texture, and association of identifiable entities that underwent metamorphosis. Many petrographers recognized that the different entities present in a coal reflected (or transmitted) the incident light to different degrees. Hoffmann and Jenkner [13] in Germany in 1932 were the first to measure quantitatively the reflectances of coal components. This pioneering work did not attract much interest until a series of papers by Seyler [14] from 1943 to 1952. Seyler proposed an interesting hypothesis of discontinuous (stepwise) reflectance series of coal components in all ranks of coal. McCartney [15] tested the applicability of Seyler's theory to a number of American coals and found that the discontinuous reflectances observed by Seyler were the result of an insufficient number of measurements and a fortuitous selection of samples furnished. Seyler's work, however, evoked much interest in the study of reflectances of coal components. McCartney, for example, noted the semiquantitative and tedious nature of the Berek photometer used by Seyler and earlier workers, and devised a detection system composed of a photomultiplier tube and electrometer, paving the way for accurate and fast reflectance measurements. The least reflecting component in a given coal was found to be exinite, followed by vitrinite, micrinite, and fusinite, in increasing order of reflectance.

Curiosity about the differences in composition and other properties of the petrographic entities to be found in a coal dates back to the 1920s. A major obstacle to systematic investigation was the size and distribution of the entities. The sizes of exinites and micrinites are usually very small, a few microns, and in many coals they are dispersed. However, the peculiarities of some coals presented opportunities to demonstrate that the different entities present in bituminous coals and lignites have differing compositions and specific gravities. In many American coals vitrinite and fusinite appear in thick lenses, and samples of them, in nearly pure forms, could be obtained with the aid of a microscope. Also, some coals contain segments very rich in exinite or resins. On comparison the compositions and chemical properties of hand-selected petrographic components were found to be different from those of the whole coal. Studies of the properties of isolated exinites date back to 1928, when Zetsche [16, 17] studied a number of exinites. Later Sprunk and co-workers [18] investigated the properties of a number of exinites found in a variety of American coals, and Macrae [19] likewise studied the exinites found in British and other coals.

Post-World War II industrial redevelopment in Europe and concern over the dwindling reserves of coking coals (in the United States and elsewhere) revived interest in a more systematic investigation of coal components. The rapid advances made in electronics and analytical instruments provided additional impetus.

It was observed that fractions of different specific gravity obtained from finely pulverized bituminous coals by successive float-and-sink operations had different petrographic compositions. Exinite tended to concentrate in the fraction

ISBN 0-201-08300-0

of least specific gravity, followed by vitrinite, micrinite, and fusinite, in the order of increasing specific gravity. Although the specific gravity of fusinite obtained from various coals did not seem to fluctuate markedly, the same was not found to be true for the other components, especially exinite. Researchers in France [20], Germany [21], the Netherlands [22], and the United States [23] succeeded in obtaining, by trial-and-error procedures, concentrates of exinite, vitrinite, and micrinite from a number of European and American coals and determined their compositions and various other physical and chemical properties.

Intense research activity flourished for a short period of time. The results of the research activity of this period are discussed in Chapters 1, 2, and 6 of Ref. 24. Table 2.4 gives an idea of the differences in composition and some physical properties of the petrographic components of a bituminous coal. It is seen that there are significant and systematic differences in their hydrogen contents, densities, and reflectances in oil. The refractive index of fusinite bucks the trend; its higher reflectivity obviously stems from a high extinction coefficient. The surface areas obtained may reflect the sizes of the homogeneous domains in the constituents because they correspond to linear dimensions ranging from 2 to 0.4 μ. It should be pointed out here that the differences in the properties of the petrographic components become less as the rank of the coal increases and are practically nonexistent in anthracites.

Table 2.4

Chemical Compositions and Some Physical Properties of Components of a High Volatile A Bituminous Coal[a]

	Composition, wt %								
Component	S	N	O	C	H	Helium Density, g/cm^3	Reflectance in Oil	Refractive Index	Surface Area[b]
Exinite	0.7	1.1	5.5	86.2	6.5	1.21	0.32	1.68	2.5
Vitrinite	1.0	1.6	8.0	84.1	5.3	1.29	0.91	1.81	2.1
Micrinite	0.6	0.7	8.0	85.9	4.8	1.32	1.61	2.01	4.6
Fusinite	0.4	0.6	4.3	91.5	3.2	1.48	2.65	1.91	9.8

[a] Ergun et al. [23].
[b] BET using nitrogen.

From a practical point of view it was hoped that a petrographic analysis and an easily determinable property of the petrographic components of coal would permit prediction of the parameters useful in coal utilization. Unfortunately, this has not proved possible. The fact that exinites are rich in hydrogen, hydrogenate easily, and oxidize with greater difficulty (under certain conditions), and that fusinites are relatively more inert, was established much earlier. The major difficulty encountered in developing a simple method for characterizing coals based

ISBN 0-201-08300-0

on their petrographic compositions and degrees of coalification proved to be the lack of correlations between the various petrographic components at different stages of metamorphosis. The properties of fusinite did not show a progressive change with coal rank. Since the major component in coals is vitrinite, its properties obviously correlated with those of the parent coal. Micrinite properties, surprisingly, showed with rank a trend toward correlation with those of vitrinites. Exinite, the hydrogen-rich component of coal, had different compositions in a coal of given rank [23].

In 1960 Shapiro and Gray [25] described a system of petrographic classification based on arbitrary reflectance limits of coal macerals. They chose to use terms such as "vitrinoid," "exinoid," "micrinoid," and "fusinoid" in place of "vitrinite," "exinite," "micrinite," and "fusinite," respectively. Shapiro and Gray subclassified each maceral as reactive or inert according to whether or not it softened and became plastic upon heating. Apparently the reflectance value (in cedar oil) separating these two classes was approximately 2.0%. For example, fusible vitrinoids were given designations from V0 to V21, while inert vitrinoids ranged from V22 to V70. This meant that there were 22 classes of the former, ranging in reflectance from 0.0 to 2.1% in 0.1% steps. Other components had similar designations, such as E0 to E15 for exinoids, all fusible, and M18 to M70 for micrinoids,

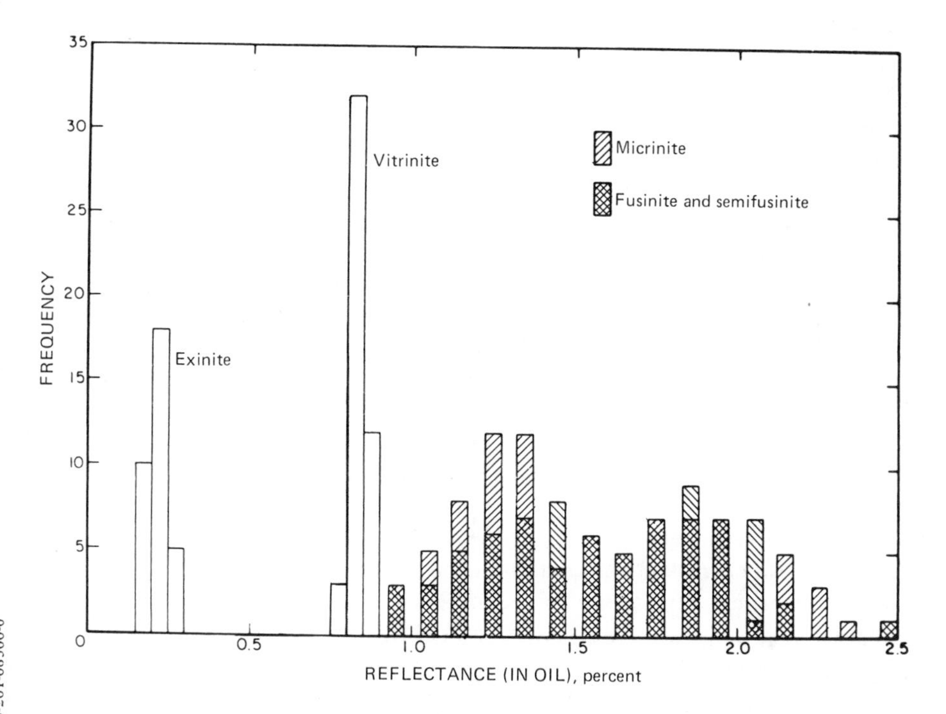

Figure 2.8. Distribution of reflectances of components in a hvAb coal, Chilton Bed, West Virginia.

ISBN 0-201-08300-0

all inert. What Shapiro and Gray had in mind was to find a correlation between coke strength and the petrographic composition of the coal (or coal blends) used in making the coke. They found it necessary to use, in an empirical way, the reflectance distributions of the components in the correlation. What van Krevelen and co-workers [22] had in mind was to seek a correlation between the reflectances (or other properties) of coal components other than vitrinite as a function of the reflectance of vitrinite, the principal coal component. They assumed that the reflectance of vitrinite would serve as a measure of the rank of coal, which is a reasonable assumption. What thwarted their effort was the lack of a correlation between vitrinite and exinite properties. Another complicating factor was the wide ranges of reflectance values of micrinites and fusinites in a given coal. In Figure 2.8 are shown the reflectance distribution of coal components in a bituminous coal [23]. It is seen that the reflectances of micronite and fusinite overlap considerably. It is not very unlikely that the reflectances of exinite, as well as those of vitrinite, in some coals show wider ranges of distribution than the one seen in Figure 2.8.

2.4. Practical Application of Coal Petrography

The significance of the petrographic heterogeneity of coals for coal utilization did not escape attention, although taking advantage of it has been largely confined to coal seam characterization and coke making. The five recognized areas of application of coal petrography are (1) description and classification of coals; (2) survey of coal resources and bed correlations; (3) blending and preparation of coals for coke making; (4) coal liquefaction; and (5) coal preparation, combustion, and gasification.

The first two items need no elaboration; they were discussed in Section 2.1. Investigation of the influence of petrographic composition on coking properties dates back to the early 1920s. Numerous researchers conducted studies over a number of years to establish a relationship between petrographic composition and coking properties [26, 27]. It was found that exinite is the most volatile component, becomes extremely fluid when heated, and yields large amounts of tars. Vitrinite fuses well in the process of coking, and this property is regarded as essential in coke making. Micrinite, on the other hand, was found to be fusible with great difficulty, especially if derived from high rank coals. On the other hand, a certain concentration of micrinite proved to be essential to obtain a coke having a high strength. If one were to draw a crude analogy between concrete and coke making, the roles of exinite, vitrinite, and inertinite would have some similarities to those of water, cement, and sand plus gravel, respectively. Suitable quality and correct proportions are the essential requirements in both cases.

A literature survey shows no significant follow-up by industry of the work done during the 1920s and early 1930s for over two decades, that is, until the late 1950s. In 1957 Ammosov et al. [28] described a method of calculating coking charges on the basis of the petrographic characteristics of coals as determined from their reflectances. Shapiro and Gray [29, 30] described a similar procedure based

ISBN 0-201-08300-0

Figure 2.1. Thin section of a coal showing an anthraxylon (vitrain) band (X200).

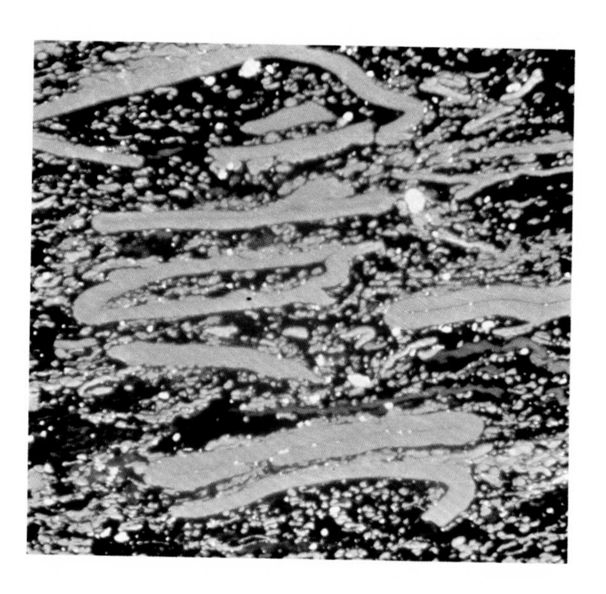

Figure 2.2. Spores in a coal as revealed in a thin section (x200).

ISBN 0-201-08300-0

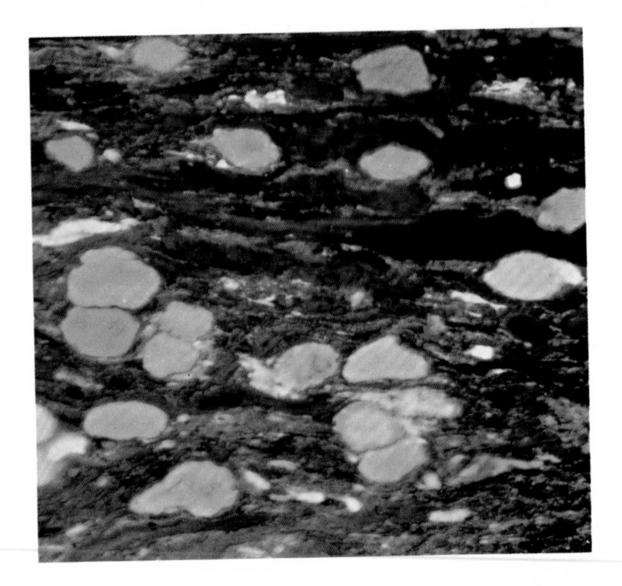

Figure 2.3. Thin section of a coal containing resin globules (x200).

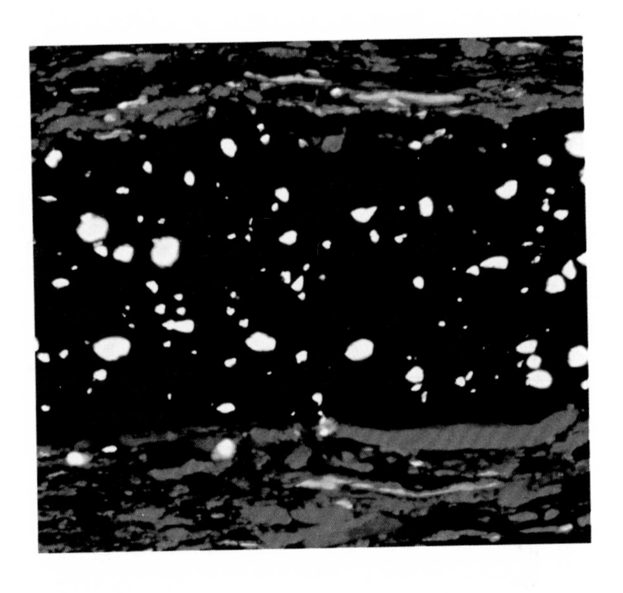

Figure 2.4. A fusain band in a coal (x200).

ISBN 0-201-08300-0

on reflectance analysis according to a classification system they had proposed earlier [25]. Harrison and co-workers [31, 32] described the application of the same general technique to different coals. The techniques and procedures adopted by the workers cited above were highly empirical and not very simple. It is unlikely that they would be generally applicable. Nevertheless, bringing the petrographic makeup into consideration in making coke from blends of a finite number of coals evidently served to establish a better basis for coke quality control than did ignoring this factor.

Another area of practical application of coal petrography is believed to be coal liquefaction by hydrogenation. The first attempt in the United States in this area dates back to 1929. Thiessen and Francis [33] reported that fusain is liquefied with great difficulty or not at all by high pressure catalytic hydrogenation. Wright and Gauger [34] concluded from their work on American coals that all banded constituents other than fusinite could be completely liquefied, but opaque attritus was more resistant. Investigations on dull (attrital) coals led to some conflicting claims, largely because of the lack of constituent (maceral) analysis.

The most serious work on hydrogenation of the petrographic constituents of a variety of coals was conducted in the late 1930s [35] batchwise in a small autoclave, using tetrahydronaphthalene as vehicle oil, and stannous sulfide as catalyst. The hydrogen pressure was 1000 psi at room temperature, and the reaction time was 3 hr. The study showed that fusains are not entirely inert; however, the oil yield from fusains, compared to the yields from whole coals or other coal constituents, was very poor. Anthraxylons (vitrinite) were found to be much more readily liquefied than fusains and were almost completely converted into a pitch. The study further indicated that the opaque attritus (micrinite, semifusinite, and perhaps finely divided fusinite) is more resistant to hydrogenation than is the translucent attritus (exinite).

The Bureau of Mines study cited above clearly showed that the petrographic composition of coal has a pronounced effect on coal liquefaction. This finding could properly be used in the selection of coal seams for liquefaction. Unfortunately, with the discovery of oil in the Arabian peninsula, the Synthetic Liquid Fuels Act was repealed and this interesting approach was not pursued. Recently, Pennsylvania State University, supported by the U.S. Department of Energy, has undertaken a research program to survey the liquefaction characteristics of American coals.

Detailed discussions of the influences of the petrographic composition of coal on liquefaction behavior are given in Chapter 5. Carbonization and combustion behaviors of coal are discussed in Chapters 2 and 3.

3. CLASSIFICATION OF COAL BY RANK AND GRADE

Designations of solid fuels as peat, lignite, subbituminous coal, bituminous coal, and anthracite undoubtedly resulted from the obvious differences in their

ISBN 0-201-08300-0

burning characteristics. Classification attempts were simply intended to place the fuels in logical order by some meaningful considerations and useful parameters. The order listed above was explained by petrographers as signifying the degree of metamorphosis of the plant debris. Since coals of differing ranks have widely different physical and chemical characteristics, the problem of classification became one of choosing a suitable property.

For a given rank of coal the amount and the nature of the mineral impurities associated with the coal vary greatly. Large coal seams containing over 30% mineral impurities are not uncommon, nor are mines containing a few percent extraneous matter. Since the calorific value of the coal purchased is reduced by the mineral matter present, the concern over grading the coal is readily appreciated.

3.1. Coal Rank and Classification by Rank

Rank characterization of coals dates back to 1827, when Regnault [36] studied the characteristics of coals ranging in rank from lignite to anthracite and went beyond that to peat and wood. In 1945 Rose [37] provided a comprehensive review of the problem of coal classification by rank.

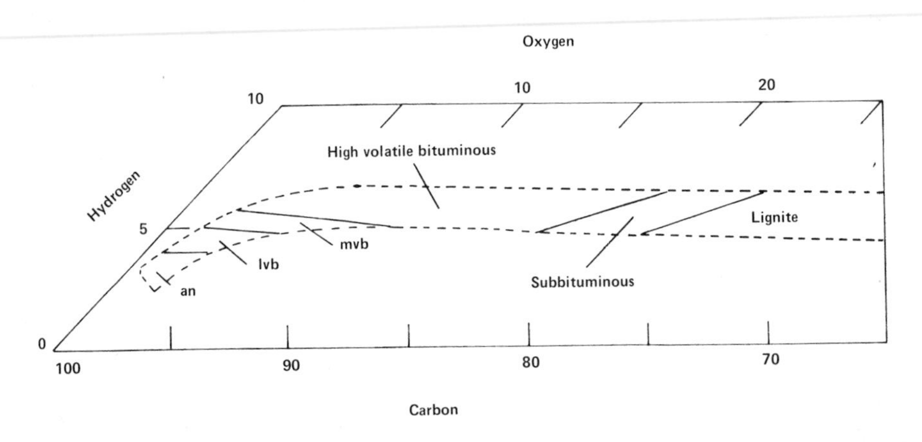

Figure 3.1. Triangular diagram of coal composition, atomic %: an – anthracite; lvb – low volatile bituminous, mvb – medium volatile bituminous.

A. ELEMENTAL COMPOSITION AS AN INDEX OF RANK

Systematic investigation of the carbon and hydrogen contents of American coals dates back to 1858, when Rogers [38] used hydrogen and carbon contents to classify Pennsylvania coals. In 1907 Grant [39] considered the inclusion of the oxygen contents and proposed a classification based on the carbon, hydrogen, and oxygen contents (everything-else-free basis) of coals. He proposed the use of a

ISBN 0-201-08300-0

triaxial plot similar to that shown in Figure 1.1. Ralston [40] systematically extended Grant's proposals, utilizing data on over 3000 coal samples of various ranks. An adaptation by Rose [37] of Ralston's plot is shown in Figure 3.1. The dotted lines in the figure represent the band within which Ralston's data fall. The demarcation lines shown in the figure between the various ranks were drawn by the author. The lines are based on the relation of ultimate analysis to the parameters used in the ASTM method. Although a case can be made for coal classification based on elemental analysis, it did not receive much support.

B. ASTM CLASSIFICATION

The original standards for classification of coals by rank (ASTM D-388) were published in 1939. The revision (ASTM D-338-66), published in 1967, remains unaltered and is shown in Table 3.1. Although this table is self-explanatory, a few remarks are in order in that it contains two parameters not yet introduced. One of them is the volatile matter of coals, which will be discussed in Section 4. It is sufficient here to state that the term "volatile matter" refers to the weight loss when dried coal samples (heated to above the boiling temperature of water) are further heated, in a covered crucible, to about 900°C. If the residue remaining is further combusted in air to produce an ash, the weight loss is referred to as the fixed carbon. Classification of medium volatile and higher rank coals is based on their volatile matter content. On a moisture- and ash-free basis (maf), the percent of fixed carbon is defined as 100 minus the percent of volatile matter; therefore it is redundant to specify both the volatile matter and the fixed carbon. By the ASTM standards the low volatile coals are distinguished from the semianthracite coals by having less than 86% fixed carbon and by being presumably agglomerating. Thus coals containing more than 86% fixed carbon (maf) are regarded as nonagglomerating. However, this hypothesis is questionable because the petrographic composition of coal is also a factor which greatly influences coal agglomerating behavior. It also appears that the fixed carbon content is not the best criterion by which to classify anthracites into three groups. As will be shown later, the graphite-like layers of the anthracite matrix range in size from about 15 to several thousand angstroms, and they appear to fall into distinct ranges. These ranges do not, however, conform to the ASTM classification shown in Table 3.1.

High volatile B bituminous (hvBb) and lower rank coals are classified according to their calorific values on a mineral matter- (or ash-) free basis, as shown in Table 3.1. In addition, their volatile matter content must be greater than 31%. The distinction between bituminous and subbituminous coals is also based on the agglomerating character of the former. Thus coals having calorific values between 10,5000 and 11,5000 Btu/lb are classified as hvCb if agglomerating and as subbituminous A if nonagglomerating. It should be mentioned that some hvCb coals do not agglomerate if oxidized in air at moderate temperatures. Dry, mineral matter-free fixed carbon (dry, MM-free *FC*) and moist, mineral matter-free calorific value

ISBN 0-201-08300-0

(moist, MM-free *Btu*) are the only parameters used in classifying coals by the ASTM method. The formulas used in calculating *FC* and *Btu* are as follows:

$$\text{Dry, MM-free } FC = \frac{(FC - 0.15S)100}{100 - (M + 1.08A + 0.55S)} \tag{3.1}$$

$$\text{Moist, MM-free } Btu = \frac{(Btu - 50S)100}{100 - (1.08A = 0.55S)} \tag{3.2}$$

where MM = mineral matter, %
Btu = calorific value, Btu/lb
FC = fixed carbon, %
M = natural moisture content, %
A = ash content, %
S = sulfur content, %

The moisture content is that which is retained in coal at 96–97% relative humidity and 30°C.

Table 3.1

Classification of Coals by Rank[a]

		Fixed Carbon Limits, % (dry, mineral matter-free basis)		Volatile Matter Limits, % (dry, mineral matter-free basis)		Calorific Value Limits, Btu/lb (moist,[b] mineral matter-free basis)		Agglomerating
Class	Group	Equal to or Greater Than	Less Than	Greater Than	Equal to or Less Than	Equal to or Greater Than	Less Than	Character
I. Anthracitic	1. Meta-anthracite	98			2			
	2. Anthracite	92	98	2	8			
	3. Semianthracite	86	92	8	14			Nonagglomerating[c]
II. Bituminous	1. Low volatile bituminous coal	78	86	14	22			
	2. Med volatile bituminous coal	69	78	22	31			
	3. High volatile A bituminous coal		69	31		14,000[d]		Commonly agglomerating[e]
	4. High volatile B bituminous coal					13,000[d]	14,000	
	5. High volatile C bituminous coal					11,500 10,500	13,000 11,500	 Agglomerating
III. Subbituminous	1. Subbituminous A coal					10,500	11,500	Nonagglomerating
	2. Subbituminous B coal					9,500	10,500	
	3. Subbituminous C coal					8,300	9,500	
IV. Lignitic	1. Lignite A					6,300	8,300	
	2. Lignite B						6,300	

[a] This classification does not include a few coals, principally nonbanded varieties, which have unusual physical and chemical properties, and which come within the limits of the fixed carbon or calorific values of the high volatile bituminous and subbituminous ranks. All of these coals either contain less than 48% dry, mineral matter-free fixed carbon or have more than 15,500 moist, mineral matter-free Btu/lb.

[b] "Moist" refers to coal containing its natural inherent moisture, but not including visible water on the surface of the coal.

[c] If agglomerating, shall be classified in the low volatile group of the bituminous class.

[d] Coals having 69% or more fixed carbon on a dry, mineral matter-free basis shall be classified according to fixed carbon, regardless of calorific value.

[e] It is recognized that there may be nonagglomerating varieties in these groups of the bituminous class, and there are notable exceptions in the high volatile C bituminous group.

ISBN 0-201-08300-0

C. BRITISH CLASSIFICATION

The British National Coal Board (BNCB) coal classification system is based on the dry, mineral matter-free volatile matter and the Gray-King coke type test. The BNCB system is shown in Table 3.2.

Table 3.2

British National Coal Board Coal Classification System

Coal Rank Code (main class)	Volatile Matter,[a] %	Gray-King Coke Type[b]	General Description
100	Under 0.1		Anthracites
200	9.1–19.5		Low volatile steam
300	19.6–32.0	A–G_9 and over	Medium volatile
400	Over 32.0	G_9 and over	High volatile, very strong caking
500	Over 32.0	G_5–G_8	High volatile, strong caking
600	Over 32.0	G_1–G_4	High volatile, medium caking
700	Over 32.0	E–G	High volatile, weakly caking
800	Over 32.0	C–D	High volatile, very weakly caking
900	Over 32.0	A–B	High volatile, noncaking

[a] On dry, mineral matter-free basis.

[b] Coals having volatile matter less than 19.6%.

D. INTERNATIONAL CLASSIFICATION

In 1956, under the auspices of the United Nations, an information classification system for hard coals by type (ECE) was devised. This system is based on the dry, ash-free volatile matter and moist, ash-free calorific value, that is, the basis of classification is similar to that of the ASTM system. However, groups and subgroups are separated by caking and coking properties. Table 3.3 lists the classification parameters and the numerical symbols representing the coal groups and subgroups.

ISBN 0-201-08300-0

Table 3.3

International Classification of Hard Coals by Type

Groups (determined by caking properties)			Code Numbers										Subgroups (determined by coking properties)		
Group Number	Alternative Group Parameters: Free-Swelling Index (crucible-swelling number)	Alternative Group Parameters: Roga Index	The first figure of the code number indicates the class of the coal, determined by volatile matter content up to 33% VM and by calorific parameter above 33% VM. The second figure indicates the group of coal, determined by caking properties. The third figure indicates the subgroup, determined by coking properties.										Subgroup Number	Alternative Subgroup Parameters: Dilatometer	Alternative Subgroup Parameters: Gray-King
3	> 4	> 45					435	535	635				5	> 140	> G_8
						334	434	534	634				4	> 50-140	G_5-G_8
						333	433	533	633	733			3	> 0-50	G_1-G_4
						332 a / 332 b	432	532	632	732	832		2	≤ 0	E-G
2	$2\frac{1}{2}$-4	> 20-45				323	423	523	623	723	823		3	> 0-50	G_1-G_4
						322	422	522	622	722	822		2	≤ 0	E-G
						321	421	521	621	721	821		1	Contraction only	B-D
1	1-2	> 5-20			212	312	412	512	612	712	812		2	≤ 0	E-G
					211	311	411	511	611	711	811		1	Contraction only	B-D
0	0-$\frac{1}{2}$	0-5		100 (I) A, B	200 II	300	400	500	600	700	800	900	0	Nonsoftening	A
Class Number →			0	1	2	3	4	5	6	7	8	9	As an indication, the following classes have approximate volatile matter contents of:		
Class Parameters	Volatile matter (dry, ash-free) →		0-3	> 3-10 (A: >3-6.5; B: >6.5-10)	> 10-14	> 14-20	> 20-28	> 28-33	> 33	> 33	> 33	> 33	Class 6 33–41% volatile matter; 7 33–44% " "; 8 35–50% " "; 9 42–50% " "		
	Calorific parameter[a] →		–	–	–	–	–	–	> 13,950	> 12,960–13,950	> 10,980–12,960	> 10,260–10,980			
Classes (determined by volatile matter up to 33% VM and by calorific parameter above 33% VM)															

[a]Gross calorific value on moist, ash-free basis (30°C, 96% relative humidity), Btu/lb.

Note 1. When the ash content of coal is too high to allow identification according to the present systems, it must be reduced by the laboratory float-and-sink method (or any other appropriate means). The specific gravity selected for floatation should allow a maximum yield of coal with 5 to 10 percent ash.

Note 2. 332a . . . >14–16% volatile matter
332b . . . >16–20% volatile matter

E. MOISTURE CONTENT AS AN INDEX OF RANK

In the ASTM classification the calorific value is used as a measure of rank for high volatile A and lower rank coal. From Table 3.1 it is obvious that the calorific

ISBN 0-201-08300-0

values of coals decrease with decrease in rank. It is also well known that the moisture contents of coals increase with decrease in rank. Recently, Ergun [41] investigated the possibility of a correlation between the calorific values of coals and their moisture contents, both on a mineral matter-free basis. Using the data recently compiled by the Bureau of Mines, [42] he obtained the plot shown in Figure 3.2. The plot represents the results of over 29,000 analyses of coal samples from 120 counties in Colorado, Montana, North Dakota, Utah, and Wyoming [42]. The reserves in these states total over 193 billion tons, that is, over 84% of the western reserves or over 50% of the total U.S. reserves. The solid line shown in Figure 3.2 is drawn according to the equation

$$CV = 19{,}000 - 1000\sqrt{12.0 + 3.1M} \tag{3.3}$$

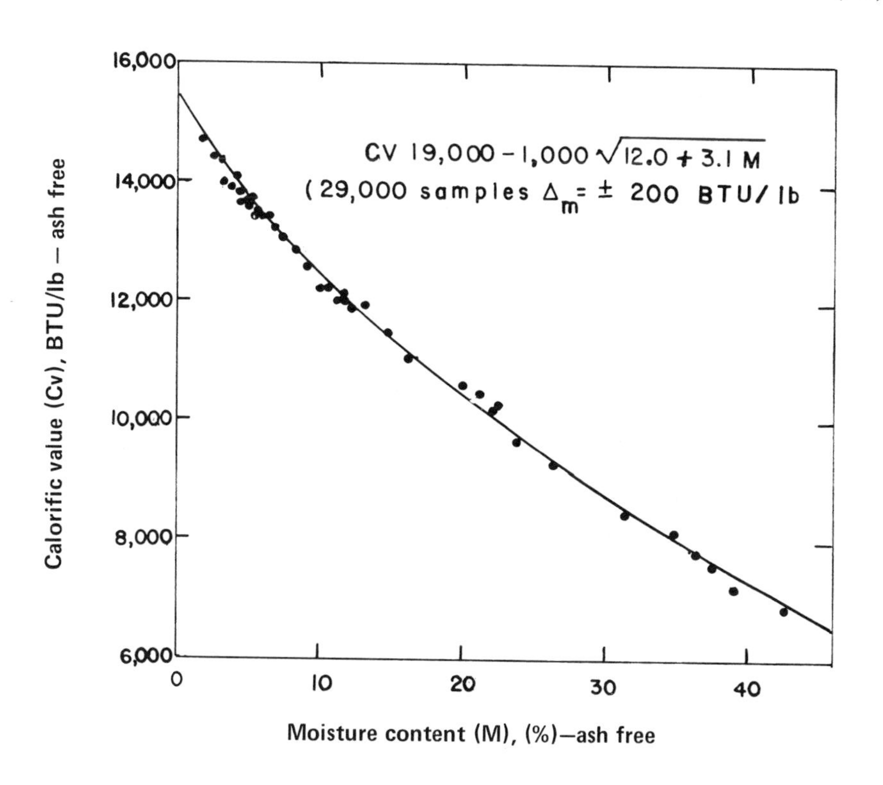

Figure 3.2. Calorific values of western coals as a function of their moisture contents.

ISBN 0-201-08300-0

where CV = calorific value. A regression analysis of the data given the Bureau of Mines IC8693 showed that formula 3.3 represents the data with a correlation coefficient over .99 and with a standard deviation under ±200 Btu for a mean calorific value of aboue 13,000 Btu/lb. Ergun also investigated the possibility of classifying high volatile A bituminous and lower rank coals by their moisture contents, their calorific values being kept within the ASTM rank classification requirements. He arrived at the moisture content demarcation values shown in Table 3.4 for coal classification.

Table 3.4

Classification of High Volatile A Bituminous and Lower Rank Coals by Their Moisture Contents[a]

	Moisture Content Range	
Rank	Lower	Upper
hvAb	–	4.3
hvBb	4.3	7.8
hvCb	7.8	15.0
sub A	15.0	21.0
sub B	21.0	25.0
sub C	25.0	33.5
lig	33.5	–

[a] Ergun [41].

Ergun also investigated the conformity to the ASTM standards of the calorific values of samples falling into each rank category by their moisture contents. The results are shown in Table 3.5. Columns 7 and 8 show the number of samples that would fall into the adjacent rank category by the ASTM standards. For example, 7442 samples classified as hvCb by their moisture contents had proper calorific values to classify them as hvCb by the ASTM standards. On the other hand, 108 samples would belong to the sub A classification by the ASTM standards. The differences encountered with the two classification criteria involved only 167 samples out of a total of over 29,000. This was rather remarkable in view of the fact that repeated calorific value determinations of high volatile C bituminous and subbituminous A coals frequently do not yield the same rank designations.

Moisture determination is much simpler and less expensive than calorific value determination. In addition to being an important parameter in coal utilization, moisture content evidently permits estimation of the calorific value of coal and serves as an index of rank.

ISBN 0-201-08300-0

Table 3.5

Classification of High Volatile Bituminous and Lower Rank Coals by Their Moisture Contents, and Conformity Comparison with ASTM Standards

Rank of Coal	ASTM Calorific Value Range		Moisture Content Range		Conformity Comparison, Number of Analyses		
	Lower	Upper	Lower	Upper	Within	Below	Above
hvAb	14,000	–	–	4.3	2,199	–	–
hvBb	13,000	14,000	4.3	7.8	9,610	–	7
hvCb	11,500	13,000	7.8	15.0	7,442	108	–
sub A	10,500	11,500	15.0	21.0	624	29	18
sub B	9,500	10,500	21.0	25.0	5,601	–	–
sub C	8,300	9,500	25.0	33.5	350	–	–
lig	–	8,300	33.5	–	3,045	–	5
Total					28,871	137	30

3.2. Coal Grade and Grade Classification

The grade of a coal depends largely on the amount and nature of the mineral impurities associated with the coal. The mineral matter content is related to the ash produced upon burning, and ash is therefore commonly used in grading coals. Other parameters which affect the grade of a coal are moisture, size, and calorific value.

A. ASTM CLASSIFICATION

The ASTM standard specifications for classification of coals by grade cover quality as determined by size, calorific value, ash content, ash softening temperature, and sulfur. Table 3.6 shows the ASTM grading of coals according to ash content, ash softening temperature, and total sulfur content. Further classification of the listed grades is possible by specifying the inorganic (pyrite) and organic sulfur contents, as will be discussed later.

B. GRADE CLASSIFICATION BY SULFUR CONTENT

At the present time one of the major obstacles to increased coal utilization is the sulfur contained in the coal mined. The current Environmental Protection Agency (EPA) regulations require that the sulfur dioxide content of the stack gases

ISBN 0-201-08300-0

not exceed 1.2 lb/million Btu generated in the industrial boilers. The Bureau of Mines recently reported that half of the coal burned by electric utilities in 1975 did not conform to environmental regulations [43]. Inasmuch as nearly 75% of the coal mined is burned in electric utility and industrial boilers [44], the sulfur reduction potentials of American coals, especially those of eastern coals, deserve special attention.

Table 3.6

Symbols for Grading Coal According to Ash, Softening Temperature of Ash, and Sulfur Where Analyses are Expressed on Basis of the Coal as Sampled

Ash[a]		Softening Temperature of Ash[b]		Sulfur[a]	
Symbol	Percent,[c] Inclusive	Symbol	Degrees Fahrenheit, Inclusive	Symbol	Percent, Inclusive
A4	0.0– 4.0	F28	2800 and higher	S0.7	0.0–0.7
A6	4.1– 6.0	F26	2600–2790	S1.0	0.8–1.0
A8	6.1– 8.0	F24	2400–2590	S1.3	1.1–1.3
A10	8.1–10.0	F22	2200–2390	S1.6	1.4–1.6
A12	10.1–12.0	F20	2000–2190	S2.0	1.7–2.0
A14	12.1–14.0	F20 minus	Less than 2000	S3.0	2.1–3.0
A16	14.1–16.0			S5.0	3.1–5.0
A18	16.1–18.0			S5.0 plus	5.1 and higher
A20	18.1–20.0				
A20 plus	20.1 and higher				

[a] Ash and sulfur shall be reported to the nearest 0.1% by dropping the second decimal figure when it is 0.01–0.04, inclusive, and by increasing the percentage by 0.1% when the second decimal figure is 0.05–0.09, inclusive. For example, 4.85–4.94%, inclusive, shall be considered to be 4.9%.

[b] Ash softening temperatures shall be reported to the nearest 10°F; for example, 2635–2644°F, inclusive, shall be considered to be 2640°F.

[c] For commercial grading of coals, ranges in the percentage of ash smaller than 2% are commonly used.

To meet the EPA requirements, the total sulfur content of the coal burned must satisfy the inequality

$$S_T \leqslant 6 \times 10^{-5} H \quad (3.4)$$

in which S_T is the weight percent of total sulfur, and H is the calorific value, Btu/lb, of coal as burned. Some coals meet these requirements as received or after removal of a portion of mineral matter by commercial washing techniques. Such coals may properly be referred to as low total sulfur (LTS) coals.

The organic sulfur contents, S_O, of some coals satisfy expression 3.4, whereas their total sulfur contents do not. The coals that satisfy the inequality

$$S_O \leqslant 6 \times 10^{-5}\ \mathrm{H} < S_T \quad (3.5)$$

may be termed low organic sulfur (LOS) coals. Some LOS coals can be cleaned

ISBN 0-201-08300-0

by physical methods to meet the EPA standards, that is, expression 3.4, with varying degrees of Btu recovery.

A considerable fraction of pyrite in many coals is embedded as very small particles (e.g., less than 50 μ in size). Finely distributed pyrite cannot, in principle, be removed by physical methods [45]. There are, however, several chemical processes under development that claim nearly complete removal of pyritic sulfur [46]. Thus LOS coals could be cleaned to meet the EPA standards physically and/or chemically if and when chemical cleaning processes prove to be technically and economically feasible.

For some chemical processes under development [46] the removal of a fraction, f, of the organic sulfur, in addition to the pyrite sulfur, is claimed. We may therefore consider a third class, the moderate organic sulfur (MOS) coals, which satisfy the inequality

$$(1 - f)S_O \leqslant 6 \times 10^{-5} HS_O \tag{3.6}$$

Such coals could be cleaned chemically or by a physical-chemical combinational approach when the processes are developed. The remaining coals, that is, those satisfying

$$(1 - f)S_O > 6 \times 10^{-5} H \tag{3.7}$$

may be classified as high organic sulfur (HOS) coals.

Table 3.7 shows the coal reserve base in the eastern United States by classification according to sulfur form and sulfur content. The figures shown are estimates based on analyses reported [47]. The table shows that the largest reserves of LTS coals in the East are in West Virginia, followed by Kentucky, Virginia, Illinois, Pennsylvania, Indiana, and Alabama. About 69% of Virginia coals fall into the LTS category, followed by Alabama (41%), Tennessee (39%), Kentucky (27%), and West Virginia (25%). The largest reserves of LOS coals are in Pennsylvania, followed by West Virginia and Illinois.

The estimate of low total sulfur coals in the eastern states shown in Table 3.7 is 12.7%. The estimate of eastern reserves containing less than 1.0% sulfur is 13.7%, according to Thomson and York [48]. Since many eastern coals containing more than 0.8% sulfur do not necessarily meet the current EPA standards, the estimates of LTS coals shown in Table 3.7 appear to be reasonable.

Estimates of low organic sulfur coals are particularly relevant to the development of coal cleaning technology. The figure shown in Table 3.7 is 29%. It is interesting to note that 33% of the coals sampled from the eastern states by Caballero et al. [49] belong to the LOS category. The reserve base estimates shown indicate that if pyrite is removed from the LOS coals the percentage of eastern coals that can be used in utility and industrial boilers will be tripled. Table 3.7 also shows that an additional 16% of the eastern coals would meet the current EPA

ISBN 0-201-08300-0

standards if, in addition to pyritic sulfur, 40% of organic sulfur could be removed by chemical cleaning processes.

Table 3.7

Reserve Base, million short tons, of Bituminous Coal in the Eastern States by Total Sulfur and Organic Sulfur Content

State	LTS	LOS	MOS	HOS	Total
Alabama	800	1,030	120	–	1,950
Illinois	1,970	11,820	9,850	42,030	65,670
Indiana	850	2,020	1,810	5,950	10,630
Kentucky	6,900	4,600	1,530	12,520	25,550
Maryland	120	580	350	–	1,050
Ohio	120	5,150	5,690	10,120	21,080
Pennsylvania	1,190	17,430	4,540	720	23,880
Tennessee	390	290	230	80	990
Virginia	2,420	1,020	30	40	3,510
West Virginia	9,900	11,880	7,120	10,690	39,590
Total	24,660	55,820	31,270	82,150	193,900
	(12.7%)	(28.8%)	(16.1%)	(42.4%)	(100%)

Table 3.8 shows the coal reserve base in western states by classification according to sulfur form and sulfur content. The accuracy of the figures shown is questionable for many states because sufficient data are not available. However, the table gives a fair indication of the reserve base by cleanability potential for many states, including Colorado, Montana, New Mexico, Oklahoma, Utah, Washington, and Wyoming. According to available data, only about 40% of the western coals would meet the current EPA standards, and about one third would require partial or complete pyrite removal.

A comparison of Tables 3.7 and 3.8 reveals that both the tonnage and the percentage of LTS coals in the western states are higher than those in the eastern states. The tonnage and percentage of LOS coals in the two regions are on comparable scales. About 8% of the western coals contain high organic sulfur, as compared with over 40% of coals in the eastern states. Nevertheless, it is evident that more than half of the western coals will require some beneficiation to meet the current EPA standards.

It should be emphasized that the reserve base estimates made in this study are based on insufficient data and are necessarily approximate [47]. A firm estimate requires acquisition of data based on a rigorous statistical analysis.

ISBN 0-201-08300-0

The amount of sulfur which can be removed by chemical or physical means is one important factor in coal utilization and conversion. The various forms of sulfur and their removal by various techniques are covered in other chapters. For example, hydrodesulfurization of coal by the use of catalyst is discussed in Chapter 5.

Table 3.8

Reserve Base, million short tons, of Bituminous Coal in the Western States by Total Sulfur and Organic Sulfur Content

State	LTS	LOS	MOS	HOS	Total
Alaska	11,070	580	–	–	11,650
Colorado	10,970	3,320	580	–	14,870
Iowa	–	–	550	2,330	2,880
Missouri	–	270	270	8,950	9,490
Montana	30,340	46,330	26,330	5,400	108,400
New Mexico	3,320	820	250	–	4,390
North Dakota	7,330	2,670	4,670	1,330	16,000
Oklahoma	460	510	210	110	1,290
Texas	–	2,450	820	–	3,270
Utah	3,030	510	190	310	4,040
Washington	1,580	270	100	–	1,950
Wyoming	26,280	18,710	8,350	–	53,340
Total	94,380	76,440	42,320	18,430	231,570
	(41%)	(33%)	(18%)	(8%)	(100%)

4. PROPERTIES OF COALS COMMONLY USED IN THEIR CHARACTERIZATION

As pointed out in the preceding sections, chemically, coal consists mainly of carbon, hydrogen, and oxygen with minor amounts of nitrogen and sulfur on an atomic basis. When a dry coal lump is crushed and dried by raising its temperature slightly over 100°C, water vapor is released and thus the coal loses weight. This weight loss is attributed to the inherent moisture content of the coal. The amount of this moisture is not trivial, for in lignite the loss is as high as 45%. Coals also contain shale, kaolin, sulfides of iron, carbonates of calcium and magnesium, alkali chlorides, and trace amounts of over 53 other elements.

Attempts to characterize coals date back to 1837. This section deals with the properties of coals commonly used in their characterization as fossil fuels and/or as sources of coke for smelting operations.

ISBN 0-201-08300-0

4.1. Proximate Analysis

The proximate analysis serves as a simple means for determining the behavior of coals when they are heated. Since an appreciable amount of water vapor is released when coals are heated to above the boiling temperature of water, the first parameter of a proximate analysis is the mositure content of coal. Another major loss occurs when coals are heated in a covered crucible or in other apparatus which prevents the oxidation of the carbon residue. In general, the weight loss becomes nearly stationary at a temperature of about 900°C. This loss is referred to as the volatile matter and constitutes the second parameter of the proximate analysis. If the remaining residue is further combusted, the residue left after the combustion is called ash, and the weight loss on combustion is referred to as fixed carbon. Fixed carbon and ash constitute the third and fourth parameters of the proximate analysis.

It is not always practical to conduct the various determinations stepwise. Thus one sample could be used for moisture content determination, another for the combined mositure and volatile matter loss, and still another for ash determination.

The proximate analysis of a coal is not a difficult operation, provided that some standard procedures are established. The moisture determination is described in ASTM D-2013, and the volatile matter and ash determinations are given in a Bureau of Mines publication [50]. The latter reference describes methods of analyzing and testing coal and coke. There are several other references that also describe these analytical procedures; however, a few factors are cited below which have drawn attention and evoked some controversy as to the meaning of the parameters determined by the proximate analysis.

A. MOISTURE

The state of water in coal and its quantitative determination have been the subject of numerous investigations. How and where a considerable amount of water is held in the coal structure was the main question. The possibility of water being held in cracks, crevices, and large pores provided a plausible answer. However, studies of water absorption isotherms and absorption swelling experiments revealed that part of the water did not seem to be behaving as liquid water [51], leading to postulation of the existence of chemisorbed water [52].

If coal is heated in vacuum at 110°C to a constant weight, a high vacuum cannot be maintained if pumping is discontinued. Moreover, mass spectrometric analyses indicate the evolution of bursts of water vapor if the temperature of the sample is raised above 110°C. Hoeppner et al. [53] found that, when lignite is heated in an inert atmosphere at temperatures of 105–175°C, water is released at measurable rates for as long as 144 hr.

All these observations indicate the complexity of the coal structure and also the need for devising a more meaningful standard procedure for moisture determination. So far, thermal methods of moisture analysis appear to be the simplest and,

ISBN 0-201-08300-0

at first sight, most meaningful. However, there is evidence that the rate of heating, the final temperature of heating, the period of heating at the final temperature, the composition of the surrounding atmosphere, the absolute pressure, the particle size, and so on all affect, to varying degrees, the amount of moisture driven away from a coal specimen; therefore one is not surprised at the number of investigations and of methods developed for determining the moisture contents of coals.

Simplicity and speed are obviously desirable features in any method devised. However, the arbitrariness of the devised thermal methods of analysis led to the development of several other schemes for moisture analysis. In one scheme, called the distillation method, coal is heated in a liquid that has a boiling point higher than that of water and is immiscible with it, for example, toluene, xylene, or kerosene; the water and liquid distillate are collected; and the volume of the water collected is measured. This method has been adopted in several national standards [54].

Other chemical techniques such as extraction, solution, and the use of chemical reagents have also been investigated. In addition, a few physical procedures such as capacitance and resistance measurements have been tried [55]. Although interesting and novel in concept, these methods have found quite limited use. A survey of the literature reveals that the objective of many of the investigations on the water contents of coals was to obtain information about the porosity and the nature of the internal surfaces of coals.

The ASTM (D-2013) procedure recommends three moisture values. First, if the coal is too wet to crush and sieve without losing moisture, it must be air dried (at 10–15°C above room temperature), which results in the value of air-dry moisture. The air-dried coal is then crushed and further dried in a forced air circulation oven at 105–110°C to constant weight. The weight lost in this operation is referred to as the residual moisture. The total moisture is the sum of the two weight losses. The total moisture is calculated from

$$\%\ \text{total moisture} = \%\ \text{air-dry} + \%\ \text{residual}\left(\frac{100 - \%\ \text{air-dry}}{100}\right) \qquad (4.1)$$

In the ASTM Standard Specifications for Classification of Coals by Rank (ASTM D-338) high volatile bituminous and lower rank coals are classified according to their calorific values on the "moist" basis (cf. Section 3.1.B). "Moist" refers to coal containing its natural moisture but not including visible water on the surface of the coal. As a means of estimating the bed moisture of coal that is either visibly wet or coal that may have lost some moisture, an equilibration method is used which is believed to restore the coal to essentially its bed moisture condition. The method consists of wetting the coal with water, draining off excess water, and equilibrating the sample at 30°C in a vacuum desiccator containing a saturated solution of potassium sulfate that maintains a relative humidity of 96%. Equilibration requires 48–72 hr, depending on the rank of the coal. The equilibrated sample can then be air dried to obtain its air-dry mositure content and further dried to calculate its residual and total moisture. The total moisture of an equilibrated sample is referred

ISBN 0-201-08300-0

to either as bed moisture or equilibrium moisture. Although more often than not overlooked, the basis of the coal moisture contents reported in the literature should be examined.

The moisture content is an important factor in the storage and utilization of coals. In the lower rank (high moisture) coals it affects slackening and weathering and is a critical factor in briquetting [56]. It is a dead weight in transportation by rail, barge, or truck and results in a reduction of heating value in combustion. As shown in Section 3.1.E, moisture content can serve as an index of rank and permits prediction of the calorific value of coal. The moisture contents of coals increase with decrease in rank. In Table 3.4 are shown the ranges of the total moisture contents (as received) of high volatile bituminous and lower rank coals. Evidently the total moisture contents of coals correlate with their calorific values, as illustrated in Figure 3.2.

B. VOLATILE MATTER

The term "volatile matter" refers to the loss of weight, corrected for moisture, when coal is heated in specified apparatus under standardized conditions. The volatile matter evolved during heating or pyrolysis of coals consists mainly of the combustible gases hydrogen, carbon monoxide, methane, and other hydrocarbons; tar vapors; and some incombustible gases such as carbon dioxide and water vapor. However, it does not include, by definition, the moisture that can be removed from coal in the moisture determination procedure. Additional moisture produced in pyrolysis is believed to stem in part from chemisorbed water and mainly from thermal decomposition of the coal substance.

The volatile matter (VM) contents of coals increase with decrease in rank. However, for low rank coals, VM content is a highly fluctuating and less predictable parameter than many other coal properties. In Figure 4.1 is shown [57] a plot of calorific values (moist, mineral matter-free basis) of 120 selected coal samples against their VM contents (dry, mineral matter-free basis). It is seen that the calorific value increases somewhat with an increase in VM content from 14.100 Btu/lb (for graphite) to a maximum of about 15,600 Btu/lb at a VM value of about 30%. Beyond a VM value of, say, 35% there appears to be no correlation between the calorific value and the VM content. Figure 4.1 clearly demonstrates that neither the calorific value nor the volatile matter can serve as a criterion of rank for the entire rank spectrum. However, the plot shows that in one region of the spectrum volatile matter is a much more sensitive parameter than calorific value, while the opposite is true for the other region of the spectrum. Therefore it is not surprising that coals containing less than 31% volatile matter are classified according to their VM contents, the cutoff points being 2, 8, 14, 22, and 31 for meta-anthracite, anthracite, semianthracite, low volatile bituminous coal, and high volatile bituminous coal, respectively (cf. Table 3.1). Lower rank coals, that is, those containing more than 31% volatile matter, are classified according to their calorific values, the cutoff values being 14,000, 13,000 11,500, 10,500, 9500, 8300, and 6300 for high volatile A bituminous, high volatile B bituminous, high volatile C

ISBN 0-201-08300-0

bituminous, subbituminous A, subbituminous B, subbituminous C, and lignite A, respectively. Coals containing less than 6300 Btu/lb are called lignite B.

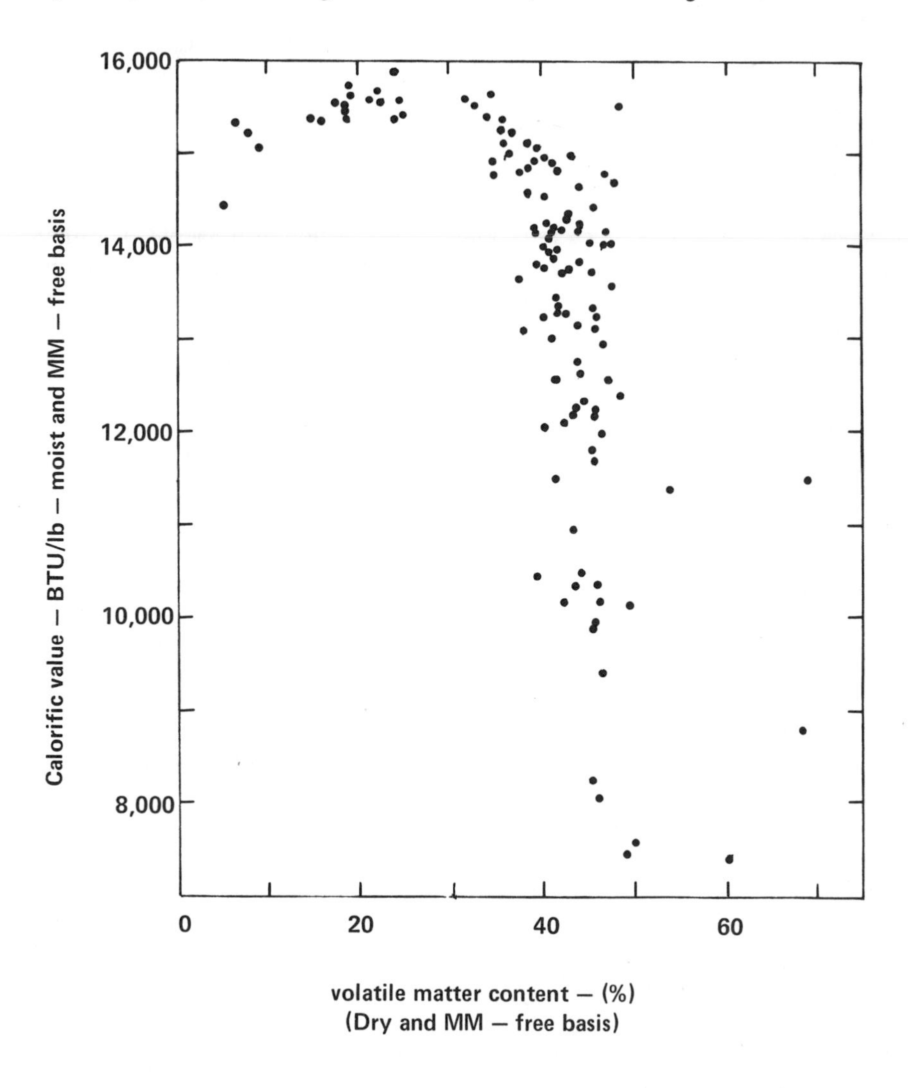

Figure 4.1. Variation of calorific value with volatile matter content.

ISBN 0-201-08300-0

In Figure 4.1 there are at least three points that seem to depart from the trend. For example, the calorific value of a sample containing 69% volatile matter is reported to be in excess of 11,500 Btu/lb. Such an anomaly could be explained readily if the coal in question contained large amounts of resin or other hydrogen-rich petrographic entities. For example, the VM contents of a resin sample obtained from a bituminous coal from Utah was 90.5%, its calorific value being in excess of 18,500 Btu/lb. However, petrographic analysis of the sample in question showed no resin and very small amounts of exinite [58]. The discrepancy could very well be the result of an error in analysis. The point to be stressed here is that abnormal properties of coals can arise from abnormal petrographic compositions.

Figure 3.2 shows an excellent correlation between the calorific values and moisture contents of coals; Figure 4.1, on the other hand, shows no discernible correlation between calorific value and VM content. Van Krevelen and Chermin [59] suggested that the VM content of coal provides a direct link with coal analysis practice and that it permits an estimation of the fraction of the aromatic carbon in coals. Since coals assume a more aromatic character with an increase in rank (and very often a decrease in hydrogen/carbon ratio is used as an indication of an increase in aromaticity), we may expect a correlation between VM content and hydrogen/carbon ratio [57]. This possibility is explored in Figure 4.2, using the Pennsylvania State data [58] and those furnished by van Krevelen and Chermin [59]. Considering the fact that both VM content and hydrogen/carbon ratio are greatly influenced by the petrographic makeup and sulfur contents of coals, the correlation observed is significant. The correlation seen in Figure 4.2 between VM content and hydrogen/carbon ratio seems to be poor compared to that in Figure 3.2 between CV and M. However, it may be pointed out that each point seen in Figure 3.2 represents the average of several hundreds of analyses, whereas the points in Figure 4.2 represent the results of a single analysis. Although an empirical parameter, VM content does not appear to be devoid of structural significance.

C. MINERAL MATTER AND ASH

Visual examinations of coal beds reveal the presence of extraneous matter as definite horizontal layers varying in thickness and extent, as surface deposits or fillings in the vertical cleats, and less frequently as intrusions of claylike or other material irregularly distributed through the coal. The horizontal layers usually contain shale and/or pyrite and vary greatly in thickness, from thin streaks to layers of several inches. They may be limited or local in extent or cover great areas, for example, the blue band of Illinois No. 6 seam. A comprehensive review of the mineral matter found in coals has been given by Thiessen [60] and Ode [61]. Some of the minerals present in coals are believed to have been introduced during the biochemical or peat stage (syngenetic minerals) and some during the metamorphic stage (epigenetic minerals) [62].

Although anthracties generally contain more mineral matter than do lignites, there is no correlation between any property of coal and its mineral matter content. For example, the mineral matter contents of high volatile A bituminous coals of Alabama range from 1.7 to over 16% [49]. Even the mineral matter content of the same coal seam (e.g., Pittsburgh) shows wide fluctuations (from 5 to over 38%) at different mines. Also, it is not uncommon to find differences of over 10% in the mineral matter contents of top and bottom benches at a given mine [49].

ISBN 0-201-08300-0

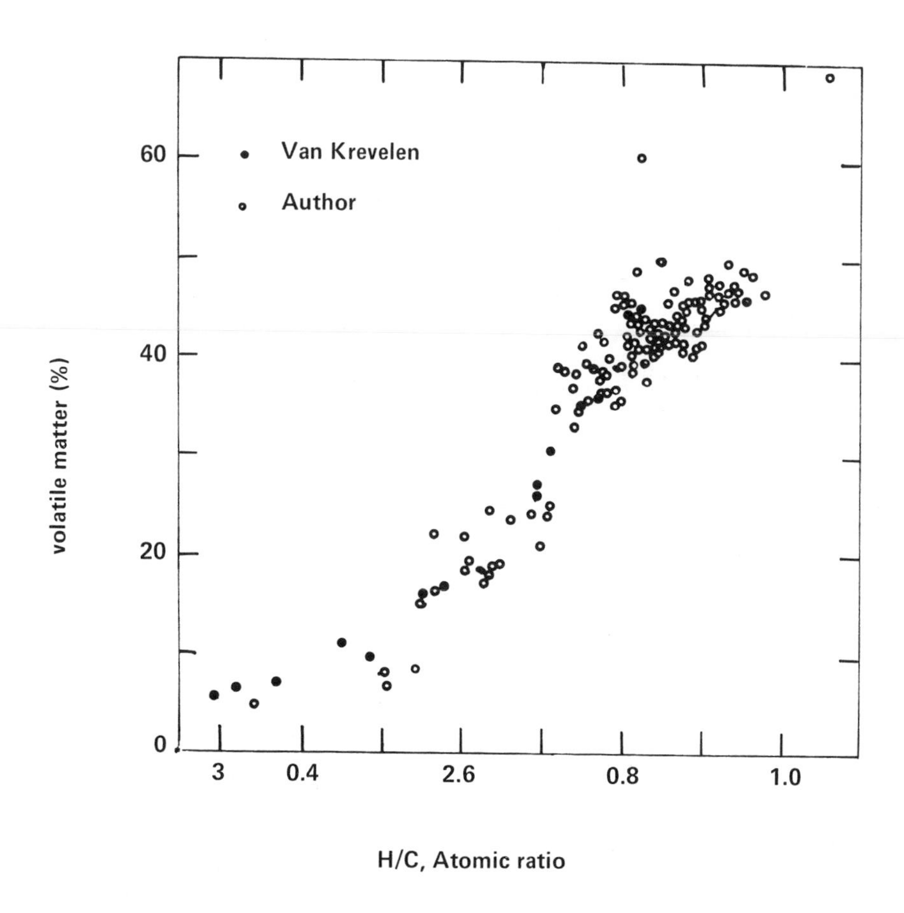

Figure 4.2. Volatile matter contents of coals as a function of H/C ratio.

Table 4.1 shows the minerals commonly associated with bituminous coals [63]. According to Nelson, more than 95% of the mineral matter associated with bituminous coals consists of species belonging to shale, kaolin, sulfide, and chloride [63].

Although the concentration of mineral matter present in coals can be determined by X-ray diffraction upon enrichment of mineral matter, earlier workers usually determined the composition of ash produced upon combustion of coal. The limits of the ash compositions of bituminous coals are shown in Table 4.2 [64-66]. It is seen that the most abundant element in coal ash is silicon oxide, followed by oxides of aluminum, iron, and calcium, suggesting that aluminum silicates are the most abundant element in bituminous coals. The reported iron and sulfur contents of German coals are high, reflecting the presence of relatively large amounts of pyrites in the original coals. The figures shown were compiled from the

ISBN 0-201-08300-0

available analytical data on coal ash in the respective countries and do not necessarily correspond to the reserve base.

Table 4.1

Minerals Associated with Bituminous Coals[a]

Group	Species	Formula
Shale	Muscovite Hydromuscovite Illite Bravaisite Montmorillonite	$(K, Na, H_3O, Ca)_2(Al, Mg, Fe, Ti)_4$ $(Al, Si)_8O_{20}(OH, F)_4$
Kaolin	Kaolinite Livesite Metahalloysite	$Al_2(Si_2O_5)(OH)_4$
Sulfide	Pyrite Marcasite	FeS_2
Carbonate	Ankerite Ankeritic calcite Ankeritic dolomite Ankeritic chalybite	$(Ca, Mg, Fe, Mn)CO_3$
Chloride	Sylvine	KCl
	Halite	$NaCl$
Accessory minerals	Quartz	SiO_2
	Feldspar	$(K, Na)_2O \cdot Al_2O_3 \cdot 6SiO_2$
	Garnet	$3CaO \cdot Al_2O_3 \cdot 3SiO_2$
	Hornblende	$CaO \cdot 3FeO \cdot 4SiO_2$
	Gypsum	$CaSO_4 \cdot 2H_2O$
	Apatite	$9CaO \cdot 3P_2O_5 \cdot CaF_2$
	Zircon	$ZrSiO_4$
	Epidote	$4CaO \cdot 3Al_2O_3 \cdot 6SiO_2 \cdot H_2O$
	Biotite	$K_2O \cdot MgO \cdot Al_2O_3 \cdot 3SiO_2 \cdot H_2O$
	Augite	$CaO \cdot MgO \cdot 2SiO_2$
	Prochlorite	$2FeO \cdot 2MgO \cdot Al_2O_3 \cdot 2SiO_2 \cdot 2H_2O$
	Diaspore	$Al_2O_3 \cdot H_2O$
	Lepidocrocite	$Fe_2O_3 \cdot H_2O$
	Magnetite	Fe_3O_4

ISBN 0-201-08300-0

Table 4.1 (Continued)

Group	Species	Formula
Accessory minerals (continued)	Kyanite	$Al_2O_3 \cdot SiO_2$
	Staurolite	$2FeO \cdot 5Al_2O_3 \cdot 4SiO_2 \cdot H_2O$
	Topaz	$(AIF)_2SiO_4$
	Tourmaline	$H_9Al_3(BOH)_2Si_4O_{19}$
	Hematite	Fe_2O_3
	Penninite	$5MgO \cdot Al_2O_3 \cdot 3SiO_2 \cdot 2H_2O$

[a] Nelson [63].

It was recognized that the ash obtained upon coal incineration is different in chemical composition and is lower in weight than the mineral matter originally present in coal. Obviously, a standard procedure had to be adopted, and an empirical formula relating the weight of the ash to the weight of the original mineral matter had to be developed. These procedures have been described in several texts [67]. The simplified empirical formulas commonly employed in calculating the mineral matter are as follows:

$$MM = 1.08 / \text{ash} + 0.55S_{total} \quad \text{(Parr)} \tag{4.2}$$

$$MM = 1.08 / \text{ash} + 0.1S_{total} \quad \text{(ASTM)} \tag{4.3}$$

If analytical data are available, the formula used is [66]

$$MM = 1.13 / \text{ash} + 0.5S_{pyritic} + 0.8CO_2 - 2.8(CO_2)S_{ash} + 2.8S_{SO_4} + 0.3C1$$

Some other aspects of the mineral elements in coal are their catalytic effects on the chemical conversion of coal and the problem of disposal and utilization of the large amounts of mineral matters and ash produced as a result of the large-scale use of coal. Both of these aspects are discussed in detail in connection with coal liquefaction in Chapter 5.

D. FIXED CARBON

The term "fixed carbon" (*FC*) refers to the weight loss upon combustion of a devolatilized coal sample. When expressed on a dry and ash-free basis, *FC* is a value complementary to volatile matter content, that is, their sum is 100. Thus Figure 4.2 also shows fixed carbon as a function of the hydrogen/carbon ratio. Although a complementary parameter, fixed carbon + mineral matter is a useful one in coking operations.

ISBN 0-201-08300-0

Table 4.2

Typical Limits of Ash Composition of Bituminous Coals[a]

Constituent	United States [64]	England [65]	Germany [66]
SiO_2	20–60	25–50	25–45
Al_2O_3	10–35	20–40	15–21
Fe_2O_3	5–35	0–30	20–45
CaO	1–20	1–10	2–4
MgO	0.3–4	0.5–5	0.5–1
TiO_2	0.5–25	0–3	–
$Na_2O + K_2O$	1–4	1–6	–
SO_3	0.1–12	1–12	4–10

[a] Ode [61].

4.2. Ultimate Analysis

Basically the ultimate analysis of coal is intended to provide the following information:

1. The moisture content on an as-received or air-dried basis under standardized conditions.
2. The mineral matter content on an as-received or moisture-free basis. More often the ash content is reported.
3. The elemental composition of the coal substance.

It is obvious that the elemental composition of coal may be reported on different bases, that is, as-received, dry, dry and ash-free, or dry and mineral matter-free. The distinction between mineral matter and ash (cf. Section 5.1 C) is often ignored.

At the present time ultimate analyses of coals are done by standardized chemical procedures and have been described in various texts [50, 61, 67]. Carbon and hydrogen, which together contribute virtually all of the energy derived from coal, are determined by burning the coal sample in oxygen in a closed system and collecting and determining quantitatively the combustion products. The ASTM standards recommend combustion at 800–850°C, whereas British standards call for a much higher temperature, 1350°C. The carbon content of coal determined by the combustion method includes the carbon of the mineral carbonates, and the hydrogen value includes the hydrogen of the moisture of the sample as well as that of the mineral hydrates. Therefore one has to be careful about the bases of ultimate

ISBN 0-201-08300-0

analyses. In the literature one often encounters a "dry basis" designation without specification as to whether air dry or total dry (cf. Section 3.1 E).

Nitrogen occurs almost exclusively in the organic matter of coal: its percentage ranges from about 1 to usually less than 2%, mainly lying between 1.0 and 1.5%. It is determined by destructive digestion of coal with a mixture of concentrated sulfuric acid, potassium sulfate, and a catalyst (mercury or a mercury salt, selenium or one of its compounds, or a mixture of the two). The procedure is called the Kjeldahl method. The nitrogen is distilled from an alkaline solution as ammonia into a standard acid solution for titration.

Sulfur occurs in coal as iron sulfide and in combination with the organic coal substance. In some weathered coals small amounts of sulfate sulfur are usually found. Several chemical methods have been developed for determining total sulfur in coals [61], but usually one of two chemical methods is used [67]. In one method the coal is oxidized in an Echka mixture (2 parts by weight of light calcinated magnesium oxide and 1 part anhydrous sodium carbonate); the resulting oxides of sulfur are retained in an alkaline solution, extracted with water, precipitated as barium sulfate, and determined gravimetrically. In the second method (British standards) coal is combusted with oxygen at high temperatures in a closed system, the sulfur oxides are absorbed in aqueous hydrogen peroxide, and the sulfuric acid formed is titrated with a standard alkali. In most schemes of coal analysis for forms of sulfur (sulfate, pyritic, and organic), the sulfate and pyritic sulfur are determined directly, and the organic sulfur is taken as the difference between the total sulfur and the sum of sulfate and pyritic sulfur. The sulfate sulfur determination involves extraction of soluble sulfates with a solution containing ethanol, water, and hydrochloric acid, and their precipitation as barium sulfate. Pyritic sulfur is determined using a sample from which sulfate sulfur has been extracted. In the pyritic sulfur determination the sulfide sulfur is removed from coal by digesting it with nitric acid, and the filtrate is then subjected to rather complicated reactions [67].

Evidently because of the lack of a simple and reliable method of determining oxygen in coal directly, the difference between 100 and the sum of the percentages of carbon, hydrogen, nitrogen, sulfur, ash, and whatever else is determined is regarded as oxygen. Numerous attempts have been made to develop chemical methods to determine oxygen directly; cf. Ref. 61. If a reliable direct method is found and if this method gives results significantly different from those obtained by difference, the discrepancy may point to the shortcomings of the methods used in the determination of other elements. Physical methods, for example, combined X-ray diffraction and X-ray fluorescence, are at such a stage that they may soon replace chemical methods.

In Table 4.3 are shown the ultimate analyses of selected coal samples representing different ranks [68]. Excluded in the table are the chlorine, sodium, and other alkali halides. Nitrogen is the least variant element in coals, ranging from 0.6 to over 1.6%. The chlorine contents of coals may range from trace amounts up to 0.3%; this element is generally more prevalent in lignites. The sulfur contents of

ISBN 0-201-08300-0

coals may range from as low as 0.3 to over 10%. Anthracites and lignites contain low sulfur; high volatile coals, especially those located in the Midwestern States, contain more sulfur.

Table 4.3

Ultimate Analyses of Selected Coal Samples of Differing Ranks[a]

Sample	Element, wt %, maf[b]				
	C	H	O	N	S
Meta-anthracite[c]	97.9	0.21	1.7	0.2	–
Anthracite[c]	95.9	0.89	1.8	0.3	1.8
Anthracite	92.8	2.7	2.9	1.0	0.6
Semianthracite	90.5	3.9	3.4	1.5	0.7
Low volatile bituminous	90.8	4.6	3.3	0.7	0.6
Medium volatile bituminous	89.1	5.0	3.6	1.7	0.6
High volatile A bituminous	84.9	5.6	6.9	1.6	1.0
High volatile B bituminous	81.9	5.1	10.5	1.9	0.6
High volatile C bituminous	77.3	4.9	14.3	1.2	2.3
Subbituminous A	78.5	5.3	13.9	1.5	0.8
Subbituminous B	72.3	4.7	21.0	1.7	0.3
Subbituminous C	70.6	4.8	23.3	0.7	0.6
Lignite	70.6	4.7	23.4	0.7	0.6

[a] Ergun [68].

[b] Moisture- and ash-free basis.

[c] Coals that yield three dimensional (hkl) crystalline X-ray patterns of graphite to a significant extent.

An inspection of Table 4.3 reveals that the carbon contents of coals decrease with decrease in rank. However, this is merely a trend; the carbon contents of coals of adjacent ranks usually overlap. A reverse trend is observed with oxygen content. Again, we should caution that the oxygen contents of adjacent-rank coals overlap considerably, although overlap is minimal in Table 4.3. The table is merely intended to give an idea of the ranges of the major elements of the coal matrix in various coals.

The ranges of the percentages of carbon, hydrogen, and oxygen shown in Table 4.3 raise the question of whether coals could be classified on the basis of their ultimate analyses. This point has already been discussed in Section 3.1A. Coal researchers have found that the hydrogen/carbon ratios of coals correlate well with many physical properties of coals [23, 69], for example, reflectance, index of

ISBN 0-201-08300-0

absorption, and specific volume. For bituminous and higher rank coals, the hydrogen/carbon ratio often has been used as a measure of rank. As pointed out in Section 4.1A, classification by ultimate analysis did not find much support.

In addition to Table 4.3, the triaxial diagram shown in Figure 3.1 gives an idea of the percentages of carbon, hydrogen, and oxygen in coals. These three elements are the most important as far as calorific value and many other properties of coals are concerned. They are not considered as environmentally objectionable materials in the combustion of coals.

4.3. Calorific Value

The calorific value of coal is a direct measure of the chemical energy stored in it and therefore is an indispensable parameter for determining the value of coal as a fuel. Its importance is further manifested by the fact that it is the parameter used in ASTM rank classification of major coal deposits in the United States. A brief description of the methods used for determination of the calorific values of coals and the methods proposed to estimate this parameter from other properties of coals are given below.

A. CALORIMETRIC DETERMINATION

Direct calorimetric determinations involve burning a weighed amount of coal in a closed vessel under a pressure of oxygen and measuring the amount of heat released. Since calorimetry is a well established chemical procedure, it is not essential to elaborate upon it. For convenience in calculations, thermochemical processes are usually conducted in near-isothermal or near-adiabatic conditions. In the former procedure heat exchange between the medium (usually water) surrounding the calorimeter (maintained at a specified temperature) and a heat sink must be provided. In the latter procedure the surrounding medium must be well insulated so that the rise in its temperature permits calculation of the heat released in the enclosed vessel. Whether operated in isothermal or adiabatic mode, the calorimeter procedures require extreme care. In the ASTM (D-2015) procedure the adiabatic method is described in detail [50].

The calorimeters provide a direct measure of heat of combustion. Although simple in operation, they are not the easiest equipment to design, construct, or operate. Therefore coal researchers have looked for correlations between the calorific values of coals and other physical-chemical properties that are easily determined. Valid correlations between calorific value and other coal properties would, of course, provide indirect means of estimating calorific values.

B. ESTIMATION FROM ULTIMATE ANALYSES

Dulong's observation regarding the additive nature of the heats of combustion of the elements present in a compound led researchers to develop empirical equations relating the calorific value of coal to its ultimate analysis. The observation,

ISBN 0-201-08300-0

that is, Dulong's law, is approximately correct in that the bonding energy is neglected. The simplest expression proposed was [70]

$$CV = 145.44X_C + 620.28(X_H - \frac{X_O}{8}) + 40.5X_S \tag{4.4}$$

In this formula CV is the calorific value, Btu/lb, on a dry, ash-free basis; the X's denote the weight percentages of the elements carbon, hydrogen, oxygen, and sulfur as designated by the subscripts; and the coefficients of the X's are the calorific values of carbon, hydrogen, and sulfur, respectively, from left to right. It is tacitly assumed that nitrogen is inert and that the oxygen of the coal is associated with hydrogen in the proper ratio to form water. Thus the excess hydrogen, together with carbon and sulfur, is available for combustion.

Selvig and Gibson [70] list nine other empirical formulas relating the calorific value of coal to its ultimate analysis. One formula, based on the most extensive investigation of British and American coals, is that of Mott and Spooner [71]:

$$CV = 144.5X_C + 610X_H - 62.5X_O + 40.5X_S \tag{4.5}$$

for coals containing less than 11% oxygen, and

$$CV = 144.5X_C + 610X_H - (65.9 - 0.31X_O)X_O + 40.5X_S \tag{4.6}$$

for coals containing more than 11% oxygen.

Schuyer and van Krevelen [72] examined the applicability of several empirical formulas relating calorific value to elemental composition for 21 coals ranging in calorific value from 13,000 to 15,800 Btu/lb. Mott and Spooner's formula yielded the least average deviation, that is, mean ±95 Btu/lb, the maximum deviation being −143 Btu/lb and the standard deviation, $\sqrt{\Sigma\Delta^2/N} = 122$ Btu/lb (±0.81%). Schuyer and van Krevelen noted that carbon and hydrogen in coal appear in both aromatic and alicyclic forms and that the bonding energies of aromatic and alicyclic compounds differ. They further postulated that the fraction of aromatic carbon can be predicted from the volatile matter content. Accordingly they proposed the following equation:

$$CV = 1408 - 14.1\text{VM} + 128X_C + 639X_H - 43.9X_O + 55.8X_S + 20.5X_N \tag{4.7}$$

Application of this equation to 21 coals yields results in somewhat better agreement (±0.75 vs. 0.81%) with the observed values than does the Mott-Spooner formula.

Ergun [73] examined the applicability of the Mott-Spooner and Schuyer-van Krevelen formulas to 121 coal samples analyzed by the Pennsylvania State University. The calorific values of the coals studied ranged from 9300 to about

ISBN 0-201-08300-0

14,900 Btu/lb on a dry basis. The Mott-Spooner formula agreed with the measured values with a standard deviation of about ±165 Btu/lb (+1.31%), the maximum deviation being +551 Btu/lb (about 4.8%). On the other hand, the Schuyer-van Krevelen equation resulted in a standard deviation of about ±253 Btu/lb (±2.0%), the maximum deviation being +705 Btu/lb. On this basis it may be concluded that the simpler Mott-Spooner equations (4.5 and 4.6) yield more accurate results that the formula proposed by Schuyer and van Krevelen, Equation 4.7. The Schuyer-van Krevelen approach could perhaps be expected to yield better agreement if its coefficients were determined from a large number of analyses. From a practical point of view, however, the Mott-Spooner formulas appear to be satisfactory if one considers the following factors:

1. Coal is an extremely heterogeneous substance. To obtain a representative sample weighing a few grams poses difficulties.
2. Calorific value determinations are subject to errors. Deviations up to 50 Btu/lb are expected between the results of duplicate analyses by the same analyst, and twice as much deviation between the results of the averages of duplicate determinations from two laboratories is not uncommon. This fact accounts for more than 60% of the resultant deviation using the Mott-Spooner formulas.
3. There are recognized variations between the results of duplicate ultimate analyses performed by the same operator. The limits of permissible variances are 0.3, 0.07, 0.05, and 0.1% for carbon, hydrogen, nitrogen, and sulfur, respectively [67].
4. Since oxygen is determined by difference, its maximum allowable variance is 0.52%. The allowable variances by the same analyst may lead to a maximum difference of 123 Btu, using the empirical equation described above.
5. Coal seams are not uniform in composition. This fact is widely recognized by the coal utilizing industry but is not appreciated by coal researchers. It is not uncommon to see in the mineral matter content of a coal extracted from a given mine a variation exceeding several percent in daily operations.

C. ESTIMATION FROM PROXIMATE ANALYSIS

In a similar manner the possibility of predicting the calorific value of coal from a proximate analysis has been explored by many investigators. Selvig and Gibson [70] listed 11 formulas developed between 1902 and 1939 for calculating calorific value from the proximate analysis. Goutal's formula [74], derived in 1902, has been the most widely used:

$$CV = 14{,}760 + a(VM) \tag{4.8}$$

where VM is the percentage of volatile matter, and a is a parameter that is a function

ISBN 0-201-08300-0

of *VM*. On a dry, ash-free basis, Goutal gave the following values of *a* as a function of *VM:*

VM	5	10	15	20	25	30	35	38	40
a	113	86	63	49	38	29	22	5.4	−3.6

Surprisingly, Equation 4.8 has yielded the best agreement with calorific value measurements for medium volatile bituminous and higher rank coals; for this range of rank the standard (ASTM) rank classification is based, accordingly, on the volatile matter content (see rank classification in Section 4.1 B).

D. ESTIMATION FROM MOISTURE CONTENT

Although it has been widely recognized that the calorific values of coals decrease and their moisture contents increase with decrease in rank, no valid correlation between calorific value and moisture content was proposed. In recent years there have been extensive surveys of U.S. coal reserves, and data regarding their mineral matter, moisture, and sulfur contents and their calorific values have been compiled [42, 48]. A statistical analysis by Ergun [41] of the data compiled by the Bureau of Mines led to the formula cited in Section 3.1:

$$CV = 19{,}000 - 1000\sqrt{12.0 + 3.1M} \qquad (4.9)$$

This formula (on a moist, mineral matter-free basis) predicted the calorific values of over 29,000 samples with a standard deviation under ±200 Btu (±1.5%). For coals containing over 12% moisture a simpler relation:

$$CV = 14{,}000 - 172M \qquad (4.10)$$

yielded a better agreement, that is, $\sqrt{\Sigma\Delta^2/N} = 118$ Btu, or 1.25%

It appears that the calorific values of coals can be predicted from their moisture contents with nearly the same accuracy as they can be predicted from ultimate analyses of the coals. In addition to being an important parameter in coal transportation and utilization, moisture content is a simpler parameter to determine than either ultimate analysis or calorific value. Equation 4.9 leads to a maximum calorific value of about 15,540 Btu/lb on a moisture- and mineral matter-free basis, a proper value for the upper rank spectrum of bituminous coals. However, the calorific values of anthracite are lower than those of high rank bituminous coals and gradually approach about 14,100 (for meta-anthracite). Therefore Equation 4.9 should not be used for anthracites.

4.4. Fusibility of Coal Ash

Although the capability exists to quantitatively analyze the mineral forms and/or the elements of the mineral matter in coals by X-ray diffraction and X-ray

ISBN 0-201-08300-0

fluorescence, such techniques are relatively new and have not been widely recognized. The composition of coal ash is commonly determined chemically by producing an ash sample from coal. Usually, a covered capsule containing a coal sample is placed in a muffle furnace and is gradually heated to 750°C in 2 hr. For ash fusibility analyses the coal sample taken is usually 3–5 g; this sample is heated to 800°C in 1½ hr.

Ash fusion tests are based on observing the temperature at which successive characteristic stages of fusion occur in a specimen of ash when heated in a laboratory furnace under specified conditions of atmosphere and rate of temperature rise. In the ASTM method (D-27) the sample of finely ground ash (−200 mesh) is further heated in an oven for 2 hr at 800°C in an oxygen atmosphere. The ash is then moistened with a 10% dextrine solution to make a thick paste and is pressed into a standard mold to form a slender, triangular pyramid, ¾ in. high (see Figure 4.3). The air-dried pyramid is then placed vertically on a ceramic base in a gas (or electrically) heated furnace. The temperature of the furnace is raised quickly to 800°C and then raised at a rate ranging from 7 to 13°C/min. To obtain a mildly reducing atmosphere, excess gas fuel is usually introduced into the burner. Four temperatures are recorded as the characteristic temperatures of the ash. The initial deformation temperature (IDT) is noted when the apex of the pyramid becomes rounded. The softening temperature (ST) is recorded when the pyramid has fused down to a nearly spherical lump having a height equal to the base. The hemispherical temperature (HT) is noted when the height of the lump becomes one half of the base. A final measurement is made of the fluid temperature (FT) when the ash fuses and spreads into a liquid layer with a height of 1 mm or less.

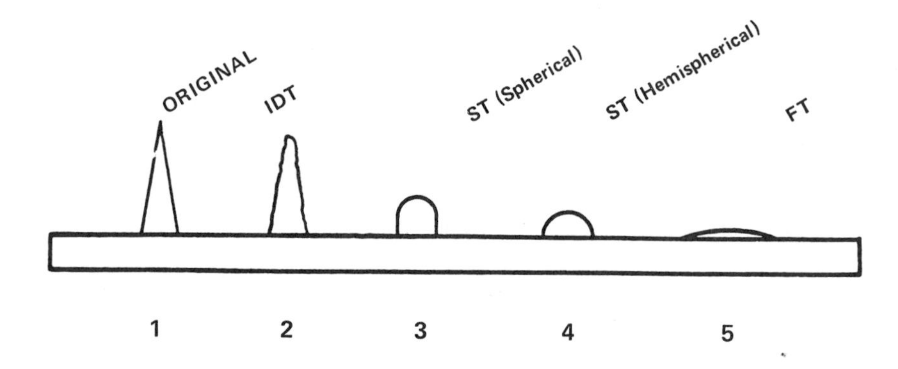

Figure 4.3. Ash fusion test – ASTM method.

ISBN 0-201-08300-0

Inasmuch as coal ash is a mixture having no well defined melting point, significance is attached to all the characteristic temperatures thus determined and to the intervals among them because they vary appreciably for different ashes. The

IDT is broadly identified with a condition of surface stickiness, the ST with plastic deformation or sluggish flow, and the FT with liquid mobility [75]. The ST value is frequently reported as the index of ash fusibility. With oxidizing conditions in the test furnace, many ashes show higher values of fusion temperatures.

Numerous studies have been conducted on the influence of alkali metals, silica, calcium, and so on on the fusion characteristics of coal ash and on the viscosity of ash at selected temperatures. Figure 4.4 shows the dependence of the IDT and ST values of ashes upon the iron contents of the ashes. The complex nature of the mineral matter evidently made it difficult to develop general correlations relating the fusion characteristics of ash to its mineral matter content.

4.5. Behavior on Heating

Regardless of type, rank, and grade, coals release moisture, gases (CO, H_2, CH_4, C_xH_y, CO_2, H_2O, H_2S), and tar vapors upon heating, and they leave a residue of char or coke behind. The compositions and relative amounts of the products depend upon the rate of heating, the final heating temperature, the pressure, the surrounding atmosphere, and, of course, the quality of the coal itself.

On the basis of their behavior, coals are broadly classified as coking and noncoking in the United States and as caking and noncaking in the United Kingdom. "Coking" refers to the behavior of the coal when it is heated rapidly, while "caking" refers to its behavior when it is heated slowly. An additional property, known as agglomerating, usually refers to the behavior of coal when heated rapidly in the volatile matter test. Thus coals are also classified broadly as agglomerating and nonagglomerating.

The above properties and classifications are usually based on empirical tests. In addition, such terms as "plasticity," "dilation," "contraction," and "agglutination" are often encountered in the literature. A very comprehensive review of the plastic, agglutinating, agglomerating, and swelling properties of coals was made by Brewer [26], and the review was updated in 1962 by Loison et al. [27].

More detailed discussions on the behavior of coal upon heating are given in other chapters in connection with coal combustion, pyrolysis, liquefaction, and gasification.

A. FREE SWELLING

In tests designed to measure the free-swelling behavior of coals, a sample (1 g) of coal is placed in a covered crucible or tube without compaction and is then heated at a fixed rate to a temperature of about 800°C. If the coal is infusible under the conditions devised, it distills without agglomeration, and the residue remains in powdered form with no visual change in appearance. Coals showing no fusion are called noncoking. If the particles of coal soften and fuse together, the mass of coal tends to swell in the crucible and resolidifies into porous coke, which is greater in volume than the original coal. The ratio of the coke to the coal volume

ISBN 0-201-08300-0

ISBN 0-201-08300-0

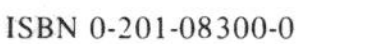

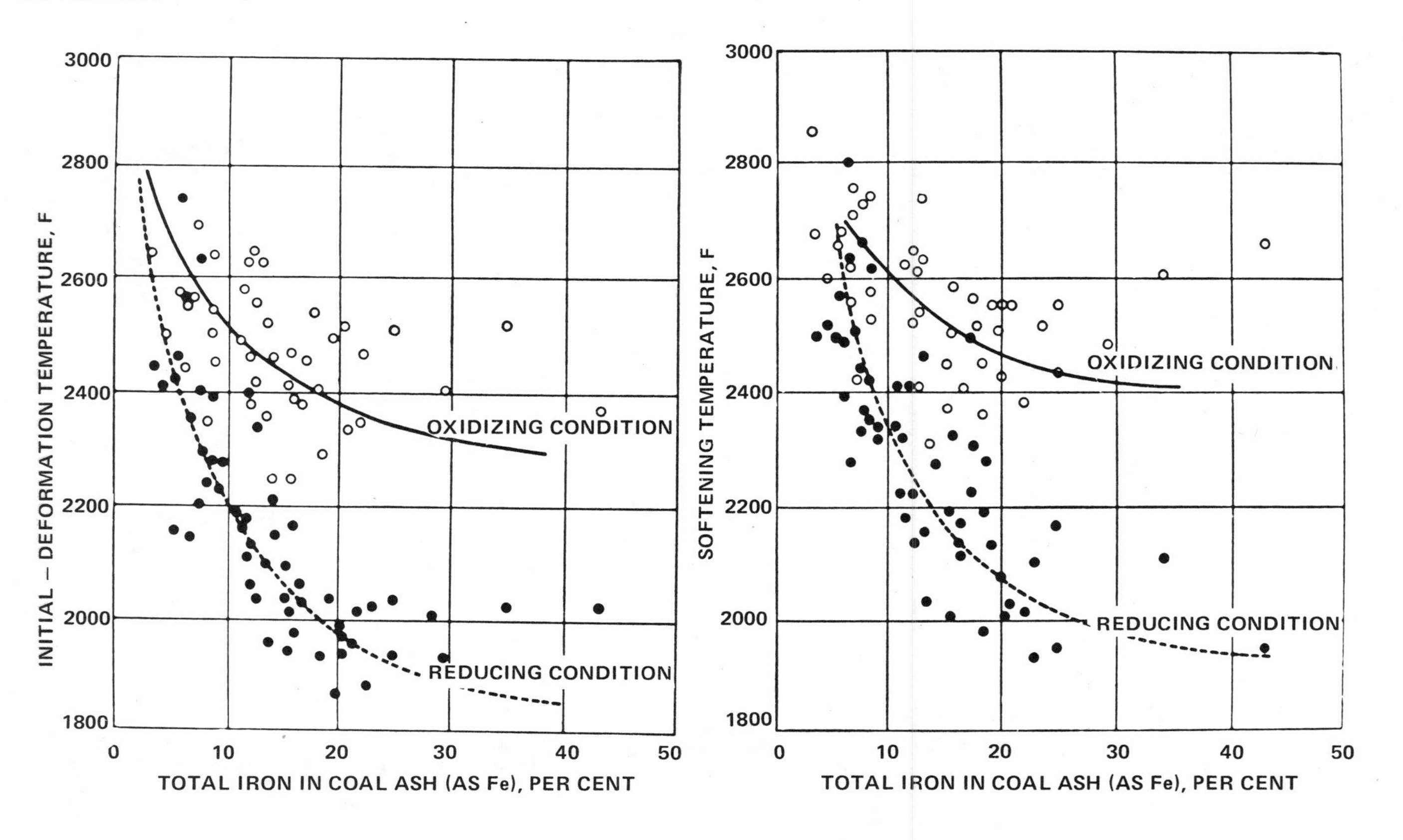

Figure 4.4. Influence of iron on coal-ash fusion temperatures.

could serve as an index of swelling. However, the common practice is to utilize a standardized crucible shape and compare the profile of the coke obtained with a series of reference profiles.

These tests make it possible to classify both weakly and medium coking coals, but they have proved to be ineffective for strongly coking coals. The tests are empirical and do not reflect any associated physical property of coal. Their simplicity in providing a measure of coal rank made them popular. The significance of the free-swelling tests has been discussed by van Krevelen and co-workers [76].

B. AGGLOMERATION

The free-swelling tests are often called agglomerating tests because of the fact that they indicate the tendency of the particles to fuse together and form a compact mass. An agglomeration test based on the free-swelling index described above is often used by the ASTM to differentiate between bituminous coals and the adjacent-rank anthracite and subbituminous coals. In this test, if the agglomerate button shows a cell structure or a swelling, or supports a 500 g weight without disintegrating, the coal used is considered to be agglomerating or bituminous.

C. DILATION

The variation in the volume of a coal sample when it is heated is called dilation if the sample used is a cut piece of coal or, more often, a pellet obtained by the compression of pulverized coal. The dilation measurement could be conducted at atmospheric or high pressures. In general, dilation is measured unidirectionally by following the variation in the length of a sample placed in a cylinder. At first, agglomeration and swelling occur during heating, followed by resolidification and further structural changes in the semicoke formed at higher temperatures. The information gained by the dilatometer tests has been categorized into behavior below the resolidification point and behavior above the resolidification point and up to 1000°C.

Many dilatometers have been devised to study the dilation of coal during heating. The most popular type is the Audibest-Arnu dilatometer. This apparatus is fully described in the publication "International Classification of Hard Coals by Type" [77].

D. PLASTICITY

Upon softening, coals do not become Newtonian fluids. Coal fluidity is a transitory phenomenon that depends upon the rate of heating, the final temperature of heating, and, if held at that temperature, upon the duration of heating. Mainly devised for assessing the coking behavior of coals, several plastometers have been designed and used to study the resistance offered to the rotation of a movable unit in the middle of a mass of powdered coal subjected to a particular rate of heating. Plastometers can be placed in two categories. The first includes the constant-torque plastometers, in which the movable unit is subjected to a constant

ISBN 0-201-08300-0

torque and the rate of rotation is measured. The second includes the variable-torque plastometers, in which the movable unit rotates at constant speed and the torque opposing the rotation is measured. Various plastometers have been designed using one or the other of the two principles cited above; these have been amply described in several texts [27]. A constant-torque Gieseler plastometer is usually used in the ASTM method of plasticity testing (ASTM D-1812).

Another technique for assessing the plasticity of coal utilizes a metal needle which under a given load penetrates the coal sample as it softens [27]. The apparatus used for this experiment is called a penetrometer. Penetrometers have been found to be more sensitive than plastometers to the softening, as well as the resolidification, temperatures in the course of a programmed heating of a compacted coal sample.

4.6. Mechanical Properties

Coal utilization involves many operations that are influenced by the mechanical properties of coal, such as mining, transportation, coarse grinding, fine grinding, injection into reaction vessels, and handling. In the course of mining and preparation, coal is subjected to a variety of forces the nature of which is not always precisely analyzed. Because shattering and grinding are common occurrences in coal handling and preparation, a few empirical tests have been devised to evaluate the degradation that coal undergoes.

A. FRIABILITY

The earliest test devised for measuring the friability of coals was the tumbler test [78]. This test employs a jar mill fitted with lifters that assist in tumbling the coal. A sample of closely sized coal is tumbled in the mill for a certain period of time at a certain number of revolutions per minute. The tumbled coal is then screened, and its average size is determined. If the average size of the tumbled coal is 75% of that of the original sample, its friability is said to be 25%. Many arbitrary parameters are included in this test, for example, size distribution of the original coal, design of the tumbler (shape, size, lifter, etc.), amount of coal charged, frequency and number of tumbles, and sieve size analysis. Since there seems to be no way to correlate the test results with any commercial coal handling operation, this test has been the subject of many investigations as to the meaning of friability tests in general. However, this early test is still used very frequently for coke friability determination.

Another test developed for determining the friability of coal is the drop-shatter test [78]. This test involves two 6 ft drops on any chosen initial size of coal and was adopted from the ASTM standards for coke. Like the tumbler test, the shatter test has many arbitrary features. The friability index is based on sieve analyses of the original and the shattered coal. The ratio of the average size after the test to that before the test multiplied by 100 is termed the size stability, and 100 minus the size stability is termed the friability. This test has also been standardized.

ISBN 0-201-08300-0

Statistical approaches to relate the strength and size distribution of broken coal to the flows and inhomogeneities in the coal have proved to be fruitful. Using the matrix representation of breakage proposed by Broadbent and Callcott [79], Berenbaum [80] analyzed the sieve-analysis size distributions obtained from the shatter tests. Utilizing a strength-size relation expressed as the probabilities of survival of pieces differing in size, Berenbaum developed expressions to predict the size degradation during transportation of coal. Brown and Hiorns [81] considered this study one of the most important advances in the 1950s from the point of view of direct practical use of fundamental studies. The inhomogeneity of a coal seam is usually caused by the variation of coal types and components in the seam. Using empirical tests, Protodyakanov showed that coal strength may vary by as much as 30% over the width of a seam and emphasized the importance of testing a sufficient number of samples representing a particular seam [82]. Pomeroy studied the model test for impact strength proposed by Protodyakanov and devised another test in which a bed of 100 g of 3/8 by 1/8 in. coal was broken by a 4 lb weight dropped 20 times from a height of 12 in. [83]. The impact strength was assessed by the percentage particles of coal having a size larger than 1/8 in. The results of the impact strength indices obtained for various coals are shown in Figure 4.5 as a function of the dry, ash-free volatile matter of the coal [84]. This plot gives some credibility to the impact strength index, although the practical use of the index remains to be demonstrated.

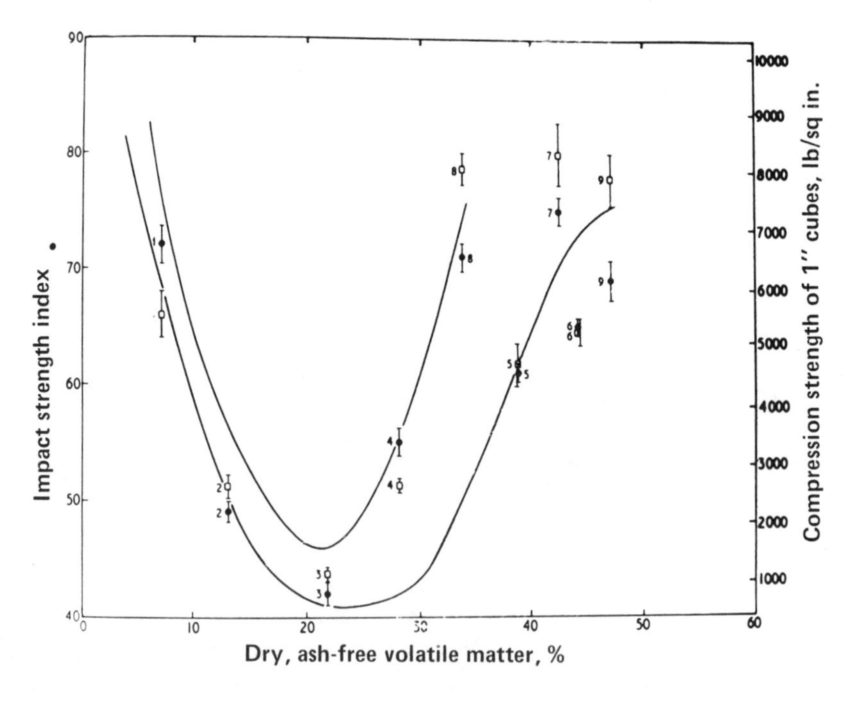

Figure 4.5. Strength and rank.

ISBN 0-201-08300-0

B. GRINDABILITY

The grindability of coal is thought to be a measure of its resistance to crushing. The tests of grindability most commonly employed are the ball-mill test and the test developed by Hardgrove [85]. Although the ball-mill test was standardized by the ASTM, it was discontinued in 1951 [81]. However, this test is still used by many researchers, and correlations have been developed between the ball-mill and Hardgrove indices by the ASTM (D-409). The Hardgrove method is generally preferred to the ball-mill method because the former is faster. The test consists of grinding a specially prepared coal sample in a laboratory mill of standard design. The factors affecting the Hardgrove grindability measurement are (1) the method of preparing the sample, (2) the method of making the test, and (3) the moisture and temperature. The arbitrariness of the various test conditions prompted many investigations into the dependence of the results on the test conditions [81].

The correlation between the grindability index and the rank of coal is shown in Figure 4.6, and it provides a general guide to the ease of grinding of a coal of a particular rank. Many argued, nevertheless, that the rank of coal is not a sufficient guide to its grindability. The curves shown in Figure 4.6, which was prepared by Brown [81], represent the limits for the range of volatile matter within which 20% of the readings obtained in an earlier survey lay; the enclosed range represents the British coals. Lignitic coals, mainly obtained from South America, lie outside the curves of Figure 4.6, as shown by the rectangular data points; and they exhibit, accordingly, high grindability (low indices).

In general, lignite and anthracite coals are more resistant to grinding (have high indices) than are bituminous coals. Both Figures 4.5 and 4.6 indicate an extremum in the respective coal property plotted for medium volatile bituminous coals and hence prove that these coals have the lowest impact strength and are the easiest to grind.

C. ELASTICITY AND STRENGTH

From a theoretical point of view, stress-strain relation, elastic modulus, internal friction, and so on are the fundamental parameters which characterize the structure of coal. Although such properties of coal have been the subject of numerous investigations, interpretation of the results has been beset by uncertainties [81]. This fact is not surprising to those familiar with the microscopic studies of coals. The presence of cracks, flaws, and other types of structure defects unquestionably affects the techniques. Unfortunately, the structure defects or the inhomogeneity of coal has eluded a statistical formulation, let alone a general correlation.

It is not implied here that the measured mechanical properties have proved to be less fruitful than any of the other measured physical properties of coals, for example, density and thermal conductivity. But from the point of view of coal structure, the mechanical properties, like many other properties, have failed to yield any specific information on coal structure and have only confirmed a trend

ISBN 0-201-08300-0

in some respects. Interpretation of the results of stress-strain and elastic module measurements of coals becomes fairly difficult because of the overwhelming influence of structural faults, which renders questionable the meaning of the results of these measurements [81].

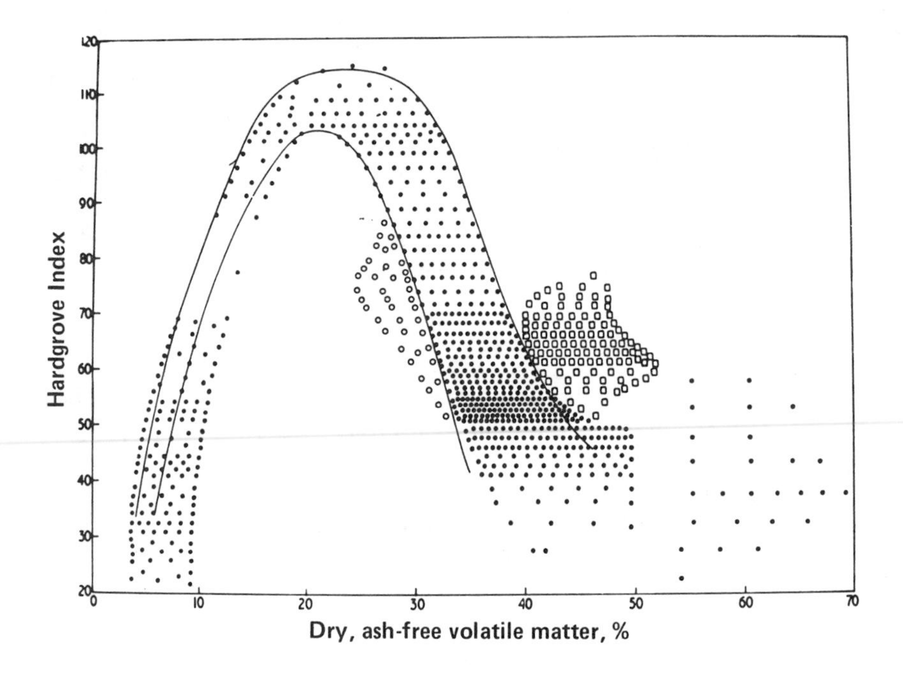

Figure 4.6. Grindability and rank.

Bengham and Maggs [86] noted that methyl alcohol absorption yielded a very large surface area and that coals swelled upon absorption. Assuming that methanol absorption is a true measure of the available internal surface area and that swelling depends upon the surface area, they derived a formula relating Young's modulus to the parameter measured:

$$E = \frac{100\rho\epsilon}{\lambda} \tag{4.11}$$

where E is Young's modulus of elasticity, ρ is the coal density, ϵ is the measured internal surface area, and λ is a constant relating the observed swelling index to the lowering of surface energy. They observed that the moduli thus calculated agreed well with those derived from compression experiments. The questionable validity of some of the assumptions made, in particular those concerning the pore size distribution and the mechanism of coal wetting, and the arbitrariness of λ limit the useful-

ISBN 0-201-08300-0

ness of Equation 4.11. Nevertheless, the Bengham-Maggs approach certainly deserves further attention.

A systematic investigation of the compression strengths of 1 in. cubes of British coals conducted by Pomeroy and Foote [84] revealed that, when compression strength is plotted against volatile matter content, the plot shows the customary minimum (see Figure 4.5) at 20–25% dry, ash-free volatile matter. The compressive strength of coals as measured in the laboratory is frequently less than the static pressure exerted on the seam by the overburden. Measurements of the compression strengths of coal samples under lateral pressures of up to 4000 psi revealed that the fracture stress at 4000 psi is higher by a factor of 3 or more than the value obtained under zero lateral pressure.

The strength of coal specimens has been measured most commonly under compression tests, partly because they are more reproducible, and partly because they have a direct application to the estimation of the load bearing capacities of coal pillars in mines.

REFERENCES

1. Thiessen, R., *U.S. Bur. Mines Inf. Circ.* 7397, 1947, 53 pp.
2. Fischer, F. and Schrader, H., *Brennstoff-Chem.*, **8**, 337–340 (1927).
3. Hendricks, T. A., Chapter 6 in *Chemistry of Coal Utilization,* Vol. 1, H. H. Lowry, Ed., John Wiley & Sons, New York, 1947.
4. Bergins, F., *Naturwissenschaften,* **16**, 1–11 (1928).
5. Siever, R., *Second Conference on the Origin and Constitution of Coal (Crystal Cliffs, Nova Scotia),* 1952, p. 341.
6. Quinn, A. W. and Glass, H. S., *Econ. Geol.*, **53**, 563 (1958).
7. Mentser, M., O'Donnell, H. J., and Ergun, S., *Fuel,* **41**, 153–161 (1962).
8. Brown, R. F., *Organic Chemistry,* Wadsworth Publishing Company, 1975, p. 814.
9. Stopes, Marie D., *Proc. R. Soc. (London),* **B90**, 470–487 (1919).
10. Thiessen, R., *Coal Age,* **18**, 1183–1189, 1223–1228, 1275–1279 (1920).
11. Thiessen, R., Sprunk, G. C., and O'Donnell, H. J., *U.S. Bur. Mines Inf. Circ.* 7021, 1938, 8 pp.; *Fuel,* **17**, 307–315 (1938).
12. Parks, B. C. and O'Donnell, J. H., *U.S. Bur. Mines Bull.* 530, 1956, 193 pp.
13. Hoffmann, E. and Jenkner, A., *Glueckauf,* **68**, 81.8 (1932); *Fuel,* **12**, 98–106 (1933).
14. Seyler, C. A., *Fuel,* **16**, 134–141 (1943); **31**, 159–170 (1952).
15. McCartney, J. T., *Econ. Geol,* **47**, 202–210 (1952).
16. Zetsche, F. et al., *Liebigs Ann.*, **461**, 89 (1928).
17. Zetsche, F., *Helv. Chim. Acta,* **14**, 59, 62, 67, 517 (1931).
18. Sprunk, G. C., Selvig, W. A., and Oda, W. H., *Fuel,* **17**, 196 (1938).
19. Macrae, J. C., *Fuel,* **22**, 117 (1943).
20. Alpern, B., *Rev. Ind. Min.*, **38**, 638 (1956).
21. Kroger, C., *Brennstoff-Chem.*, **37**, 192 (1956).

ISBN 0-201-08300-0

22. Durmans, H. N. M., Huntjens, F. J., and van Krevelen, D. W., *Fuel,* **36**, 321 (1957).
23. Ergun, S., McCarney, J. T., and Mentser, M., *Econ. Geol.,* **54**, 1068 (1959).
24. Lowry, H. H., *Chemistry of Coal Utilization,* Supple. Vol., John Wiley & Sons, New York, 1963.
25. Shapiro, N. and Gray, R. J., *Proceedings of the Illinois Mining Institute, 68th Year,* 1960, pp. 83–97.
26. Brewer, R. E., in Chapter 6 of *Chemistry of Coal Utilization,* Vol. 1, H. H. Lowry, Ed., John Wiley & Sons, New York, 1947.
27. Loison, R. et al., *ibid.,* Supple. Vol., 1963, Chapter 4.
28. Ammosov, I. L. et al., *Kuks i Khim.,* **12** (1957).
29. Shapiro, N., Gray, R. J., and Eusner, G. R., *Proc. AIME,* **20**, 89–112 (1964).
30. Shapiro, N. and Gray, R. J., *J. Inst. Fuel,* **37**, 234–242 (1964).
31. Harrison, J. A., *Proceedings of the Illinois Mining Inst., 69th Year,* pp. 17–43, 1961.
32. Harrison, J. A., Jackson, H. W., and Simon, J. A., *Ill. State Geol. Surv. Circ.* 366, 1964, 20 pp.
33. Thiessen, R. and Francis, W., *U.S. Bur. Mines Tech. Paper* 446, 1929, 27 pp.
34. Wright, C. C. and Gauger, A. W., *American Mining Congress Yearbook,* 1937, pp. 381–383.
35. Fisher, C. H. et al., *Ind. Eng. Chem.,* **31**, 190–195 (1939); *Fuel,* **18**, 196–203 (1939); *Ind. Eng. Chem.,* **31**, 1155–1161 (1939); *Fuel,* **18**, 132–141 (1939).
36. Regnault, V., *Ann. Chim. Phys.,* **66**, 337–365 (1837); *Ann. Mines,* **12**, 161–240 (1837).
37. Rose, H. J., Chapter 2 of *Chemistry of Coal Utilization,* Vol. 1, H. H. Lowry, Ed., John Wiley & Sons, New York, 1947, p. 47 et seq.
38. Rogers, H. D., *The Geology of Pennsylvania,* Vol. 2, J. B. Lippincott, Philadelphia, 1858, pp. 988–997.
39. Grant, F. F., *Econ. Geol.,* **2**, 225–241 (1907).
40. Ralston, D. C., *U.S. Bur. Mines Tech. Paper* 93, 1915, 41 pp.
41. Ergun, S., Manuscript in preparation.
42. Hamilton, P. A., White, D. H., Jr., Matson, T. K., *U.S. Bur. Mines Inf. Circ.* 8693, 1975, pp. 282–322.
43. "Effects of Air Quality Requirements on Coal Supply," EPRI, U.S. Bur. Mines Contract J0155164, 1976.
44. "National Energy Outlook," Federal Energy Administration, 1976.
45. McCartney, J. T., O'Donnell, H. J., and Ergun, S., *U.S. Bur. Mines Rep. Invest.* 7231, 1969.
46. Ergun, S. et al., "An Analysis of Chemical Coal Cleaning Processes," Bechtel Corp., U.S. Bur. Mines Contract J0166191, 1977.
47. Ergun, S., "An Analysis of Technical and Economic Feasibility of Combination of Physical and Chemical Cleaning of Coal Prior to Combustion," Lawrence Berkeley Laboratory, University of California, November 1977.
48. Thomson, R. D. and York, H. E., *U.S. Bur. Mines Inf. Circ.* 8680, 1975, 537 pp.
49. Caballero, J. A., Johnson, M. T., and Deurbrouck, A. W., *U.S. Bur. Mines Rep. Invest.* 8118, 1976, 323 pp.

ISBN 0-201-08300-0

50. "Methods of Analyzing and Testing Coal and Coke," *U.S. Bur. Mines Bull.* 638, 1967.
51. Lecky, J. A., Hall, W. K., Anderson, R. B., *Nature,* **168**, 24–25 (1951).
52. Mukherjee, P. N., Basak, G. N., and Lehiri, A., *Fuel,* **30**, 215–216 (1951).
53. Hoeppner, J. J., Fowkes, W. W., and McMurtrie, R., *U.S. Bur. Mines Rep. Invest.* 5215, 1956, 25 pp.; *Fuel,* **36**, 469–474 (1957).
54. Deutscher Normenauschuss, DIN 51718-1950; Polish Standard Comm., PN-57 C-04315; British Standards Inst., British Standard 1016, 1957, Part 1.
55. Badzioch, S. and Cornford, G. B., *BCURA Bull.,* **16**, 77–89 (1952).
56. Gauger, A. W., Chapter 17 of *Chemistry of Coal Utilization,* Vol. 1, H. H. Lowry, Ed., John Wiley & Sons, New York, 1947.
57. Ergun, S., Manuscript in preparation.
58. Spackman, William, "Characteristics of American Coals in Relation to Their Conversion into Clean Energy Fuels," Pennsylvania State University, Interim Report, 1976.
59. Van Krevelen, D. W. and Chermin, H. A. G., *Fuel,* **33**, 338–347 (1954).
60. Thiessen, G., Chapter 14 of *Chemistry of Coal Utilization,* Vol. 1, H. H. Lowry, Ed., John Wiley & Sons, New York, 1947.
61. Ode, W. H., Chapter 5 of *Chemistry of Coal Utilization,* Supple. Vol., H. H. Lowry, Ed., John Wiley & Sons, New York, 1963.
62. Gumz, W., Kirsch, H., and Mackowsky, Marie T., *Schlackenkunde,* Springer Verlag, Berlin, 1958, pp. 105–123.
63. Nelson, J. B., *BCURA Bull.,* **17**, 43 (1953).
64. Selvig, W. A. and Gibson, F. H., *U.S. Bur. Mines Bull.* 567, 1956.
65. King, J. G. and Crassley, H. E., Gr. Brit. Dept. Sci. Ind. Res., *Fuel Res., Survey Paper* 28, 1933, 20 pp.
66. Rosin, P. and Fehling, R., *Ber. Reichskohlenrat* 54/55, 1935.
67. Abernathy, R. F., Ergun, S., Friedel, R. A., McCarthy, J. T., and Wender, I., in *Encyclopedia of Industrial Analysis,* F. F. Snell and L. S. Ettre, Eds., Wiley-Interscience, New York, 1970, Vol. 10; *Coal and Coke,* pp. 209–262.
68. Ergun, S., *U.S. Bur. Mines Bull.* 648, 1968, 35 pp.
69. Ergun, S., Mentser, M., and Howard, H. C., *Fuel,* **38**, 495–499 (1959).
70. Selvig, W. A. and Gibson, F. N., Chapter 4 of *Chemistry of Coal Utilization,* Vol. 1, H. H. Lowry, Ed., John Wiley & Sons, New York, 1947.
71. Mott, R. A. and Spooner, C. E., *Fuel,* **19**, 226–231, 242–251 (1940).
72. Schuyer, J. and van Krevelen, D. W., *Fuel,* **33**, 348–353 (1954).
73. Ergun, S., Manuscript in preparation.
74. Goutal, M., *Compt. Rend.,* **135**, 477–479 (1902); *J. Soc. Chem. Ind.,* **21**, 1267 (1902).
75. Ely, F. G. and Bernhart, D. H., in *Chemistry of Coal Utilization,* Vol. 2, H. H. Lowry, Ed., John Wiley & Sons, New York, 1947, p. 826.
76. Van Krevelen, D. W., Huntjens, F. J., and Dormans, H. N. M., *Fuel,* **35**, 462–475 (1956).
77. United Nations Publ., 1956 II.E.4, E/ECE/247, E/ECE/Coal/110, 1956, 52 pp.
78. Yancey, H. F. and Greer, M. R., Chapter 5 of *Chemistry of Coal Utilization,* Vol. 1, H. H. Lowry, Ed., John Wiley & Sons, New York, 1947.

ISBN 0-201-08300-0

79. Broadbent, S. R. and Callcott, T. G., *Phil. Trans. R. Soc. (London),* **A249,** 99–123 (1956).
80. Berenbaum, R., *J. Inst. Fuel,* **34**, 367–374 (1961).
81. Brown, R. L. and Hiorns, F. J., Chapter 3 of *Chemistry of Coal Utilization,* Supple. Vol., H. H. Lowry, Ed., John Wiley & Sons, New York, 1963.
82. Protodyakanov, M. M., *Izv. Akad. Nauk. SSR,* No. 2, 283–298 (1953).
83. Pomeroy, C. D. J., *J. Inst. Fuel,* **30**, 50–54 (1957).
84. Pomeroy, C. D. and Foote, P., *Colliery Eng.,* **37**, 146–154 (1960).
85. Hardgrove, R. M., *Trans. Am. Soc. Mech. Engrs.,* **54**, 37–46 (1932).
86. Bengham, D. H. and Maggs, F. A. P., *Proc. Conf. on Ultra-fine Structure of Coals and Cokes,* BCURA, London, 1944, pp. 118–130.

ISBN 0-201-08300-0

2. Rates of Coal Pyrolysis and Gasification Reactions

C.Y. Wen and S. Dutta

1. INTRODUCTION

Currently, coal constitutes only one fifth of all the sources of energy consumed in this country. The balance is made up of natural gas, domestic oil, imported oil, and hydro, nuclear, and geothermal sources. The expansion of coal production and the development of technologies for more efficient coal combustion and coal conversion to heat and gaseous and liquid fuels are one of the prime objectives in meeting the energy requirement. Publications on coal combustion and gasification are numerous. Researchers of various interests and backgrounds have contributed to the understanding of the complex problems of coal conversion processes. Nevertheless, the uncertainty regarding the kinetics of coal reactions and the physical and chemical behaviors of coal during chemical conversion still presents a major problem to the designer of coal combustion or gasification units. Concurrently with the emphasis on greater and more efficient utilization of coal, there has arisen a pressing need for a better understanding of the rates of coal pyrolysis, gasification, and combustion reactions.

Coal undergoes a series of complex physical and chemical changes upon heating. The nature of these changes, though primarily dependent on the type of coal, is greatly modified by the temperature, heating rate, soaking period, pressure, and gaseous environments. Reactor type, sample size, particle size, and hydrodynamic conditions are other factors which also affect these changes.

Except in conditions of extremely rapid heating and at high temperatures, pyrolysis or thermal decomposition is the first step through which the coal particles undergo major physical and chemical changes. During the pyrolysis of coal a substantial weight loss occurs because of the evolution of volatile matters; the amount

C. Y. Wen and E. Stanley Lee, (Eds.), Coal Conversion Technology, ISBN 0-201-08300-0

ISBN 0-201-08300-0

and the composition depend on coal type, coal size, and the conditions prevailing in the gasifier or combustor. After the pyrolysis a series of consecutive and parallel reactions, involving both the residual solid char and the gaseous products of the primary decomposition stage, takes place.

The following are the main reactions that need to be considered in the design of a coal combustor or gasifier:

Solid-Gas Reactions

$$C^*(s) + O_2(g) \rightarrow CO_2(g)$$
$$2C^*(s) + O_2(g) \rightarrow 2CO(g)$$
$$C^*(s) + 2H_2(g) \rightarrow CH_4(g)$$
$$C^*(s) + CO_2(g) \rightarrow 2CO(g)$$
$$C^*(s) + H_2O(g) \rightarrow CO(g) + H_2(g)$$

(C^* = carbon in char.)

Gas-Gas Reactions

$$2CO(g) + O_2(g) \rightarrow 2CO_2(g)$$
$$CO(g) + H_2O(g) \rightarrow CO_2(g) + H_2(g)$$
$$CO(g) + 3H_2(g) \rightarrow CH_4(g) + H_2O(g)$$
$$CH_4(g) + 3/2O_2(g) \rightarrow CO(g) + 2H_2O(g)$$
$$\text{Hydrocarbons}(g) + O_2(g) \rightarrow CO(g) + xH_2O(g)$$
$$\text{Higher hydrocarbons}(g) \xrightarrow{\text{cracking}} y(\text{lower hydrocarbons})(g) + zC(s)$$
$$\text{Unsaturated hydrocarbons}(g) + H_2(g) \rightarrow \text{saturated hydrocarbons}(g)$$
$$\text{Unsaturated hydrocarbons}(g) \xrightarrow{\text{polymerization}} \text{higher hydrocarbons}(g \text{ or } \ell)$$

The extents and the rates of the above reactions are determined by the reaction conditions, primarily temperature, pressure, gaseous environment, and the hydrodynamic conditions present in the system. The phenomena of rapid pyrolysis and hydrogasification, which are not yet understood fully, have become of considerable interest in recent years because of their potential applications.

The initial stage of pyrolysis and the char-gas reactions are the subjects of discussion in this chapter. Gas-gas reactions, which are covered in other chapters, will not be discussed here. The purpose of this chapter is to assess the available data and correlations in a systematic way and to present these subjects in a consolidated

ISBN 0-201-08300-0

and useful form that can be applied to combustor and gasifier designs. In this chapter the subjects of coal/char pyrolysis and hydrogasification will be discussed in greater detail than the other char-gas reactions.

2. PYROLYSIS

2.1. Physical and Chemical Changes in Coal Pyrolysis

A series of consecutive and parallel reactions occurs during the rapid stage of pyrolysis of coal. The effect of initial heating is to release the occluded gases and moisture from coal. Although most of the uncombined water is driven off below 105°C, this water is not completely removed until the temperature reaches around 300°C. The phenomena that take place thereafter, as the temperature rises, are schematically shown in Figure 2.1. For caking coals, as temperature increases, the coal particle softens to form a metastable plastic intermediate, metaplast, which depolymerizes to yield the primary volatile products and the semicoke. During this process the particle swells to a degree, depending on its swelling index and the heating conditions. The semicoke decomposes slowly on further heating. The British Standard swelling index (BSSI) can be roughly correlated to the hydrogen (H) and carbon (C) contents of coal by the following formula (Lowry [75]):

$$\mathrm{BSSI} = \frac{1.4}{(H - 4.0)} \times [(C - 94) + 11(H - 4.0)]$$

At a particular heating rate the temperatures of the start of softening, plasticity, and swelling and the temperature of peak rate of decomposition are lower for coals with higher volatile matter contents. For a particular coal these temperatures increase with increase in heating rate [108].

Typical swelling characteristics of coals in inert atmosphere are illustrated in Figure 2.2. The swelling indices of coals of one particle size or their swelling behavior under one condition of temperature, heating rate, pressure, and gaseous environment, however, does not tell how they are going to behave in an entirely different situation.

Table 2.1 shows the possible chemical reactions which take place during pyrolysis in an inert atmosphere, at conditions of moderate heating rates and temperatures, and at atmospheric pressure. At conditions of long vapor residence time and/or extremely rapid heating rates and temperatures (e.g., in flash heating), considerable amounts of acetylene, other unsaturated hydrocarbons, and cracked carbon are formed by the vapor cracking of the primary products of gaseous hydrocarbons.

ISBN 0-201-08300-0

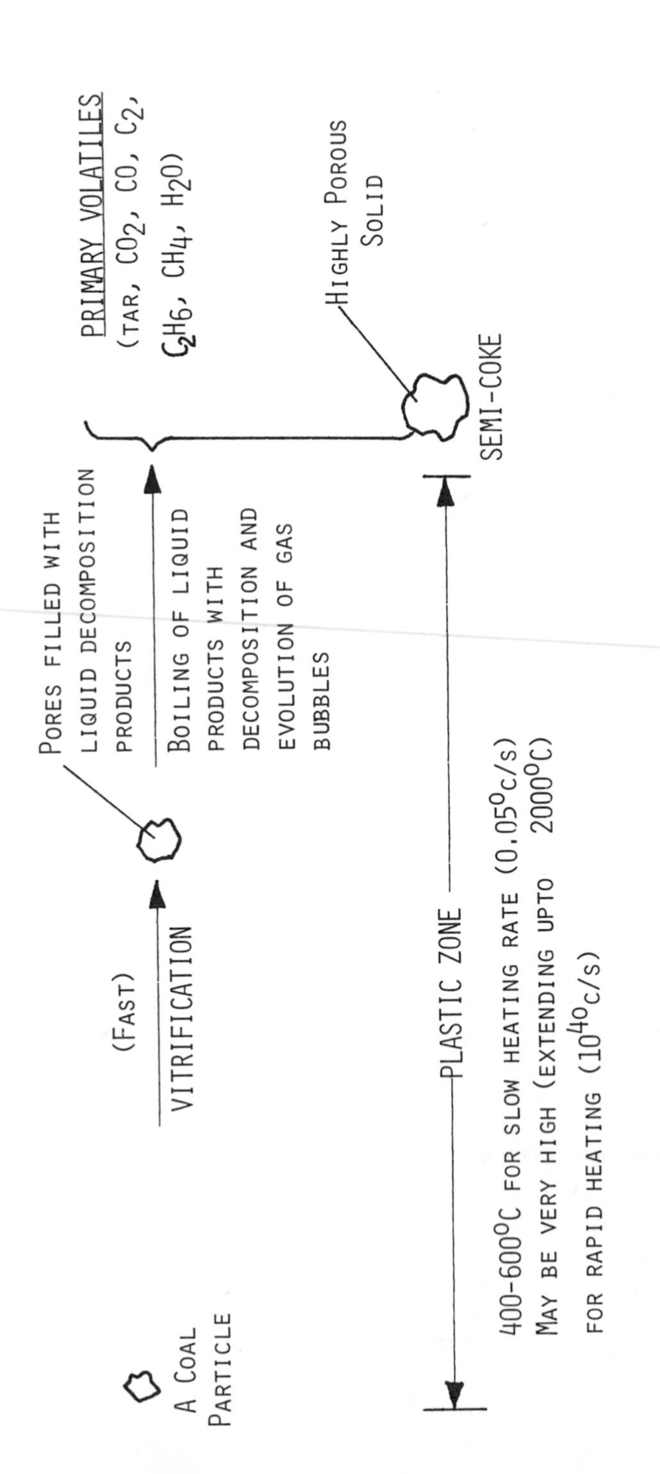

ISBN 0-201-08300-0

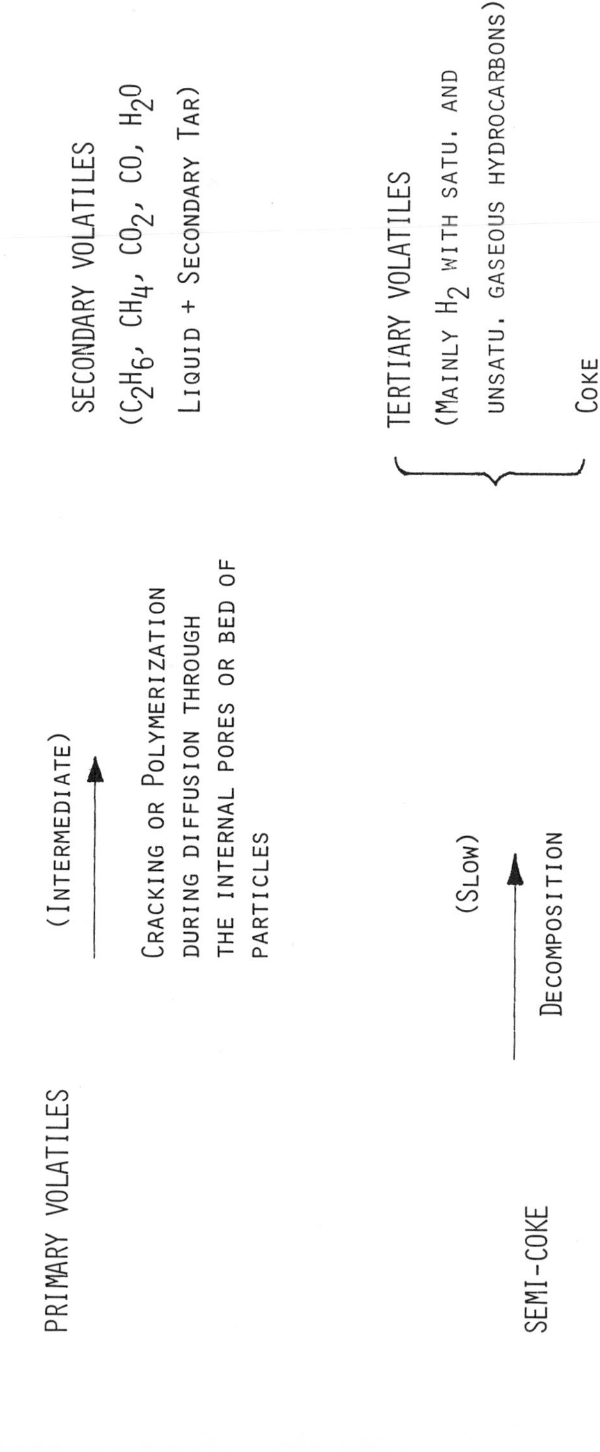

Figure 2.1. Suggested mechanism for pyrolysis of a single swelling coal particle.

ISBN 0-201-08300-0

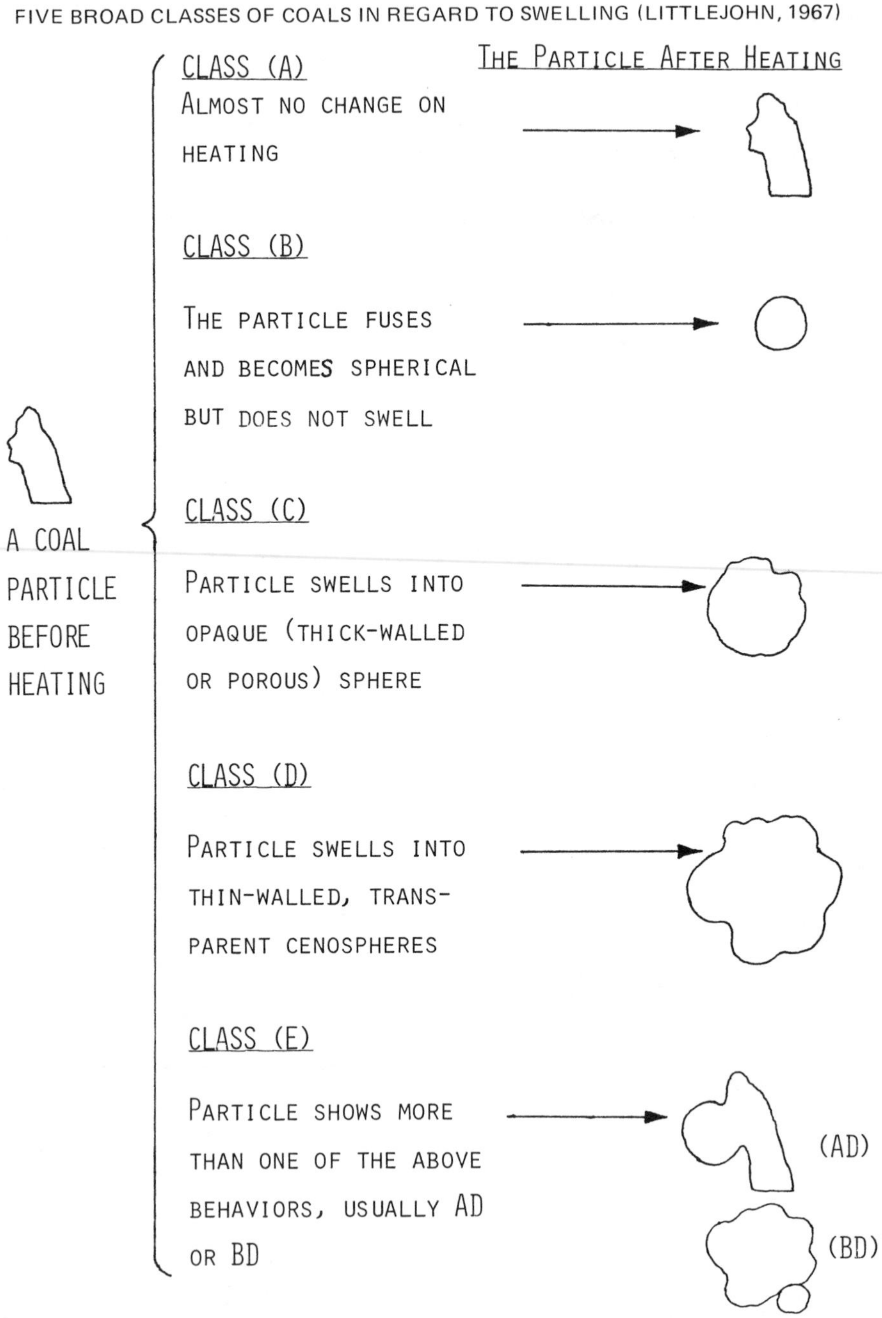

FIVE BROAD CLASSES OF COALS IN REGARD TO SWELLING (LITTLEJOHN, 1967)

A COAL PARTICLE BEFORE HEATING

THE PARTICLE AFTER HEATING

CLASS (A)
ALMOST NO CHANGE ON HEATING

CLASS (B)
THE PARTICLE FUSES AND BECOMES SPHERICAL BUT DOES NOT SWELL

CLASS (C)
PARTICLE SWELLS INTO OPAQUE (THICK-WALLED OR POROUS) SPHERE

CLASS (D)
PARTICLE SWELLS INTO THIN-WALLED, TRANS-PARENT CENOSPHERES

CLASS (E)
PARTICLE SHOWS MORE THAN ONE OF THE ABOVE BEHAVIORS, USUALLY AD OR BD

(AD)

(BD)

Figure 2.2. Swelling of coal particles in inert atmosphere. Five broad classes of coals in regard to swelling (Littlejohn [72]).

ISBN 0-201-08300-0

Table 2.1

The Chemical Processes of Coal Pyrolysis

Product	Source	Process
1. Tar + liquid	Weakly bonded ring clusters	Distillation + decomposition
2. CO_2	Carboxyl groups	Decarboxylation
3. CO (<500°C)	Carbonyl groups and ether linkages	Decarbonylation
4. CO (>500°C)	Hetero-oxygens	Ring rupture
5. H_2O	Hydroxyl groups	Dehydroxylation
6. $CH_4 + C_2H_6$	Alkyl groups	Dealkylation
7. H_2	Aromatic C-H bonds	Ring rupture

Volatiles are released from coal in approximately the following order: H_2O, CO_2, CO, C_2H_6, CH_4, tar + liquid, H_2.

2.2. Volatiles Yield, Product Distribution and Kinetics of Coal Pyrolysis

A knowledge of the kinetics of coal pyrolysis is essential in predicting the yield and product distribution in a coal gasifier. An understanding of coal pyrolysis is very important in view of the potential of the process to take adavntage of (a) the phenomena of rapid pyrolysis and (b) possibility of obtaining higher yields of gaseous and liquid hydrocarbons by the application of pressure and hydrogen atmosphere. Rapid pyrolysis drew the attention of investigators after it was first demonstrated by Hawk et al. [48], Loison and Chauvin [73], and Eddinger et al. [24] and confirmed by Kimber and Gray [68] that volatile matters significantly higher than those indicated by the proximate analysis can be obtained from coal by rapid heating. On the other hand, the pilot plant studies of Garrett Research and Development [94] have demonstrated that about 30–35% conversion of coal to oil is possible in pyrolysis by the rapid heating of coal at comparatively lower temperatures, and with minimum residence time of the volatile matters. It has again been shown by the pilot plant studies of Union Carbide [3] that production of both the gaseous and liquid hydrocarbons is improved significantly under high partial pressure of hydrogen.

Depending on the desired yields, the product distribution, and the processing conditions required, both the heating rates and the temperature zones of pyrolysis can be classified roughly into four categories, as is shown in Table 2.2. Slow heating is the classical procedure that has been practiced for a number of years in the carbonization industry to make coke for steel production. The conventional laboratory assay methods like those used in the Fischer assay and ASTM proximate

ISBN 0-201-08300-0

analysis also fall in this category. Low, intermediate, or high temperatures are used in the carbonization, depending on the desirability of yields of coke or volatile matters. The object of rapid heating, as has already been indicated, is to get as much of the products as possible in the volatile forms. Higher yields of liquid and gaseous hydrocarbons and correspondingly smaller coke or char production are the essence of this technique. Flash heating, a comparatively recent interest, is aimed at the production of acetylene and other unsaturated hydrocarbons. Table 2.3 shows the suggested temperature programming mode (i.e., the heating rate, temperature, and residence times) to be used for the desired volatile product.

Table 2.2

A. Classification of Heating Rates

	Heating Rate, $^{o}C/s$	Heat-up Time to $1000^{o}C$ for $\sim 100\ \mu$ in Size
1. Slow heating	$\ll 1$	20 min
2. Intermediate heating	$5 \sim 100$	$10\ s \sim 4$ min
3. Rapid heating	$500 \sim 100{,}000$	$10\ ms \sim 2\ s$
4. Flash heating	$> 10^6$	< 1 ms

B. Classification of Temperature Zones for Carbonization of Coal

1. Low temperature carbonization $\sim 500^{o}C$
2. Intermediate temperature carbonization $\sim 750^{o}C$
3. High temperature carbonization $\sim 1000^{o}C$
4. Very high temperature carbonization $> 1200^{o}C$

Figure 2.3 shows qualitatively the effects of heating rate and temperature on (a) the relative total yields of volatiles, (b) the relative ratios of liquid (liquid and tar) to gaseous volatiles, and (c) the gas composition. In this figure it is assumed that the solid is held at the final temperatures until complete decomposition, while

ISBN 0-201-08300-0

the vapor residence time is kept at the minimum. Thus at a particular heating rate both the total yield of volatiles and the ratio of gaseous to liquid hydrocarbons increase with increase in temperature. On the other hand, if the heating rate is increased, with the final temperature kept constant at a low level (around 500°C), the yield of total volatiles increases but the ratio of gaseous to liquid hydrocarbons decreases. However, at a much higher level of temperature (around 1000°C) and a rapid heating rate both the total yield and the ratio of gaseous to liquid hydrocarbons increase again. Because of certain distinct differences in the heating techniques involved and the yield structures, the subjects of slow, rapid, and flash pyrolysis are discussed separately in the following sections.

Table 2.3

Programmed Temperature Pyrolysis (PTP)

	Suggested Temperature Programming Mode			
Desired Volatile Product	Heating Rate	Temperature of Carbonization	Solid Residence Time	Volatile Residence Time
1. Tar	Rapid	Low (~500°C)	Long	Short
2. Liquid	Rapid	Intermediate (~750°C)	Long	Long
3. Gas	Rapid	High (>1000°C)	Long	–[a]
4. CH_4	Rapid	~600°C	Long	–
5. H_2	Rapid	1000~1100°C	Long	–
6. C_2H_2 + unsaturates	Flash	>1200°C	Long	Intermediate
7. CO	–	Intermediate (~750°C)	Long	–

[a] Dash means that the effect is either uncertain or insignificant.

ISBN 0-201-08300-0

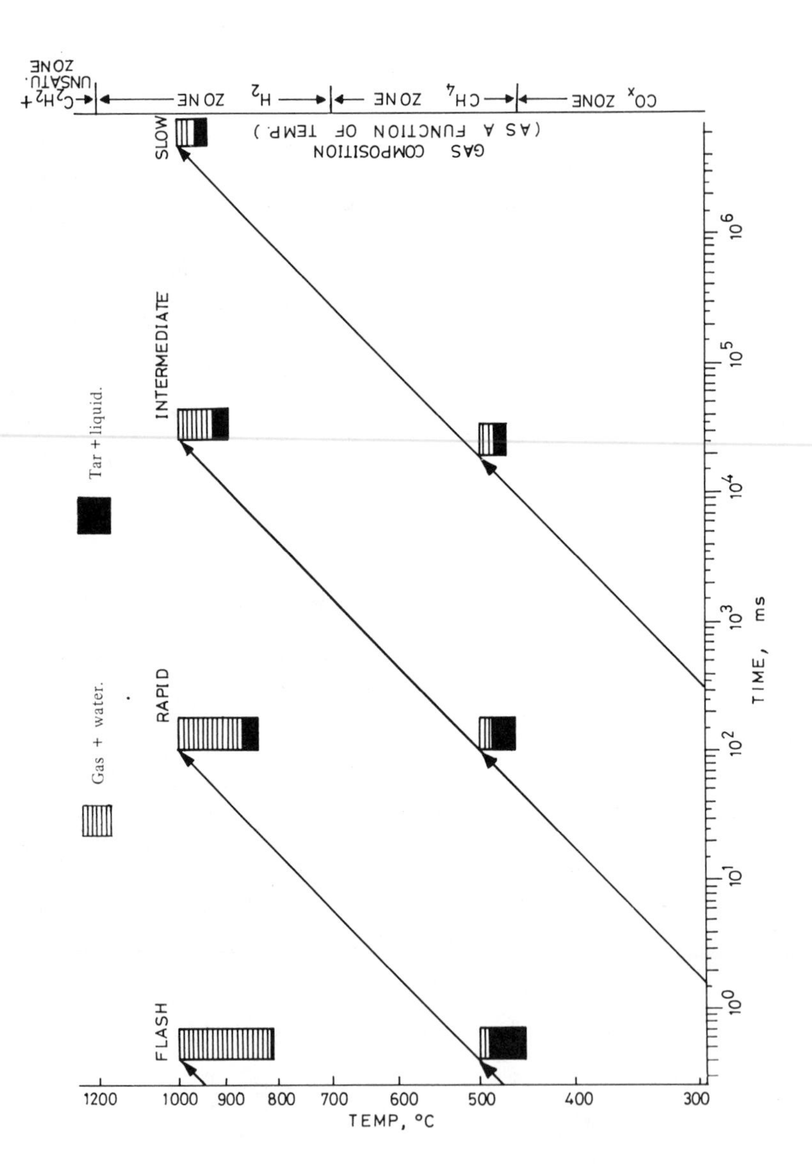

Figure 2.3. Relative yields and product distributions of pyrolysis in inert atmosphere as functions of temperature, time, and heating rate.

ISBN 0-201-08300-0

A. SLOW PYROLYSIS

In this chapter the term "slow pyrolysis" will mean pyrolysis by slow heating. Similarly the terms "rapid pyrolysis" and "flash pyrolysis" will mean pyrolysis by rapid and flash heating, respectively.

When coal or char is heated at a slow or an intermediate rate (<100°C/s), the time required for almost complete devolatilization of the coal or char is significantly less than the time required for heating the particles to the final temperatures. Therefore, in pyrolysis by slow or intermediate heating, the ultimate yields and quality of volatiles and chars, rather than the rate of pyrolysis, are important. Figure 2.4 shows a few typical volatiles yield versus temperature curves as observed in the Fischer assay of some coal samples [39, 40]. In this figure, as well as in the subsequent ones, the percent yield of volatiles (maf basis) has been normalized by the proximate volatile matter (VM) content (maf basis) of the respective coal as determined by the ASTM method.

Table 2.4 shows the useful correlations available for calculating the total loss of volatile matter and the rate of pyrolysis at slow, intermediate, and rapid heating rates. The curves in Figure 2.4 were based on the equation of Gregory and Littlejohn [45], modified as indicated under "Remarks" in Table 2.4. The calculated curves for three proximate VM value (proximate VM values mentioned are on a moisture- and ash-free basis, unless otherwise specified), which approximately cover the range of volatile matter contents of most of the coals, are presented.

Figures 2.5 and 2.6 show the volatiles composition and gas composition as functions of temperature as found in the Fischer analysis by Goodman et al. [39, 40] of eight American coals including peat, lignite, and bituminous and subbituminous coals. These figures also include the vacuum pyrolysis data for a few British coals studied by Fitzgerald and van Krevelen [32]. The National Coal Board Coal Rank Code numbers have been abbreviated as NCB C.R.C. numbers on these and subsequent figures. Figure 2.5 shows that, whereas the amount of tar and liquid does not increase beyond a temperature of about 600°C, the amount of gas and water continues to rise. Among the gaseous products, as the temperature rises, the total amount of CO and CO_2 decreases, while the amount of hydrogen increases. The total amount of CH_4 and C_2H_6 reaches a peak value at around 500°C and decreases on further increase in temperature.

The simple rate equation, Equation 2 in Table 2.4, proposed by Wen et al. [113], was applied by Dutta et al. [23] to the analysis of coal and char pyrolysis performed in a thermobalance. The rates of pyrolysis and the volatiles yields as functions of time and temperatures were experimentally determined during the heat-up periods of a few coal and char samples. Temperatures were simultaneously recorded during these heat-up periods as a function of time. Values of A' and B' were found to be 0.33 s^{-1} and 2.5 kcal/mol (10.5 kJ/mol), respectively, for the char samples. For the two coals an activation energy of about 7 kcal/mole (29.3 kJ/mol) was found. This study shows that coal or char pyrolysis can be presented approximately by a single Arrhenius type equation which is useful for design purposes.

ISBN 0-201-08300-0

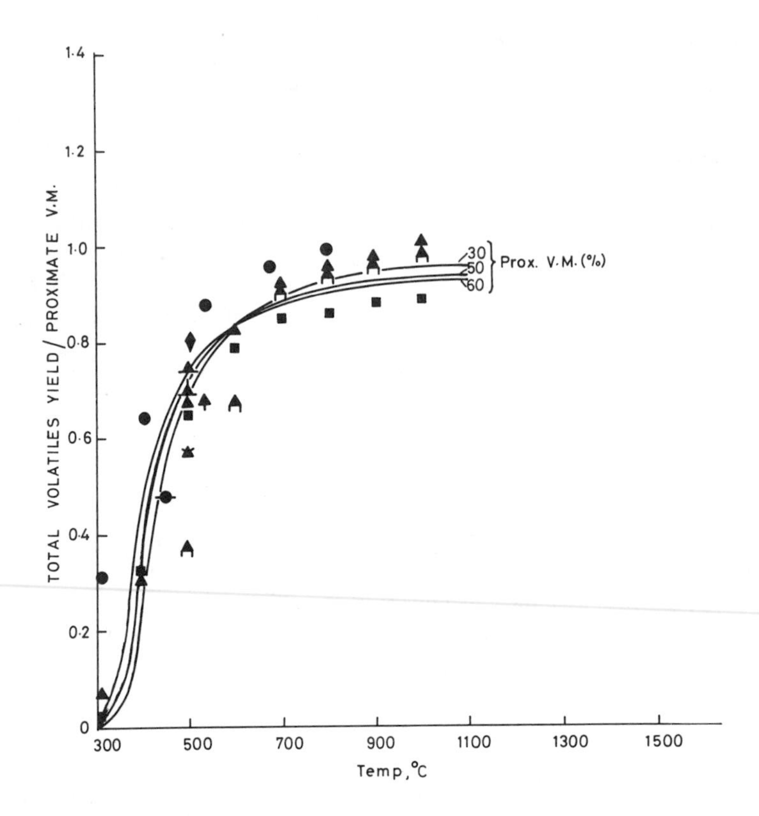

Figure 2.4. Volatiles loss as a function of temperature in slow pyrolysis. Experimental data from Goodman et al. [39, 40] on Fischer assays and from Kasurichev [65].

Ireland peat (66.7);* French lignite (50.7); Texas lignite (50.1); Texas lignite char produced at 510°C (31.5); Elkol subbitu (44.5); Ecuador subbitu (45.4); Mexico HVbb (57.8); Wyoming subbitu (47.7); Utah HVbb (46.8); Moscow District coal (50.2). *The numbers within parentheses would indicate percent proximate VM of coals in this and the following figures.

Pitt [86] proposed that pyrolysis be considered as a large number of independent chemical reactions involving the original coal molecule. Depending on the strength, the ruptures of different covalent bonds of coal molecules take place at different temperatures with different rates. He therefore suggested the use of a distribution function of activation energy, instead of using an overall activation energy like A', as in Equation 2. It is thus necessary to know at least one more

ISBN 0-201-08300-0

kinetic parameter, the standard deviation σ, for the activation energy distribution function, in addition to the mean activation energy and the preexponential factor for describing the kinetic behavior of coal pyrolysis. As will be seen in the following section on rapid pyrolysis, such an approach has recently been adopted by several investigators.

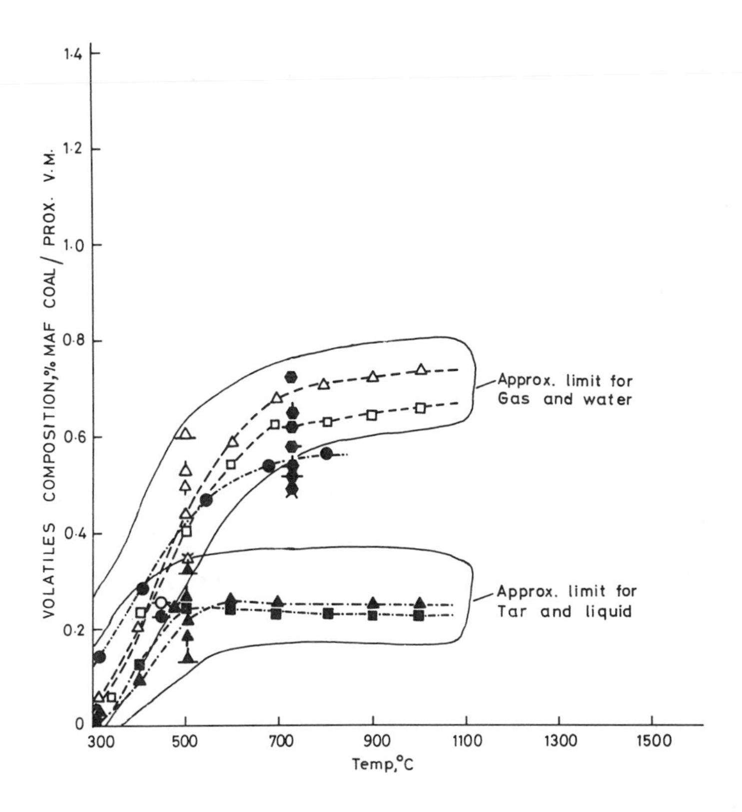

Figure 2.5. Product distribution in slow pyrolysis. Solid points: tar + liquid. Open points: gas + water. The data points (except ● and the hexagonal points) are from Fischer assays (Goodman et al. [39, 40]).

Ireland peat (66.7); French lignite (50.7); Texas lignite (50.1); Ecuador subbitu (45.4); Mexico HVbb (57.8); Wyoming subbitu (47.7); Utah HVbb (46.8); □ ■ Elkol subbitu (44.5); ● Moscow District coal (50.2).

The seven hexagonal solid points are for pyrolysis in vacuum of seven British Coals (Fitzgerald and van Krevalen [32]). The NCB C.R.C. numbers for these coals (from top to bottom) and VMs are 100b (8.2), 201b (13.7), 901 (26.5), 301 (23.3), 802 (41.9), 701 (35.7), and 401a (30.6).

ISBN 0-201-08300-0

Table 2.4

Correlations for Pyrolysis of Coal/Char in Inert Atmosphere

Author	Heating Rate	Temperature Range and Pressure (gaseous atmosphere)	Correlation[a] for Total Volatiles Yield, V, or Rate	Equation No.	Remarks
Gregory and Littlejohn [45]	Slow or intermediate	500–1100°C, atmospheric (N_2)	$V = VM - R' - W$ where $R' = 10^{(11.47 - 3.961 \log_{10} T + 0.005\,VM)}$, $W = 0.20\,(VM - 10.9)$	(1)	The following correlation showed better agreement with a few data, tested, on Fischer assay: $V = 1.1\,VM - R' - W$.
Wen et al. [113]	Slow, intermediate, or rapid	550–1500°C, atmospheric (N_2)	$\frac{dX}{dt} = A'e^{-B'/RT}(f - X)$	(2)	Here f is the final conversion, and X is the conversion at any time t.
Juntgen and van Heek [63]	Slow, intermediate	Up to 1000°C, atmospheric (N_2)	$\frac{dV'}{dT} = \frac{K_0 V_0}{m'} \exp\left[-\frac{E}{RT} - \frac{K_0 R T^2}{mE} \cdot e^{-(E/RT)}\right]$	(3)	Here V' represents the volume of any particular gas released at time t, and not the total volatiles; m' is the heating rate, dT/dt.
Howard and Essenhigh [53]	Rapid	200–1550°C, atmospheric (air)	$\frac{dV}{dt} = K_0 e^{-(E/RT)}(V_0 - V)$	(4)	

ISBN 0-201-08300-0

ISBN 0-201-08300-0

Table 2.4 (Continued)

Badzioch and Hawksley [11]	Rapid (25,000–50,000°C/s)	Up to 1000°C atmospheric (N_2)	$V = Q \cdot VM\,(1 - D) \cdot [1 - \exp(-Ae^{-B/T}t)]$ (5)	By far the best available useful correlation, applicable to bituminous coals and semianthracite with carbon contents of 79–92%. Correlations for the constants of the equations are provided by the authors.
Anthony and Howard [3]	Slow, intermediate, or rapid	Up to 1000°C, 0.001–100 atm (He, H_2)	$V = V^* [1 - \int_0^\infty \exp(-\int_0^t k\, dt) f(E)\, dE]$ (6) where $k = k_0\, e^{(-E/RT)}$ $V^* = V^*_{nr} + V^{**}_r/(1 + K_c P)$ $f(E) = [\sigma(2\pi)^{1/2}]^{-1} e^{-(E-E_0)^2/2\sigma^2}$	The only model that includes the pressure effect on pyrolysis in inert atmosphere, but requires that seven parameters be determined for each coal/char.
Solomon et al. [98, 99, 100]	Rapid	300–1250°C, vacuum	$\frac{dW'}{dt} = K_o e^{(-E/RT)}(W'_o - W')$ (7)	Here W' and W'_o represent the weight of any particular gas or tar released at time t = t and at $t = \infty$, respectively.

[a] See "Notations,"

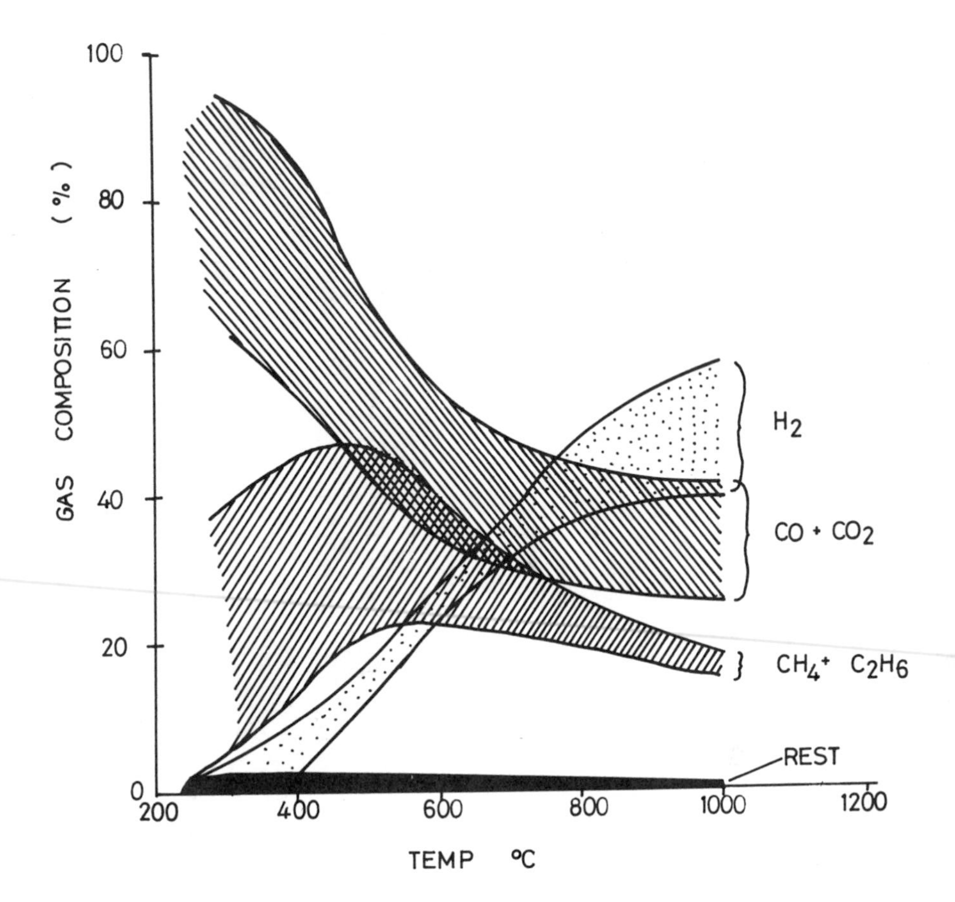

Figure 2.6. Gas composition as a function of temperature in slow pyrolysis. The bands show the spread of data for the coals in Figures 2.4 and 2.5.

A problem which is more baffling than predicting the total yield is the prediction of product distribution from coal pyrolysis. A fundamental approach to this problem was made by Hanbaba et al. [47] and Juntgen and van Heek [63] in their work on gas release from coal pyrolysis as a function of the rate of heating. They proposed Equation 3 in Table 2.4, which is based upon experiments with heating rates ranging from 10^{-2} to more than 10^{5} ^{0}C/min. This equation is, therefore, applicable to pyrolysis by both slow and fast heating rates.

More recently Campbell and Stephens [17] applied Equation 3 to the release rate of the major gases during the pyrolysis of a subbituminous coal at a heating

ISBN 0-201-08300-0

rate of 3.33°C/min at temperatures between 110 and 1000°C. By fitting the experimental rate versus temperature curves, the following values of the rate parameters were obtained for the release of the major gaseous products.

Gas	K_0 s^{-1}	E, kcal/mol*
H_2	20	22.3
CH_4	1.67×10^5	31.0
CO_2	550	19.5
CO	55	18.0
C_2H_6	1.67×10^6	33.4
C_3H_8	7.3×10^6	35.0
C_2H_4	2.3×10^6	33.4

*1 kcal/mol = 4.187 kJ/mol.

Juntgen and van Heek [63] on the other hand, obtained values of 1.67×10^9 s^{-1} and 42.1 kcal/mol (176.3 kJ/mol) for K_0 and E, respectively, for C_2H_6 release, using a German coal (prox. VM: 19.5%) in their experiment.

Equation 3 or a modification of it is attractive for representing the kinetic behaviors of coal pyrolysis, especially in relation to the product distribution, although this equation fails to determine the composition and the rate of release of the liquids and tars in the volatiles. However, products like CH_4, C_2H_4, and C_2H_2 are apparently not the results of single decomposition reactions. Moreover, the reactions leading to these products also vary with the heating rate, temperature level, or coal type. Further investigations with a wide variety of coals, heating rates, and temperature levels are expected to throw more light on this subject.

B. RAPID PYROLYSIS

The phenomenon of rapid pyrolysis has the potential of becoming one of the most effective ways to utilize hydrocarbons contained in coal. To achieve a rapid rate of pyrolysis, pulverized coal burners, fluidized beds, free-fall type reactors, entrained beds, and cyclone beds are often utilized.

An experimental method for the kinetic studies of rapid pyrolysis was developed by Sapatina et al. (Badzioch [10]). They dropped 147 to 227 μm coal

ISBN 0-201-08300-0

particles vertically through an electrically heated refractory tube. The time of fall, which was about 0.45 s, was determined by high-speed cinematography. The temperature range used in this study was limited, however, to a maximum of only 600°C. As already mentioned, the results of investigations provided by Loison and Chauvin [73] and Eddinger et al. [24] and confirmed by Kimber and Gray [68] established the fact that a significantly higher amount of volatile matter than that given by proximate analysis is attainable in rapid pyrolysis. Kimber and Gray used a transport reactor attaining a heating rate of 10^5 to 10^6 °C/s, and a solid residence time of only 45 to 110 ms. Later Friedman et al. [33] and Badzioch and Hawksley [11] also used transport reactors in their studies, and attained solid residence times of only 4 to 100 ms. Mentser et al. [77, 78] used a cylindrical stainless steel mesh (400 mesh) carrying coal particles, which was directly heated by electric current. A heating rate of about 10^4 °C/s and a solid residence time of 60 to 140 ms were thus attained. More recently Anthony and Howard [3] and Solomon [98, 99] used almost the same technique as that of Mentser et al. in their investigations. Figure 2.7, the legend of which is given in Table 2.5, shows typical results of these investigations and indicates that volatiles evolution up to 50% higher than the proximate VM yield is attainable in rapid pyrolysis. However, as shown in the figure, Mentser et al. found for the four samples they studied a sharp decrease in volatile loss in the temperature range 700 to 1000°C. Such apparently unexpected behavior has not been reported by any other investigator.

Badzioch and Hawksley [11] analyzed a large number of data obtained for 10 bituminous coals and 1 semianthracite and arrived at the empirical formula 5 shown in Table 2.4. The average values and the approximate correlations for the constants of Equation 5 are given in Table 2.6, where C represents the carbon content of the coal on an maf basis. The approximate correlations for the constants, as functions of carbon contents of the high swelling coals, have been obtained by regression analyses of the curves given by Badzioch and Hawksley [11].

Figure 2.7 shows a few calculated curves based on the equation of Badzioch and Hawksley. Curve A is for weakly swelling coals (BSSI $\leqslant 4.0$), whereas curves B are for a few highly swelling coals (BSSI > 4.0). The experimental data points shown in Figure 2.7 are those of investigators other than Badzioch and Hawksley. In view of the extreme rapidity and complexity of the process, the calculated curves should be considered as fairly good approximations of most of the experimental observations. There are two principal reasons for the discrepancies between calculated and experimental yields. The calculations were done by assuming that the processes occurred isothermally for a period of 100 ms. In actual processes this time includes both the heat-up time and the isothermal periods or represents only the heat-up periods needed to reach the final temperatures. Equation 5, therefore, gives somewhat higher yields than were observed, partly also because of the type of reactor, the sample size, and the heating technique used by the various investigators. The calculated curves in Figure 2.7 can be improved, however, to represent the observed results more closely, by doing the nonisothermal calculations for the

ISBN 0-201-08300-0

actual heat-up periods, when temperature is a known function of time. Comparison of the calculated curves with the experimental observations also indicates that Equation 5 can be modified by taking into consideration the fact that the temperature of "threshhold of decomposition" is shifted to higher temperatures with a decrease in the VM contents of the coals.

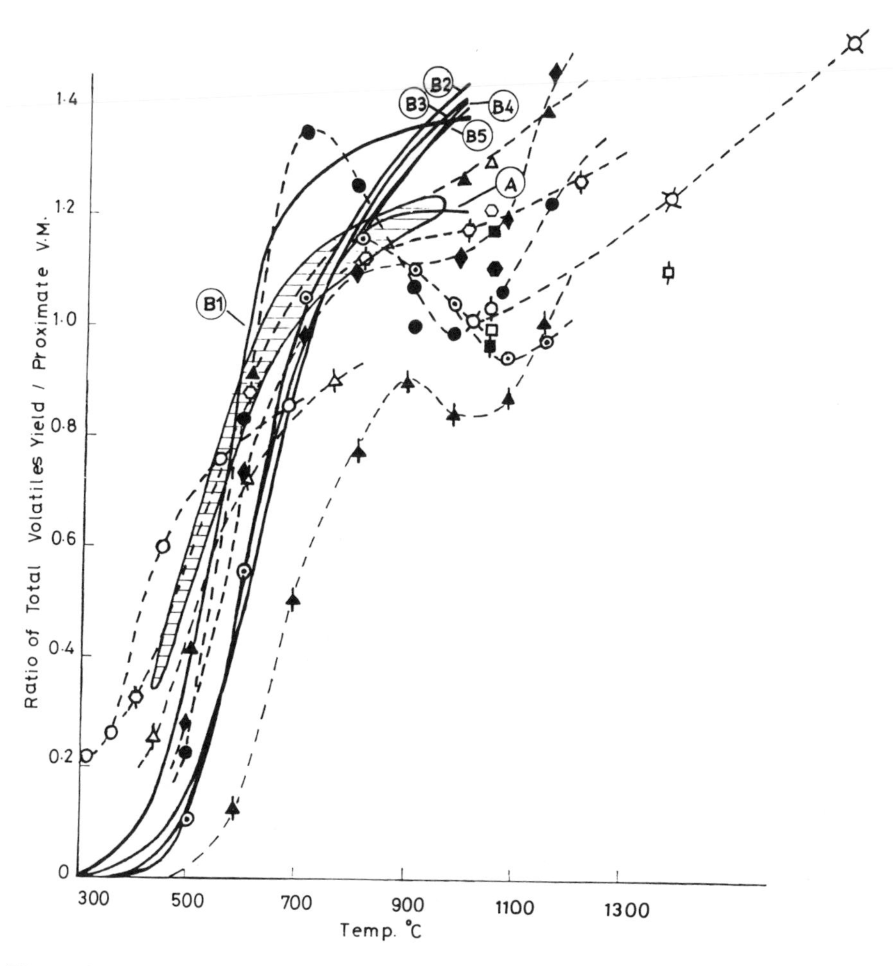

Figure 2.7. Effect of temperature on total volatiles yield from pyrolysis by rapid heating. Experimental curves for different coals obtained by various investigators are listed in Table 2.5. Calculated curves (solid lines) according to Badzioch and Hawksley [11].

ISBN 0-201-08300-0

Neither the above correlations nor the correlations developed afterwards are able to represent the peculiar nature of some of the temperature versus yield curves, as observed by Mentser et al. [77, 78], which show a decrease of yield as the tem-

Table 2.5

Legend to Figure 2.7

Investigator	Coal (VM)	Swelling Index	Experimental Points	Calculated Curve
Pitt [86]	NCB C.R.C. No. 902 (46.2)	Low	○	A
Loison and Chauvin [73]	Maigre Oignies, No. 2 (10.2)	Low	□	A
	Bergmannsglück (19.6)	Low	△	A
	Emma (22.8)	Low	⬡	A
	Faulauemont (38.9)	Low	⦶	A
Kimber and Gray [68]	NCB C.R.C. No. 902 (38.3)	Low	⧖	A
	NCB C.R.C. No. 902 (40.0)	Low	⊡	A
Friedman et al. [33]	Elkol (42.7)	Low	⟁	A
Mentser et al. [77]	Illinois No. 2 HVcb, vitrain (48.4)	Low	◉	A
Anthony and Howard [3]	Pittsburgh HVb (46.2)	Low	⌬	A
Loison and Chauvin [73]	Lens Lievin No. 10 (29.6)	High	⬢	B-5
	Flenus de Bruay (33.1)	High	■	B-2
	Wendel III (37.1)	High	⧯	B-2
Mentser et al. [77]	Pocahontas No. 3 LVb, vitrain (20.5)	High	⍋	B-4
	Pittsburgh HVab, vitrain (35.7)	High	●	B-1
Mentser et al. [78]	Lower Kittanning (Pa.) MVb, vitrain (25.5)	High	▲	B-4
	Rock-Springs No. 7½, vitrain (37.9)	High	◆	B-3
Solomon [99]	Lower Kittanning HVa (53.0); Hazard No. 7, HVa (45.5); Middle Kittanning HVb (40.0); Ohio No. 2, HVc (40.0); Pittsburgh HVa (40.0)	–	The shaded area	–

ISBN 0-201-08300-0

perature increases from 700 to 1000°C.

Following the idea of Pitt [86], Anthony and Howard [3] proposed Equation 6 in Table 2.4 for representing the rate of coal pyrolysis at any gas pressure in an inert atmosphere. The following values of the rate parameters were determined from experimental data by Anthony et al. [4] for two coal samples at atmospheric pressure.

Parameter	Montana Lignite (VM: 37.4%, original coal weight basis)	Pittsburgh Seam Bituminous Coal (VM: 39.81%, original coal weight basis)
V^*	40.63	37.18
E_0, kcal/mol (kJ/mol)	48.72 (204.0)	36.89 (154.4)
σ, kcal/mol (kJ/mol)	9.38 (39.3)	4.18 (17.5)
k_0, s^{-1}	1.07×10^{10}	2.91×10^9

According to Anthony et al. [4], the primary volatiles formed by decomposition within the coal particles are of two categories: reactive, V_r^{**}, and nonreactive, V_{nr}^{**}. The nonreactive volatiles escape from the particles as such (i.e., $V_{nr}^{**} = V_{nr}^*$), whereas part of the reactive volatiles (V_r^{**}) deposits inside the particles by polymerization and/or cracking. Thus only part (V_r^*) escapes out. This is believed to be the main reason for the higher volatiles yield in rapid heating than in slow heating. In rapid heating the residence time of the primary decomposition products within the coal particles is shortened, lessening the chance of secondary reactions within the coal matrix. This mechanism will be discussed in more detail in Section 3.5 on the coal/char hydrogen reaction.

Solomon [99] and Solomon and Colket [100] applied Equation 7 in Table 2.4 to the release rate of both gases and tar from 13 different coals (12 bituminous coals and one lignite) at near-isothermal conditions in a temperature range of 300–1250°C and for residence times of 3 to 180 sec. The specific rate constants and activation energies obtained for the major products from all the coals are as follows:

Product	K_0, s^{-1}	E, kcal/mol
H_2	3.6×10^3	25.4
Hydrocarbons	290	15.9
CO_2	33	9.7
CO (1st stage)	7×10^3	20.6
CO (2nd stage)	1.1×10^5	34.6
H_2O	27	9.9
Tar	81	11.6

ISBN 0-201-08300-0

Table 2.6

Constant	Semianthracite	Weakly Swelling Coals	High Swelling Coals
Q, –	1.41	1.41	$1.249 + (C - 78)0.0364$
D, –	0.65	0.14	$e^{-K_1(T-K_2)}$ for $T > K_2$ (For $T \leqslant K_2$, $D = 1.0$)
			$K_1 = -0.2676 + 0.6189 \times 10^{-2}C - 0.3533 \times 10^{-4}C^2$
			$K_2 = -112{,}100 + 2526C - 14.14C^2$
A, s^{-1}	1.79×10^7	1.34×10^5	$-0.3885 \times 10^8 + 0.90 \times 10^6 C - 0.515 \times 10^4 C^2$
B, $^\circ$K	14,100	8900	8900

Experiments of Suuberg et al. [102] on Montana lignite indicate more than one stage in the evolutions of CO_2, CO, CH_4, C_2H_4, and tar in yield versus temperature diagrams. The first stage occurs between 400 and 600°C, and the second stage between 700 and 850°C. The carbon oxides tend to show a third step above 1000°C. Previously only CO was found to be evolved in two stages.

Figure 2.8 shows the effect of pressure on coal pyrolysis in inert atmosphere as reported by Anthony and Howard [2].

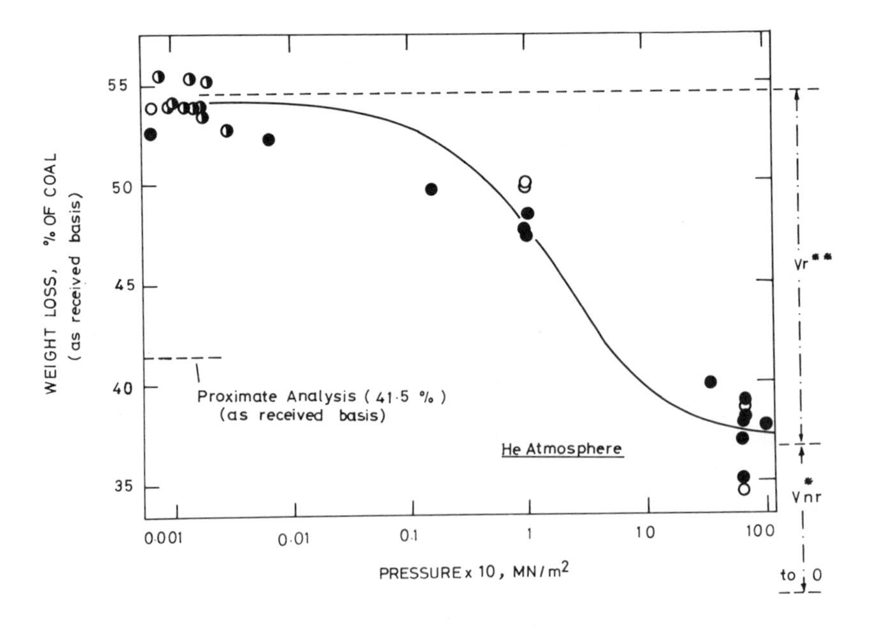

Figure 2.8. Effect of pressure on pyrolysis in inert atmosphere (Anthony and Howard [3]. Sample: Ireland mine coal. Final holding temperature: 1000°C. Experimental time: 5–20 s. Mean particle diameter: 70 μ. Nominal heating rate: 10,000°C/s (O), 3000°C/s (◑), and 700°C/s (●).

ISBN 0-201-08300-0

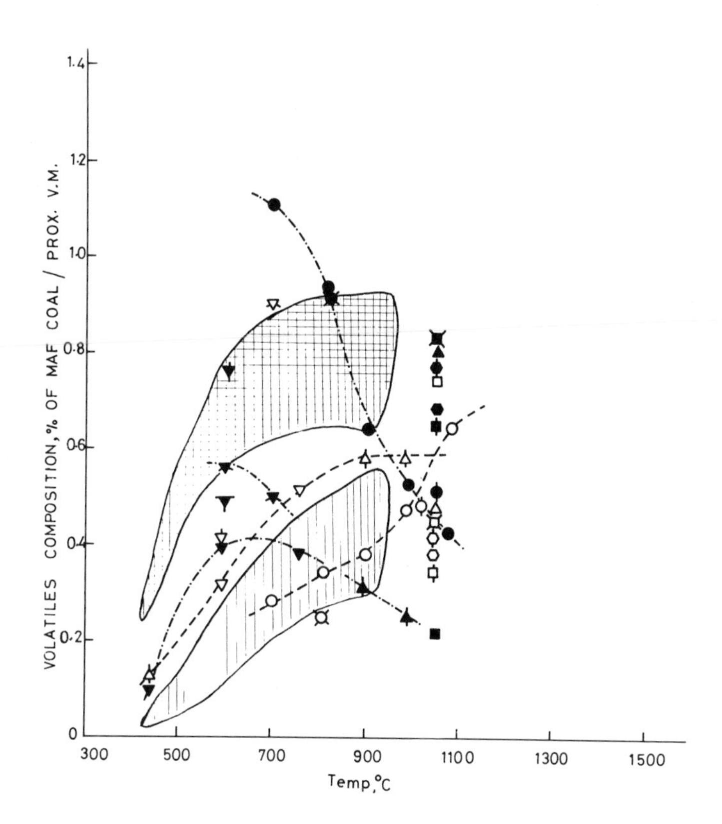

Figure 2.9. Product distribution as a function of temperature in rapid pyrolysis. [cross-hatched], Solid points: tar + liquid. [vertical hatched], Open points: gas + water. □ ■ Flenus de Bruay (33.1); △ ▲ Bergmannsgluck (19.6); Emma (22.8); □ ■ Maigre Oignies No. 2 (10.2); ⬡ ⬢ Len-Lievin No. 10 (29.6); Wendel, III (37.1); Faulguemont (38.9), Loison and Chauvin [73]; ▽ ▼ Elkol (42.7); Federal No. 1 (39.5); Orient No. 3 (37.0); ▼ Kopperston No. 2 MVb (33.1); Friedman et al. [33]; Pocahontas No. 3 LVb Vitrain (20.5); ○ ● Pittsburgh HVab Vitrain (35.7); Illinois No. 2 HVcb Vitrain (48.4), Mentser et al. [77].

Shaded: The coals listed in Table 2.5 under Solomon [98] indicating rapid pyrolysis show a similar trend to that observed in slow pyrolysis.

Figure 2.9 shows the observed product distribution with respect to tar, liquid, gas, and water in rapid pyrolysis for a few coals, as a function of temperature. In this figure the shaded areas cover the data for five bituminous coals observed under identical and isothermal conditions by Solomon [98, 99]. These data show a trend identical with those observed in slow pyrolysis (Figure 2.5). Thus in rapid pyrolysis also the amounts of tar and liquid tend to approach some limiting values after

ISBN 0-201-08300-0

about 700°C, while the production of gas and water continues to rise. The data points and the curves shown in Figure 2.9 are those observed by various other investigators under widely different experimental conditions.

Figures 2.10(a) and 2.10(b) show the gas composition and the water/gas ratio (by volume) in the product for five bituminous coals as observed by Solomon [99]. This figure indicates that the production of hydrogen sharply rises at the expense of hydrocarbons with an increase in pyrolyzing temperature. The percentage of CO does not change appreciably, while the percentage of CO_2 declines.

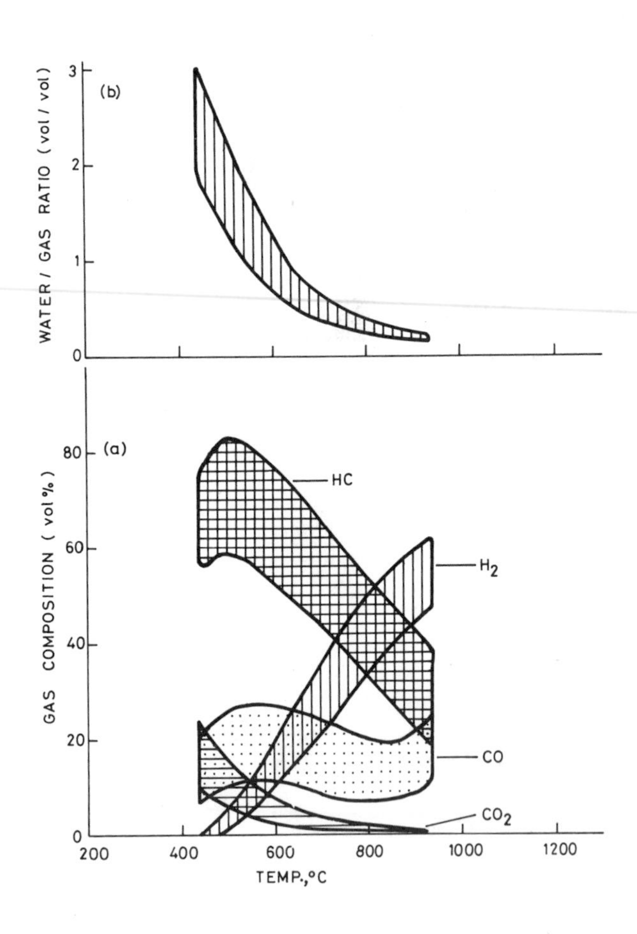

Figure 2.10. Gas composition (a) and water/gas volumetric ratio (b) as a function of temperature in rapid pyrolysis. Based on the data of Solomon [98, 99] on five bituminous coals. Residence time: 20 s.

ISBN 0-201-08300-0

Figures 2.11(a) and 2.11(b) show the compositions and the average molecular weights of the hydrocarbons as a function of temperature, for three bituminous coals and one lignite. The volumetric compositions in Figure 2.11(a) are calculated from the (weight percent) data of Solomon [98], with the following assumptions:

1. C-2 hydrocarbons consist of 50% C_2H_6 and 50% C_2H_4 (mol. wt. 29).
2. C-3 hydrocarbons consist of 100% C_3H_8 (mol. wt. 44).
3. C-4+ hydrocarbons consist of 50% C_4H_{10} and 50% C_5H_{12} (mol. wt. 65).

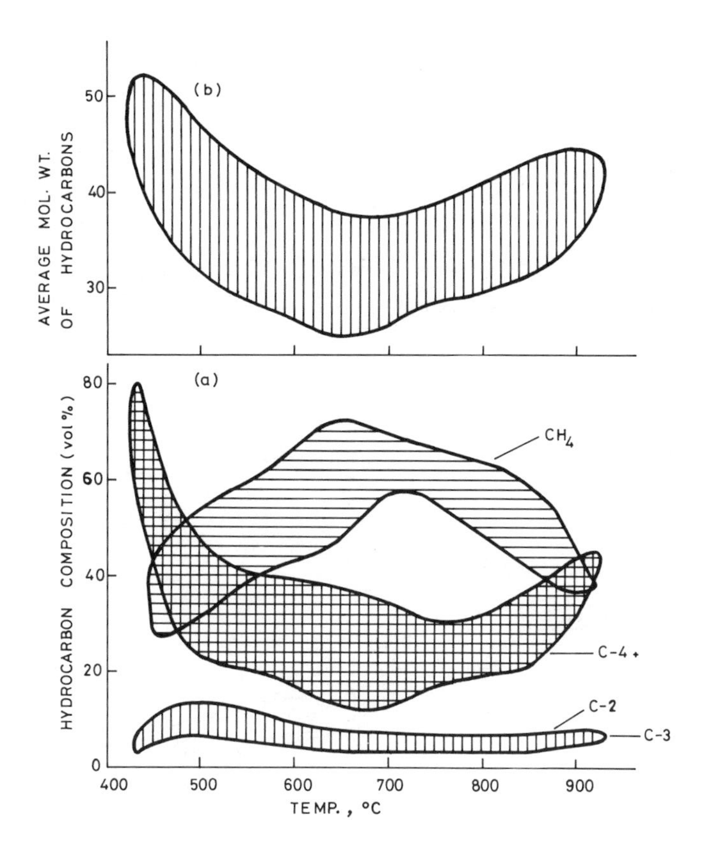

Figure 2.11. Hydrocarbon composition (a) and average molecular weights of hydrocarbons (b) as a function of temperature in rapid pyrolysis. Based on the data of Solomon [98, 99] on three bituminous coals and 1 lignite. Residence time: 20 s.

ISBN 0-201-08300-0

Figures 2.12(a) and 2.12(b) show the elemental compositions of char and tar as observed by Solomon [99] for 12 bituminous coals. Figure 2.12(a) indicates that, although the oxygen and hydrogen contents of char drop sharply with an increase in temperature, there is no appreciable change in the carbon and nitrogen contents. Tars are found to be virtually identical with their parent coals in carbon, oxygen, and nitrogen contents but richer in hydrogen. Tars become enriched with hydrogen because of being very similar in composition to their parent coals but having lower molecular weights. The sulfur contents of both char and tar vary most wildly with the nature of the parent coal, as shown in Figures 2.12(a) and 2.12(b), and are thus highly unpredictable.

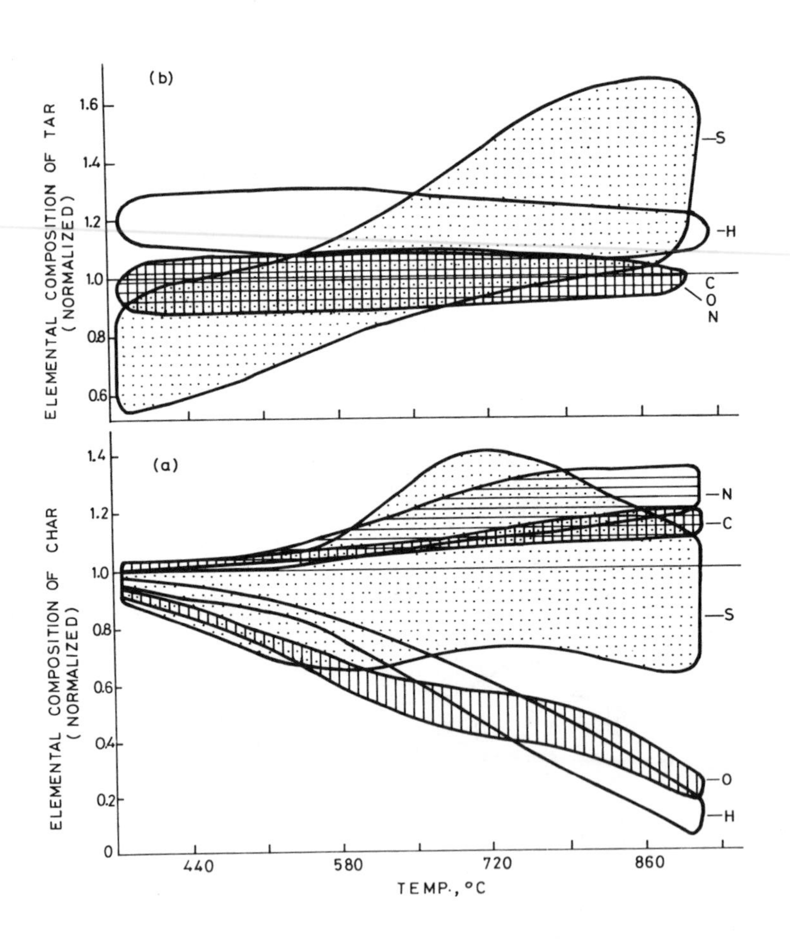

Figure 2.12. Elemental compositions of char (a) and tar (b) normalized to the compositions in the parent coal, as a function of temperature in rapid pyrolysis. Based on the data of Solomon [99] on 12 bituminous coals. Residence time: 20 s.

ISBN 0-201-08300-0

Nitrogen and sulfur are the two principal pollutants from coal. The nature of the structures of nitrogen and sulfur compounds in coals and the reactions leading to these evolutions during pyrolysis and subsequent stages of char and tar decompositions are of great importance. A great many studies have therefore been made in these areas. It is believed in general that coal nitrogen is almost entirely contained in the tightly bound ring structures of coal. The nitrogen contents and the structures of the nitrogen compounds of tar and the parent coal are very similar, since the nitrogen-containing ring structures are evolved from coal without cleavage, during the primary stages of pyrolysis. The tar nitrogen and the other nitrogen in char evolve in various forms by ring ruptures when subjected to higher temperatures. Sulfur, in contrast to nitrogen, occurs in coal in several forms, which can be broadly classified as inorganic and organic. Aside from a small amount of sulfates the principal constitutent of inorganic sulfur is iron pyrite (FeS_2). The decomposition of FeS_2 takes place at fairly low temperature (~400°C), producing FeS, sulfur, H_2S, and organic sulfur compounds. The compound FeS decomposes at much higher temperatures (~900°C) to form Fe, releasing sulfur and sulfur compounds. Organic sulfur compounds are generally classified into three categories, namely, loosely bound sulfur, called organic-I; tightly bound sulfur, called organic-III; and a category intermediate between these two forms, called organic-II. The loosely bound forms of sulfur decompose at low temperature (~400°C), while the tightly bound forms decompose at temperatures above 500°C. Some of the tightly bound organic sulfur compounds and some of the intermediate category are formed by incorporation of pyritic sulfur into the ring structures. The extreme variations in the sulfur contents of tar and char produced from coal pyrolysis, as seen in Figures 2.12(a) and 2.12(b), are due to wide differences in the compositions of the various forms of sulfur in coals. Solomon [99] obtained the following first-order rate constants for the char decompositions leading to nitrogen and sulfur losses, for all of the 12 bituminous coals and one lignite studied:

Rate Constant	Element	Frequency Factor, sec^{-1}	Activation Energy, kcal/mol
k_N	Nitrogen	3.6	15.0
k_{py}	Pyritic sulfur	6×10^2	16.7
k_I	Organic-I sulfur	100	11.6
k_{II}	Organic-II sulfur	10	11.6
k_{III}	Organic-III sulfur	2.2×10^9	56.1

The behavior of coal undergoing rapid pyrolysis is apparently similar to that of organic polymers at high heating rates. Thus at high heating rates the coal molecules tend to devolatilize without substantial change, while at low heating rates

ISBN 0-201-08300-0

they undergo considerable decomposition and repolymerization reactions. The tar produced by rapid heating at intermediate temperatures (750 to 900°C) has therefore been found to be very similar in composition to coal [122], and usually heavier than the tar produced in slow heating.

The kinetic model applicable to the release rates of simple gaseous products by pyrolysis at both slow and rapid heating that was proposed by Juntgen and van Heek [63] was discussed in the preceding section dealing with pyrolysis by slow heating. Reidelbach and Summerfield [90] proposed a more complicated kinetic model for the calculation of the pyrolysis products in terms of tar, liquid, char, and gases. A scheme of reactions that involves as many as seven decomposition reactions, including two secondary ones in the gaseous phase, was used to develop this model. Although the model requires the experimental determination of a large number of parameters, such a detailed mechanistic approach was made in the belief that a similar treatment is necessary for predicting the complex yield structure of coal pyrolysis under all conditions.

Most of the experimental results and the correlations available for pyrolysis are applicable to coal particles below approximately 100 μ in size, and for sample sizes ranging from micrograms to a maximum of a few grams. The yield, product distribution, and duration of decomposition reactions will be affected if the particle size and/or the sample size is large. This is due to two principal reasons. First, there will be a temperature gradient within the individual particles and/or the bed of particles, and, second, the residence time of the primary volatile decomposition products within the particles and/or bed will increase, raising the chance of secondary decomposition reactions. Very little experimental evidence is available to make a proper assessment of these effects.

C. FLASH PYROLYSIS

Systematic investigation on the flash pyrolysis of coal started with the works of Rau and Seglin [88, 89] and Sharkey et al. [96]. Rau and Seglin attained a solid residence time of 0.65 ms by using a GE FT 524, a xenon filled, quartz helix flash tube which heated a 10 mg sample of coal powder, 5–10 μ in size. The percentage of coal volatilized could not be determined, but it was believed to be higher than the proximate VM yield. However, the interesting finding was that the product gases contained 45–67% H_2, only 10–25% of CO_x, CH_4, and C_2H_6, and 20–30% of "remaining." This remaining portion contained about 70% C_2H_2 in addition to other unsaturated hydrocarbons. Sharkey et al. used a flash heating technique similar to that of Rau and Seglin, as well as laser irradiation. In both techniques a solid residence time of about one ms was attained. The product gases contained 45 to 69% H_2, about 20% CO_x, CH_4, and C_2H_6, and 10 to 33% of "remainder." The remainder contained about 70% of C_2H_2.

Granger and Ladner [44] used the xenon flash tube technique for heating several coals with different incident energies. They reported the detailed product yields and compositions from these experiments. Figures 2.13, 2.14, and 2.15 show

ISBN 0-201-08300-0

these results; Figure 2.15 also contains the data of Durmosch et al. [21]. The temperatures shown in these figures, as well as those reported by all the investigators on flash pyrolysis, are estimated values based on approximate calculations from the incident energies. Later Durmosch et al. [21] used microwave and Woodburn et al. [120] used a shock tube for the flash heating of coal and reported the product characteristics. Durmosch et al. attained an extremely high temperature of 1750 to 1900°C and, using a solid residence time of 20 to 40 s, obtained a product gas that contained as much as 85 to 89% H_2, only 2 to 5% of CO_x, CH_4, and C_2H_6, and 7 to 12% of "remainder."

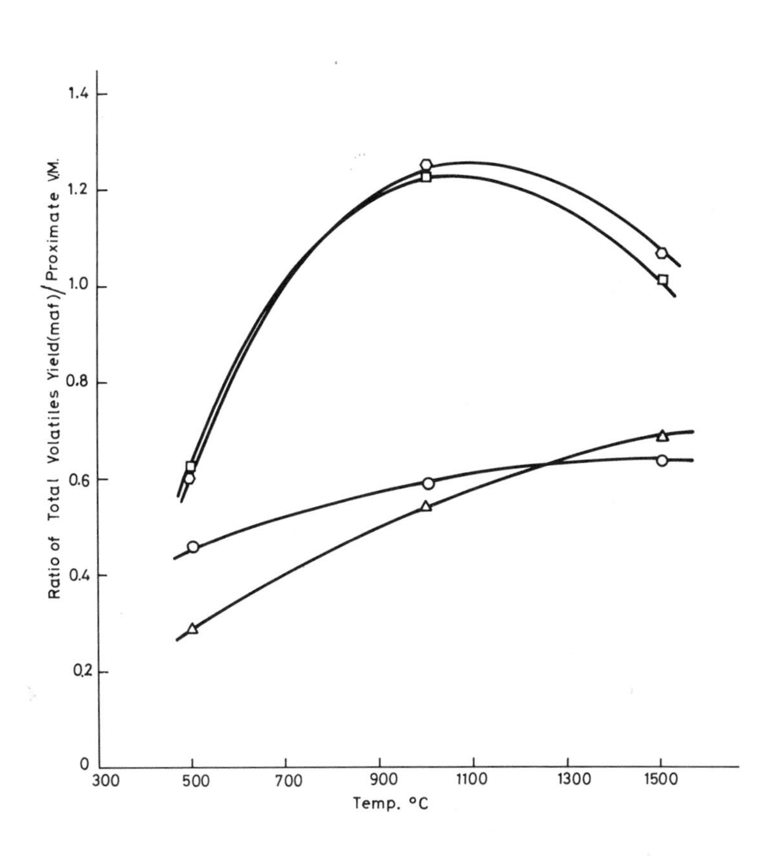

Figure 2.13. Total volatiles yield as a function of temperature in flash pyrolysis. Based on the data of Granger and Ladner [44]. O NCB C.R.C. No. 902 (33.2); △ NCB C.R.C. No. 601 (32.8); □ NCB C.R.C. No. 301b (25.6); ⬡ NCB C.R.C. No. 203 (18.0).

ISBN 0-201-08300-0

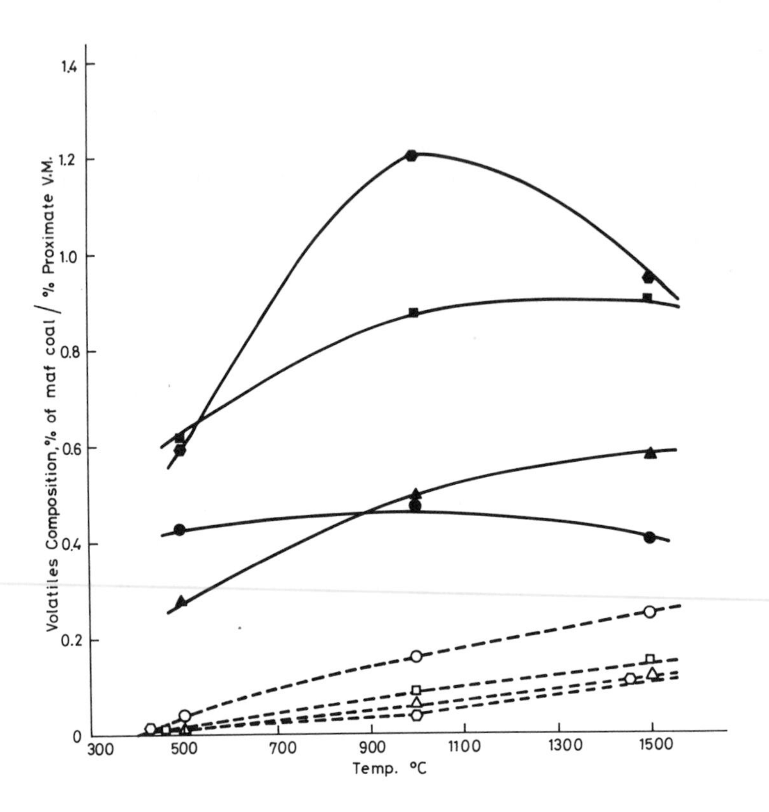

Figure 2.14. Product distribution as a function of temperature in flash pyrolysis. Based on the data of Granger and Ladner [44]. Solid points: tar + liquid. Open points: gas + water.

○ ● NCB C.R.C. No. 902 (33.2); △ ▲ NCB C.R.C. No. 601 (32.8); □ ■ NCB C.R.C. No. 301b (25.6); ⬡ ⬢ NCB C.R.C. No. 203 (18.0).

The data available on the yields and product distributions from flash pyrolysis of coal are insufficient to draw a generalized conclusion. The data of Rau and Seglin and Woodburn et al. on the gas composition could not be shown in Figure 2.15, since no estimates of the particle temperatures were available. However, a high yield of acetylene and other unsaturated hydrocarbons has been found to be a characteristic feature of this technique in most of these experiments. A mathematical model for pyrolysis of a coal particle in an inert gas is presented as an Addendum of this chapter on page 158.

ISBN 0-201-08300-0

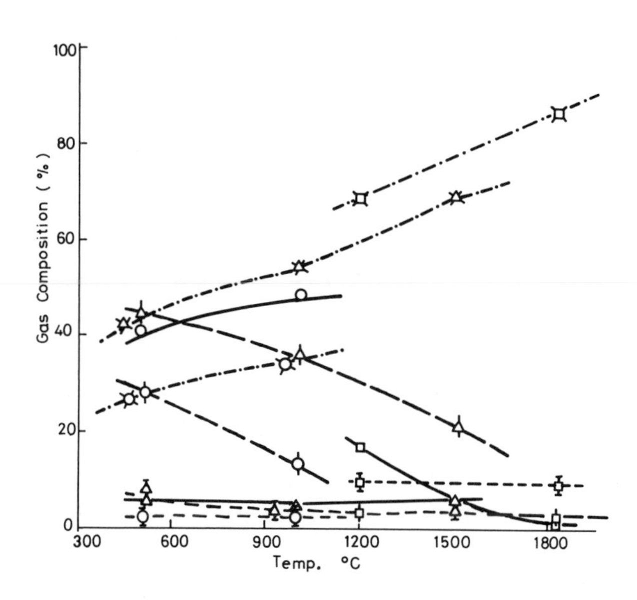

Figure 2.15. Gas composition as a function of temperature in flash pyrolysis.

Gas	Symbol	Coal (VM)	Symbol	Coal (VM)	Symbol	Coal
$CO + CO_2$	○	NCB C.R.C. No. 902 (33.2)	△	NCB C.R.C. No. 601 (32.8)	□	German K-23 (23.5)
$CH_4 + C_2H_6$		NCB C.R.C. No. 902 (33.2)		NCB C.R.C. No. 601 (32.8)		German K-23 (23.5)
H_2		NCB C.R.C. No. 902 (33.2)		NCB C.R.C. No. 601 (32.8)		German K-23 (23.5)
Rest		NCB C.R.C. No. 902 (33.2)		NCB C.R.C. No. 601 (32.8)		German K-23 (23.5)

Granger and Ladner [44] — Durmosch et al. [21]

3. CHAR-GAS REACTIONS

The phenomena involved in the char production due to the initial rapid pyrolysis are shown schematically in Figure 3.1. In the presence of oxygen, generally the evolution of volatile matters takes place first, followed by the burning of these volatiles. However, in combustion at extremely high temperatures and at

ISBN 0-201-08300-0

rapid heating rates the coal particle can burn directly at its outside surface, before devolatilization takes place. More details of the phenomenon of simultaneous release and combustion of coal volatiles are given in Chapter 3. In most cases the char produced in this stage is distinctly different from its parent coal in size, shape, and pore characteristics, whether it is produced in the presence or the absence of oxygen. The char thus produced takes part in the second-stage char-gas reactions, namely, the char-H_2, char-O_2, char-H_2O, and char-CO_2 reactions.

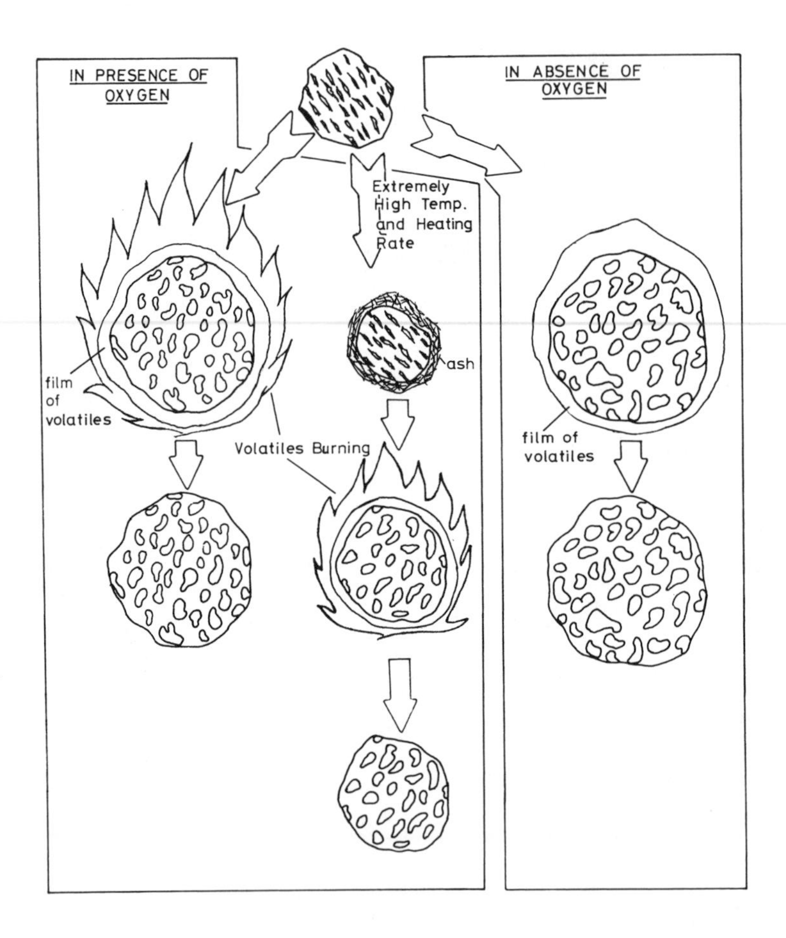

Figure 3.1. Pyrolysis of coal.

3.1. General Characteristics of Char-Gas Reactions

The char-gas reactions that take place during the second stage following the pyrolysis reaction may be classified into two distinct categories, namely, volumetric reactions and surface reactions. Char-gas reactions, whether of the volumetric or surface type, take place on the external or internal surface of the char. Thus

ISBN 0-201-08300-0

diffusion is an important step in heterogeneous char-gas reactions. In a volumetric reaction the reacting gas diffuses into the interior of the particles, and the reaction zone spreads throughout the body of the solid. As the reaction proceeds, product solid or ash layer may build up at the outside surface of the particles as the "reacting zone" continues to shrink. In a surface reaction, on the other hand, the reacting gas can hardly penetrate into the interior of the solid particles, and the reaction is therefore confined to the surface of the "shrinking core of unreacted solid." In such a case, therefore, the reaction interface is sandwiched between the inner unreacted core of solid and the outer product solid layer (ash).

Figures 3.2(a) and 3.2(b) show typical concentration profiles of carbon and reacting gas in surface reaction and volumetric reaction as a function of carbon conversion. Generally speaking, surface reaction occurs when the chemical reaction is very fast, such as combustion reaction, and diffusion is the rate controlling step.

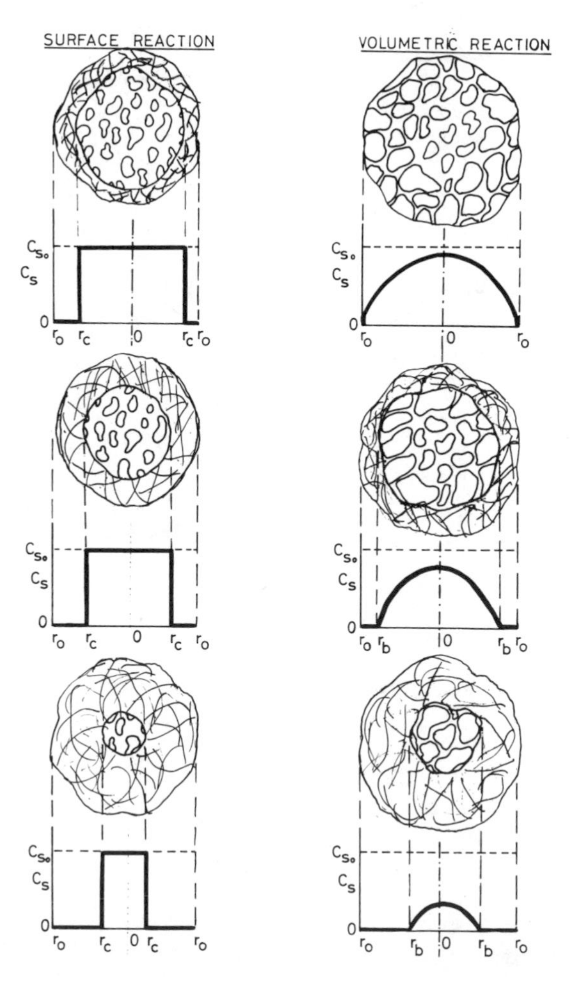

Figure 3.2(a). Carbon concentration profile.

ISBN 0-201-08300-0

Volumetric reaction, on the other hand, is characteristic of slow reactions in porous solids.

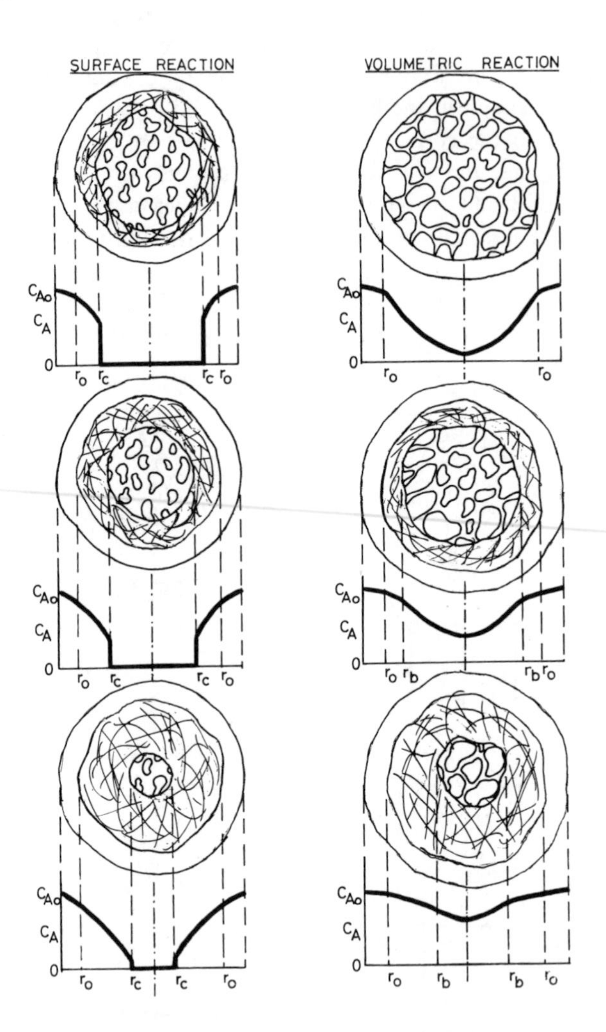

Figure 3.2(b). Gas concentration profile.

3.2. General Rate Expressions for Char-Gas Reactions

A. VOLUMETRIC REACTION

When a gaseous species reacts with char, the species must diffuse to and be adsorbed on the reactive sites of the char surface before the reaction takes place

ISBN 0-201-08300-0

and the products are formed on the carbon surface. The products must subsequently desorb from the active sites and diffuse out of the particles. Several rate expressions of the Langmuir-Hinshelwood type, particularly for char-CO_2 and char-steam reactions, have been proposed by various investigators on the basis of the concepts of adsorption and desorption of gases on solid surfaces. These rate expressions are of the following general form:

$$\text{Rate} = \frac{k_1' [A]^{x'}}{k_2' + k_3' [A] + k_4' [B]} \tag{3.1}$$

for a reaction of the type

$$C^*(s) + x'A(g) \rightarrow B(g)$$

where k_1', k_2', k_3', and k_4' represent the rates of the various steps of adsorption, desorption, and reaction involved in the proposed reaction mechanism, and [A] and [B] are the concentrations of the reactant and product gases, respectively. From a mechanistic point of view rate equations based on adsorption mechanisms are attractive if they indeed represent the true mechanism. These equations may also be very important when the effects of various gases on the individual reactions are expected to be significant. However, the direct applicabilities of most of these equations for design purposes are limited because of the requirement of more than one arbitrary rate constant. Thus applying the rate equation of the char-hydrogen reaction requires the knowledge of two rate constants, and applying the rate equations of the char-carbon dioxide and char-steam reactions requires the knowledge of at least three rate constants. Each of these constants is in turn a function of temperature. Thus the requirements are doubled when the temperature effect is included in these rate equations. Determination of these rate constants is extremely difficult, and the validity of these mechanisms is often questionable. Nevertheless, attempts have been made by many investigators to determine the rate constants of these mechanistic rate equations. However, wide variations are observed in the reported values.

For most practical purposes, empirical rate expressions which involve the use of only one or two rate constants to account for the effect of concentration on individual reactions are more useful. The rate constants of these empirical expressions are determined comparatively easily from experimental time-conversion data. However, extrapolation beyond the range of experimentation should be avoided. The general rate expression for a volumetric reaction can be expressed in the following form:

$$-\frac{dC_S}{dt} = k_v \cdot \alpha_v(x, T) \cdot C_A^n \cdot C_S^m \tag{3.2}$$

or

$$\frac{dx}{dt} = k_v \cdot \alpha_v(x, T) \cdot C_A^n \cdot C_{S_0}^{m-1} \cdot (1-x)^m \tag{3.3}$$

ISBN 0-201-08300-0

where x is the conversion of carbon in the char-gas reaction, k_v is the volumetric reaction rate constant having a dimension of $L^{3(m+n-1)}/mol^{m+n-1} \cdot \theta$, and $\alpha_v(x, T)$ represents the relative available pore surface area of particles and is a function of carbon conversion and temperature. This term represents the ratio of the available pore surface area of the particles at any stage of conversion to that at zero conversion level.

As will be seen later, the general equation can be simplified to the following form for most char-gas reactions following the volumetric reaction mode:

$$\frac{dx}{dt} = k_v \cdot \alpha_v(x) \cdot C_A^n (1-x)^m \tag{3.4}$$

by assuming that $\alpha_v(x, T) = \alpha_v(x)$.

The assumption in Equation 3.4 is valid for narrow temperature ranges, when the effect of conversion on the relative available pore surface area is much more significant than the effect of temperature.

B. SURFACE REACTION

In a surface reaction the rate is proportional to the surface area of the reaction interface and can be expressed as follows:

$$\frac{dx}{dt} = k_s \cdot S_{g_{ex}} \cdot C_A^n \cdot C_{S_0}^m \tag{3.5}$$

where $S_{g_{ex}}$ $(= S_{ex}/w_o)$ is the geometric surface area of the shrinking interface per unit original weight of a particle, and k_s is the surface reaction rate constant, having a dimension of $L^{3(m+n)-2}/(mol)^{m+n-1} \cdot \theta$. In Equation 3.5 the carbon concentration is constant and is equal to the original carbon concentration of the char. This is in contrast to the situation in a volume reaction (see Equation 3.3), where the carbon concentration decreases with conversion. This contrast has also been shown clearly in the carbon concentration profiles for the two reaction modes in Figure 3.2(a).

In the char-oxygen reaction, for example, Equation 3.5 can be expressed in a much simpler form as follows:

$$\frac{dx}{dt} = k_s' \cdot S_{g_{ex}} \cdot P_{O_2}$$

where P_{O_2} is the partial pressure of oxygen, and the time-conversion relation is expressed as

$$t = \frac{r_0 \rho_s}{k_s' P_{O_2}} \left[1 - (1-x)^{1/3}\right] \tag{3.6}$$

ISBN 0-201-08300-0

3.3. Changes in Coal/Char Pore Characteristics during Gasification Reactions and Their Effects on Rates

In a volumetric reaction the rate of a char-gas reaction is proportional to the available pore surface area of the particles. The rate expression for a volumetric reaction contains a relative surface area term, α_v, which is a function of conversion and temperature, as shown in Equations 3.2 and 3.3. Therefore a discussion on the pore characteristics of chars in relation to their parent coals is necessary.

Coal is a porous solid with porosities ranging from 2 to 20%, depending on the type or seam. In general, the porosity of coal has been found to be roughly related to its rank. Thus, the porosity is minimum (2 to 5%) in coals with 15 to 25% proximate volatile matters (maf basis) and with carbon contents of 87 to 91% (maf basis) [31]. The lump densities of coals vary from 0.9 to 1.5 g/cm^3.

The shape and size distribution of pores vary widely from coal to coal. The pores may be regular-cylindrical shaped, cone shaped, slit type, channel type, or a combination of two or more of these varieties. The pore sizes can be broadly classified into three categories, as shown in Table 3.1.

Table 3.1

Broad Pore Size Classification for Coals and Chars

Sl Number	Type	Size Range	Common Measurement Technique	Remarks
1	Macropores	500 Å to a few microns (50 nm to a few microns)	Mercury porosimetry	Account for the bulk of pore volumes in most coals and chars.
2	Mesopores or transitional pores	20–500 Å (2–50 nm)	Partly mercury porosimetry and partly BET methods	May contribute to both the pore volumes and pore surface areas of coal/char, depending on the pore size distribution. These pores serve as the principal passages for reactant and product gases to diffuse in and out of the individual particles during solid-gas reactions.
3	Micropores	< 20 Å (< 2 nm)	BET methods	Constitute the bulk of the pore surface areas measured by BET methods using CO_2 or N_2, in many high surface area coals/chars (surface areas >25 m^2/g).

Because of the wide variations in pore sizes, shapes, and internal structures, the pore surface areas as measured by the BET method show a wide range of values

ISBN 0-201-08300-0

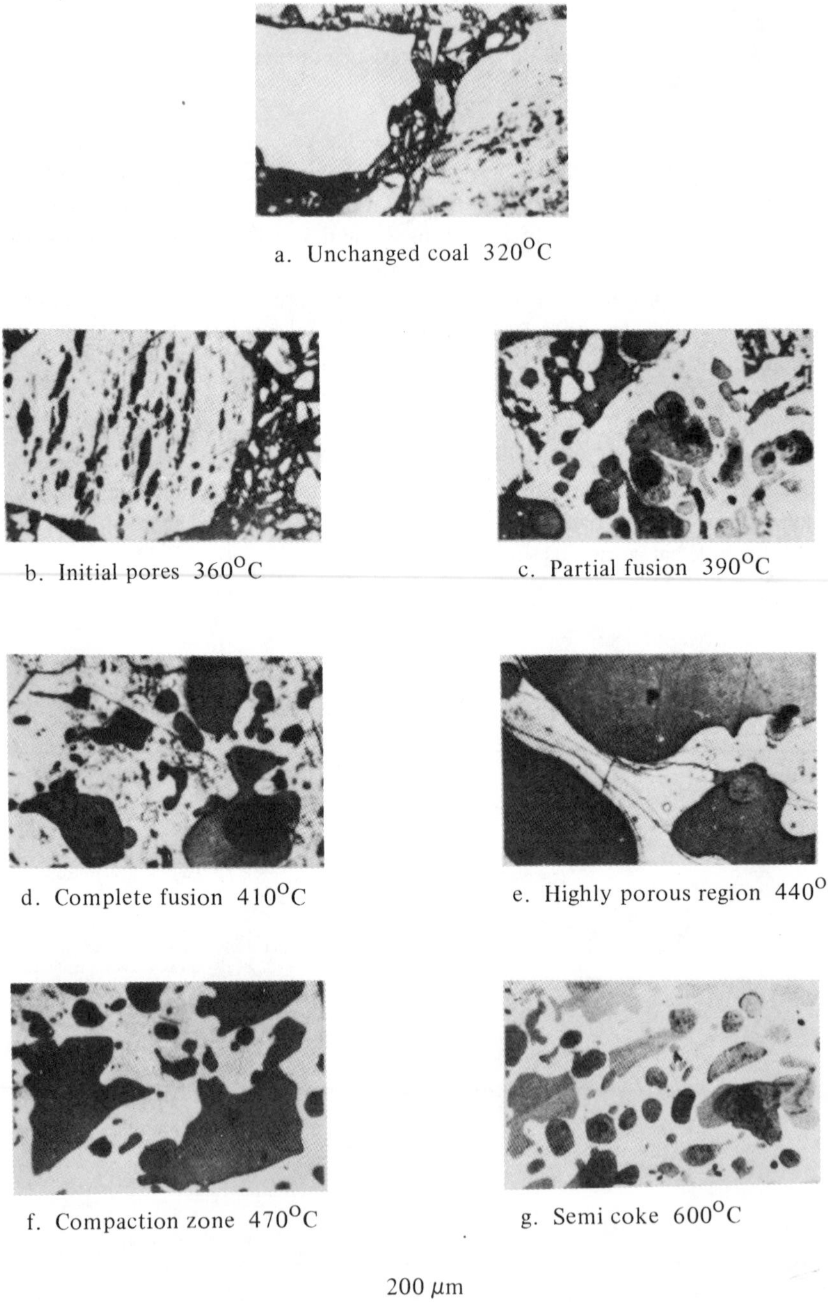

a. Unchanged coal 320°C

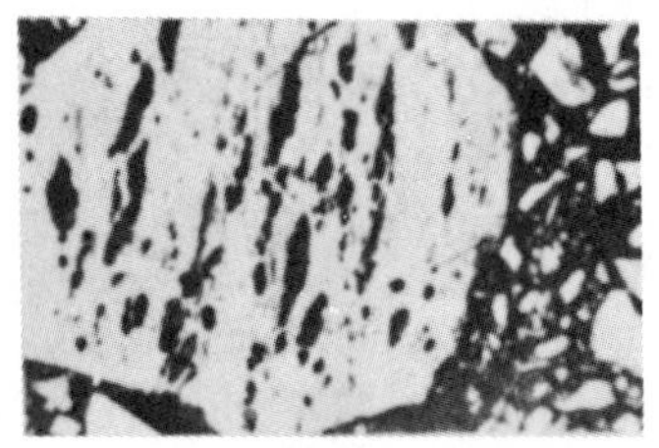

b. Initial pores 360°C

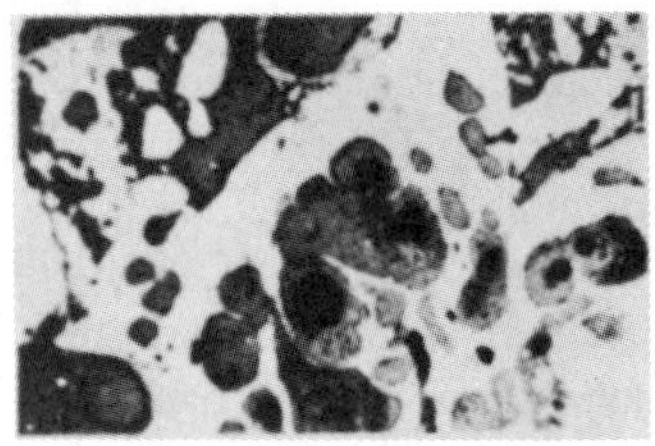

c. Partial fusion 390°C

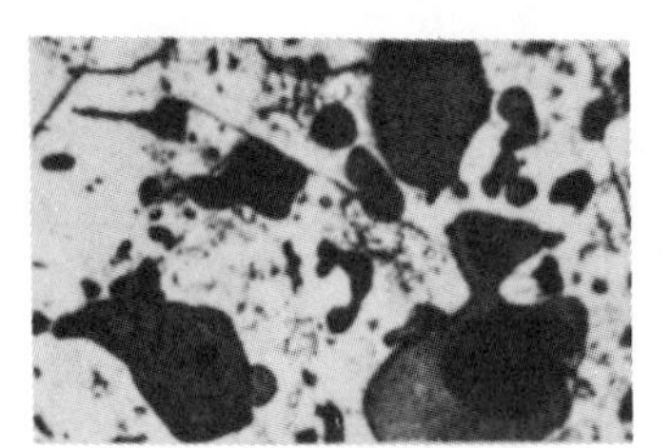

d. Complete fusion 410°C

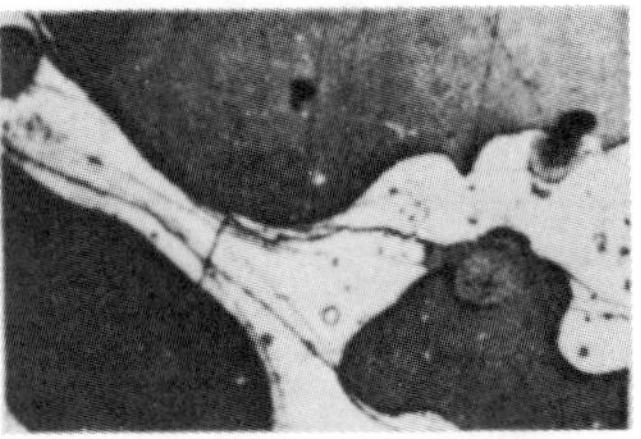

e. Highly porous region 440°C

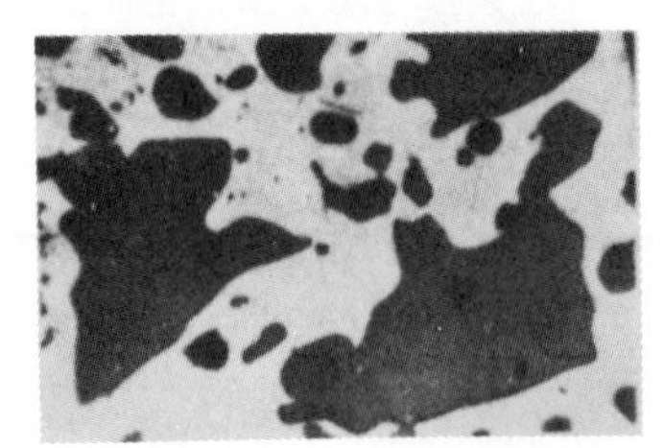

f. Compaction zone 470°C

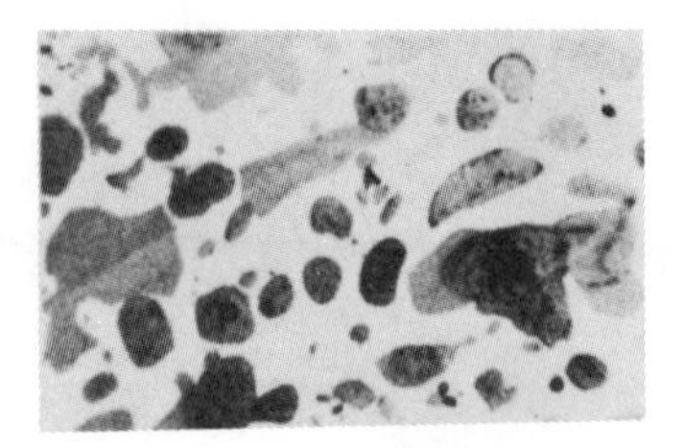

g. Semi coke 600°C

200 μm

Figure 3.3. Formation and development of the pore structure during carbonization of Blackhall coal. Reproduced from Hays et al. [50], *Fuel,* **55**, p. 299 (1976) by permission of the publishers, IPC Business Press Ltd.©.

ISBN 0-201-08300-0

(less than 1 to as high as 90 m^2/g), when N_2 is used as an adsorbate. When CO_2 serves as the adsorbate, the surface area values have been found to have range from 100 to 400 m^2/g, for a wide variety of American coals (anthracite to lignite) tested (Gan et al. [36]). More details on the pore characteristics of coal/char are given in Chapter 1.

As already discussed in this chapter, coal undergoes the first and quite often the most drastic changes in its physical characteristics during the stage of pyrolysis. In most cases, especially with high-swelling coals, pyrolysis brings about an enormous change in pore characteristics. The photographs in Figure 3.3 exemplify one such change for a single coal particle. The type and degree of change in coal/char characteristics during pyrolysis depend not only on the coal type, but also on the particle size, temperature, heating rate, gaseous environments, and hydrodynamic conditions of the process. In most cases the char produced from coal pyrolsis shows a much larger surface area than does its parent coal because of a significant increase in porosity. Thus nitrogen surface areas as large as 200–400 m^2/g can be seen in many chars obtained from coals having original surface areas of 10–20 m^2/g.

As the char produced by the initial stage of pyrolysis starts reacting with the gases in the second stage of gasification, the sizes of the particles and/or their internal pore structures undergo subsequent change. The nature and the degree of these changes depend mainly on the following factors:

1. The initial pore characteristics of the char.
2. The reaction involved, that is, whether it is a char-CO_2 or a char-O_2 reaction, for example.
3. The temperature of the reaction.
4. The extent of conversion.
5. The char particle size.

Many char samples show very high pore surface areas (>50 m^2/g) when measured by the BET method using N_2 or CO_2. However, in most cases a significant portion of these measured areas is found to be occupied by a very large number of finer micropores. When these chars undergo reactions with various gases at high temperatures, most of the measured surface areas may not become available, because the finer pores may not be accessible to the reacting gases under the reaction conditions. In such cases the reaction rate will be proportional only to the portion of pore surface area available (accessible) to the reacting gases at the reaction temperature and pressure. The amount of available surface area, however, may change significantly during reactions, as demonstrated by Patel et al. [84], Berger et al. [12], Dutta et al. [23], and Dutta and Wen [22].

The electron microscopic photographs in Figure 3.4 show the initial structures of three char samples before reaction and the changes in pore structure of one of these samples with conversion, as it reacts with CO_2. These photographs show only the changes in macropore structure in a char with reaction. However, a similar change occurs in the accessible meso- and micropore networks inside the char particles.

ISBN 0-201-08300-0

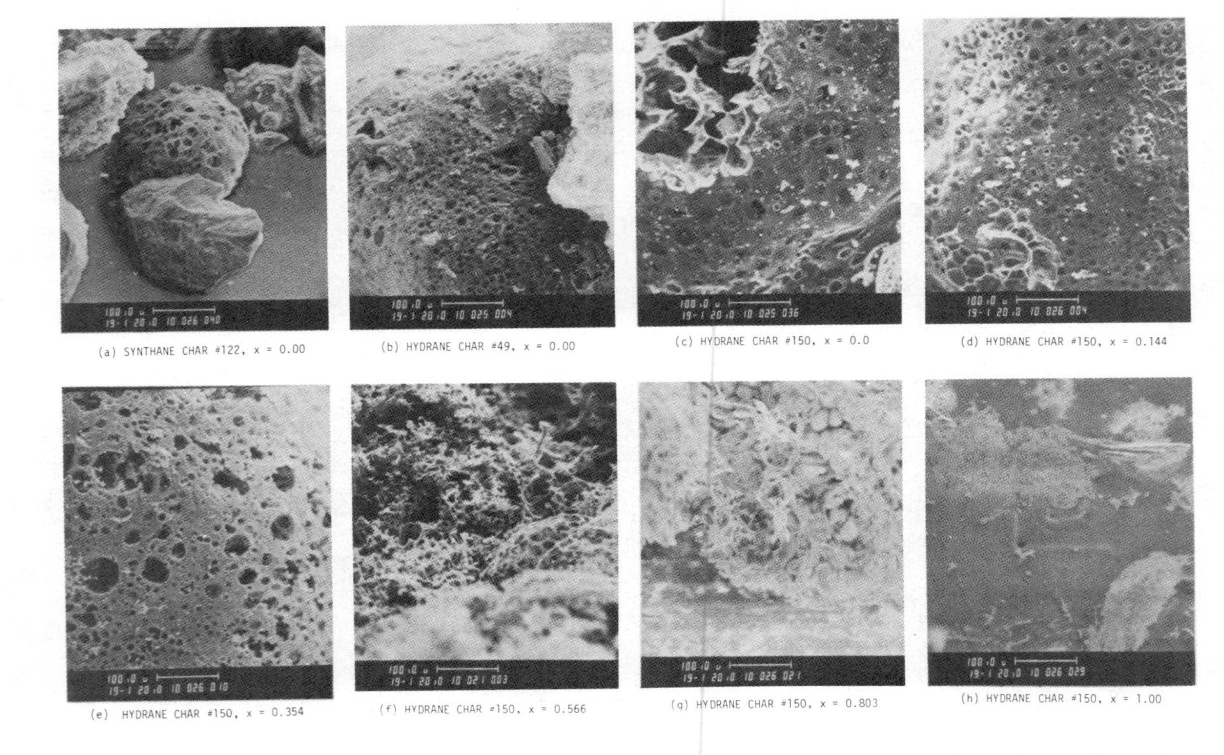

Figure 3.4. Surfaces of three char samples (a-c) before reaction (average particle sizes) as viewed through scanning electron microscope at a magnification of 190: (a) Synthane char No. 122, $x = 0.00$; (b) Hydrane char No. 49, $x = 0.00$; (c) Hydrane char No. 150, $x = 0.0$. Surfaces of one char sample (Hydrane char No. 150) at various conversion levels as viewed through a scanning electron microscope at a magnification of 190; (d) $x = 0.144$; (e) $x = 0.354$; (f) $x = 0.566$; (g) $x = 0.803$; (h) $x = 1.00$.

ISBN 0-201-08300-0

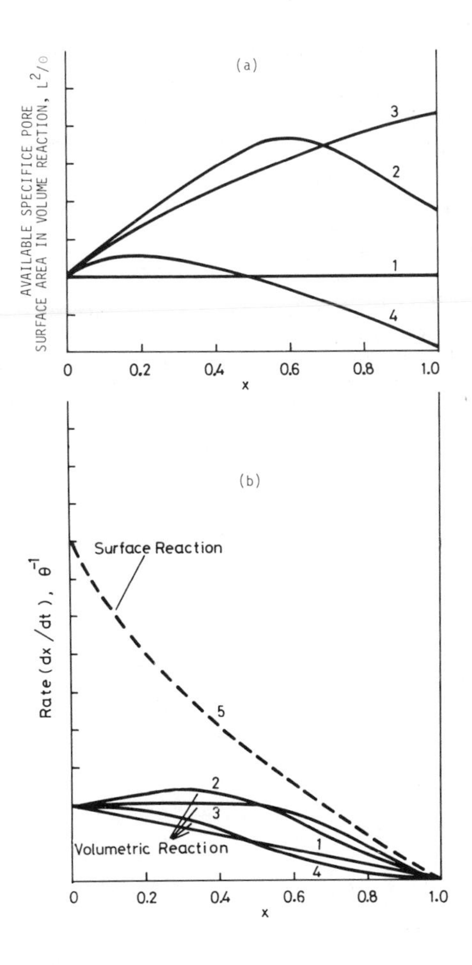

Figure 3.5. Effect of the change in pore characteristics during reaction on the rate curves.

It has been observed by Kawahata and Walker [66] and Dutta et al. [23] that the change in pore surface area with conversion depends not only on carbon conversion, but also on the temperature level of reaction. However, the effect of temperature has been found to be much less than that of conversion in many cases.

In volume reactions, when the rate is dominated by chemical reaction rate, four distinct cases may arise in regard to the changes in pore characteristics of

ISBN 0-201-08300-0

particles during reaction, as shown in Figure 3.5(a):

1. There is no appreciable change in available pore surface area: curve 1.
2. Surface area increases to a maximum and then decreases: curve 2.
3. Surface area increases continuously: curve 3.
4. Surface area decreases below the initial value after a small increase at the beginning: curve 4.

In case 1 the rate of a char-gas reaction is a linear function of conversion according to Equation 3.4 and as shown by curve 1 in Figure 3.5(b). In cases 2, 3, and 4 the function $\alpha_v(x)$ in Equation 3.4 changes with conversion, and the resulting rate curves deviate from linearity,as shown by curves 2, 3, and 4 in Figure 3.5(b).

When there is a surface reaction and the rate is controlled by the rate of surface reaction, the specific interface area term $S_{g_{ex}}$ in Equation 3.5 can be represented in terms of conversion as follows:

$$\frac{dx}{dt} = k_s' \cdot \frac{3r_c^2}{r_0^3 \rho_s} \cdot P_A$$

$$= 3k_s' \cdot \frac{(1-x)^{2/3}}{r_0 \rho_s} \cdot P_A \qquad (3.7)$$

since

$$S_{g_{ex}} = \frac{4\pi r_c^2}{(4/3)\pi r_0^3 \rho_s} = \frac{3r_c^2}{r_0^3 \rho_s} \quad \text{and} \quad r_c = r_0(1-x)^{1/3}$$

The rate of surface reaction is thus proportional to $(1 - x)^{2/3}$ and follows a course similar to that shown by curve 5 in Figure 3.5(b).

3.4. Catalytic Effects in Char-Gas Reactions

The discussions in preceding sections have been confined to gasification reactions without paying due attention to the possible catalytic or inhibiting effects of the mineral matters present in coals and chars. It has long been known, and lately been demonstrated by several investigators, that a number of inorganic elements present in coal/char have potential effects on the rates of gasification reactions. Such effects may be significant on reactions occurring both on the coal/char surface and in the gas phase. Reactions in the gas phase are affected by ash particles, whereas those on or inside the reacting coal/char particles are affected by the dispersed minerals in the coal/char body.

In general, the alkali, alkaline earth, and transition metals have been found to be the most effective catalysts. The catalysts believed to be most effective for the four important char-gas reactions are shown in Table 3.2.

ISBN 0-201-08300-0

Table 3.2

	Reaction	Effective Catalysts
1.	Char-oxygen	Fe, Co, Ni
2.	Char-steam	K, Na, Ni
3.	Char-carbon dioxide	K, Na, Li, Ni, Co, Fe, Ca
4.	Char-hydrogen	K, Ni

When such a catalytic effect is significant, the rate expressions given in preceding sections may or may not represent the intrinsic reaction rates, depending on the presence or absence of catalysts. The rate constants k_v and k_s of Equations 3.2–3.5 may include the catalytic effects. To isolate such effects, k_v and k_s in Equations 3.2–3.5 should be replaced as follows:

$$k_v = Z_v k_{vt} \text{ and } k_s = Z_s k_{st} \tag{3.8}$$

where k_{vt} and k_{st} are the true rate constants of a reaction, and Z_v and Z_s represent the effect due to catalysis. The correct values of true rate constants are extremely difficult to determine, however, since the presence of a trace of impurity (solid or gaseous) may have a significant effect on the measured rate in many cases. For all practical purposes, therefore, rate expressions of the forms of Equations 3.2–3.5 are more useful. There are some indications [110] that catalysts or impurities affect not only the preexponential factor, k_{v0}, but also the activation energy, E, of carbon-gas reactions. Thus for carbon-CO_2 and carbon-O_2 reaction systems the values of E have been found to decrease in the presence of catalytic minerals in coal.

The catalytic effects on the values of Z_v and Z_s for a given catalytic mineral with respect to a particular reaction depend on four factors:

1. The chemical form of the catalyst.
2. The physical form of the catalyst.
3. The amount of catalyst.
4. The temperature of reaction.

Thus Fe, Co, and Ni have been found to be the most effective catalysts when they are present in their elemental states or are transformed to the elemental states during reaction. Potassium and sodium have been found to be most effective in the form of carbonates and least effective as phosphates. Among the oxides of iron and other transition metals, the stoichiometrically deficient oxides are believed to be

ISBN 0-201-08300-0

the better catalysts in C-CO_2 and C-H_2O reactions. Thus FeO or Fe_3O_4 is a better catalyst than Fe_2O_3 in such reactions. Among the salts of these metals, the organic salts like oxalates, acetates, and citrates show catalytic effects superior to those of the inorganic salts. This is due to the fact that the former group of salts yields finer subdivision and dispersion of the metal ions inside the body of the reacting solid particles. The catalytic activity decreases with the increase in size of the dispersed catalyst particles.

The activity increases with an increase in the amount of catalyst (or impurity). It soon reaches a "saturation point," however, above which a larger amount does not have an appreciable effect.

Experiments of Exxon Research and Engineering Company (Kalina [64]) indicate that treatment of coals with solutions of Na_2CO_3 and/or K_2CO_3 not only catalyzes the char-steam reaction but also significantly reduces the agglomerating tendency of caking coals. The experiments were conducted at about 700°C and with up to 15% potassium based on the carbon in the coal. The rate of gasification was found to be almost proportional to the concentration of the impregnated potassium. The work of Battelle's Columbus Laboratories [19] also demonstrates that impregnation of CaO into coal before gasification can prevent agglomeration of coal and greatly increase the coal/char reactivity and hydrocarbon yields in the gasifier. These studies indicate definite potentials for catalytic coal gasification in commercial applications.

It is believed that impurities decrease the CO/CO_2 ratio in the C-O_2 reaction, because the impurities catalyze the secondary CO → CO_2 reaction, without significantly affecting the primary reaction, C → CO.

It is known that CO and H_2 are inhibitors of C-CO_2 and C-H_2O reactions. However, this may be true only as far as the reactions are uncatalyzed. If these reactions are catalyzed, particularly by oxides of Ni, Co, or Fe, CO and H_2 may instead act as promoters. They may reduce the oxides of Ni, Co, and Fe and, if the temperatures are favorable, may bring them to their metallic states, which are the most effective catalysts for gasification reactions. For the same reason steam may also act as a promoter for these reactions, since it produces H_2 and CO with carbon or char. A detailed review on the mechanism and related aspects of catalysis of carbon gasification has been presented by Walker et al. [110]. A discussion regarding effect of CaO on char-CO_2 reaction is presented in 3.6.

3.5. Coal/Char-Hydrogen Reaction

The reaction of hydrogen with coal or char is often called hydrogasification or hydrocarbonization. This is one of the important coal gasification reactions primarily aimed at upgrading the hydrocarbon content of the product gas. This reaction is the key to the economic production of SNG (synthetic natural gas) and liquid hydrocarbons.

ISBN 0-201-08300-0

A. REACTION MECHANISM

It has long been known that methane can be synthesized by directly reacting carbon with hydrogen. However, the reaction mechanism of hydrogenation of coal is much more complicated than that operating in the formation of CH_4 from pure carbon. In coal-hydrogen reactions CH_4 is not the only product, particularly during the first stage of rapid hydrogasification. In this period, pyrolysis and hydrogenation reactions (hydrogenation of both volatiles and char) take place simultaneously. Zielke and Gorin [124] proposed the following reactions steps as a possible mechanism for CH_4 production from coal/char during hydrogasification; (A) represents a postulated basic unit of ring structures of an average molecule in coal or char:

CH ═ CH
═ C C — $+ H_2$ $\underset{k_2}{\overset{k_1}{\rightleftharpoons}}$ CH_2 — CH_2
C — C ═ C C —
C — C

(A) Saturation (B)

CH_2 — CH_2 CH_3 CH_3
═ C C — $+ H_2$ $\xrightarrow{k_3}$ ═ C C —
C — C C — C

(B) Ring Rupture (C)

CH_3 CH_3 H H
═ C C — $+ 2H_2$ $\xrightarrow{k_4}$ ═ C C — $+ 2CH_4$
C — C C — C

(C) Dealkylation (D)

ISBN 0-201-08300-0

In this mechanism saturation of the aromatic rings occurs during hydrogenation, followed by ring rupture and dealkylation. The dealkylation step was assumed to be much faster than the other steps. The rate equation proposed by Zielke and Gorin [124] is as follows:

$$\text{Rate} = \frac{k_1 k_3 (\text{A}) (\text{H}_2)^2}{k_2 + k_3 (\text{H}_2)} \tag{3.9}$$

According to Equation 3.9, the order of the char-H_2 reaction will tend to a value of 2 at very low hydrogen partial pressure and a value of unity at very high hydrogen partial pressure.

B. EXPERIMENTAL TECHNIQUES USED IN HYDROGASIFICATION STUDIES

Goring et al. [41, 42], Zielke and Gorin [124], Hiteshue et al. [51, 52], Birch et al. [13], and Pyrcioch and Linden [87] used fluidized beds for experimental studies of hydrogasification of coal/char. In subsequent years several other methods were used for the experimental investigation of hydrogasification, with the aim of reducing the heat-up time of the coal particles and of gaining the combined benefits of both rapid pyrolysis and high pressure hydrogasification. Feldkirchner and Linden [29] used a vertical semiflow reactor. Moseley and Patterson [80] employed both a horizontal transport reactor and a horizontal tubular reactor heated by a series of moving gas burners. Later Moseley and Patterson [81] also used a vertical transport reactor in their studies. Coates et al. [20] and Johnson [60] employed entrained flow reactors. Anthony et al. [4] and Squires et al. [101] applied a fast heating technique which involved direct resistance heating of the metallic sample holder. Besides these fast heating techniques, Johnson [59] used a thermobalance to obtain more accurate data for a kinetic model of hydrogasification. Johnson's thermobalance, along with the experimental technique, is described in Chapter 4.

ISBN 0-201-08300-0

C. EFFECTS OF TEMPERATURE AND HYDROGEN PARTIAL PRESSURE ON THE RAPID AND SLOW STAGES OF HYDROGASIFICATION BY SLOW HEATING

Figures 3.6 to 3.9 represent typical time-conversion data of the above investigators for different coals at different temperatures and hydrogen partial pressures. Here experiments conducted in fluidized beds, semiflow reactors, or thermobalances are categorized as slow heating experiments. These figures also show the product-volatile compositions where available. As is seen in Figure 3.6, which is based on the data of Feldkirchner and Linden for a semiflow reactor, there is an extremely rapid initial stage of gasification in which 20 to 40% of the coal carbon is gasified in a few seconds. The remaining fraction of carbon gasifies comparatively slowly during the subsequent stage.

In Figure 3.7 the data of Hiteshue et al. [51, 52] from fluidized bed experiments are plotted. Although, data at very short residence times cannot be obtained in such experiments, the available results point to a phenomenon identical to that described above. Thus in the first 100 sec the reaction rates are found to be much faster than those in the subsequent periods. The initial stage of rapid conversion is obviously due to the contribution of the fast pyrolysis process occurring simultaneously with hydrogasification. Thus, as Figure 3.9 shows, this initial stage of rapid conversion or the so-called first stage of hydrogasification is absent when a completely devolatilized char is used.

In Figure 3.8 the time-conversion curve for coal B (proximate VM = 31.9%) in hydrogen atmosphere is compared with that in nitrogen atmosphere at the same temperature and pressure. The first stage of gasification, which is virtually complete in the initial few seconds, gasifies more than 50% of carbon in hydrogen atmosphere, but only about 30% in nitrogen atmosphere. Higher hydrogen partial pressure improves the carbon conversion in the first stage for two principal reasons. First, hydrogen gasifies part of the solid char. Second, it renders part of the primary volatile decomposition products unreactive, thereby helping that portion escape unchanged from the solid particle.

ISBN 0-201-08300-0

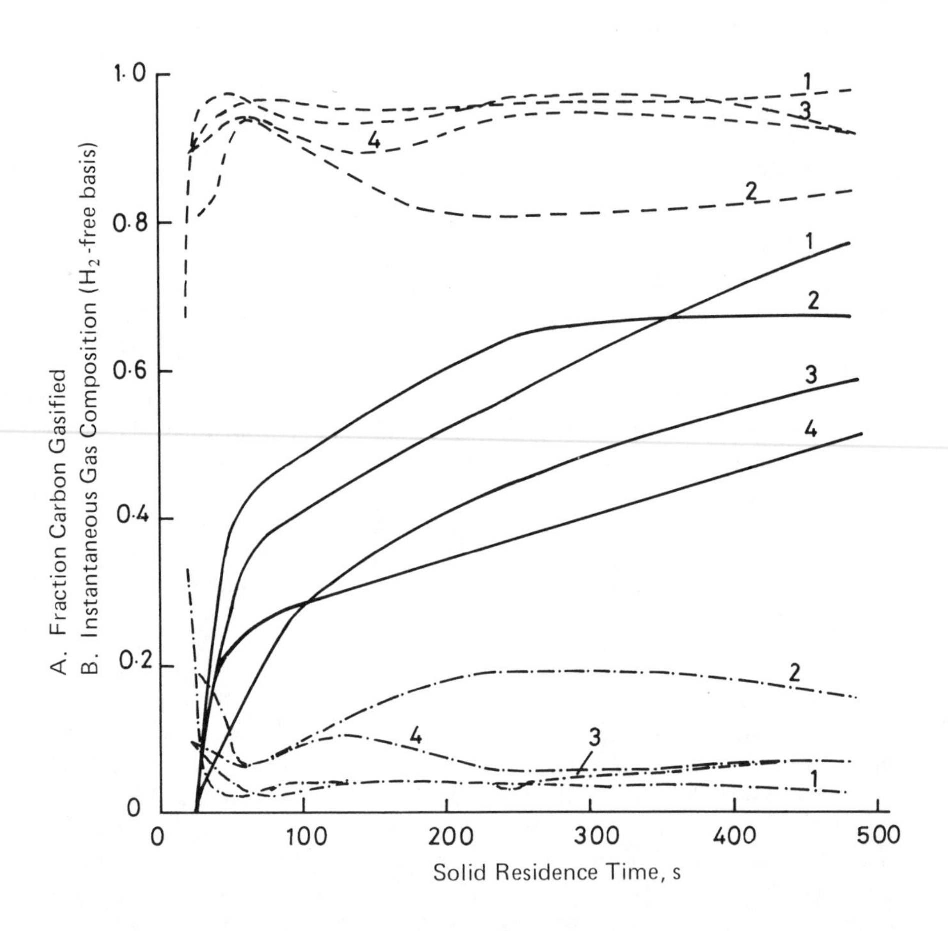

Figure 3.6. Effects of coal/char type and residence time on conversion and gas composition in hydrogasification by slow heating. Based on the data of Feldkirchner and Linden [29]. Temperature: 930–950°C. Pressure: 10.4 MN/m^2. Ambient gas: H_2.

A. —— Fraction of carbon gasified. B. —·— $CO + CO_2$, — — $CH_4 + C_2H_6$.

1. Concoal Montour No. 10 bitu. Coal (37.0). 2. North Dakota lignite (44.2). 3. Med. vol. anthracite (5.7). 4. Concoal char from coal 1 (19.1).

ISBN 0-201-08300-0

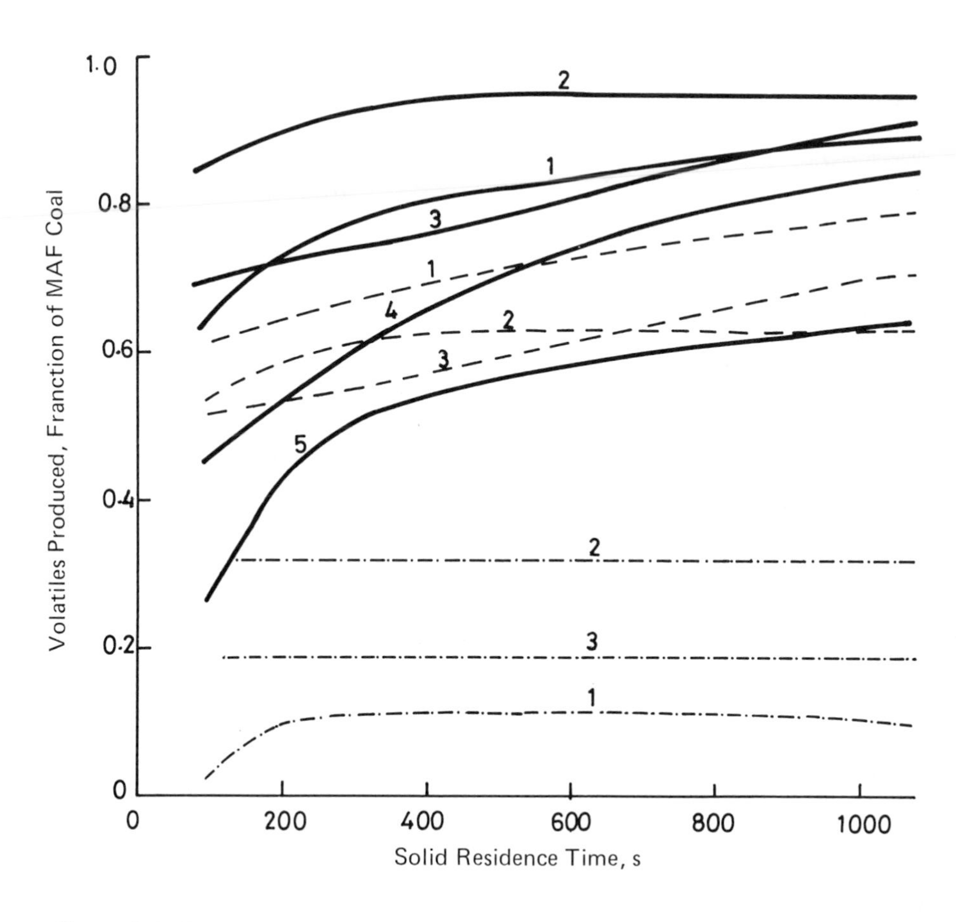

Figure 3.7. Product distribution as a function of coal type and solid residence time in hydrogasification by slow heating. Based on the data of Hiteshue et al. [51, 52]. Temperature: 800°C. Pressure: 41.4 MN/m^2. Ambient gas: H_2.

—— Total yield. –·– Tar + liquid. – – Gas + water. Volatiles produced include absorbed (reacted) hydrogen.

1. Rock-Springs (Wyoming) HVcb with 1% Mo (48.7, calc.). 2. Texas lignite (75.2, calc.). 3. Pittsburgh HVb (42.2, calc.). 4. Wyoming char from coal 1. 5. Med. vol. anthracite.

ISBN 0-201-08300-0

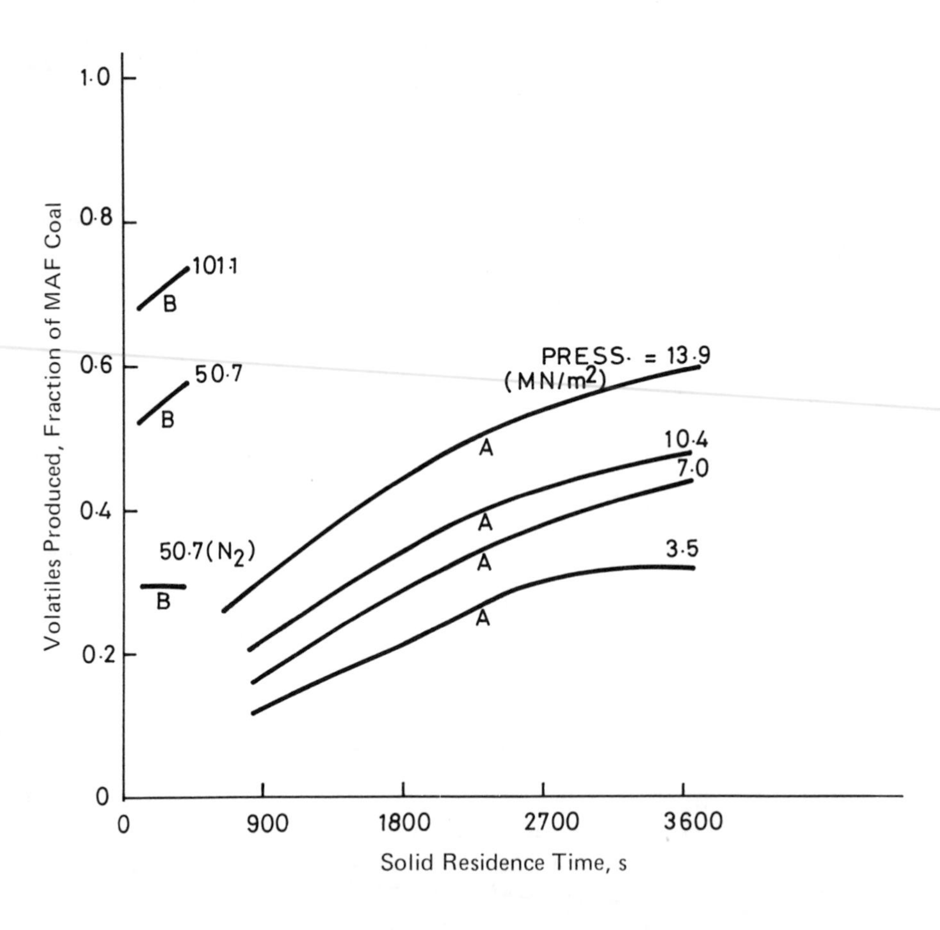

Figure 3.8. Effect of pressure and residence time on the second stage of hydrogasification in a fluidized bed. A. Concoal char from Montour No. 10 bitu coal by low temperature fluidized bed carbonization (19.1); temperature: 760°C; Pyrcioch and Linden [87]. B. Air-pretreated Pittsburgh No. 8 coal char (31.9); temperature: 927°C; ambient gas: H_2 (except as indicated otherwise) Johnson [59].

ISBN 0-201-08300-0

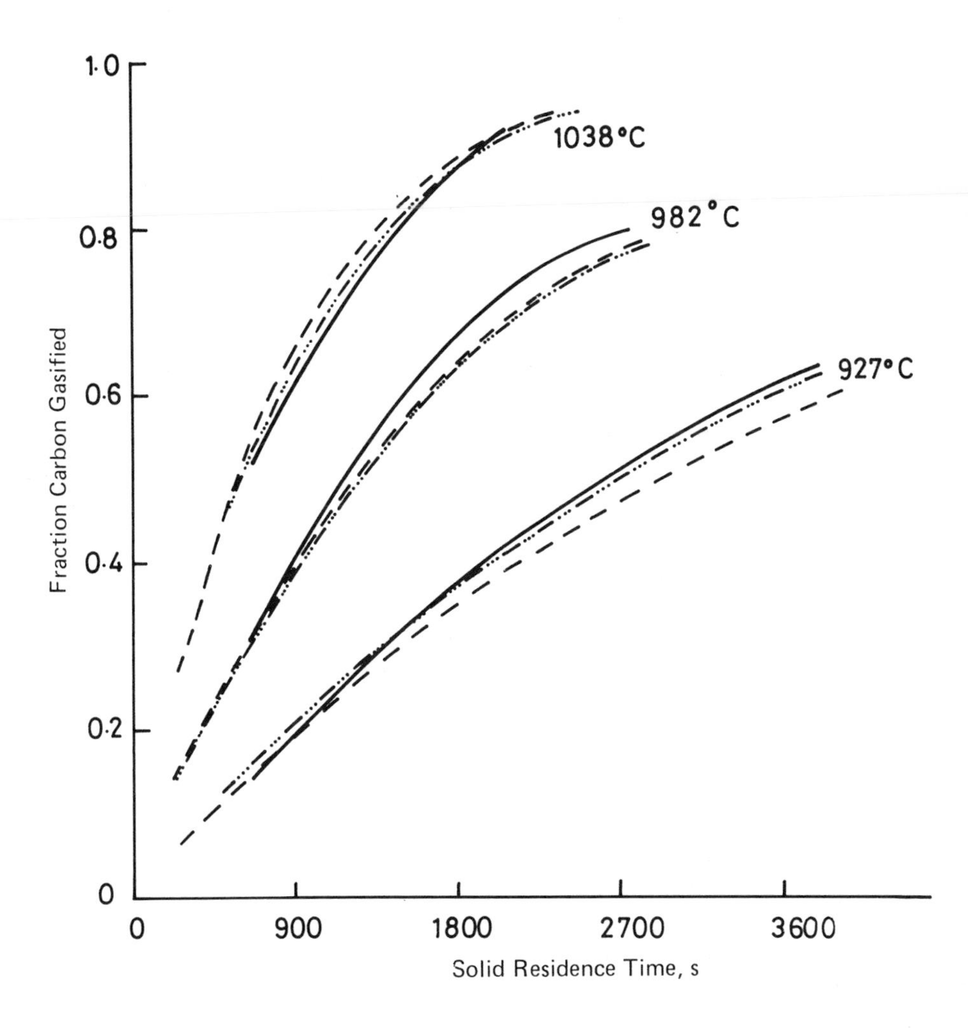

Figure 3.9. Effect of temperature and residence time on the second stage of hydrogasification in a thermobalance. Based on the data of Johnson [59]. Pressure: 3.5 MN/m^2. Ambient gas: H_2. Air-pretreated Pittsburgh No. 8 coal char (0.0).

—— Experimental curves. — — Calculated curves with f_0 = 0.80, according to Johnson correlation [59]. —···— Calculated curves with $k'_{v0} = 2.27 \times 10^3\ s^{-1}\ (MN/m^2)^{-1}$ and E = 172 kJ/mol, according to equations of Wen and Huebler [114].

ISBN 0-201-08300-0

D. KINETIC EQUATIONS FOR HYDROGASIFICATION

Wen and Huebler [114] proposed the following equations for the kinetics of first-stage and second-stage hydrogasification.

First stage:

$$\frac{dX}{dt} = k_v'(f - X)(C_{H_2} - C_{H_2}{}^*) \tag{3.10}$$

where X is the overall carbon conversion in the first stage at any time, t, and f is the overall carbon conversion at the end of the first stage, both based on initial carbon content. The point where the slope of the time-conversion curve changes sharply at the initial stage can be considered as the approximate boundary between the first and second stages in hydrogasification. The value of k_v', the first-stage rate constant, has to be determined experimentally for each coal/char. Approximate values are as follows:

$$k_v' \approx 0.95 \times 10^{-3}\ \text{m}^3/\text{mol} \cdot \text{s for raw coal}$$

$$\approx 0.90 \times 10^{-5}\ \text{m}^3/\text{mol} \cdot \text{s for pretreated char}$$

Also in Equation 3.10 C_{H_2} and $C_{H_2}{}^*$ are the hydrogen concentrations in the system and at equilibrium, respectively. The value of $C_{H_2}{}^*$ has to be calculated from the hydrogen-char reaction equilibrium constant, which is a function of the carbon conversion of the char (Wen and Huebler [114]).

Second stage:

$$\frac{dx}{dt} = k_v(1 - x)(C_{H_2} - C_{H_2}{}^*) \tag{3.11}$$

where x is the conversion due to the second stage alone, and k_v is the rate constant for the second stage, or

$$x = \frac{X - f}{1 - f}$$

Johnson [59] proposed the following empirical correlations for the first and second stages of hydrogasification, based on the experimental data that he gathered using a thermobalance.

First stage:

$$\int_0^X \frac{e^{\gamma X^2}\, dX}{(1 - X)^{2/3}} = 0.0092 f_R P_{H_2} \tag{3.12}$$

ISBN 0-201-08300-0

where X = carbon conversion in the first stage

f_R = relative reactivity factor of the coal/char for the first-stage gasification

γ $\simeq$ 0.97 for H_2 or H_2-CH_4 mixtures

$\simeq$ 1.70 for H_2-steam mixtures

Equation 3.12 is applicable to temperatures above 815°C and heating rates of 15 to 100°C/s. In this equation it is assumed that the first-stage conversion is independent of temperature above 815°C.

Second stage:

$$\frac{dx}{dt} = f_L k_I (1 - x)^{2/3} e^{-\gamma' x^2} \qquad (3.13)$$

where x = carbon conversion due to the second stage alone

f_L = $f_0 e^{8467/T_0}$

T_0 = maximum temperature to which char was exposed before gasification, °R

f_0 = relative reactivity factor of coal/char for second-stage gasification

$$k_I = \frac{P_{H_2}^2 e^{2.6741-33076/T} (1 - P_{CH_4}/P_{H_2}^2 K_E)}{1 + P_{H_2} e^{-10.452+19976/T}}$$

K_E = carbon-hydrogen-methane equilibrium constant

T = gasification temperature, °R

P_{H_2}, P_{CH_4} = partial pressures of hydrogen and methane in the bulk gas, atm

$$\gamma' = \frac{52.7 P_{H_2}}{1 + 54.3 P_{H_2}}$$

Table 3.3 shows values of f_0, the relative reactivity factor for the second stage, according to Johnson's correlation, and of the rate constant, k_v, according to the equation of Wen and Huebler, for a few coals and chars. In Table 3.3 the data of Pyrcioch and Linden [87], obtained at a comparatively low temperature (760°C), show that the reactivity of char decreases as the first-stage carbon conversion increases. Similar observations have been made by other investigators. The reactivity apparently reduces to a constant value as the volatile matter of the coal/

ISBN 0-201-08300-0

Table 3.3

Rate Parameters of the Second Stage of Hydrogasification

Investigator	Coal/Char Type	Proximate VM	C	H	O	Temperature, °C	Pressure, MN/m^2	Approximate First-Stage Carbon Conversion	Relative Reactivity Factor, f_o (Johnson [59])	Rate Constant k_v, s^{-1} $(MN/m^2)^{-1}$ (Wen and Huebler [114])
Hiteshue et al. [51]	Wyoming HVcB impregnated with 1% Mo	48.7[a]	77.6	5.6	14.3	800	41.4	0.495	1.7	0.281×10^{-4}
Hiteshue et al. [52]	Texas lignite	57.2[a]	68.2	5.5	22.4	800	41.4	0.645	1.1	0.185×10^{-4}
	Pittsburgh HVb	42.2[a]	81.4	5.5	9.6	800	41.4	0.532	1.4	0.097×10^{-4}
	Wyoming char prepared by heating in He for 2 h at 600°C		91.9	2.3	4.0	800	41.4	0.352	2.2	0.196×10^{-4}
	Anthracite MV (Pa.)		92.3	2.6	3.8	800	41.4	0.400	0.8	0.200×10^{-4}
Pyrcioch and Linden [87]	Concoal char from Montour No. 10 bit. coal by low-temp. fluidized bed	19.1	84.5	3.6	10.9	760	3.5	0.032	8.0	0.227×10^{-4}
		19.1	84.5	3.6	10.9	760	7.0	0.049	6.0	0.159×10^{-4}
		19.1	84.5	3.6	10.9	760	10.4	0.075	5.0	0.119×10^{-4}
		19.1	84.5	3.6	10.9	760	13.9	0.135	4.0	0.112×10^{-4}

ISBN 0-201-08300-0

ISBN 0-201-08300-0

Table 3.3 (Continued)

Feldkirchner and Linden [29]	Montour No. 10 bit. coal	37.0	82.8	5.5	10.0 (O + N)	940	10.4	0.365	1.4	0.20×10^{-3}
	North Dakota lignite	44.2	69.6	4.8	25.1	940	10.4	0.420	1.8	0.25 xx 10^{-3}
	Med. volatile anthracite	5.7	93.0	2.7	3.3	940	10.4	0.300	1.0	0.13×10^{-3}
	Concoal char from Pittsburgh Montour No. 10 bit. coal	19.1	84.5	3.7	10.8	940	10.4	0.240	0.7	0.095×10^{-3}
Johnson [59]	Air-pretreated Pittsburgh No. 8 coal char	0.0	100.0[b]	0.0[b]	0.0[b]	927	3.5	0.00		
		0.0	100.0[b]	0.0[b]	0.0[b]	982	3.5	0.00	0.8	$k_{v0} = 2.27 \times 10^{+3}$ $(MN/m^2)^{-1}s^{-1}$
		0.0	100.0[b]	0.0[b]	0.0[b]	1038	3.5	0.00		$E = 41.1$ kcal/mol (172 kJ/mol)

[a] Calculated from empirical formula.
[b] Assumed.

char is almost completely driven off. In Figure 3.9 time-conversion curves calculated according to the correlations of both Wen and Huebler and of Johnson are plotted for one set of data with the values of the rate parameters and reactivity factors shown in Table 3.3. It is seen that both the Wen and Heubler equation and Johnson's correlation are adequate for the second stage of hydrogasification.

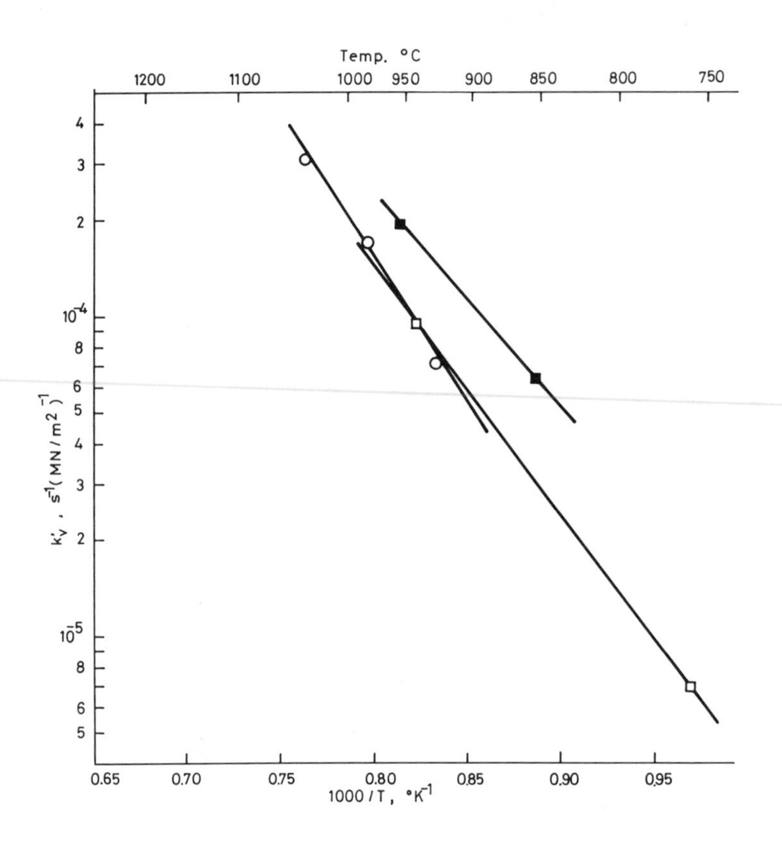

Figure 3.10. Arrhenius plots for the second stage of hydrogasification. ■ Australian Yallourn coal (51.7); Birch et al. [13]; pressure: 4.2 MN/m^2; gas: H_2. □ Concoal Montour No. 10 bitu. coal (37.0); Pyrcioch and Linden [87]; pressure: 10.4 MN/m^2; gas: H_2. ○ Air-pretreated Pittsburgh No. 8 coal char (0.0); Johnson [59]; pressure: 3.5 MN/m^2; gas: H_2.

In Figure 3.10 the Arrhenius plots are made according to the equation by Wen and Huebler for the second stage, for the data of Birch et al. [13], Pyrcioch and Linden [87], and Johnson [59]. The apparent activation energy of the coal/char hydrogen reaction, in the second stage, is found to lie between 30 and 41

ISBN 0-201-08300-0

kcal/mol (125.6 and 171.7 kJ/mol). Gilliland and Harriott [37] obtained activation energies of 36 to 55 kcal/mol (150.7 to 230.3 kJ/mol) for the carbon-hydrogen reaction, for various forms of carbon deposited on solid catalysts. An activation energy of 40 kcal/mol (167.5 kJ/mol) for the char-hydrogen reaction in the second stage is recommended. A few other data not included in Figure 3.10 show activation energies below 25 kcal/mol (104.7 kJ/mol). The apparent reasons for lower activation energies in these observations are as follows. In the fluidized bed experiments of Pyrcioch and Linden [87] and Lewis et al. [71] and in other U.S. Bureau of Mines and Institute of Gas Technology experiments, the data were taken at comparatively lower temperatures (650 to 800°C). Devolatilization of the chars is not complete at these low temperatures, as the data of Pyrcioch and Linden [87] show in Table 3.3. In an Arrhenius plot covering such a low temperature range, the first and the second stage of hydrogasification were not isolated. In Figure 3.10 the low temperature rate data of Pyrcioch and Linden [87], as shown in Table 3.3, have been modified to eliminate the effect of the first stage. In rapid hydrogasification experiments of most of the recent investigators (presented in Section 3.5F) the apparent reason for the low values of the calculated activation energy is the uncertainty regarding the isothermal temperatures and their exact durations in those experiments.

E. PRODUCT COMPOSITIONS IN HYDROGASIFICATION BY SLOW HEATING

In Figure 3.6 the instantaneous gas compositions are shown for four coal samples in fluidized bed experiments. This figure shows that, during the very initial periods, a substantial amount of CO_x is formed, whereas CH_4 and C_2H_6 are the main components (~95%) of the gaseous volatiles subsequently. Figure 3.7 shows that the evolution of tar and liquid is virtually complete during the first stage of hydrogasification, whereas the gaseous volatiles continue to evolve during the subsequent stage.

F. HYDROGASIFICATION BY RAPID HEATING

(i) *Effects of Temperature and Hydrogen Partial Pressure.* Figures 3.11 and 3.12 show the effects of temperature and hydrogen partial pressure on the yields and on the compositions of volatiles in hydrogasification by rapid heating. The techniques used by Moseley and Patterson [80], Coates et al. [20], and Johnson [60], utilizing transport and entrained flow reactors, and the direct electrical heating techniques of Anthony et al. [4] and Squires et al. [101] fall in the category of rapid heating. The data of Anthony and Howard [2] in Figure 3.12 show a minimum in the conversion versus pressure curve. This is due to two competing effects of pressure in the first stage of hydrogasification: hydrogen partial pressure increases the rates of hydrogenation reactions, but gas pressure suppresses pyrolysis.

Anthony and Howard [3] proposed a mechanistic model for the kinetics of hydrogasification, shown schematically in Figure 3.13. In this model the total

ISBN 0-201-08300-0

volatile loss is calculated, as a function of time, temperature, and pressure, according to the following equation:

Total volatiles loss = total nonreactive volatiles formed, V_{nr}^{**} + part of reactive volatiles formed, V_r^{*} + total reactive char gasified

$$= V_{nr}^{**} + V_r^{**} \Bigg/ \left(1 + \frac{k_1''}{K_c + k_2'' P_{H_2}}\right) + k_3'' P_{H_2} \tag{3.14}$$

where V_{nr}^{**}, V_r^{**}, k_1'', K_c, k_2'', and k_3'' are the constants for a particular coal sample. Temperature effect is taken into account by the following assumptions:

1. Gaussian distribution of activation energy, E, with a mean value of E_0 and standard deviation of σ.

2. $E_{\text{nonreactive volatiles}} > E_1 > E_{\text{reactive volatiles}}$, where E_1 represents the maximum activation energy of the reactive volatiles and the minimum activation energy of the nonreactive volatiles.

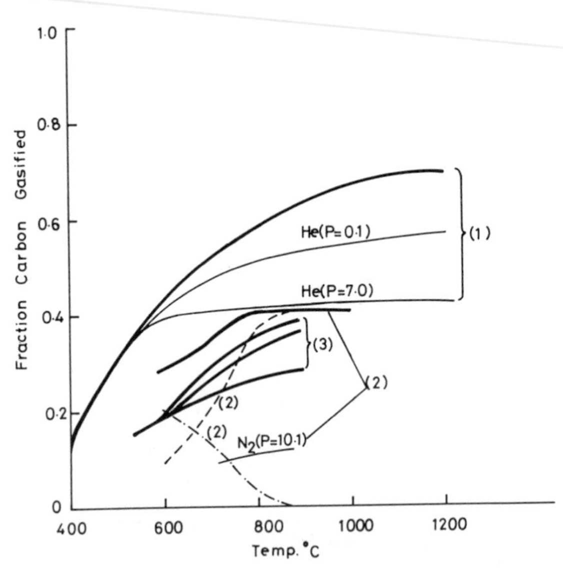

Figure 3.11. Effect of temperature and coal type on hydrogasification by rapid heating.——— Fraction of carbon gasified in inert atmosphere.——— Same in H_2 atmosphere. —·— Same to tar + liquid. — — Same to gas + water.

1. Pittsburgh No. 8 HVb coal (46.2); hydrogen pressure: 7.0 MN/m^2; solid residence time: 5–20 s; heating rate: 750°C/s; Anthony and Howard [2]. 2. Illinois No. 6 coal (40.2); hydrogen pressure: 10.1 MN/m^2; solid residence time: 3–4 s; heating rate: 650°C/s; Squires et al. [101]. 3. Montana lignite (45.9); hydrogen pressure: 1.8 MN/m^2; solid residence time: 5–14 s; heating rate: unknown; Johnson [60].

ISBN 0-201-08300-0

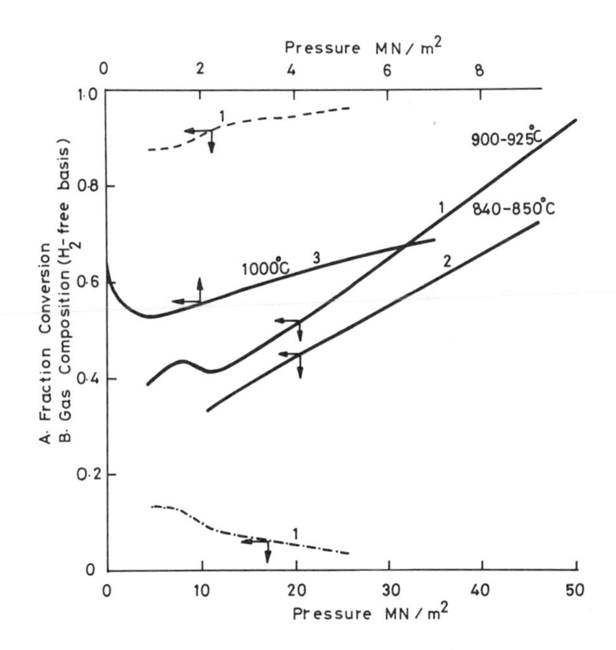

Figure 3.12. Effect of pressure on hydrogasification by rapid heating. A.—— Fraction of carbon gasified in coals 1 and 2.—— Fraction of volatiles produced, maf basis, in coal 3. B. —— $CH_4 + C_2H_6$, —·— $CO + CO_2$.

1. Kingsburry coal, NCB C.R.C. No. 902, HVb (37.5); solid residence time: 1 s, heating rate: unknown; Moseley and Patterson [80]. 2. Same as 1. 3. Pittsburgh No. 8 HVb coal (46.2); solid residence time: 5–20 s; heating rate: 750°C/s; Anthony and Howard [2].

Anthony and Howard have successfully explained their experimental data on one bituminous coal and one lignite with the help of the above model. However, this model requires that nine parameters be determined experimentally for each coal/char sample. More recently Russell et al. [93] have also proposed an analytical model to relate trends to the chemical and physical rate processes competing within coal particles in rapid hydropyrolysis.

The experiments of Squires et al. [101] show (Figure 3.11) that there is practically no liquid or tar in the volatiles at temperatures above 900°C, when the gasification was done in a hydrogen partial pressure of 100 atm (10.1 MN/m^2) and at a rapid rate. At 1000°C all the carbon gasified was found to appear as almost pure CH_4. The data of Moseley and Patterson [80], as presented in Figure 3.12, show that the percentage of CO_x decreases, and that of CH_4 and C_2H_6 increases, in the gaseous products as the hydrogen partial pressure is raised.

(ii) *Effects of Solid and Gas Residence Times.* Graff et al. [43] studied the effects of solid soaking time (the solid residence time at the final steady tempera-

ISBN 0-201-08300-0

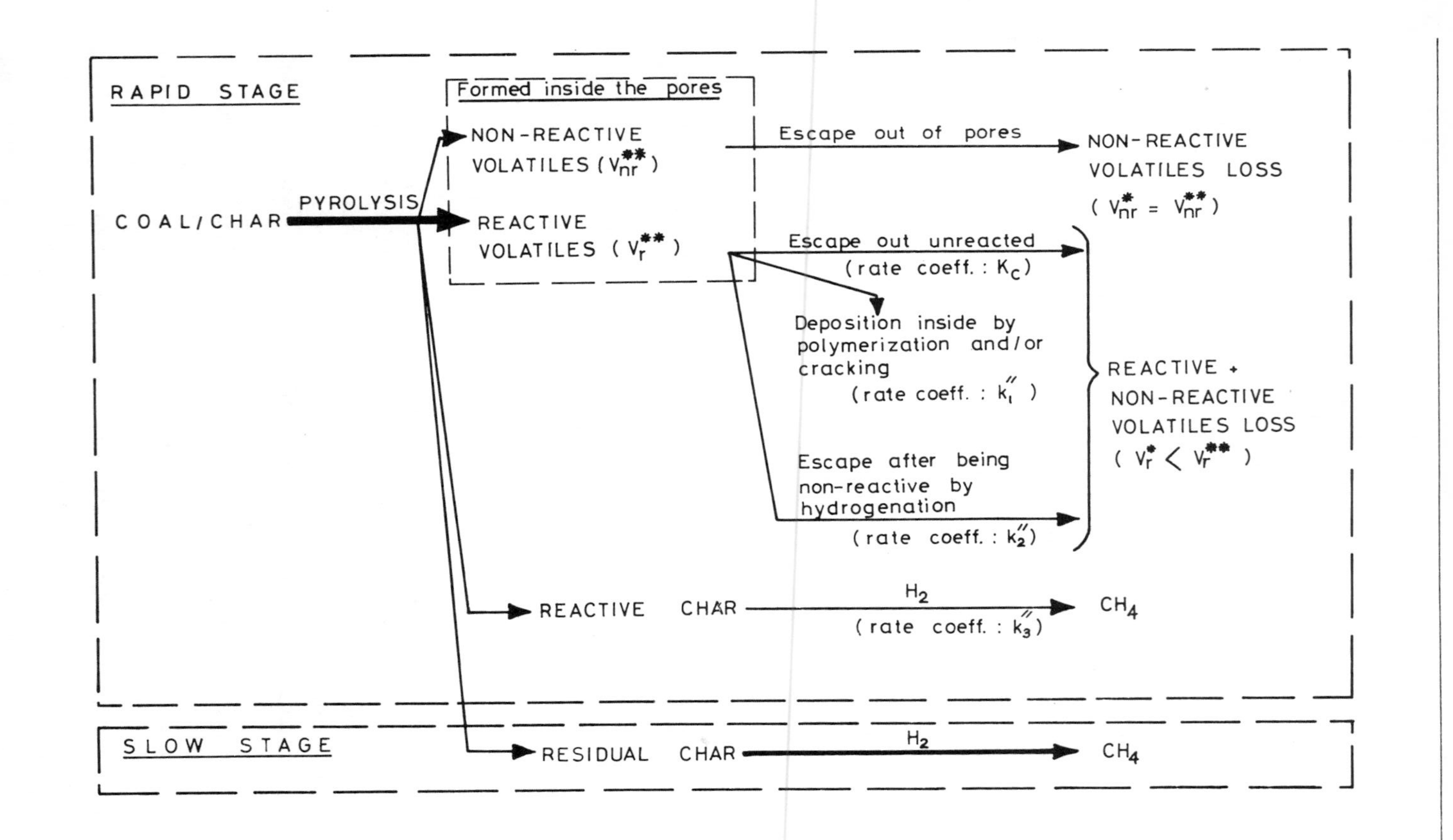

Figure 3.13. Kinetic model for coal pyrolysis and hydrogasification. Anthony and Howard [3].

ISBN 0-201-08300-0

ture) and gas residence time on the volatile products of hydrogasification. Their results, as shown in Figure 3.14, indicate that, although an increase in soaking time favors the conversion, cracking and/or polymerization of the volatile products, particularly of tar and liquid, takes place with longer gas residence times, leading to a partial loss of such products.

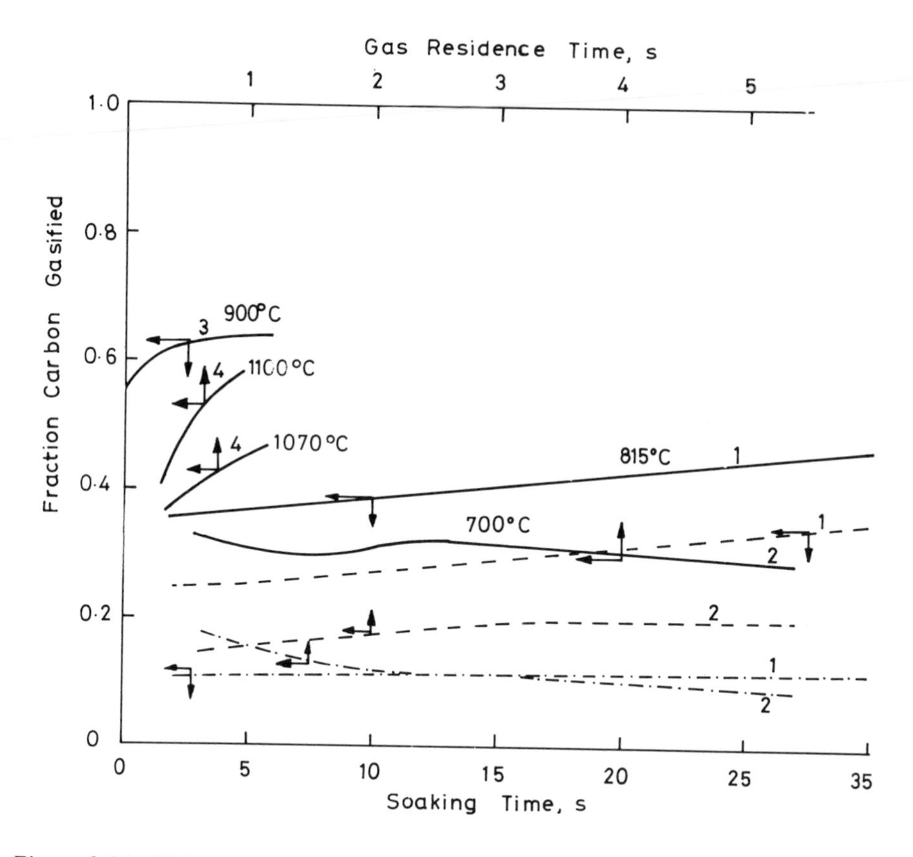

Figure 3.14. Effects of soaking time and gas residence time on hydrogasification by rapid heating.——Fraction of carbon gasified. —·— Same to tar + liquid. —— Same to gas + water.

1. Illinois No. 6 coal (40.2); hydrogen pressure: 10.1 MN/m^2; heating rate: 650°C/s; Graff et al. [43]. 2. Same as 1. 3. Pittsburgh No. 8 HVb coal (46.2); hydrogen pressure: 7.0 MN/m^2; heating rate: 750°C/s; Anthony and Howard [2]. 4. Kingsbury HVb coal char (20.8); hydrogen pressure: 35.2 MN/m^2, heating rate: unknown; Moseley and Patterson [80].

ISBN 0-201-08300-0

(iii) *Effect of Heating Rate.* Studies of Graff et al. [43] indicate that a higher heating rate favors higher carbon conversion, as is shown in Figure 3.15.

(iv) *Effects of Particle Size and Sample Size.* Particle size and sample size are two important factors, among others, that have considerable effects on the yields and product distribution in hydrogasification. However, the experimental data available are not adequate to draw quantitative conclusions on these effects. Figure 3.16 shows two sets of data on the effects of particle size on hydrogasification.

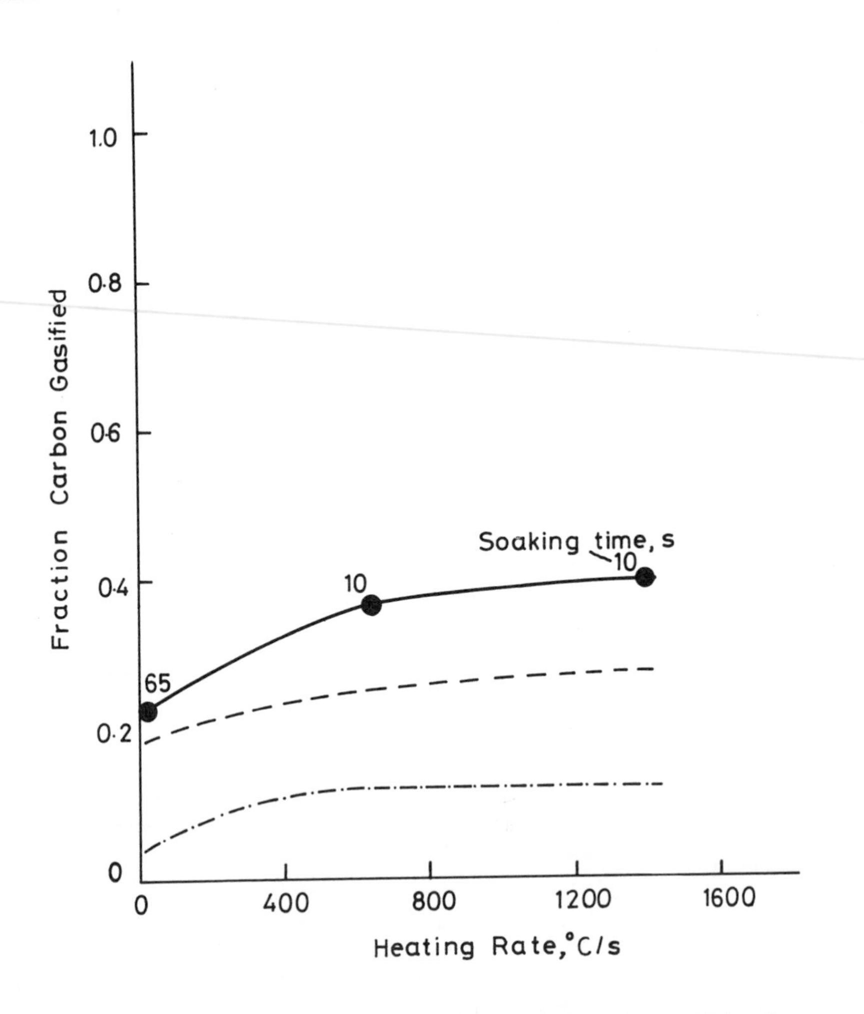

Figure 3.15. Effect of heating rate on hydrogasification by rapid heating. —— Fraction of carbon gasified. –– Same to gas + water. –·– Same to tar + liquid. Illinois NO. 6 HVb coal (40.2); hydrogen pressure: 10.1 MN/m^2; temperature: 785°C; Graff et al. [43].

ISBN 0-201-08300-0

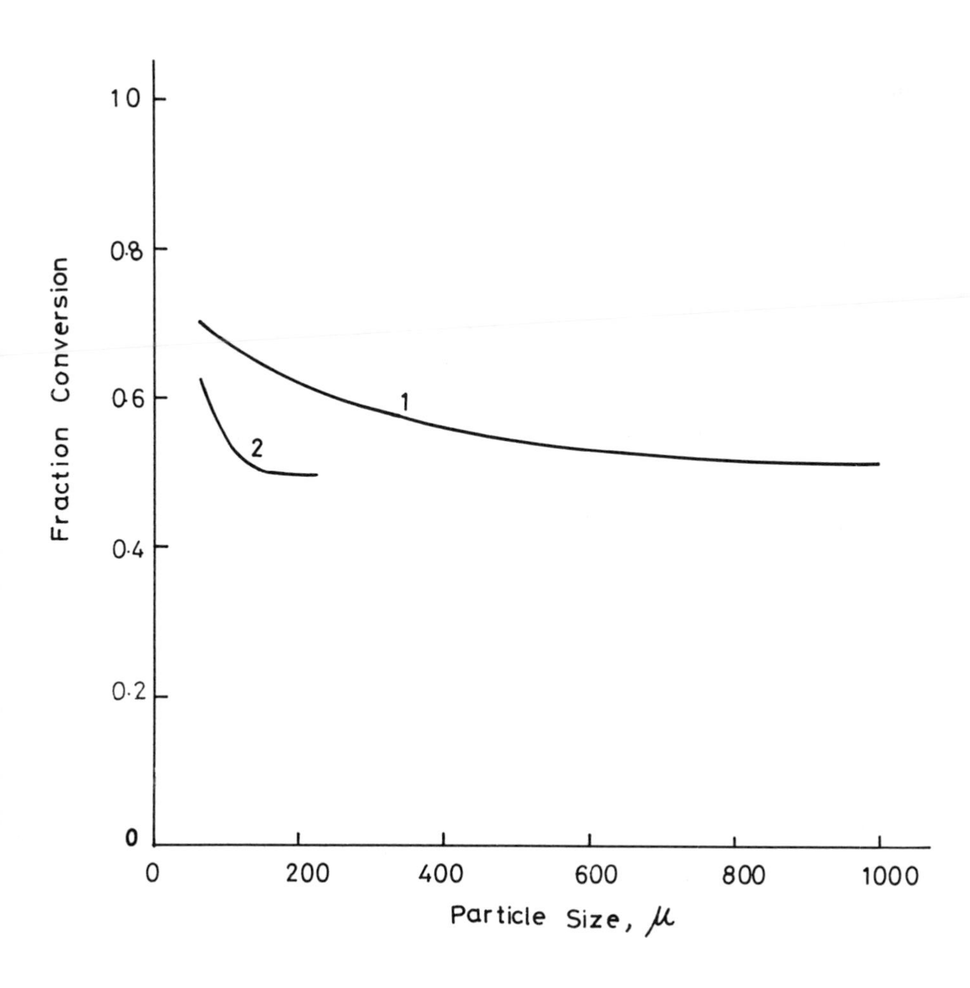

Figure 3.16. Effect of particle size on hydrogasification by rapid heating.

1. Pittsburgh No. 8 HVb coal (46.2). The fraction conversion means the volatiles produced as fraction of maf coal. Hydrogen pressure: 7.0 MN/m^2; temperature: $1000^{o}C$; heating rate: $700^{o}C/s$; solid residence time: 5–20 s; Anthony and Howard [2]. 2. Kingsbury HVbB coal, NCB C.R.C. No. 902 (37.5). Fraction conversion means the fraction of carbon gasified. Hydrogen pressure: 17.6 MN/m^2; temperature: 900–925°C; heating rate: unknown; solid residence time $\approx$ 1 s; Moseley and Patterson [80].

ISBN 0-201-08300-0

3.6. Char-Carbon Dioxide Reaction

The char-CO_2 reaction is often used to test the reactivities of different types of char produced from different processes and from different parent coals. It is a relatively slow reaction, is easy to measure, and, as will be seen, has similarity to the char-steam reaction. For smaller char particle sizes (less than about 300 μ) and lower temperatures (below about 1000°C), the char-CO_2 reaction is normally controlled by the chemical reaction rate and occurs nearly uniformly throughout the interior surfaces of the char particles.

Several mechanisms have been proposed for the char-CO_2 reaction. One of them, proposed by Walker et al. [109], is as follows:

$$C_f + CO_2(g) \underset{k_6}{\overset{k_5}{\rightleftarrows}} C(O) + CO(g)$$

$$C(O) \xrightarrow{k_7} CO(g)$$

The rate expression according to the above mechanism is as follows:

$$\text{Rate} = \frac{k_5(CO_2)}{1 + \frac{k_5}{k_7}(CO_2) + \frac{k_6}{k_7}(CO)} \tag{3.15}$$

According to Equation 3.15, CO would have an inhibiting effect on the char-CO_2 reaction. Moreover, the order of the reaction with respect to the CO_2 concentration would tend to unity at low partial pressure and to zero at high partial pressure of CO_2. The rate expression in the form of Equation 3.15 is, however, difficult to apply to gasifier designs for the reasons mentioned earlier. The empirical reaction rate can be expressed as follows (Dutta et al. [23]):

$$\frac{dx}{dt} = \alpha_v k_v C_A^n (1 - x) \tag{3.16}$$

where C_A = CO_2 concentration

$$\alpha_v = \frac{\text{available pore surface area per unit weight at any stage of conversion}}{\text{initial available pore surface area per unit weight}}$$

$$= 1 \pm 100 x^{\upsilon\beta} e^{-\beta x}, 0 \leqslant \upsilon \leqslant 1,$$

υ, β = constants characteristics of a given char and reaction conditions

In Equation 3.16 the rate is expressed as proportional to the partial pressure of CO_2 raised to the power n, where $0 \leqslant n \leqslant 1$. The values of n vary, depending on the experimental conditions. Walker et al. [109], Turkdogan et al. [104], Turkdogan and Vinters [105], Wen and Wu [119], and Fuchs and Yavorsky [34] reported the effect of CO_2 concentration on the carbon-CO_2 reaction rate; their

ISBN 0-201-08300-0

results are shown in Figure 3.17. As can be concluded from this figure, the rate of the char-CO_2 reaction with respect to CO_2 concentration can be considered as first order up to one atmosphere pressure but approaches zero order at high pressures (above 15 atm).

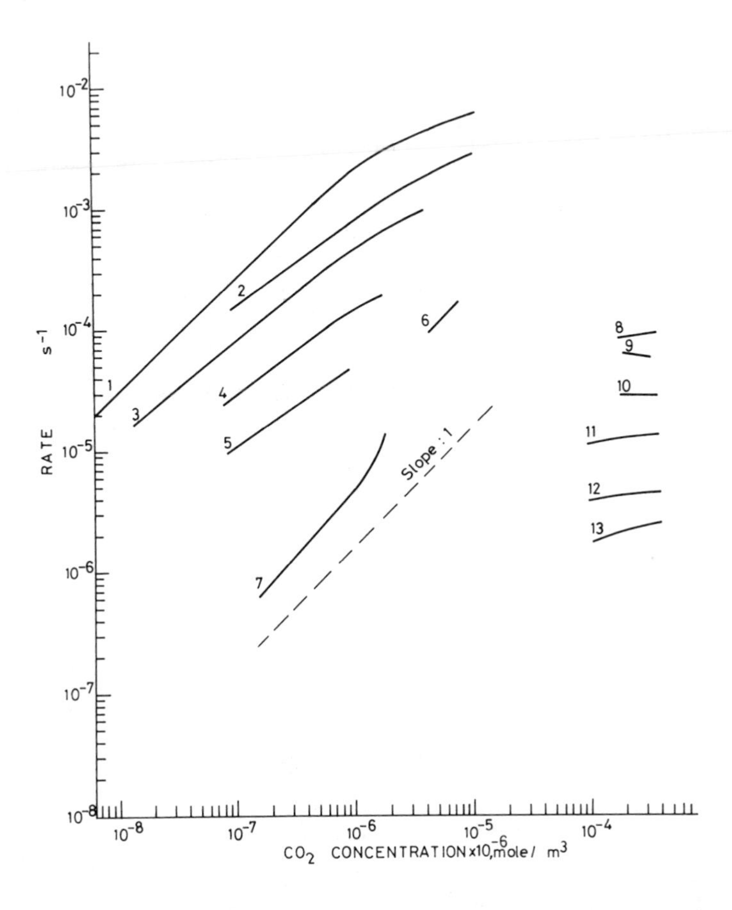

Figure 3.17. Effect of CO_2 concentration on the rate of gasification of carbons, at different temperatures, as observed by various investigators (see the legend in Table 3.4).

The term α_v is included in Equation 3.16 to take account of the fact that each coal/char sample is found to have its own characteristic rate-conversion curve. The different rate curves are apparently due to the different ways in which the pore characteristics of the samples change with conversion.

Figure 3.18 shows the Arrhenius plots. The rate constant k_v of Equation 3.16 is applicable only when the chemical reaction rate controls the process. At higher

ISBN 0-201-08300-0

Table 3.4

Legend to Figure 3.17

Curve Number	Investigators	Temperature, °C	Type of Carbon and Size	Diluent	Remarks
1	Turkdogan and Vinters [105]	1100	Coconut charcoal, -30 + 40 mesh	Ar	Initial rate
2	Turkdogan and Vinters [105]	1000	Coconut charcoal, -30 + 40 mesh	Ar	Initial rate
3	Turkdogan and Vinters [105]	1300	Electrode graphite, -30 + 50 mesh	Ar	Initial rate
4	Turkdogan and Vinters [105]	1200	Electrode graphite, -30 + 50 mesh	Ar	Initial rate
5	Turkdogan and Vinters [105]	1100	Electrode graphite, -30 + 50 mesh	Ar	Initial rate
6	Wen and Wu [119]	1087	Active charcoal and cement mixture, spheres 1.14–2.53 cm in diameter	N_2	Initial rate
7	Walker et al. [109]	1305	Spectroscopic carbon rod, 5.1 cm long and 1.27 cm in diameter	N_2	Initial rate
8	Fuchs and Yavorsky [34]	800	Hydrane char from Pittsburgh coal, -60 + 100 mesh	He	Average rate

ISBN 0-201-08300-0

Table 3.4 (Continued)

9	Fuchs and Yavorsky [34]	792.5	Hydrane char from Pittsburgh coal, -60 + 100 mesh	He	Average rate
10	Fuchs and Yavorsky [34]	750	Hydrane char from Pittsburgh coal, -60 + 100 mesh	He	Average rate
11	Turkdogan et al. [104]	1000	Electrode graphite spheres, 2.22 cm in diameter	None	Initial rate
12	Turkdogan et al. [104]	950	Electrode graphite spheres, 2.22 cm in diameter	None	Initial rate
13	Turkdogan et al. [104]	900	Electrode graphite spheres, 2.22 cm in diameter	None	Initial rate

ISBN 0-201-08300-0

temperatures, however, diffusion resistance within the solid particles may become appreciable, and therefore an effectiveness factor must be introduced for such cases. This effectiveness factor, η, is defined here as follows:

$$\eta = \frac{De \left.\dfrac{dC_A}{dr}\right|_{r_0} \cdot 4\pi r_0^2}{\left(\dfrac{4}{3}\pi r_0^3\right)\left[\alpha_v' k_v C_A C_{S0} (1-x)\right]} \tag{3.17}$$

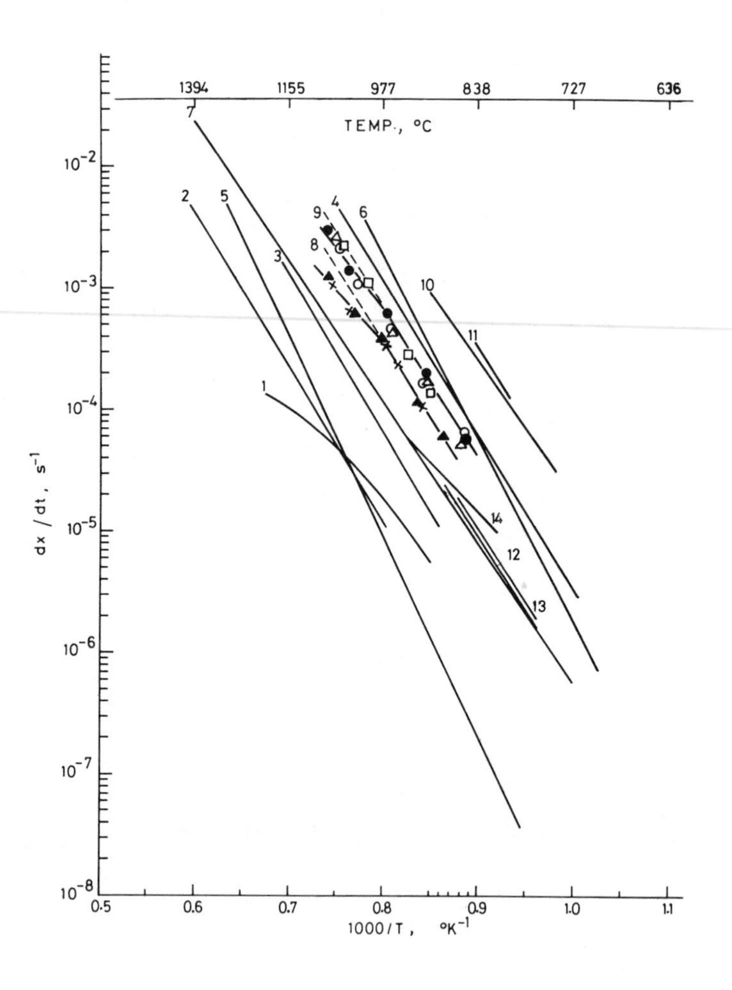

Figure 3.18. Arrhenius plots for the gasification of various forms of carbon in CO_2 (see the legend in Table 3.5).

ISBN 0-201-08300-0

This equation takes account of the fact that the effective diffusivity, D_e, may also change during reaction. This change may be expressed as

$$D_e = D_{e0}\, g(x)$$

where D_{e0} is the effective diffusivity of a particular char at zero conversion, $g(x)$ is a certain function of conversion x, and $\alpha_v' = \alpha_v/g(x)$. Equation 3.17 can be solved by appropriate boundary conditions as follows:

$$\eta = \frac{3}{M}\left(\frac{1}{\tanh M} - \frac{1}{M}\right) \tag{3.18}$$

where $M = \phi_0\ [(1 - x)\alpha_v']^{1/2}$ and $\phi_0 = r_0 \sqrt{k_v C_{S0}/D_{e0}}$. For chemical reaction control, $\phi_0 \simeq 0$. Thus the rate of gasification, including the influence of intraparticle diffusion, is expressed as

$$\frac{dx}{dt} = \eta \alpha_v k_v C_A(1 - x) \tag{3.19}$$

Thus, when diffusion controls the rate of conversion, the change in effective diffusivity, D_e, must be known as a function of conversion, in addition to the change in the relative reactivity factor, α_v.

A few experimental observations indicate that the reactivity of a char in the char-CO_2 reaction, under chemical reaction control, depends more on its coal seam than on the gasification scheme used for its production. The rate of the char-CO_2 reaction has been found to have little relation to the total surface areas of pores of the char particles measured by the nitrogen adsorption method. There is indication that only a fraction of the total nitrogen surface area which is occupied by pores above a certain size (about 15 Å in radius) is available for reaction (see Dutta et al. [23]).

In Figure 3.19 the reactivities of several chars have been plotted versus the oxygen contents of their parent coals. This figure tends to indicate that the reactivity of a char increases with the oxygen content of its parent coal. However, such a conclusion cannot be generalized until further investigations are made, particularly on the catalytic effects of the mineral matters in coals.

Studies of Muralidhara [83] on a few samples of lignite, bituminous and subbituminous coals, and anthracite, of particle size 50–75 μ and with CaO and oxygen up to 4% and 3.5%, respectively, indicate that the coal/char reactivity to CO_2 is

ISBN 0-201-08300-0

Table 3.5

Legend to Figure 3.18

Curve Number	Investigators	Type, Size, and Shape of Carbon	Remarks
1	Yoshida and Kunii [123]	Graphite sphere, 1.5 cm in diameter	Initial rates
2	Ergun [25]	Ceylon graphite, -10 + 200 mesh	Initial rate in a fluidized bed
3	Ergun [25]	Activated graphite, -10 + 200 mesh	Initial rate in a fluidized bed
4	Ergun [25]	Activated carbon, -10 + 200 mesh	Initial rate in a fluidized bed
5	Turkdogan and Vinters [105]	Electrode graphite particles, -10 + 40 mesh	Initial rates
6	Turkdogan and Vinters [105]	Coconut charcoal particles, -10 + 40 mesh	Initial rates
7	Austin and Walker [9]	Graphitized carbon cylinder, 5.1 cm long and 1.27 cm in diameter	Calculated initial rates
8,9	Dutta et al. [23]	○ – Illinois coal No. 6 ● – Synthane char No. 122 △ – Hydrane char No. 49 □ – IGT char No. HT155 ▲ – Hydrane char No. 150 X – Pittsburgh HVab coal all of size -35 + 60 mesh	Rates at 20% conversion level

ISBN 0-201-08300-0

Table 3.5 (Continued)

10	Fuchs and Yavorsky [34]	Hydrane char from Pittsburgh coal, -60 + 100 mesh	Average rate in a fluidized bed at 16–32 atm partial pressure of CO_2 with He as diluent
11	Fuchs and Yavorsky [34]	Synthane char from Illinois coal No. 6, -60 + 100 mesh	Average rate in a fluidized bed at 32 atm partial pressure of CO_2 with He as diluent
12	Tyler and Smith [106]	A test electrode particles, 0.9 mm	Rates at 23% conversion level
13	Tyler and Smith [106]	Commercial aluminum smelting electrode (anode) particles, 0.9 mm	Rates at 29% conversion level
14	Tyler and Smith [106]	Petroleum coke particles, 0.9 mm	Rates at 21% conversion level

ISBN 0-201-08300-0

dependent on both the CaO and the O_2 content of char. This dependency was expressed as follows:

$$\left(\frac{dx}{dt}\right)_{x=0} = 151.4 \times 10^6 e^{-35,000/T}$$

$$+ 63.1 \times 10^6 e^{-27,000/T} C_{CaO} (1 + 241.4 C_{O_2}), (T \text{ in } {}^oK)$$

where the rate, $(dx/dt)_{x=0}$, is in reciprocal seconds (sec^{-1}), and C_{CaO} and C_{O_2} are the weight fractions of CaO and oxygen, respectively, in char. Experiments of Aggarwal [1], conducted at temperatures up to 1100°C, indicate that CO has a poisoning effect on the char-CO_2 reaction so long as CO concentration is low. At high CO concentration, ($CO/CO_2 > 2.7$), iron oxides in coal are reduced to metallic iron which is an excellent catalyst for CO_2 gasification [110].

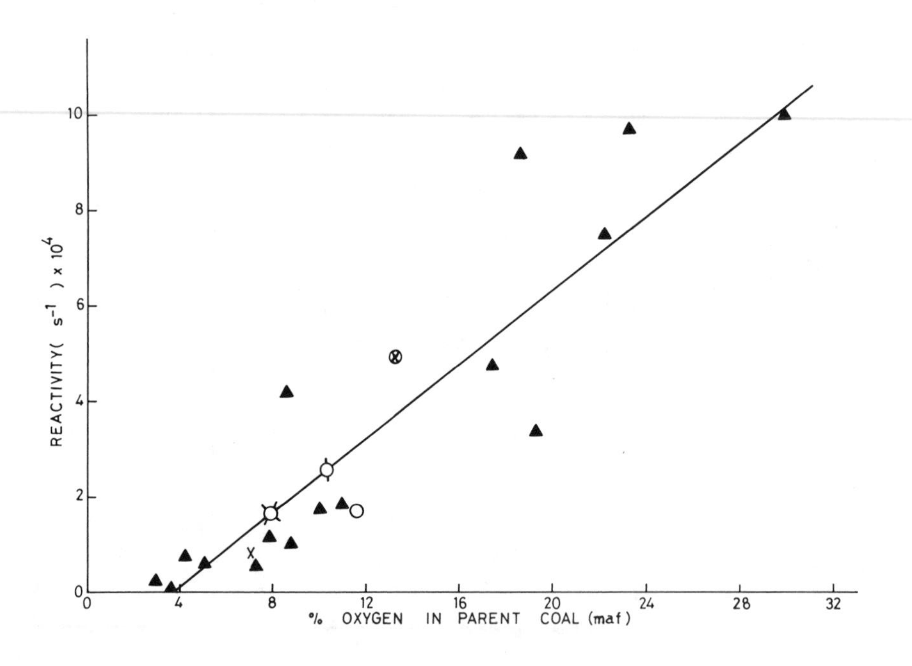

Figure 3.19. Reactivities of coals in CO_2 atmosphere during the second-stage reaction at 900°C (Dutta et al. [23]).

ISBN 0-201-08300-0

3.7. Char-Oxygen Reaction

The burning of char or the char-oxygen reaction is the fastest of the char-gas reactions taking place in a combustor or gasifier. In most cases this reaction takes place at the external surface of the char particles and is controlled by ash-layer diffusion. However, if the temperature and/or the particle size increases, the reaction may proceed toward the gas-film diffusion control regime. On the other hand, if the temperature and/or the particle size decreases substantially, the reaction may proceed toward the chemical reaction control regime, and may take place uniformly throughout the internal pore surfaces of the particles. Field [30] showed that in pulverized coal combustors, for particles below 50 μm, the combustion is chemical reaction controlled, whereas for particles larger than 100 μm the combustion is diffusion dominated. Mulcahy and Smith [82] reported that for particles larger than 100 μm the burning rate is gas-film diffusion controlled at temperatures above 1200°K. According to the studies of Mulcahy and Smith [82], Sergent and Smith [95], Hamor et al. [46], and Smith and Tyler [97], for coal/char particles about 90 μm in size, the burning rate is in the chemical reaction control regime up to temperatures about 750°K, whereas for smaller particles (~20 μm) the chemical reaction control regime extends up to a temperature as high as 1600°K.

For the conditions of ash-layer and gas-film diffusion controls, the char-O_2 reaction rate can be expressed as follows (Field et al. [31]):

$$\frac{dx}{dt} = \frac{P_{O_2}}{1/k_{\text{diff}} + 1/k_s} \tag{3.20}$$

where dx/dt = rate per unit of external particle surface

P_{O_2} = partial pressure of oxygen in free stream

k_{diff} = diffusional reaction rate coefficient

k_s = surface reaction rate coefficient

For small particles in fluidized beds, where the relative velocity of solid particles compared to that of the gas, is not large, the following formula can be used for the approximate calculation of k_{diff} (Field et al. [31]):

$$k_{\text{diff}} = \frac{0.292\,\Psi\,D'}{d_p T_m}$$

where Ψ = a mechanism factor

T_m = mean temperature in the film boundary between the solid particles and the bulk gas, °K

$= (T_s + T)/2$

T = temperature of the bulk gas, °K

T_s = particle surface temperature, °K

D' = diffusion coefficient of oxygen

$= 4.26\,(T/1800)^{1.75} \cdot 1/P'$, cm²/s

ISBN 0-201-08300-0

P' = total pressure, atm

d_p = particle diameter, cm

Since the char-oxygen reaction or the combustion of char is highly exothermic and extremely fast, an accurate estimation of the particle surface temperature, T_s, is difficult. This temperature can be 400 to 600°C higher than the bulk gas temperature, T, in many combustors and gasifiers. An inaccurate estimate of T_s may thus lead to serious errors in kinetic calculations. If it is assumed that there is no temperature gradient within the particles and that the particles are suspended in a stagnant gas, the value of T_s can be calculated from the following energy balance equation, through iterative computation (Field et al. [31]):

$$\frac{p_{O_2} \cdot Q}{1/k_{\text{diff}} + 1/k_s} = \frac{2\lambda(T_s - T)}{d_p} + \epsilon\sigma(T_s^4 - T^4) \tag{3.21}$$

where Q = $[7900(2/\Psi - 1) + 2340(2 - 2/\Psi)]\ 4187$, J/kg

λ = $\lambda_0[(T_s + T)/(2T_0)]^{0.75}$

λ_0 = thermal conductivity of the surrounding gas at the reference temperature, T_0

σ = Stefan-Boltzmann constant

ϵ = emissivity of particles

The radiation loss (the last term of Equation 3.21) can be ignored if the bed temperature is below 1100°C.

The mechanism factor, Ψ, takes a value of 2 when CO is the direct product of the char-O_2 reaction and a value of 1 when CO_2 is the direct product. Several investigations were made to determine the value of Ψ as a function of temperature, particle size, and carbon type. No quantitative conclusions can be drawn from these studies, but they showed that smaller particle size and/or higher temperatures favor CO formation, whereas larger particle size/or lower temperatures favor CO_2 formation. The following formulas are suggested for the value of Ψ, based on the above investigations:

For $d_p \leqslant 0.005$ cm

$$\Psi = \frac{2Z + 2}{Z + 2}$$

where $Z = 2500e^{-6249/T_m}$

ISBN 0-201-08300-0

For $0.005 \text{ cm} < d_p \leqslant 0.1 \text{ cm}$

$$\Psi = \frac{1}{Z + 2}\left[(2Z + 2) - \frac{Z(d_p - 0.005)}{0.095}\right], (d_p \text{ in cm})$$

For $d_p > 0.1$ cm

$$\Psi = 1.0$$

The value of the mechanism factor is thus determined by the value of Z, which represents the ratio of CO to CO_2 formed on the surface of char particles. According to the expression for Z given above (proposed by Arthur [6]), CO is the principal product of char-oxygen surface reaction above a temperature of about 900°C for small particle sizes. However, the next issue in dispute is the site of CO oxidation. Two basic mechanistic models, known as the so-called single-film model and the double-film model, are available. These are represented as Model I and Model II (Caram and Amundson [18]) in Figure 3.20. In the single-film model oxygen diffuses through the stationary boundary layer to the surface of the char particle, where it reacts to form CO and/or CO_2. Unreacted CO may escape into the bulk of the well mixed region. According to the double-film model, the char reacts with CO_2, and not with oxygen, to product CO, which burns in a thin flame front in the boundary layer. The CO is exhausted within the boundary layer, and oxygen does not reach the char surface. Generally, small particles ($<100\ \mu$) burn according to the single-film model; large particles (>2 mm), according to the double-film model (Caram and Amundson [18]). Avedesian and Davidson [7] used a model similar to Model II in their studies of carbon particle combustion in a fluidized bed.

However, these two models may represent only the two extreme cases of char burning, and not a generality. The actual mechanism of char burning may be more complicated because of competing factors like char particle size, char reactivity, temperature level, and bulk oxygen concentration. In general, therefore, oxygen and CO_2 can both reach the char surface and can react simultaneously. Such a general situation is depicted in Model III of Figure 3.20, as proposed by Essenhigh in Chapter 3 of this book. In this model the temperature and concentration profiles have been extended into the depth of the particles. Under conditions of pure diffusion control, the combustion of CO can be fast enough to consume all the oxygen before it reaches the char surface, and the char reacts with CO_2 alone. Under conditions of chemical reaction control, on the other hand (for example, for particles below 100 μm in size), CO_2 and oxygen will have almost equal opportunities to

ISBN 0-201-08300-0

reach the char surface. However, as Field et al. [31] and Borghi et al. [15] have pointed out, the char-CO_2 reaction is too slow compared to the char-O_2 reaction, and therefore the latter is the principal reaction on the char surface. In essence, therefore, there is a possibility of having two ignition phenomena, one corresponding to char surface ignition and the other corresponding to CO flame ignition in the boundary layer. For highly reactive chars or at high char surface temperatures this could sometimes lead to combustion oscillations due to periodic "choking off" of the oxygen supply to the char surface, as indicated by Kurylko and Essenhigh [70] and Caram and Amundson [18].

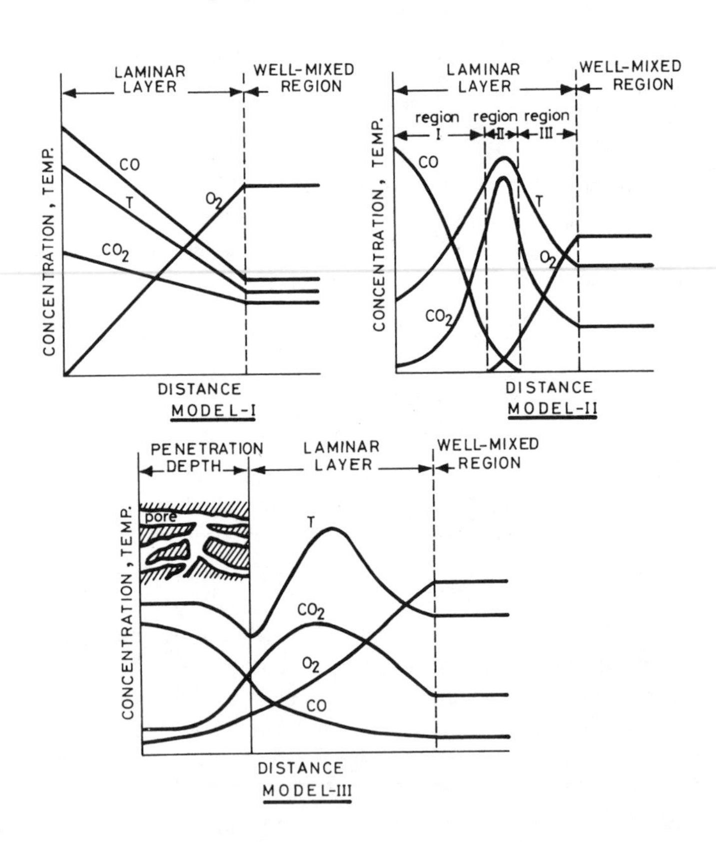

Figure 3.20. Schematic concentration and temperature profiles around burning char particles in different models.

For gasification conditions where the Nusselt number is far greater than 2 (i.e., for high gas velocities and large particle size), estimations of k_{diff} and T_s become more difficult. Field et al. [31] have discussed the methods for these estimations and various related topics of coal combustion.

ISBN 0-201-08300-0

The surface reaction rate coefficient, k_s, can be calculated from the following equation (Field et al. [31]):

$$k_s = k_{s0} e^{-17{,}967/T_s} \tag{3.22}$$

where $k_{s0} = 8710$ g/cm^2 · s · atm [8.71×10^5 kg/m^2 · s · (MN/m^2)].

The value of k_{s0} varies with the type of coal or carbon. An approximate value for char is indicated above.

For the conditions in the chemical reaction control regime, the char-O_2 reaction can be expressed as (Dutta and Wen [22])

$$\frac{dx}{dt} = \alpha_v k_v P_{O_2}(1 - x) \tag{3.23}$$

where dx/dt is the rate per unit mass of the solid reactant, and α_v is a reactivity factor which is a function of conversion and represents the relative available pore surface area defined in Section 3.6 on the char-CO_2 reaction.

In Equation 3.23 the rate of reaction is assumed to be proportional to the partial pressure of oxygen, P_{O_2}. The same assumption was made by many earlier investigators in treating the kinetics of the coal/char-oxygen reaction. However, the "true" order of the char-oxygen reaction is still being debated, and fractional orders covering almost the entire range between zero and unity have been reported by various investigators (e.g., Lewis et al. [71]; Armiglio and Duval [5]; Magne and Duval [76]; Smith and Tyler [97]; Tyler et al. [107]; and Glassman [38]).

Smith and Tyler [97] argued that the order of reaction with respect to oxygen partial pressure, as implied by Equation 3.20, should be regarded as an apparent order, n_a, and that implied by Equation 3.23 should be regarded as the true order, n_t, for the char-oxygen reaction. The two terms are related to each other as follows:

$$n_a = \frac{n_t + 1}{2}$$

The experimental data of Smith and Tyler [97] tend to indicate that the true order, n_t, of the char-oxygen reaction is between 0.2 and 0.3 in the chemical reaction control regime, at oxygen partial pressure between 0.05 and 0.2 atm (5 to 20 kPa) and at temperatures below about 490°C. The study was made with a sample of low rank brown Australian coal char of particle sizes 20 to 90 μm. At the same oxygen partial pressures, but at a higher temperature of about 625°C or above, the true order of reaction, n_t, appeared to be close to zero. Under these conditions the char-oxygen reaction is predominantly controlled by pore diffusion, and the apparent order of reaction, n_a, is close to 0.5. Dutta and Wen [22] found n_a to be close to unity for P_{O_2} in the range 0.002 to 0.2 atm (0.2 to 2 kPa) when diffusion is controlling. The sample used was Synthane char No. 122 (obtained

ISBN 0-201-08300-0

from Illinois No. 6 coal by the Synthane process) of particle size -35 + 60 mesh, and the temperature was 910°C.

Smith and Tyler [97] proposed the following single rate expression, which encompasses both the chemical control regime and the pore diffusion control regime:

$$R_s = 1.34\, e^{-32{,}600/RT_s} \quad (\text{g/cm}^2 \cdot \text{s})$$

for the char samples studied by them, over the temperature range of 370 to 1550°C and for a P_{O_2} of 0.1 to 1 atm (10 to 100 kPa). Here R_s is the rate of carbon combustion per unit of total surface area (the total pore surface area measured by the CO_2 adsorption method). When the pore diffusion is controlling at higher temperatures, the reaction occurs predominantly on the external surfaces of particles and the reaction rate, $R_{a,c}$, is usually expressed in terms of the external (geometric) surface area. The two rates, R_s and $R_{a,c}$, are related to each other approximately as follows (Smith and Tyler [97]):

$$R_{a,c} = 394\epsilon \left[(P_{O_2} R_s R)\sqrt{M/(\tau T)}\right]^{0.5} \quad (\text{g/cm}^2 \cdot \text{s})$$

where P_{O_2} is the partial pressure of oxygen (in atm), and R is the gas constant (in cm^3 · atm/g-mol · °K). The two equations given above should be used with caution, since they are based on a limited number of data.

Although the differences between the reactivities of different chars are small when ash-layer or gas-film diffusion controls the overall rate, such differences can be appreciable in the chemical reaction control regime, as is seen in Figure 3.21. Thus the devolatilized Illinois coal No. 6 is about 10 times more reactive than the devolatilized Hydrane char No. 49, obtained from the same coal. Figure 3.21 also shows that the reactivity of a char in the chemical reaction control regime for the char-O_2 reaction depends more on the degree of gasification of the char than on its parent coal. This is in contrast to the observations in char-CO_2 reaction studies, where the reactivity of a char was found to depend more on its coal seam than on the degree of char gasification. The relative reactivity of a char therefore appears to be determined not only by its own physical characteristics, but also by the reaction it undergoes.

ISBN 0-201-08300-0

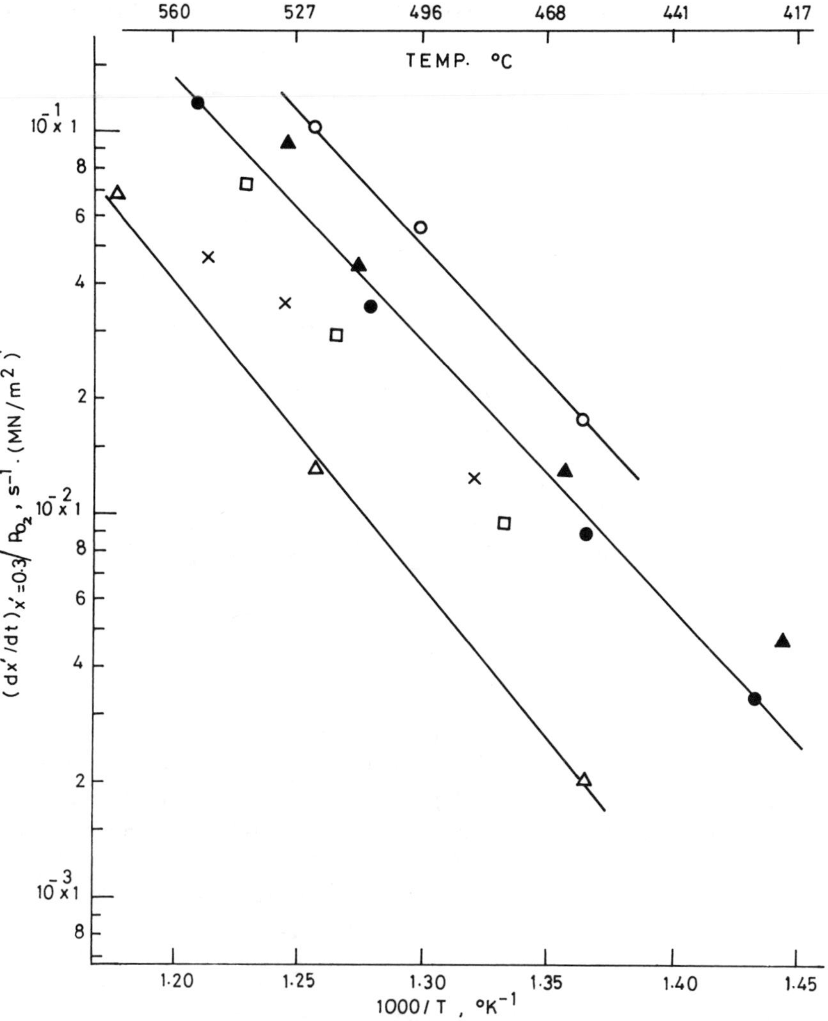

Figure 3.21. Arrhenius plots for six coal/char samples for char-oxygen reaction in reaction control regime. ○ No. 6; ● No. 4; △ No. 5; □ No. 3; × No. 2; ▲ No. 1. The numbers correspond to the serial numbers of the coals and chars of Figure 3.22.

ISBN 0-201-08300-0

Figure 3.22 shows that the changes in available pore surface area or in the relative reactivity factor, α_v, with conversion due to the char-O_2 reaction are much greater than those due to the char-CO_2 reaction. Because of such drastic modification of pore characteristics during the char-O_2 reaction, the reactivities of the chars in an oxygen atmosphere show less dependence on their parent coals.

A more comprehensive discussion on the combustion of coal is presented in Chapter 3.

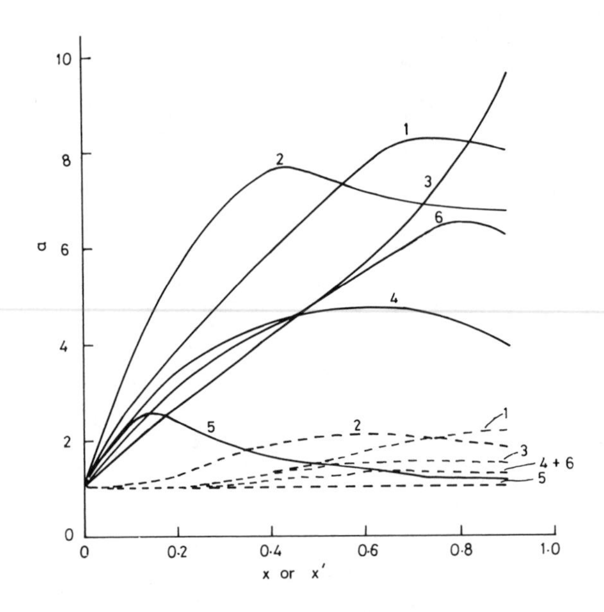

Figure 3.22. Relative available pore surface area (a) versus conversion for devolatilized coals and chars. — For reaction in O_2. -- For reaction in CO_2. 1. Devolatilized hydrane char No. 150 (parent coal: No. 2). 2. Devolatilized Pittsburgh HVab coal. 3. Devolatilized IGT char No. HT155 (parent coal: No. 6). 4. Devolatilized synthane char No. 122 (parent coal: No. 6). 5. Devolatilized hydrane char No. 49 (parent coal: No. 6). 6. Devolatilized Illinois coal No. 6.

3.8. Coal/Char-Steam Reaction

The carbon-steam reaction has been one of the most important reactions in industrial practice for many years, and a large amount of literature is available on this subject. However, there are many inconsistencies in the earlier works, especially in regard to the mechanism of this reaction. It is now established that CO and H_2 are the primary and principal products of this reaction and that hydrogen has a strong retarding effect on it (Johnstone et al. [61]; Klei et al. [69]).

ISBN 0-201-08300-0

Most of the earlier investigators used Langmuir type adsorption equations to express the rate of this reaction. Walker et al. [109] proposed the following mechanism and rate expression for the carbon-steam reaction:

$$C_f + H_2O(g) \underset{k_9}{\overset{k_8}{\rightleftharpoons}} H_2(g) + C(O)$$

$$C(O) \xrightarrow{k_{10}} CO(g)$$

$$\text{Rate} = \frac{k_8(H_2O)}{1 + \dfrac{k_8}{k_{10}}(H_2O) + \dfrac{k_9}{k_{10}}(H_2)} \qquad (3.24)$$

Ergun and Mentser [27], on the other hand, proposed the following mechanism and rate expression:

$$C_f + H_2O(g) \underset{k_9}{\overset{k_8}{\rightleftharpoons}} H_2(g) + C(O)$$

$$C(O) \xrightarrow{k_{10}} CO(g)$$

$$C(O) + CO \underset{k_{12}}{\overset{k_{11}}{\rightleftharpoons}} CO_2(g) + C_f$$

$$\text{Rate} = \frac{k_8(H_2O) + k_{11}(CO_2)}{1 + \dfrac{k_8}{k_{10}}(H_2O) + \dfrac{k_9}{k_{10}}(H_2) + \dfrac{k_{12}}{k_{11}}(CO)} \qquad (3.25)$$

According to Equation 3.24 only H_2 has an inhibiting effect on the char-H_2O reaction, whereas Equation 3.25 has both CO and H_2 inhibiting the reaction.

Wen [112] proposed the following simpler equation based on the available rate data:

$$\frac{dx}{dt} = k_v\left(C_{H_2O} - \frac{C_{H_2}C_{CO}RT}{K}\right)(1-x) \qquad (3.26)$$

where k_v represents the rate constant, and K the equilibrium constant, of the reaction. Figure 3.23 shows the Arrhenius plots for k_v for various carbon types, based on the rate data of several investigators. On the basis of this figure, an average activation energy of 35 kcal/mol (146 kJ/mol) is suggested for the char-steam reac-

ISBN 0-201-08300-0

tion. Figure 3.23 does not include a few sets of data obtained by other investigators, for example, Rossberg [91], Ergun and Mentser [26], Montet and Myers [79], and Klei et al. [69], who reported very high values of activation energy (60 to 80 kcal/mol or 251 to 335 kJ/mol) for this reaction.

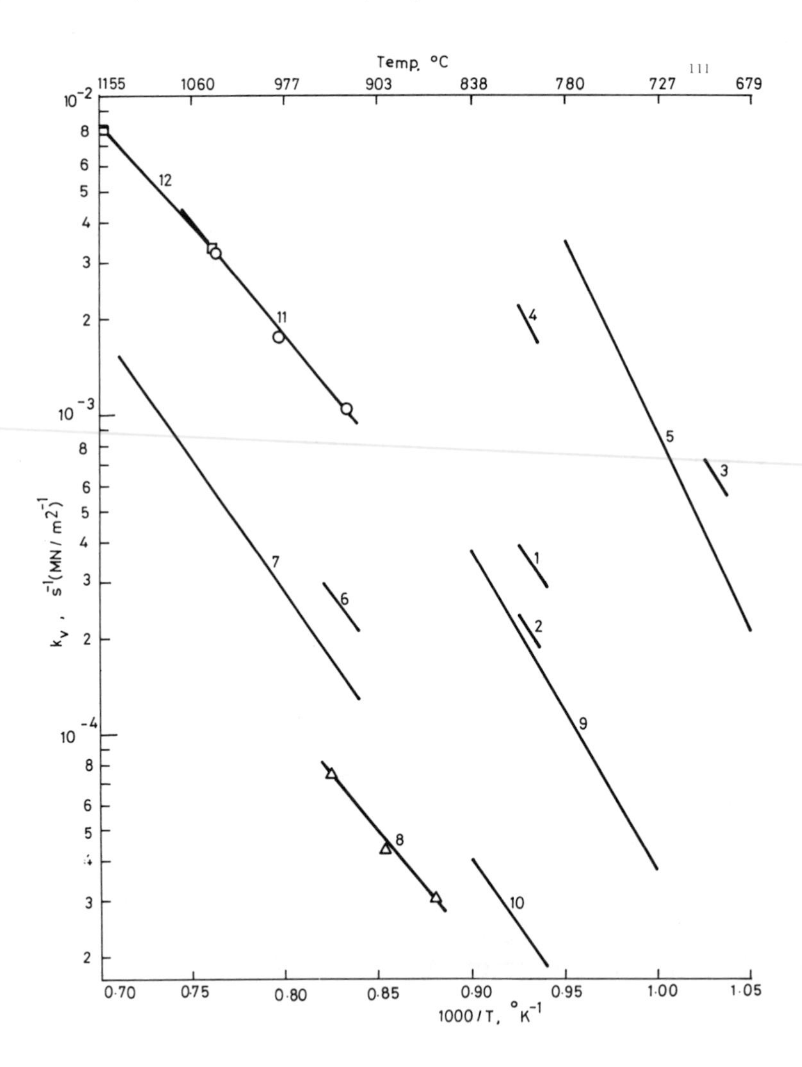

Figure 3.23. Arrhenius plots for various forms of carbon, coal, and char in carbon-steam or char-steam reaction (see the legend in Table 3.6).

ISBN 0-201-08300-0

Table 3.6

Legend to Figure 3.23

Line	Investigators	Material	Total Pressure, MN/m^2
1,2	Jolly and Pohl [62]	Coke	0.1
3,4	Gadsby et al. [35]	Nut char, coal char	0.1
5	Long and Sykes [74]	Coconut shell charcoal	0.02–0.1
6	Feldkirchner and Linden [29]	Low temperature bituminous coal char	10.4
7	Feldkirchner and Huebler [28]	Low temperature bituminous coal char	7.0
8	Johnstone et al. [61]	Cylindrical porous graphite rod	0.1
9,10	Blackwood and McGrory [14]	Coconut carbon	0.1–5.0
11	Johnson [59]	Air-pretreated HVab Pittsburgh No. 8 coal	3.5
12	Jensen [58]	Coal minerals from Kentucky No. 9 coal	0.1

The relative reactivities of various forms of carbon in the reaction with steam are almost similar to those found in the reactions with oxygen and CO_2. Thus, as seen in Figure 3.23, graphitized carbons tend to be the least reactive, and coconut shell charcoals the most reactive, forms of carbon. Coal/chars show intermediate reactivities. Again, reactivities between various coal/chars vary, depending on the parent coals and pretreatment conditions. However, very little study has been made on the relative reactivities of different coal/chars in the char-steam reaction.

Johnson [59] observed that steam has little effect on the rapid stage of coal/char gasification. For gases containing steam and hydrogen he suggested a value of 1.7 for the kinetic parameter γ in Equation 3.12 for this stage. For the char-steam reaction in the second stage, Johnson proposed a rate expression similar to Equation 3.13 for the second stage of the char-hydrogen reaction, but with a different expression for the rate constant. In his studies on coal/char gasification by steam

ISBN 0-201-08300-0

and hydrogen, he found it necessary to consider a third independent reaction:

$$H_2 + H_2O + 2C \rightleftharpoons CO + CH_4 \quad (3.27)$$

Thus, for steam-hydrogen mixtures, the rate constant k_I in Equation 3.13 has to be replaced by k_T, where

$$k_T = k_I + k_{II} + k_{III}$$

and k_I, k_{II}, and k_{III} represent the rate constants for the char-hydrogen reaction, the char-steam reaction, and reaction 3.27, respectively. Johnson gave correlations for k_{III} and the parameter γ' of Equation 3.13 for char-steam and char-hydrogen-steam reactions.

It appears that, as in the case of the char-hydrogen reaction, a simple equation such as Equation 3.26 is adequate for the second stage of the char-steam reaction.

Recent investigations of Jensen [58], Klei et al. [69], and Kayembe and Pulsifer [67] have demonstrated that the carbon-steam or char-steam reaction is chemical reaction controlled for smaller carbon/char particles (roughly $<500\ \mu$) and at temperatures up to about 1000–1200°C. At these conditions the reaction occurs uniformly throughout the interior of the pore surfaces of the solid particles.

The order of the char-steam reaction varies with steam concentration in a way similar to that in which the order of the char-CO_2 reaction varies with CO_2 concentration. Thus the order of the char-steam reaction with respect to steam concentration is approximately unity up to unit partial pressure of steam, but tends to become zero as the steam partial pressure rises significantly (Walker et al. [109]).

It is believed that the mineral matters in coal/char affect the rates of the various char-gas reactions. Although the studies on such catalytic effects are not conclusive, the investigation of Kayembe and Pulsifer [67] tends to indicate that calcium oxide and iron(III) oxide, which possibly catalyze the char-CO_2 reaction, may inhibit the char-steam reaction. These workers found that the char-steam reaction is catalyzed by alkaline salts like potassium and sodium carbonates.

3.9. Comparison of the Rates of Char-Gas Reactions

Pyrolysis of coal or char is normally the first set of reactions to occur. The rate of pyrolysis, which depends greatly on the operating conditions, is rather rapid compared to the rates of other char-gasification reactions. The char-oxygen reaction, which occurs concurrently with and/or following the pyrolysis stage, is the fastest among the char-gas reactions. The rates of pyrolysis and of the four char-gas reactions, as observed for several coals, are shown in Figure 3.24. The Arrhenius plots for char-O_2 and char-CO_2 reactions, shown at atmospheric pressure (0.1 MN/m^2), are based on experimental data at atmospheric pressure. The Arrhenius plots for char-H_2 and char-H_2O reactions, shown at atmospheric pressure, were

ISBN 0-201-08300-0

calculated using correlations obtained from data for higher pressures. The char-H_2 reaction can be assumed, with a reasonable degree of accuracy, to be first order. However, as discussed previously, such an assumption for char-H_2O and char-CO_2 reactions is not valid at higher pressure, as shown in Figure 3.24. Thus, although the Arrhenius plot for the char-H_2O reaction at atmospheric pressure shown in Figure 3.24 is only an extrapolation, the figure provides an order of magnitude comparison of various char-gasification reaction rates. As can be seen, at atmospheric pressure the char-H_2 reaction is slow. However, as the pressure is increased, the orders of both the char-CO_2 and char-H_2O reactions tend to approach zero, and therefore at considerably higher partial pressures of the reacting gases the char-H_2 reaction rate may become much greater than the rates of the other two reactions.

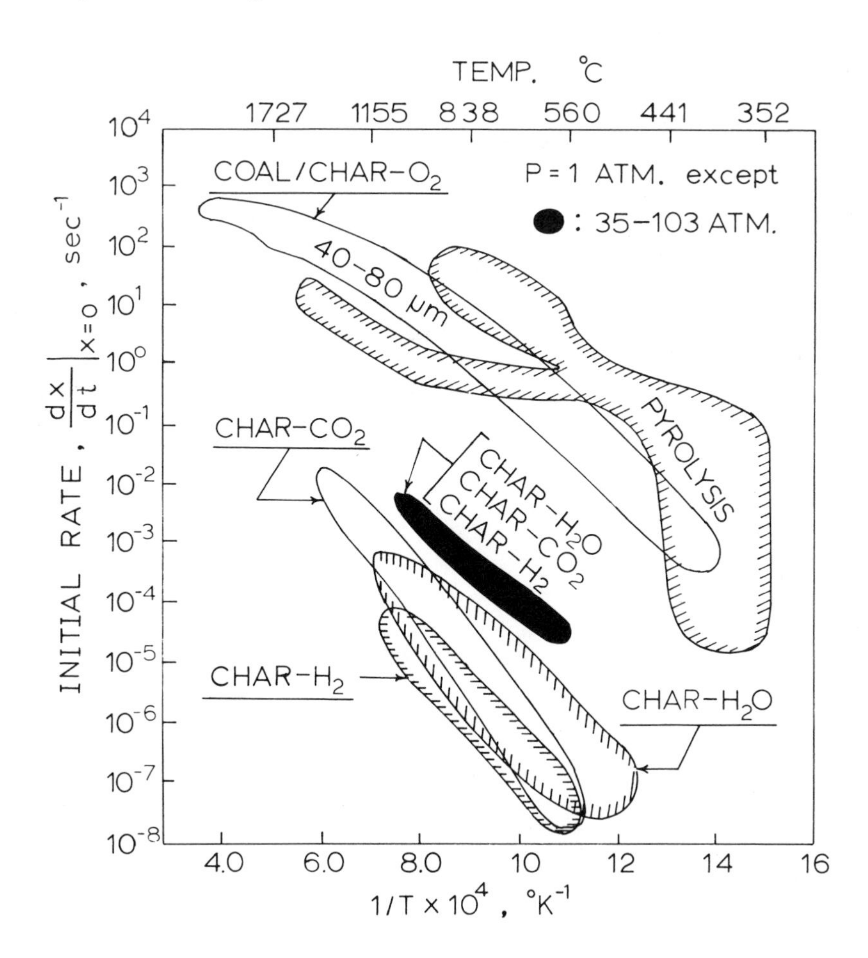

Figure 3.24. Comparison of initial rates of pyrolysis, combustion, and gasification of coal/char.

ISBN 0-201-08300-0

The approximate activation energies of the four char-gas reactions are as follows:

Reaction	Activation Energy, kJ/mol
Char-O_2	149
Char-H_2O	146
Char-CO_2	247
Char-H_2	172

3.10. Phenomenological Models for Solid-Gas Reactions [111,116]

A brief discussion of phenomenological models for solid-gas reactions is presented in this section. These models may be applied to char gasification reactions.

A. VOLUMETRIC REACTION MODEL

As shown in Figure 3.2 a reaction proceeds according to the volumetric reaction model, two zones appear: an outer zone in which the solid is totally exhausted (diffusion zone), and an inner, shrinking zone in which the reaction continues to take place (reaction zone). A finite length of time is required for the formation of the ash layer to begin at the outside surface of the reacting particles. This period of reaction is termed the first-stage reaction. The material balances for reactants during this stage of reaction in a spherical particle are represented by the following volumetric reaction equations:

$$\overline{D}_{eA}\left(\frac{d^2\overline{C}_A}{dr^2} + \frac{2}{r}\cdot\frac{d\overline{C}_A}{dr}\right) - ak_v\overline{C}_A^n C_S^m = 0 \tag{3.28}$$

$$\frac{dC_S}{dt} = -k_v\overline{C}_A^n C_S^m \tag{3.29}$$

Boundary conditions:

$$\overline{D}_{eA}\cdot\frac{d\overline{C}_A}{dr} = k_{mA}(C_{A0} - \overline{C}_A) \quad \text{at } r = r_o$$

$$\frac{d\overline{C}_A}{dr} = 0 \quad \text{at } r = 0 \tag{3.30}$$

$$C_S = C_{S0} \quad \text{at } t = 0$$

ISBN 0-201-08300-0

The period following the first stage, when an ash layer begins to form, is termed the second-stage reaction. During this period two sets of simultaneous material balance equations are needed to describe the system. The first set of equations, the material balance for the inner reaction zone, is the same as Equations 3.28–3.30. The second set of equations represents the material balance for the outer diffusion zone in the exhausted product layer and is as follows:

$$D_{eA}\left(\frac{d^2C_A}{dr^2}+\frac{2}{r}\cdot\frac{dC_A}{dr}\right)=0 \qquad (3.31)$$

$$C_S = 0 \qquad (3.32)$$

Boundary conditions:

$$D_{eA}\cdot\frac{dC_A}{dr}=k_{mA}(C_{A0}-C_A) \quad \text{at } r=r_o \qquad (3.33)$$

$$C_A=\bar{C}_A \quad \text{at } r=r_b$$

where r_b is the boundary between the inner reaction zone and the outer diffusion zone.

B. UNREACTED CORE SHRINKING MODEL

The material balance equations for the unreacted core model are identical to those used for the diffusion zone in the second stage of the volumetric reaction model, except for the surface reaction term appearing as one of the boundary conditions. These equations are expressed as follows:

$$D_{eA}\left(\frac{d^2C_A}{dr^2}+\frac{2}{r}\cdot\frac{dC_A}{dr}\right)=0 \quad \text{at } r_o>r>r_c \qquad (3.34)$$

Boundary conditions:

$$D_{eA}\left(\frac{dC_A}{dr}\right)=k_{mA}(C_{A0}-C_A) \quad \text{at } r=r_o$$

$$D_{eA}\left(\frac{dC_A}{dr}\right)=ak_sC_A^nC_{S0}^m \quad \text{at } r=r_c \qquad (3.35)$$

$$-D_{eA}\left(\frac{dC_A}{dr}\right)=aC_{S0}\frac{dr_c}{dt} \quad \text{at } r=r_c$$

ISBN 0-201-08300-0

C. SOLUTIONS OF VOLUMETRIC REACTION AND UNREACTED CORE SHRINKING MODELS AND THEIR INTERRELATIONSHIPS

Volumetric reaction and unreacted core models have been studied by several investigators (Rossberg and Wicke [92]; Petersen [85]; Walker et al. [109]; Yagi and Kunii [121]; Ausman and Watson [8]; Calvelo and Cunningham [16]; Hawtin and Murdoch [49]; Wen [111]; Wen and Wang [116]; Ishida and Wen [54, 55, 56, 57]; Wen and Wei [117, 118]; and Szekely et al. [103]).

General solutions of Equations 3.28 to 3.35 are not available. However, analytical solutions of these equations have been obtained under specific conditions. Wen et al. [111, 116, 117, 118] have given such solutions for spherical particles undergoing reactions which are first or zero order with respect to gas concentration but are independent of solid concentration. Szekely et al. [103] have provided solutions that are applicable to flat or cylindrical solid shapes as well.

(i) *Solutions for Volumetric Reaction Model.* For a volumetric reaction that is first order with respect to gaseous concentration but independent of solid concentration, Equations 3.28–3.31, with the accompanying boundary conditions, yield, for a spherical particle, the following solutions for the concentration profiles of the gaseous reactants and solid conversion:

First stage:

$$\frac{\bar{C}_A}{C_{A_0}} = \frac{1}{\theta_{vc}} \cdot \frac{\sinh(\bar{\phi}_v \xi)}{\xi \sinh \bar{\phi}_v} \tag{3.36}$$

where

$$\xi = \frac{r}{r_o}, \quad \bar{\phi}_v = r_o \sqrt{\frac{a k_v C_{S0}}{\bar{D}_{eA}}}$$

and θ_{vc}, the length of time (dimensionless) for the duration of the first-stage reaction (constant rate period), can be given as

$$\theta_{vc} = 1 + \frac{\bar{D}_{eA}}{k_{mA} r_o} (\bar{\phi}_v \coth \bar{\phi}_v - 1) \tag{3.37}$$

$$\frac{C_S}{C_{S_0}} = 1 - \frac{\sinh(\bar{\phi}_v \xi)}{\xi \sinh \bar{\phi}_v} \cdot \frac{\theta_v}{\theta_{vc}} \tag{3.38}$$

where

$$\theta_v = k_v C_{A0} t \tag{3.39}$$

and

$$x = \frac{3}{(\bar{\phi}_v)^2} \cdot (\bar{\phi}_v \coth \bar{\phi}_v - 1) \cdot \frac{\theta_v}{\theta_{vc}} \tag{3.40}$$

ISBN 0-201-08300-0

If gas-film diffusion resistance is negligible, the concentration profiles and conversion at the end of the first stage are obtained from Equations 3.36, 3.38, and 3.40 by substituting $\theta_v = \theta_{vc} = 1$.

Second stage:

$$\frac{\bar{C}_A}{C_{Ab}} = \frac{\xi_b \cdot \sinh(\bar{\phi}_v \cdot \xi)}{\xi \cdot \sinh(\bar{\phi}_v \cdot \xi)} \text{ for } 0 \leqslant r \leqslant r_b \tag{3.41}$$

$$\frac{C_A}{C_{Ab}} = \frac{\xi_b}{\xi} \cdot \frac{1 - \xi + \xi/N_{Sh}}{1 - \xi_b + \xi_b/N_{Sh}} + \frac{1 - \xi_b/\xi}{1 - \xi_b + \xi_b/N_{Sh}} \quad \text{for } r_b \leqslant r \leqslant r_o \tag{3.42}$$

$$\bar{C}_A = C_A = C_{Ab} \text{ at } r = r_b \tag{3.43}$$

$$\frac{C_{Ab}}{C_{A0}} = \frac{1}{1 + \dfrac{\bar{D}_{eA}}{D_{eA}}\left(1 - \xi_b + \dfrac{\xi_b}{N_{Sh}}\right)[\bar{\phi}_v \xi_b \coth(\bar{\phi}_v \xi) - 1]} \tag{3.44}$$

$$\frac{C_S}{C_{S0}} = 1 - \frac{\xi_b \sinh(\bar{\phi}_v \xi)}{\xi \sinh(\bar{\phi}_v \xi_b)} \tag{3.45}$$

$$x = 1 - \xi_b{}^3 + \frac{3\xi_b}{(\bar{\phi}_v)^2}[\bar{\phi}_v \xi_b \coth(\bar{\phi}_v \xi_b) - 1] \tag{3.46}$$

and

$$\theta_v = 1 + \left(1 - \frac{\bar{D}_{eA}}{D_{eA}}\right) \ln \frac{\xi_b \sinh \bar{\phi}_v}{\sinh(\bar{\phi}_v \xi_b)} + \frac{\phi_v^2}{6}(1 - \xi_b)^2.$$

$$(1 + 2\xi_b) + \frac{\bar{D}_{eA}}{D_{eA}}(1 - \xi_b)[\bar{\phi}_v \xi_b \coth(\bar{\phi}_v \xi_b) - 1]$$

$$+ \frac{\phi_v{}^2}{3N_{Sh}}(1 - \xi_b{}^3) + \frac{\xi_b}{\bar{N}_{Sh}}[\bar{\phi}_v \xi_b \coth(\bar{\phi}_v \xi_b) - 1] \tag{3.47}$$

where $\bar{C}_A$, C_A, and C_{Ab} represent the gaseous reactant concentrations in the reaction zone, in the ash layer, and at the boundary, r_b, between these two zones, respectively; and $N_{Sh} = k_{mA} r_o / D_{eA}$, $\bar{N}_{Sh} = k_{mA} r_o / \bar{D}_{eA}$, $\xi_b = r_b / r_o$, $\phi_v = r_o \sqrt{a k_v C_{S0} / D_{eA}}$.

ISBN 0-201-08300-0

The time required for the reaction to complete both the first and the second stage, θ_v^*, is

$$\theta_v^* = 1 + \frac{\bar{\phi}_v^{\,2}}{6}\left(1 + \frac{2}{N_{Sh}}\right) + \left(1 - \frac{\bar{D}_{eA}}{D_{eA}}\right) \ln \frac{\sinh \bar{\phi}_v}{\bar{\phi}_v} \tag{3.48}$$

When $D_{eA} = \bar{D}_{eA}$, the volumetric reaction model is called a homogeneous model. In a homogeneous reaction system the diffusion of gaseous reactant A through the solid becomes the rate controlling factor, as the value of ϕ_v $(= \bar{\phi}_v)$ is increased. Equations 3.41, 3.42, 3.46, and 3.47 are then reduced to

$$C_A = 0 \text{ for } 0 \leqslant r \leqslant r_b \tag{3.49}$$

$$\frac{C_A}{C_{A0}} = \frac{1 - \xi_b/\xi}{1 - \xi_b + \xi_b/N_{Sh}} \text{ for } r_b \leqslant r \leqslant r_o \tag{3.50}$$

$$x = 1 - \xi_b^{\,3} \tag{3.51}$$

and

$$\frac{\theta_v}{\theta_v^*} = 1 - 3(1 - x)^{2/3} + 2(1 - x) \tag{3.52}$$

These results (Equations 3.49 to 3.52) are exactly the same as those obtained from the unreacted core shrinking model under the pseudo-steady-state assumption when the ash diffusion is the rate controlling factor. On the other hand, when the chemical reaction is very slow and becomes the rate controlling factor, ϕ_v $(= \bar{\phi}_v)$ approaches zero and

$$\bar{C}_A = C_{A0} \text{ for } 0 \leqslant r \leqslant r_0 \tag{3.53}$$

$$\frac{C_S}{C_{S0}} = 1 - \frac{\theta_v}{\theta_v^*} \tag{3.54}$$

$$x = \frac{\theta_v}{\theta_v^*} \tag{3.55}$$

(ii) *Solutions for Unreacted Core Shrinking Model.* For a reaction that is first order with respect to gaseous concentration but independent of solid concen-

ISBN 0-201-08300-0

tration, Equation 3.34 with the boundary conditions of Equation 3.35 readily yields the following solutions for a spherical particle with constant radius r_o:

$$\frac{C_A}{C_{A0}} = \frac{\left(1 + \frac{D_{eA}}{ak_s C_{S0} r_c}\right)\frac{1}{r_c} - \frac{1}{r}}{\left(1 + \frac{D_{eA}}{ak_s C_{S0} r_c}\right)\frac{1}{r_c} - \left(1 - \frac{D_{eA}}{k_{mA} r_o}\right)\frac{1}{r_o}} \tag{3.56}$$

$$\theta_s = 1 - \xi_c + \frac{\phi_c}{2}(1 - \xi_c^2) + \frac{\phi_c}{3}\left(\frac{1}{N_{Sh}} - 1\right)(1 - \xi_c^3) \tag{3.57}$$

$$x = 1 - \xi_c^3 \tag{3.58}$$

where

$$\phi_c = \frac{r_o a k_s C_{S0}}{D_{eA}}, \quad \theta_s = \frac{k_s C_{A0} t}{r_o}, \quad \xi_c = \frac{r_c}{r_o}$$

The time for complete conversion, when $t = t^*$ and $r_c = 0$, is obtained from Equation 3.57 as

$$t^* = \frac{r_o a C_{S0}}{C_{A0}}\left(\frac{1}{3k_{mA}} + \frac{r_0}{6D_{eA}} + \frac{1}{ak_s C_{S0}}\right) \tag{3.59}$$

The concentration profiles of solid and gas within a reacting particle were qualitatively shown earlier, for the two models, in Figures 3.2(a) and 3.2(b).

If for a volumetric reaction $\overline{D}_{eA}$ is much smaller than D_{eA} (i.e., $\overline{D}_{eA} << D_{eA}$), Equations 3.47 and 3.46 become

$$\frac{\theta_v}{\overline{\phi}_v} = 1 - \xi_b + \frac{\phi_v}{2}\sqrt{\frac{\overline{D}_{eA}}{D_{eA}}}(1 - \xi_b^2) + \frac{\phi_v}{3}\sqrt{\frac{\overline{D}_{eA}}{D_{eA}}}\left(\frac{1}{N_{sh}} - 1\right)(1 - \xi_b^3) \tag{3.60}$$

$$x = 1 - \xi_b^3 \tag{3.61}$$

ISBN 0-201-08300-0

Equations 3.60 and 3.61 are identical with Equations 3.57 and 3.58. We can thus relate the surface based quantities k_s, ϕ_c, and θ_s to the volume based quantities k_v, ϕ_v, and θ_v as follows:

$$\phi_c = \phi_v \sqrt{\frac{\overline{D}_{eA}}{D_{eA}}}, \quad \theta_s = \frac{\theta_v}{\overline{\phi}_v}, \quad \text{and} \quad k_s = \sqrt{\frac{\overline{D}_{eA} k_v}{a C_{S0}}}$$

The phenomenological similarity between the volumetric reaction model and the unreacted shrinking core model at $\overline{D}_{eA} << D_{eA}$ can be realized for the following reason. When $\overline{D}_{eA}$ becomes very small and hence $\overline{\phi}_v$ becomes very large (say greater than 100), the gaseous reactant concentration $\overline{C}_A$ at r_b approaches zero, according to the volumetric reaction model, and the reaction zone becomes very narrow. Under this condition there is no penetration of gas into the reaction core, and hence the two models become identical.

Thus, when the diffusion through the product layer is the rate controlling step, the volumetric reaction model becomes the shrinking core model. Diffusion is likely to become the rate controlling step if the chemical reaction is very fast, if the reaction occurs at very high temperature, and particularly if the particle diameter is large. This situation is often observed for combustion of char in a gasifier.

Thus the volumetric reaction model is more versatile in applicability and can be applied to most char-gas reactions.

Figure 3.25 shows the characteristic behaviors of the volumetric reaction and the surface reaction in various temperature regions. As shown in Figure 3.25(a), the concentration profiles of the reacting gas A and the solid are both flat in the lowest temperature region, *V*, where the rate controlling step is the intrinsic chemical reaction rate of the solid. In the highest temperature region, I, the concentration of gaseous reactant at the boundary between the reaction zone and the product layer (or the gas layer, if no solid product is formed) is practically zero, and the rate controlling step is diffusion. In region III the concentration gradient of the gas, C_A, is very steep within the particle, and the reaction is confined to a narrow band at the interface between the product ash layer and the reacting zone.

Comparison of Figures 3.25(a) and 3.25(b) reveals that as the temperature increases, approaching zones I, II, and III, the volumetric reaction model and the shrinking core model become identical as diffusion becomes the rate controlling step. The slopes (E_{ks}/R) of the Arrhenius plots (lines NN') of both models, under such conditions, become equal to $E_{kv}/2R$, where E_{kv} is the activation energy of the volumetric reaction at the conditions of region V in Figure 3.25(a). As the temperature is increased beyond region II, the activation energy approaches a still lower value of E_D, which is due to the diffusion control.

Wen and Wei [118] have provided solutions, based on the unreacted core shrinking model, for determining the selectivity of one reaction over the others in a system where more than one reaction takes place simultaneously. Selectivity criteria have been developed for independent, parallel, and consecutive reactions with orders up to unity with respect to gaseous concentration.

ISBN 0-201-08300-0

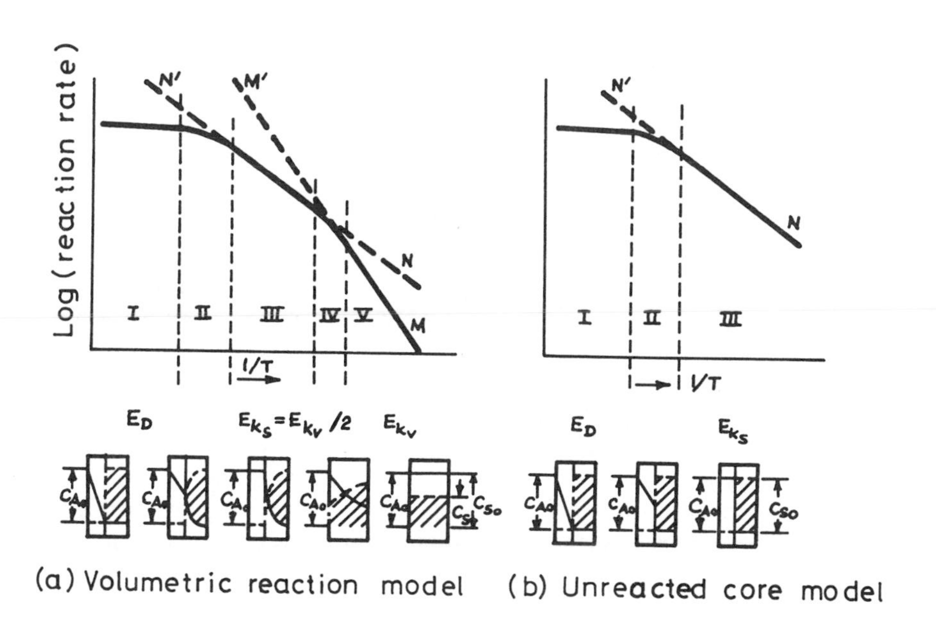

Figure 3.25. Schematic diagram representing characteristic behavior of solid-gas reaction systems under various temperature regions.

When the chemical reaction is accompanied by a heat effect, not only the concentration gradient, but also an appreciable temperature gradient, may exist within the particles. Such a situation can easily arise in a highly exothermic reaction like the char-oxygen reaction or in an endothermic reaction like the char-steam reaction. In such cases it may not be possible to neglect the effect of the temperature gradient on the reaction rates. In the case of exothermic reactions thermal instability and abrupt transition of the rate controlling step may also take place during the course of the reaction. Wen [111], Ishida and Wen [55], Ishida et al. [57], and Wen and Wei [118] have discussed these problems and given some of their solutions for nonisothermal systems involving both single and simultaneous reactions.

3.11. Application of Solid-Gas Reaction Models to Char-Gas Reactions

The rate of the char-oxygen reaction is faster than the rates of the other char-gas reactions by a factor of about 10^5, at the same concentration of all the gases, as is seen in Figure 3.24. The char-oxygen reaction is mostly a diffusion-limited surface reaction, particularly for large particles, and the unreacted core shrinking model can be applied to this reaction. The char-hydrogen, char-steam, and char-carbon dioxide reactions are comparatively much slower and take place with considerable gas penetration inside the pores of the particles. The volumetric reaction

ISBN 0-201-08300-0

model is therefore the proper representation of these reactions. In a system of mixed gas streams, the char-oxygen reaction will take place on the outside surface of the shrinking particle interface, whereas the other three char-gas reactions are expected to occur inside the pores of the particles by penetration of the corresponding reactant gases.

The analysis of the mixed reaction system can be greatly simplified, however, if it is noted that, because of the extremely fast rate of the oxygen reaction, which takes place on the outside surfaces of the particles, the penetration of the other reactant gases into the particles and the extent of other char-gas reactions will be negligible as long as any significant amount of oxygen is present in the gas stream. At a very low oxygen concentration, and at high concentrations of the other reactant gases, the rates of the other char-gas reactions may become comparable to the rate of the char-oxygen reaction. However, in such a situation most of the available oxygen will react with the product gases (CO, H_2, and CH_4) of the other reactions diffusing out of the particles, before it can reach the particle surface for reaction. The design procedure is therefore simplified if it is assumed that, until all the oxygen is depleted, the char-oxygen reaction is the only solid-gas reaction that takes place in the system. This is followed by the other char-gas reactions occurring simultaneously. These reaction steps, as well as the nature of the change in the pore structure of a char particle as the reactions proceed, are shown schematically in Figure 3.26.

The ash produced from most coals (with ash contents less than about 15%) is highly porous and fragile. In most situations, particularly in entrained and fluidized beds, the ash formed probably does not remain on the outside surface of the reacting particles, but becomes detached from the unreacted char by attrition. Therefore, depending on the solid-gas contacting arrangement in a gasifier, the reacting char may be considered to exist without an ash layer.

MATERIAL BALANCE FOR CHAR GASIFICATION REACTIONS

(i) *Char-O_2 Reaction: Unreacted Core Shrinking Model with No Ash Layer*

O_2 balance

$$k_{mA}(C_{A0} - C_A) = ak_sC_A \quad \text{at } r = r_c \tag{3.62}$$

Carbon balance

$$-C_{S0}\left(\frac{dr_c}{dt}\right) = k_sC_A \tag{3.63}$$

Initial condition

$$r_c = r_0 \text{ at } t = 0$$

ISBN 0-201-08300-0

ISBN 0-201-08300-0

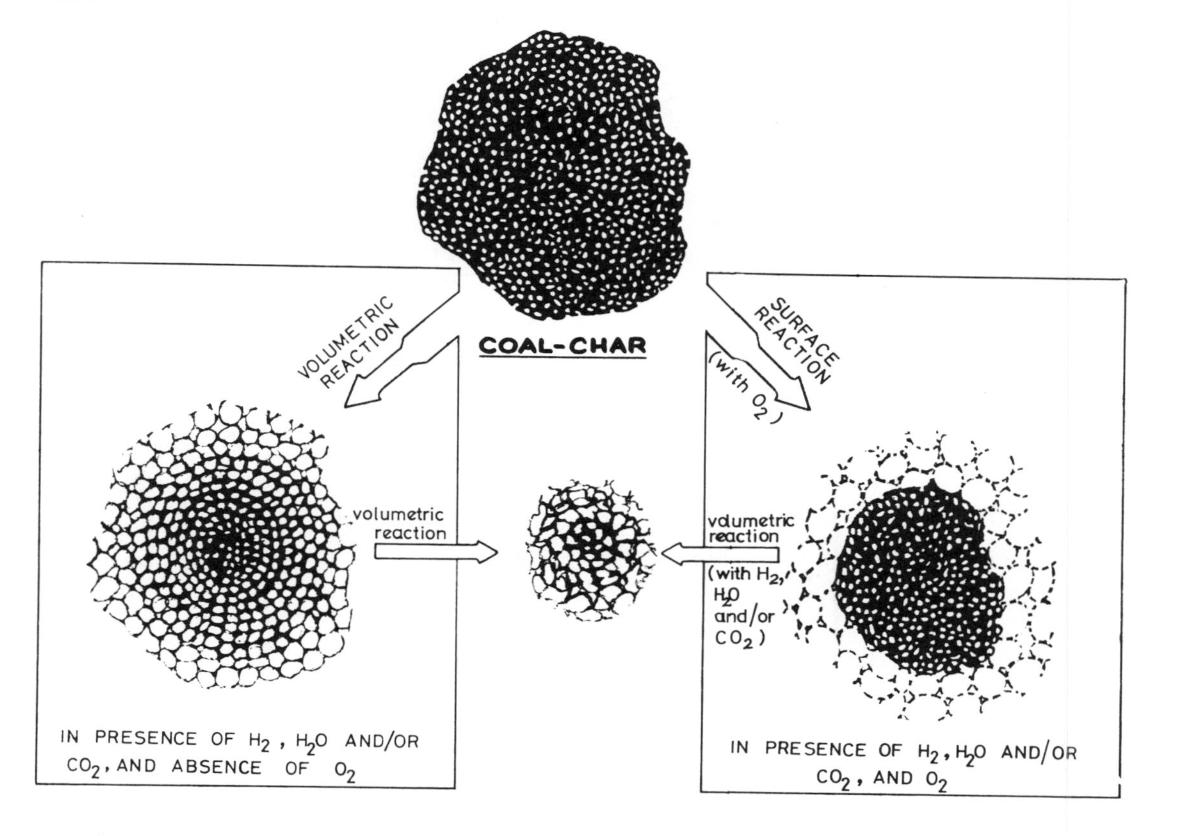

Figure 3.26. Reaction of coal-char.

Rate expression generally used: derived from above equations

$$\text{Rate} = \frac{P_{A0}}{1/k_s' + 1/k_{\text{diff}}} \quad (\text{mol/L}^2 \cdot \theta) \tag{3.64}$$

where $P_{A0} = C_{A0}RT$, $k_s' = k_s/RT$, $k_{\text{diff}} = k_{mA}/aRT$

(ii) *Char-H_2, Char-H_2O, and Char-CO_2 Reactions: Volumetric Reaction Model with No Ash Layer*

H_2 balance

$$D_{eH}\left(\frac{d^2 C_H}{dr^2} + \frac{2}{r} \cdot \frac{dC_H}{dr}\right) - ak_{vH}C_H C_S = 0 \tag{3.65}$$

where D_{eH}, C_H, and k_{vH} refer to H_2.

Boundary conditions

$$D_{eH}\left(\frac{dC_H}{dr}\right) = k_{mH}\,(C_{H0} - C_H) \text{ at } r = r_c \tag{3.66}$$

$$\frac{dC_H}{dr} = 0 \text{ at } r = 0 \tag{3.67}$$

H_2O and CO_2 balance

Equations 3.65–3.67 with D_{eH}, C_H, k_{vH} and k_{mH} replaced by corresponding quantities for H_2O and CO_2.

Carbon balance

$$\frac{dC_S}{dt} = -C_S\,(k_{vH}C_H + k_{vW}C_W + k_{vD}C_D) \tag{3.68}$$

Initial condition

$$C_S = C_{S0} \text{ at } t = 0$$

Subscripts H, W, and D in Equation 3.68 refer to H_2, H_2O, and CO_2, espectively.

ISBN 0-201-08300-0

The value of the stoichiometric coefficient, a, in Equation 3.65, for the char-H_2 reaction, depends on the hydrogen content of the char. For a devolatilized char the value of this coefficient increases from about 1.2 at zero carbon conversion level to about 2 near complete conversion (Wen et al. [115]). For char-H_2O and char-CO_2 reactions the value of a is close to unity. In the case of the char-O_2 reaction, the product can be either CO_2 or CO or both. The distribution of CO and CO_2 or the mechanism factor Ψ, discussed in connection with the diffusional reaction rate coefficient, k_{diff}, in Section 3.7 on the char-O_2 reaction, is the reciprocal of the stoichiometric coefficient a.

The rates of the char-H_2, char-CO_2, and char-H_2O reactions are expressed by Equations 3.11, 3.16, and 3.26, respectively, in preceding sections. All these rate equations take a simplified form:

$$\frac{dx}{dt} = k_v C_A (1 - x) \tag{3.69}$$

if the following assumptions are made:

1. The equilibrium hydrogen concentration ($C^*_{H_2}$) is negligible compared to the bulk concentration of H_2 (C_{H_2}).
2. $\alpha_v \approx 1.0$.
3. The inhibiting effect of H_2 and CO on the char-H_2O reaction rate is negligible.

Equation 3.69 then takes the following form, when expressed in terms of carbon concentration:

$$\frac{dC_S}{dt} = -k_v C_A C_S \tag{3.70}$$

This simplified rate expression has been used for carbon balance in Equation 3.68.

4. REMARKS

The phenomena of pyrolysis, particularly those of rapid and of flash pyrolysis, are not yet well understood. Very little is known regarding the effects of coal type and particle size on the yields and product distributions obtained from these processes. These aspects of pyrolysis need extensive investigations. An

ISBN 0-201-08300-0

attempt to quantitatively describe pyrolysis of a coal particle is made in Addendum of this chapter.

Detailed investigations are required also on the relative reactivities of various forms of coal and char in regard to each of the four char-gas reactions.

For the char-gasification reaction, models follow two modes of char depletion: (a) char without ash-layer formation during gasification due to attrition, and (b) char with the ash layer remaining intact at the outside surface of the reacting char. The actual process, however, may not fall into either of these extreme categories. The char particle may deplete in size gradually with a part of the ash layer remaining at the outside surface, or may collapse suddenly into small fragments after maintaining the initial size up to a certain conversion level. The actual reaction rates, as well as the elutriation characteristics in a bed of char particles (which also depend on the hydrodynamic conditions in the system), will therefore depend on the size change and decrepitation characteristics of a particular coal or char sample. Very little is known about this aspect at present.

In coal gasification most of the char-gas reactions mentioned earlier can be expected to take place simultaneously, some to a greater, and others to a lesser, extent. In addition, a number of gas phase reactions such as the water-gas shift reaction will also take place along with the char-gas reactions. However, very little is known regarding the effects of these reactions on individual reaction rates, when they take place simultaneously. The mineral matter or ash of coal often plays an important role as catalyst for many of these reactions. More research is needed to clarify these aspects.

NOTATIONS

A = A constant of Equation 5 of Table 2.4, s^{-1}

A' = Overall preexponential factor for the rate of pyrolysis in Equation 2 of Table 2.4, s^{-1}

a = Stoichiometric coefficient of the solid-gas reaction, $aA(g) + S(s) \rightarrow$ products, –

B = A constant of Equation 5 of Table 2.4, ^{o}K

B' = Overall activation energy for the rate of pyrolysis in Equation 2 of Table 2.4, kJ/mol

C_A = Gas concentration; C_{As}, same at the outside surface of a char particle; C_{A0}, same in the bulk gas, mol/m^3

ISBN 0-201-08300-0

$\overline{C}_A$	=	Concentration of gaseous component A in the reaction zone within solid particles in the volumetric reaction, mol/m^3
C_{Ab}	=	Concentration of gaseous component A at the reaction interface in the volumetric reaction, mol/m^3
C_f	=	Free active sites on carbon surface
$C(O)$	=	Oxygen atom chemisorbed on carbon
C_S	=	Carbon concentration of char; C_{S0}, same at zero conversion, mol/m^3
D	=	A constant of Equation 5 of Table 2.4, –
D_e	=	Effective diffusivity in solid; D_{e0}, same at zero conversion, mol/m·s
$D_{eA}, \overline{D}_{eA}$	=	Effective diffusivity of gaseous component A in the ash layer and reaction zone, respectively, mol/m·s (Effective diffusivity can be roughly estimated by $\epsilon^n D_m$ where ϵ is porosity of ash layer or reaction zone, n is 2 ∿ 3 and D_m is molecular diffusivity of gas A.)
E	=	Activation energy, kJ/mol
E_0	=	Mean activation energy, kJ/mol
f	=	Final fraction of conversion of coal due to pyrolysis in inert atmosphere, or overall fraction of carbon conversion of coal during the first stage of hydrogasification, –
K_0	=	Rate constant for the release of a particular component of gaseous volatiles, including tar due to pyrolysis, s^{-1}
K_c	=	Overall mass transfer coefficient of the primary volatiles formed within the pores of coal particles because of pyrolysis, s^{-1}
k, k_0	=	Pyrolysis rate constants, s^{-1}
k'_1, k'_2, k'_3, k'_4 $k_1 \ldots k_{12}$	=	Constants of Langmuir-Hinshelwood type rate expressions for char-H_2, char-CO_2, and char-H_2O reactions
k''_1	=	Overall rate constant for decomposition reaction in pyrolysis in Figure 3.13, s^{-1}
k''_2	=	Overall rate constant for hydrogenation of reactive volatiles in Figure 3.13, s^{-1} (MN/m^2)$^{-1}$

ISBN 0-201-08300-0

k_3''	=	Rate constant for hydrogenation of reactive char in Figure 3.13, $(MN/m^2)^{-1}$
k_{mA}	=	Mass transfer coefficient of gaseous reactant A at gas film, m/s
k_s, k_{s0}	=	Surface reaction rate constant, $m^{3(m+n)-2}/(mol)^{m+n-1} \cdot s$ unless otherwise specified
k_s'	=	Surface reaction rate constant for a reaction first order with respect to gas concentration, $mol/m^2/(MN/m^2) \cdot s$
k_v, k_{v0}	=	Volumetric rate constant of second-stage char-gas reactions, $m^{3(m+n-1)}/(mol)^{m+n-1} \cdot s$ unless otherwise specified
k_v'	=	Rate constant of Equation 3.10, $m^3/mol \cdot s$
M	=	Molecular weight of oxygen
m	=	Order of reaction with respect to solid reactant concentration, –
m'	=	Heating rate, dT/dt, °K/s
n	=	Order of reaction with respect to gaseous reactant concentration; n_a, apparent order; n_t, true order, –
P	=	Total pressure, MN/m^2
P_A	=	Partial pressure of gaseous reactant; P_{A0}, same in bulk gas, MN/m^2
P_{O_2}, P_{H_2}, P_{CH_4}	=	Partial pressure of O_2, H_2, and CH_4, respectively, MN/m^2 unless otherwise specified
Q	=	A constant of Equation 5 in Table 2.4, –
R	=	Gas constant, kJ/mol · °K
r	=	Radius of a char particle; r_o, outside radius; r_c, radius at the reaction interface in unreacted core shrinking model; r_b, radius at the reaction interface in volumetric reaction model, m
$R_{a,c}$	=	Reaction rate coefficient, expressed per unit of external (geometric) surface area of particles, $g/cm^2 \cdot s$
R_s	=	Reaction rate coefficient, expressed per unit of total surface area of particles, $g/cm^2 \cdot s$
S_{ex}	=	Geometric surface area of the shrinking interface of reacting solid particles, m^2
S_{gex}	=	Geometric surface area of the shrinking interface of reacting solid particles per unit of initial weight, m^2/mol

ISBN 0-201-08300-0

T = Temperature, °K unless otherwise specified

t = Time, s

V = Yield of total volatiles (percent of moisture- and ash-free coal), –

V_0 = Volume of any particular component of gaseous volatiles released because of pyrolysis at time $t = \infty$, m^3

V^* = Ultimate yield of total volatiles (percent) from coal due to pyrolysis, in hydrogen or inert atmosphere, –

V' = Volume of any particular component of gaseous volatiles evolved because of pyrolysis, m^3

V_{nr}^{**} = Nonreactive volatiles (percent) formed up to $t = \infty$ (potential ultimate yield of reactive volatiles), –

V_{nr}^{*} = Ultimate yield of nonreactive volatiles (percent) from coal due to pyrolysis in inert atmosphere, –

V_r^{**} = Reactive volatiles (percent) formed up to $t = \infty$ (potential ultimate yield of reactive volatiles), –

V_r^{*} = Reactive volatiles (percent) lost from particle up to $t = \infty$, –

VM = Proximate volatile matter content of coal (percent) (moisture- and ash-free basis), –

X = Fraction of conversion based on initial carbon content due to pyrolysis, or due to the first stages of hydrogasification and char-steam reactions, –

x = Fraction of carbon conversion of char in the second stage alone due to char-gas reactions, –

Z_s, Z_v = Catalytic activity factor, –

Greek Letters

α_v = Relative pore surface area function, –

γ = A constant of Equation 3.12, –

ϵ = Porosity of solid, unless otherwise specified, –

ρ_s = Particle density of solid, mol/m^3

σ = Standard deviation in the Gaussian distribution of activation energy, kJ/mol

τ = Tortuosity of pores in solid, –

ISBN 0-201-08300-0

APPENDIX TO CHAPTER 2

A Mathematical Model for Pyrolysis of Coal*

Pyrolysis of coal, which occurs in all coal conversions, is perhaps the most difficult to model mathematically. A mathematical model of a single-particle coal pyrolysis in an inert atmosphere is presented on the basis of the following hypothesis.

As the temperature of the coal particle is increased, thermal decomposition takes place within the particle, resulting in the production of tar [122] and gases. As discussed previously, the product undergoes secondary reactions such as:

(a) cracking to form lighter gases from the relevant functional groups, and
(b) polymerization and deposition while the tar diffuses through the internal pores of the particle.

The assumptions used to formulate the single-particle model are as follows:

(a) pseudo-steady-state concentration profiles,
(b) negligible increase in internal pressure, and
(c) equal binary diffusivities.

This model combines the chemical reactions and the transport processes occurring during pyrolysis.

A. CHEMICAL REACTIONS:

Three simultaneous reactions occurring during pyrolysis in an inert atmosphere are devolatilization, cracking, and deposition. For convenience, the products of pyrolysis are classified as char, tar, and gas. Char is defined as the undistillable material that remains in the form of a solid. Tar is defined as the distillable liquid, which has a molecular weight larger than C_6. Gas is defined as those components lighter than C_6, that is, CO, CH_4, CO_2, C_2H_6, H_2O, etc. Both tar and gas occur in the form of vapor when coal is pyrolyzed. During pyrolysis, all of the chemical reactions are assumed to be first-order with respect to the concentration of the reactants. The temperature effect on the rate constants can be expressed by an

*Wen, C.Y. and Chen, L. H., "A model for coal pyrolysis," Paper presented at ACS National Meeting, Washington, D.C., Sept. 9-14, 1979.

ISBN 0-201-08300-0

Arrhenius form. The chemical reactions and the rate expressions for the pyrolysis of a coal particle are summarized as follows:

- *Devolatilization*

$$\text{Coal} \xrightarrow{k_1} X_1 \cdot \text{tar} + (1 - X_1) \cdot \text{char}$$

$$\text{Rate} = k_{10} \cdot \exp(-E_1/RT) \cdot C_{\text{coal}}$$

- *Cracking*

$$\text{Tar} \xrightarrow{k_2} \text{product gas}$$

$$\text{Rate} = k_{20} \cdot \exp(-E_2/RT) \cdot C_{\text{tar}}$$

- *Deposition*

$$\text{Tar} \xrightarrow{k_3} \text{char}$$

$$\text{Rate} = k_{30} \cdot \exp(-E_3/RT) \cdot C_{\text{tar}}$$

The rates of formation of tar, product gas, and inert gas can then be expressed as:

$$R_{\text{tar}} = X_1 \cdot k_1 \cdot C_{\text{coal}} - (k_2 + k_3) \cdot C_{\text{tar}}$$

$$R_{\text{gas}} = k_2 \cdot C_{\text{tar}} \text{ and } R_{\text{inert gas}} = 0$$

B. TRANSPORT PROCESSES:

Both mass and heat transfer affect the pyrolysis of a single coal particle. This is particularly significant for large particles.

Mass transfer

The coal particle can be considered as a porous sphere, which retains its integrity while pyrolysis reaction proceeds. The conservation equation for the gaseous species, i, inside the particle having a mass concentration, C_i, can be formulated as: [93]

$$\frac{1}{r^2}\frac{\partial}{\partial r}(r^2 \cdot N_i) = R_i$$

where R_i is the rate of generation of the species i due to the chemical reactions.

ISBN 0-201-08300-0

N_i is the mass flux of the species i and can be expressed as the sum of the rate of diffusion in the radial direction plus bulk flow through the pores. Thus,

$$N_i = -D_{\text{eff},i} \frac{\partial C_i}{\partial r} + W_i \sum_j N_j$$

W_i, the weight fraction of the species i in the gas phase, can be expressed as:

$$W_i = C_i / \sum_j C_j$$

The solid concentration, C_i, which is necessary for calculating the reaction rate R_i, can be obtained from the material balance equation as:

$$\frac{dC_i}{dt} = R_i$$

The conservation equation for the gaseous species i across the gas film can be written as:

$$N_i = k_{gi} \; [C_{i,\text{s}} - C_{i,\text{b}}]$$

where $C_{i,\text{s}}$ and $C_{i,\text{b}}$ are the concentrations of the gas species i at the particle surface and at the bulk gas stream outside, respectively.

Here k_{gi} is the mass-transfer coefficient across the gas film and can be estimated from an appropriate mass-transfer correlation.

Heat transfer

The energy-balance equation for the particle is derived by taking into account convective, radiative, and conductive heat transfer with the heating devices and the heat of reaction of the pyrolysis process:

$$C_{ps} \cdot \rho_s \cdot \frac{dT}{dt} = \frac{3}{r_0} h_c(T_w - T) + \frac{3\sigma Fe}{r_0} (T_w{}^4 - T^4) + \frac{3ka}{r_0^2} (T_w - T) + \Sigma_i (-\Delta H_i) R_i$$

ISBN 0-201-08300-0

where a represents the fraction of the surface area of the particle in contact with the heating element, and T_w is the temperature of the heating element.

The mass-transfer equations and the heat-transfer equations discussed above can be solved numerically based on the following initial and boundary conditions:

$$t = 0, \quad C_i = C_{io}, \quad \text{and } T = T_o$$

$$r = 0, \quad N_i = 0, \quad \text{and } r = r_o, C_i = C_{is}$$

C. WEIGHT-LOSS CALCULATIONS

The observable phenomena for pyrolysis are particle weight loss, WL, and product distribution. The amount of tar and gas formed at any time, t, can be estimated by integrating the mass flux for tar and gas generated by the particle for a given time interval:

$$(\text{WL})_i = 4\pi r_0^2 \int_0^t N_i(t, r_0) \cdot dt$$

The total weight loss of the particle can be calculated either by the addition of the weight loss of tar and gas or by the subtraction of the weight of unreacted coal and char remaining in the particle during pyrolysis, from the original weight of the coal. Thus,

$$\begin{aligned}\text{Total weight loss} &= \sum_i (\text{WL})_i \\ &= (\text{Original weight of coal}) - (\text{Weight of unreacted coal}) - \\ &\quad (\text{Weight of char formed})\end{aligned}$$

D. DETERMINATION OF RATE CONSTANTS

The pyrolysis data of Anthony and Howard [3, 4] for bituminous coal and those of Suuberg et al. [102] for lignite were used to determine the reaction-rate constants for the devolatilization step and the deposition step. The cracking-reaction rate constants for bituminous and subbituminous coal and lignite were chosen on the basis of the product-distribution data of Solomon [98, 99]. The

ISBN 0-201-08300-0

reaction-rate constants obtained for different ranks of coals are tabulated in Table 1A.

Table 1A.

Reaction-Rate Constants for Coal Pyrolysis Model

Reaction-rate constant	Bituminous coal	Subbituminous coal	Lignite
k_{10} (1/sec)	1.1×10^5	7.5×10^4	5.1×10^4
E_1 (cal/g-mole)	21,200	18,700	16,200
k_{20} (1/sec)	9.7×10^9	3.5×10^{10}	8×10^{10}
E_2 (cal/g-mole)	29,000	27,750	26,500
k_{30} (1/sec)	5.3×10^4	2.5×10^4	1.1×10^3
E_3 (cal/g-mole)	7,000	5,500	4,000

A comparison between the calculated results and the experimental data for the weight-loss history in pyrolysis of bituminous coal is shown in Figure 1A. The comparison of the product distribution of tar and gas is shown in Figure 2A for bituminous coal. Figure 3A shows the application of the model with the predetermined reaction-rate constants for bituminous coals with various volatile matters. The amount of tar formed in the devolatilization step, X_1, can be correlated with

ISBN 0-201-08300-0

the volatile matter content (based on dry, ash-free coal) for various types of coal as follows:

Bituminous coal: $X_1 = 1.30\ (\text{V.M.}) + 0.025$

Lignite: $X_1 = 0.95\ (\text{V.M.}) + 0.025$

This model is useful for predicting the amounts of tar and gas produced during pyrolysis in coal gasifiers or combustors.

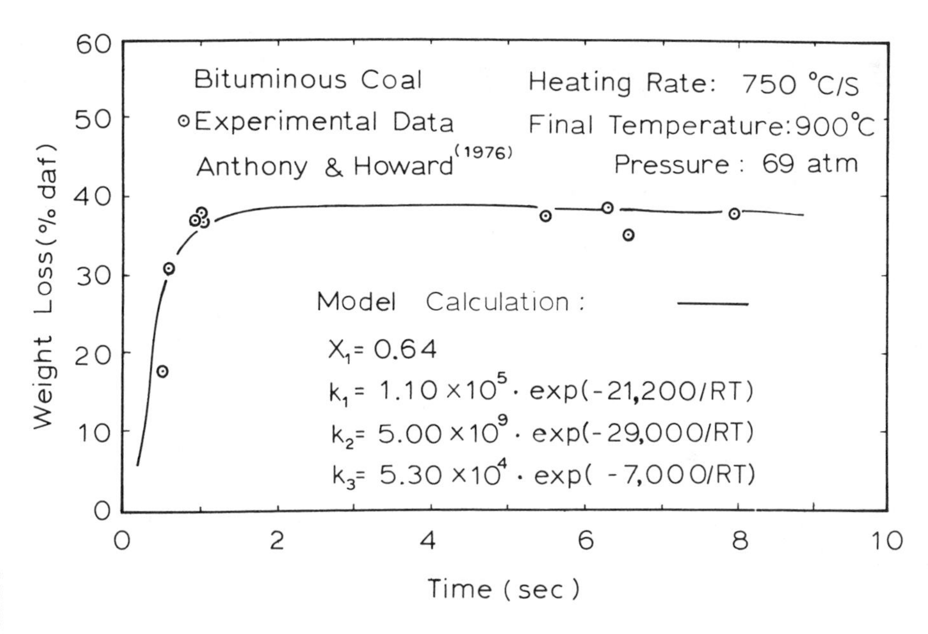

Figure 1A. Weight loss during bituminous coal pyrolysis.

ISBN 0-201-08300-0

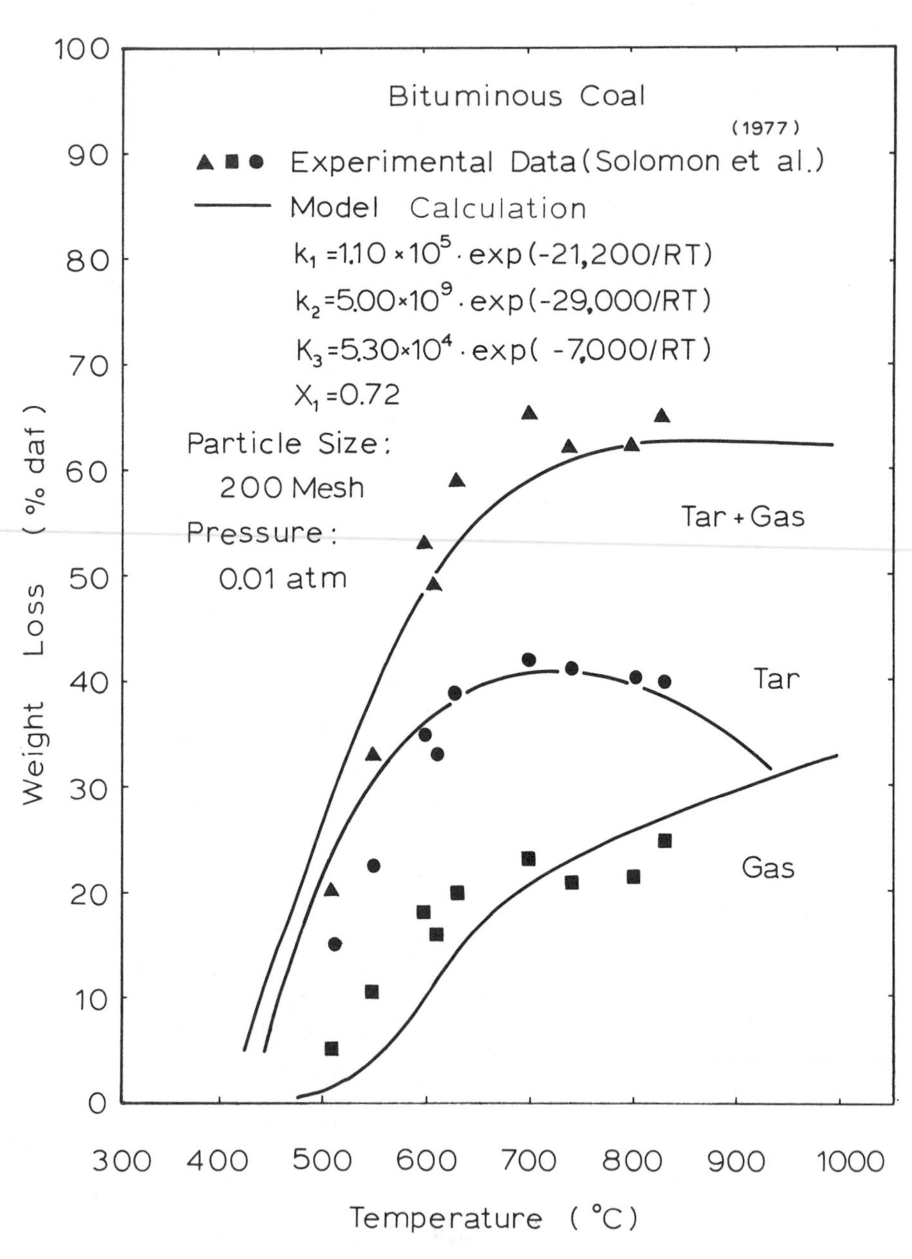

Figure 2A. Product distribution as a function of temperature in bituminous coal pyrolysis (V.M. = 0.53).

ISBN 0-201-08300-0

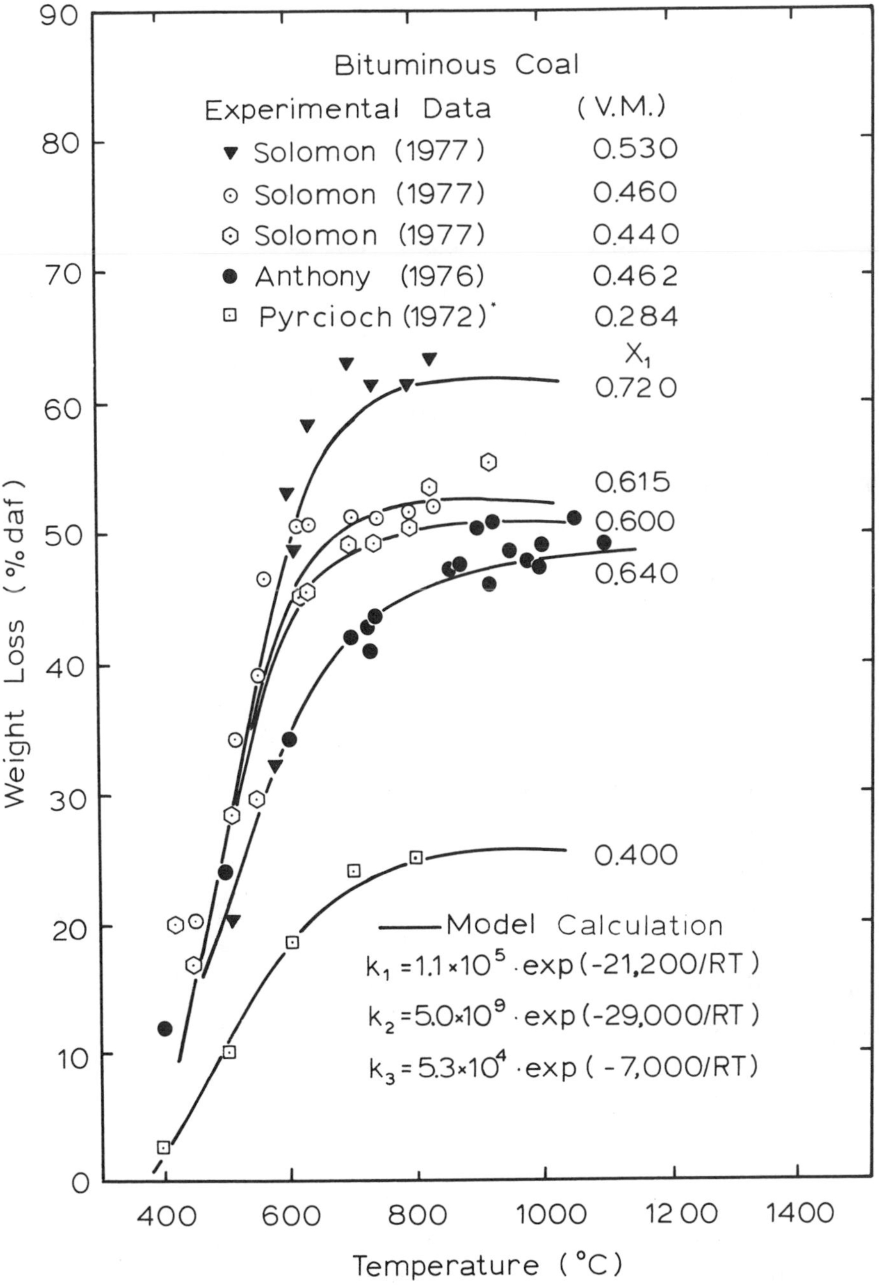
Bituminous Coal
Experimental Data (V.M.)
▼ Solomon (1977) 0.530
⊙ Solomon (1977) 0.460
⬡ Solomon (1977) 0.440
● Anthony (1976) 0.462
⊡ Pyrcioch (1972)* 0.284
X1
0.720
0.615
0.600
0.640
0.400
—Model Calculation
k1 = 1.1×10^5 · exp(-21,200/RT)
k2 = 5.0×10^9 · exp(-29,000/RT)
k3 = 5.3×10^4 · exp(-7,000/RT)
Weight Loss (% daf)
Temperature (°C)
0
10
20
30
40
50
60
70
80
90
400
600
800
1000
1200
1400

Figure 3A. Effect of temperature on the weight loss in bituminous coal pyrolysis (particle size ≅ 70 μm, pressure ≅ 0.01 to 35 atm., heating rate ≅ 700°C/s).

ISBN 0-201-08300-0

Nomenclature

a	Fraction of the contact surface area of the particle with heating elements, –
C_{ij}	Mass concentration of species i at position j, g/cm^3
C_{pi}	Heat capacity of species i, cal/g^{O}K
D_{eff}	Effective diffusivity, cm^2/sec
D_p	Particle diameter, cm
e	Emissivity of the particle, –
F	Geometric factor related with radiation heat transfer, –
h_c	Convective heat-transfer coefficient, cal/cm^2 ^{O}K · sec
k	Thermal conductivity of the coal particle, cal/cm ^{O}K · sec
k_g	Mass-transfer coefficient in the gas film, cm/sec
N_i	Mass flux of gas species i, g/cm^2 · sec
r, r_0	Radius of the particle, cm
R_i	Rate of generation of species i, g/cm^3 · sec
t	time, sec
T	particle temperature, ^{O}K
T_w	Temperature of heating devices, ^{O}K
V.M.	Weight fraction of volatile matter in daf coal, –
W_i	Weight fraction of species i in the gas phase, –
$(WL)_i$	Weight loss of coal due to formation of gaseous species, i, –
X_1	Weight fraction of coal that converted to tar during devolatilization step, –
ΔH_i	Heat of reaction for formation of i, cal/g
ρ_i	Density of species i, g/cm^3
σ	Stefan–Boltzmann constant, cal/cm^2 ^{O}K^4 sec

REFERENCES

1. Aggarwal, A. K. Ph.D. Dissertation, West Virginia University, 1978.
2. Anthony, D. B. and Howard, J. B., Private communication, 1975.
3. Anthony, D. B. and Howard, J. B., *AIChE J.*, **22** (4), 625 (1976).
4. Anthony, D. B., Howard, J. B., Hottel, H. C., and Meissner, H. P., *Proceedings of the 15th Symposium on Combustion (International)*, Combustion Institute, Pittsburgh, Pa., 1974, p. 1303.

ISBN 0-201-08300-0

5. Armiglio, H. and Duval, X., *J. Chem. Phys.*, **64**, 916 (1967).
6. Arthur, J. R., *Trans. Faraday Soc.*, **47**, 164 (1951).
7. Avedesian, M. M. and Davidson, J. F., *Trans. Inst. Chem. Eng.*, **51**, 121 (1973).
8. Ausman, J. M. and Watson, C. C., *Chem. Eng. Sci.*, **17**, 323 (1962).
9. Austin, L. G. and Walker, P. L., Jr., *AIChE J.*, **9**, 303 (1963).
10. Badzioch, S., The British Coal Utilization Research Association Monthly Bulletin, Vol. XXV, No. 8, 1961, p. 285.
11. Badzioch, S. and Hawksley, P. B. W., *Ind. Eng. Chem. Process Des. Dev.*, **9**, 52 (1970).
12. Berger, J., Siemieniewska, T., and Tomkow, K., *Fuel*, **55**, 9 (1976).
13. Birch, T. J., Hall, K. R., and Urie, R. W., *J. Inst. Fuel*, **33**, 422 (1960).
14. Blackwood, J. D. and McGrory, F., *Aust. J. Chem.*, **11**, 16 (1958).
15. Borghi, G., Sarofim, A. F., and Beer, J. M., Paper 34C, 70th Annual AIChE Meeting, New York, November 1977.
16. Calvelo, A. and Cunningham, R. E., *J. Catal.*, **16**, 397 (1970).
17. Campbell, J. H. and Stephens, D. R., *Preprints, Am. Chem. Soc., Div. Fuel Chem.*, **21** (7), 94 (1976).
18. Caram, H. S. and Amundson, N. R., *Ind. Eng. Chem. Fund.*, **16**, 171 (1977).
19. Chauhan, S. P., Feldman, H. F., Stanbaugh, E. P., and Oxley, J. H., *Preprints, Am. Chem. Soc., Div. Fuel Chem.*, **20** (4), 207 (1975).
20. Coates, R. L., Chen, C. L., and Pope, B. J., *Coal Gasification, Advances in Chemistry Series* 131, American Chemical Society, Washington, D.C., 1974, p. 92.
21. Durmosch, R., Islam, K. U., and Oleret, H., *Chem. Ing. Tech.*, **46**, 961 (1974).
22. Dutta, S. and Wen, C. Y., *Ind. Eng. Chem. Process Des. Dev.*, **16**, 31 (1977).
23. Dutta, S., Wen, C. Y., and Belt, R. J., *Ind. Eng. Chem. Process Des. Dev.*, **16**, 20 (1977).
24. Eddinger, R. T., Friedman, L. D., and Rau, E., *Fuel*, **45**, 245 (1966).
25. Ergun, S., *J. Phy. Chem.*, **60**, 480 (1956).
26. Ergun, S. and Mentser, M., *U.S. Bur. Mines Bull.* 598, 1962.
27. Ergun, S. and Mentser, M., *Chem. Phys. Carbon*, **1**, 203 (1965).
28. Feldkirchner, H. L. and Huebler, J., *Ind. Eng. Chem. Process Des. Dev.*, **4**, 134 (1965).
29. Feldkirchner, H. L. and Linden, H. R., *Ind. Eng. Chem. Process Des. Dev.*, **2**, 153 (1963).
30. Field, M. A., *Combust. Flame*, **13**, 237 (1969).
31. Field, M. A., Gill, D. W., Morgan, B. B., and Hawksley, P. G. W., *Combustion of Pulverized Coal*, BCURA, 1967.
32. Fitzerald, D. and van Krevelen, D. W., *Fuel*, **38**, 17 (1959).
33. Friedman, L. S., Rau, E., and Eddinger, R. T., *Fuel*, **47**, 149 (1968).
34. Fuchs, W. E. and Yavorsky, P. M., *Preprints, Am. Chem. Soc., Div. Fuel Chem.*, **20** (3), 115 (1975).
35. Gadsby, J., Hinshelwood, C. N., and Sykes, K. W., *Proc. R. Soc. (London)*, **A187**, 129 (1946).
36. Gan, H., Nandi, S. P., and Walker, P. L., Jr., *Fuel*, **51**, 272 (1972).
37. Gilliland, E. R. and Harriott, P., *Ind. Eng. Chem.*, **46**, 2195 (1954).

ISBN 0-201-08300-0

38. Glassman, I., Paper 34e, 70th Annual AIChE Meeting, New York, November 1977.
39. Goodman, J. B., Gomez, S., Parry, V. F., and Landers, W. S., *U.S. Bur. Mines Bull.* 530, 1953.
40. Goodman, J. B., Gomez, S., and Parry, V. F., *U.S. Bur. Mines Rep. Invest.* 5383, 1958.
41. Goring, G. E., Curran, G. P., Tarbox, R. P., and Gorin, E., *Ind. Eng. Chem.*, **44**, 1051 (1952).
42. Goring, G. E., Curran, G. P., Zielke, C. W., and Gorin, E., *Ind. Eng. Chem.*, **45**, 2586 (1953).
43. Graff, R. A., Dobner, S., and Squires, A. M., *Fuel,* **55**, 113 (1976).
44. Granger, A. F. and Ladner, W. R., *Fuel,* **49**, 17 (1970).
45. Gregory, D. R. and Littlejohn, R. F., *BCURA Mon. Bull.,* **29** (6), 173 (1965).
46. Hamor, R. J., Smith, I. W., and Tyler, R. J., *Combust. Flame,* **21**, 153 (1973).
47. Hanbaba, P. H., Juntgen, H., and Peters, W., *Brennstoff-Chem.,* **49**, 368 (1968).
48. Hawk, C. O., Schlesinger, M. D., and Hiteshue, R. W., *U.S. Bur. Mines Rep.* 6264, 1963.
49. Hawtin, P. and Murdoch, R., *Chem. Eng. Sci.,* **19**, 819 (1964).
50. Hays, D., Patrick, J. W., and Walker, A., *Fuel,* **55**, 299 (1976).
51. Hiteshue, R. W., Anderson, R. B., and Schlesinger, M. D., *Ind. Eng. Chem.*, **49**, 2008 (1957).
52. Hiteshue, R. W., Anderson, R. B., and Schlesinger, M. D., *Ind. Eng. Chem.*, **52**, 577 (1960).
53. Howard, J. B. and Essenhigh, R. H., *Ind. Eng. Chem. Process Des. Dev.,* **6**, 74 (1967).
54. Ishida, M. and Wen, C. Y., *AIChE J.,* **14**, 311 (1968).
55. Ishida, M. and Wen, C. Y., *Chem. Eng. Sci.,* **23**, 125 (1968).
56. Ishida, M. and Wen, C. Y., *Chem. Eng. Sci.,* **26**, 1031 (1971).
57. Ishida, M., Wen, C. Y., and Shirai, T., *Chem. Eng. Sci.,* **26**, 1043 (1971).
58. Jensen, G. A., *Ind. Eng. Chem. Process Des. Dev.,* **14**, 314 (1975).
59. Johnson, J. L., *Coal Gasification, Advances in Chemistry Series* 131, American Chemical Society, Washington, D.C., 1974, p. 145.
60. Johnson, J. L., *Preprints Am. Chem. Soc., Div. Fuel Chem.,* **20** (3), 61, 145 (1975).
61. Johnstone, H. F., Chen, C. Y., and Scott, D. S., *Ind. Eng. Chem.,* **44**, 1564 (1952).
62. Jolly, L. J. and Pohl, A., *J. Inst. Fuel,* **26**, 33 (1953).
63. Juntgen, H. and van Heek, K. H., *Fuel,* **47**, 103 (1968).
64. Kalina, T., "Exxon Catalytic Coal Gasification Process – Development Program," *Tech. Prog. Rep.* FE-2369-4-5-6-7, 1976.
65. Kasurichev, A. P., "Technological Utilization of Fuel for Energy," *Izd. Akad. Nauk SSR (Moscow),* 1960, p. 82 [translated by S. Badzioch (1961)].
66. Kawahata, M. and Walker, P. L., Jr., *Proceedings of the 5th Carbon Conference,* Vol. II, 1963, p. 251.
67. Kayembe, N. and Pulsifer, A. H., *Fuel,* **55**, 211 (1976).
68. Kimber, G. M. and Gray, M. D., *Combust. Flame,* **11** (4), 360 (1967).

ISBN 0-201-08300-0

69. Klei, H. E., Sahagian, J., and Sundstrom, D. W., *Ind. Eng. Chem. Process Des. Dev.*, **14**, 470 (1975).
70. Kurylko, L. and Essenhigh, R. M., *Proceedings of the 14th Symposium (International) on Combustion*, Combustion Institute, Pittsburgh, Pa., 1972, p. 1375.
71. Lewis, J. B., Connor, P., and Murdoch, R., *Carbon*, **2**, 311 (1964).
72. Littlejohn, R. F., *J. Inst. Fuel*, **40**, 128 (1967).
73. Loison, R. and Chauvin, R., *Chem. Ind.*, **91** (3), 269 (1964).
74. Long, F. J. and Sykes, K. W., *Proc. Ro. Soc. (London)*, **A193**, 377 (1948).
75. Lowry, H. H. (Chairman), *Chemistry of Coal Utilization*, Vol. I, John Wiley & Sons, New York, 1945, p. 187.
76. Magne, P. and Duval, X., *Bull. Soc. Chim.*, **5**, 1585 (1971).
77. Mentser, M., O'Donnell, H. J., and Ergun, S., *Preprints, Am. Chem. Soc., Div. Fuel Chem.*, **14** (5), 94 (1970).
78. Mentser, M., O'Donnell, H. J., Ergun, S., and Friedel, R. A., *Coal Gasification, Advances in Chemistry Series* 131, 1974, p. 1.
79. Montet, G. L. and Myers, G. E., *Carbon*, **6**, 627 (1968).
80. Moseley, F. and Patterson, D., *J. Inst. Fuel*, **38**, 13 (1965).
81. Moseley, F. and Patterson, D., *J. Inst. Fuel*, **40**, 523 (1967).
82. Mulcahy, M. F. R. and Smith, I. W., *Rev. Pure Appl. Chem.*, **19**, 81 (1969).
83. Muralidhara, H. S., Ph.D. Dissertation, West Virginia University, 1978.
84. Patel, R. L., Nandi, S. P., and Walker, P. L., Jr., *Fuel*, **51**, 47 (1972).
85. Petersen, E. E., *AIChE J.*, **3**, 443 (1957).
86. Pitt, G. J., *Fuel*, **41**, 267 (1962).
87. Pyrcioch, E. J. and Linden, H. R., *Ind. Eng. Chem.*, **52**, 290 (1960).
87a. Pyrcioch, E. J., Feldkirchner, H. L., Tsaros, C. L., Johnson, J. L., Bair, W. G., Lee, B. S., Schora, F. C., Huebler, J., and Linden, H. R., "Production of Pipeline Gas by Hydrogasification of Coal," IGT Res. Bull., No. 39, Chicago (1972).
88. Rau, E. and Seglin, L., *Preprints, Am. Chem. Soc., Div. Fuel Chem.*, **7**, 154 (1963).
89. Rau, E. and Seglin, L., *Fuel*, **43**, 147 (1964).
90. Reidelbach, H. and Summerfield, M., *Preprints, Am. Chem. Soc., Div. Fuel Chem.*, **20** (1), 161 (1975).
91. Rossberg, M., *Z. Elektrochem*, **60**, 952 (1956).
92. Rossberg, M. and Wicke, E., *Chem. Ing. Tech.*, **28**, 181 (1956).
93. Russell, W. B., Saville, D. A., and Greene, M. I., *AIChE J.*, **25**, 65 (1979).
94. Sass, A., *Chem. Eng. Prog.*, **70** (1), 72 (1974).
95. Sergent, G. D. and Smith, I. W., *Fuel*, **52**, 52 (1973).
96. Sharkey, A. G., Jr., Shultz, J. L., and Friedel, R. A., *Nature*, **202**, 988 (1964).
97. Smith, I. W. and Tyler, R. J., *Combust. Sci. Technol.*, **9**, 87 (1974).
98. Solomon, P. R., "The Evolution of Pollutants during the Rapid Devolatilization of Coal," Rep. R76-952588-2, United Technologies Research Center, East Hartford, Conn., 1977.
99. Solomon, P. R., "The Evolution of Pollutants during the Rapid Devolatilization of Coal," Rep. NSF/RA-770422, NTIS PB 278496/AS, 1977.
100. Solomon, P. R. and Colket, M. B., Private communication, 1978.

ISBN 0-201-08300-0

101. Squires, A. M., Graff, R. A., and Dobner, S., *Science,* **189**, 793 (1975).
102. Suuberg, E. M., Peters, W. A., and Howard, J. B., *Ind. Eng. Chem., Process Des. Dev.,* **17**, 37 (1978).
103. Szekely, J., Evans, J. W., and Sohn, H. Y., *Gas-Solid Reactions,* Academic Press, New York, 1976.
104. Turkdogan, E. T., Koump, V., Vinters, J. V., and Perzak, T. F., *Carbon,* **6**, 467 (1968).
105. Turkdogan, E. T. and Vinters, J. V., *Carbon,* **7**, 101 (1969).
106. Tyler, R. J. and Smith, I. W., *Fuel,* **54**, 99 (1975).
107. Tyler, R. J., Wouterhood, M. J., and Mulcahy, M. F. R., *Carbon,* **14**, 271 (1976).
108. Van Krevelen, D. W., Huntjens, F. J., and Dormans, H. N. M., *Fuel,* **35**, 462 (1956).
109. Walker, P. L., Jr., Rusinko, F., Jr., and Austin, L. G., *Advances in Catalysis,* Vol. 11, Academic Press, New York, 1959, p. 133.
110. Walker, P. L., Jr., Shelef, M., and Anderson, R. A., *Chem. Phys. Carbon,* **4**, 287 (1968).
111. Wen, C. Y., *Ind. Eng. Chem.,* **60**, 34 (1968).
112. Wen, C. Y., "Optimization of Coal Gasification Processes," *R & D Res. Rep.* 66, Vol. 1, 1972, Chapter 4, p. 74.
113. Wen, C. Y., Bailie, R. C., Lin, C. Y., and O'Brien, W. S., *Coal Gasification, Advances in Chemistry Series* 131, American Chemical Society, Washington, D.C., 1974, p. 9.
114. Wen, C. Y. and Huebler, J., *Ind. Eng. Chem. Process Des. Dev.,* **4**, 142, 147 (1965).
115. Wen, C. Y., Mori, S., Gray, J. A., and Yavorsky, P. M., *AIChE Symp. Ser.,* **73** (161), 86 (1977).
116. Wen, C. Y. and Wang, S. C., *Ind. Eng. Chem.,* **62**, 30 (1970).
117. Wen, C. Y. and Wei, L. Y., *AIChE J.,* **16**, 848 (1970).
118. Wen, C. Y. and Wei, L. Y., *AIChE J.,* **17**, 272 (1971).
119. Wen, C. Y. and Wu, C. N. T., *AIChE J.,* **22**, 1012 (1976).
120. Woodburn, E. T., Everson, R. C., and Kirk, A. R. M., *Fuel,* **53**, 38 (1974).
121. Yagi, S. and Kunii, D., *5th International Symp. on Combustion,* Reinhold, N.Y., 1955, p. 231.
122. Yellow, P. C., *BCURA Mon. Bull.,* **29** (9), 285 (1965).
123. Yoshida, K. and Kunii, D., *J. Chem. Eng. (Jap.),* **2**, 170 (1969).
124. Zielke, C. W. and Gorin, E., *Ind. Eng. Chem.,* **47**, 820 (1955).

ISBN 0-201-08300-0

3. Coal Combustion

Robert H. Essenhigh

1. INTRODUCTION

The combustion of coal as it is carried out in existing furnaces and boilers today is a practical art that still owes more to past empirical experience than it does to scientific understanding. The process of combustion is now reasonably well understood in broad outline and in some detail, but prediction of the behavior of a given system from first principles or mathematical models is still not possible. The overall performance of a coal burning unit can be reasonably well predicted from a combination of simple theoretical considerations and past experience, but the empirical content of the bases for prediction must be fully recognized by practioners in the art of coal firing and equipment design. Research is, of course, under way to fill in the gaps in information and understanding, but from the point of view of presenting a coherent account of the behavior of coal-fired systems it is still best to approach the task from the operational point of view. In this chapter, therefore, the approach will be to describe existing coal-fired systems, the essentials of the equipment used, the understanding that we now have of how those experimental systems behave, and our present knowledge of the underlying fundamentals.

Note on units. Although consistent units, preferably SI, throughout are desirable, engineering experience given in units other than FPS is uninterpretable by many readers. Therefore FPS units have been used where this practice has been customary in the past. Usually SI or related units are included in parentheses (also see Table 4.2).

2. COAL BURNING SYSTEMS AND APPLICATIONS

2.1. System Function

The function of any boiler or furnace (or kiln, still, stove, pot, etc.) is to burn fuel to heat some material or process stream (water, steam, glass, refractory, metal,

C. Y. Wen and E. Stanley Lee, Editors, Coal Conversion Technology, ISBN 0-201-08300-0

ISBN 0-201-08300-0

etc.). The function of the burner is to introduce the fuel and air, or other oxidants/reactants such as oxygen and steam, and to mix them to maintain a stable flame. The function of the combustion chamber is to contain the flame to permit maximum transfer of heat, governing thermal efficiency, as well as to provide sufficient space for complete combustion (combustion efficiency) and to permit proper removal of the products of combustion and ash (pollution control). There are design constraints: the combustion chamber must be large enough to contain the flame; its relative dimensions must fit the flame shape, which is controlled by the fuel used and the burner design; flame stability can be affected by the nature of the combustion chamber walls, whether they are hot as in most furnaces or cold as in most boilers; the furnace temperature can also affect corrosion or slag attack on the walls or on the stock in the furnace; the design must permit proper effluent control with respect to environmental regulations.

2.2. Firing Methods

The principal methods of firing coal are listed in Table 2.1 and are cross-listed under "Applications" in Table 2.2. Most are discussed in some greater detail in Section 3.

Table 2.1

Firing Methods[a]

System	Burner/Combustion Chamber	Application
1. Fixed bed (grate) (size: lump; up to 10 cm)	1.1 Hand stoking	Boilers
	1.2 Overfeed (sprinkler stoker)	Boilers
	1.3 Cross-feed (traveling grate)	Boilers and process
	1.4 Underfeed: (i) Single retort	Boilers and process
	(ii) Multiple retort	Boilers
	1.5 Vibragrate	Boilers
2. Fluid bed (size: crushed; 1–5 mm)	2.1 Ballasted } { (i) Nonrecycle	Boilers
	2.2 Unballasted } { (ii) Recycle (fast)	Boilers
3. Entrained or disperse phase (size: ground <100 μm)	3.1 Straight shot (process) burners	Process furnaces
	3.2 Grid and multijet	Boilers (shell)
	3.3 U and W flame	Boilers (water tubes)
	3.4 Tangential ("external" turbulence)	Boilers
	3.5 "Flare" ("self-turbulating")	Boilers and process
	3.6 Cyclone	Boilers
	3.7 Reciprocating engines	Direct power
	3.8 Gas turbines	Direct power
4. Combined firing	4.1 Pulverized coal over grate firing	Boilers
	4.2 Oil and/or gas over grate firing	Boilers
	4.3 Coal/oil dispersions	Boilers and process

[a] In principle, many of the systems can also be operated as gasifiers by firing fuel-rich.

ISBN 0-201-08300-0

Table 2.2

Applications and Methods of Firing Coal

	Applications	Firing Methods
1.	Boilers – steam raising (i) Shell (fire/smoke tube) (ii) Water tube	Fixed and entrained bed; fluid bed (ballasted) under development; Ignifluid; combined
2.	Process furnaces Ferrous and nonferrous melting, reheat, heat treatment, forge; glass melting; refractories firing; ore and lime roasting; etc.	Entrained; fixed in some applications; gasifiers; possibly coal/oil
3.	Direct power (i) Reciprocating engines (ii) Gas turbines	Entrained; gasifiers ("power" gas); possibly coal/oil
4.	Special reaction systems (i) Stirred reactor (research) (ii) Plug flow flame (research) (iii) Explosion flame	Entrained

Coal may be fired in three primary ways: in a fixed bed on a grate, in a fluid bed, or pulverized and entrained. The three systems may be distinguished by their solid particle kinematics, as illustrated schematically in Figure 2.1 [1]. The combination of particle and gas motion generates three mixing descriptions that are significantly different.

1. In the fixed (or broken) bed the particles and gas are essentially in counterflow (or cross flow, see Section 3.2). In principle, therefore, particle histories, which are needed for theoretical models, can be uniquely determined and/or predicted.

2. In the fluid bed the closest description of the particle behavior is "well stirred," and particle histories are less easily determined.

3. In the entrained bed the particles move mostly with the gas. Practical entrained systems generally contain regions of backmix flow and regions of plug or piston flow. There are then potential problems of particle history determinations (as with the fluid bed) in the backmix region. In entrained flow, however, the particles and surrounding gas tend to move together, which effectively they do not

ISBN 0-201-08300-0

in the fluid bed. The overall or net behavior of the particles with respect to the gas is parallel flow.

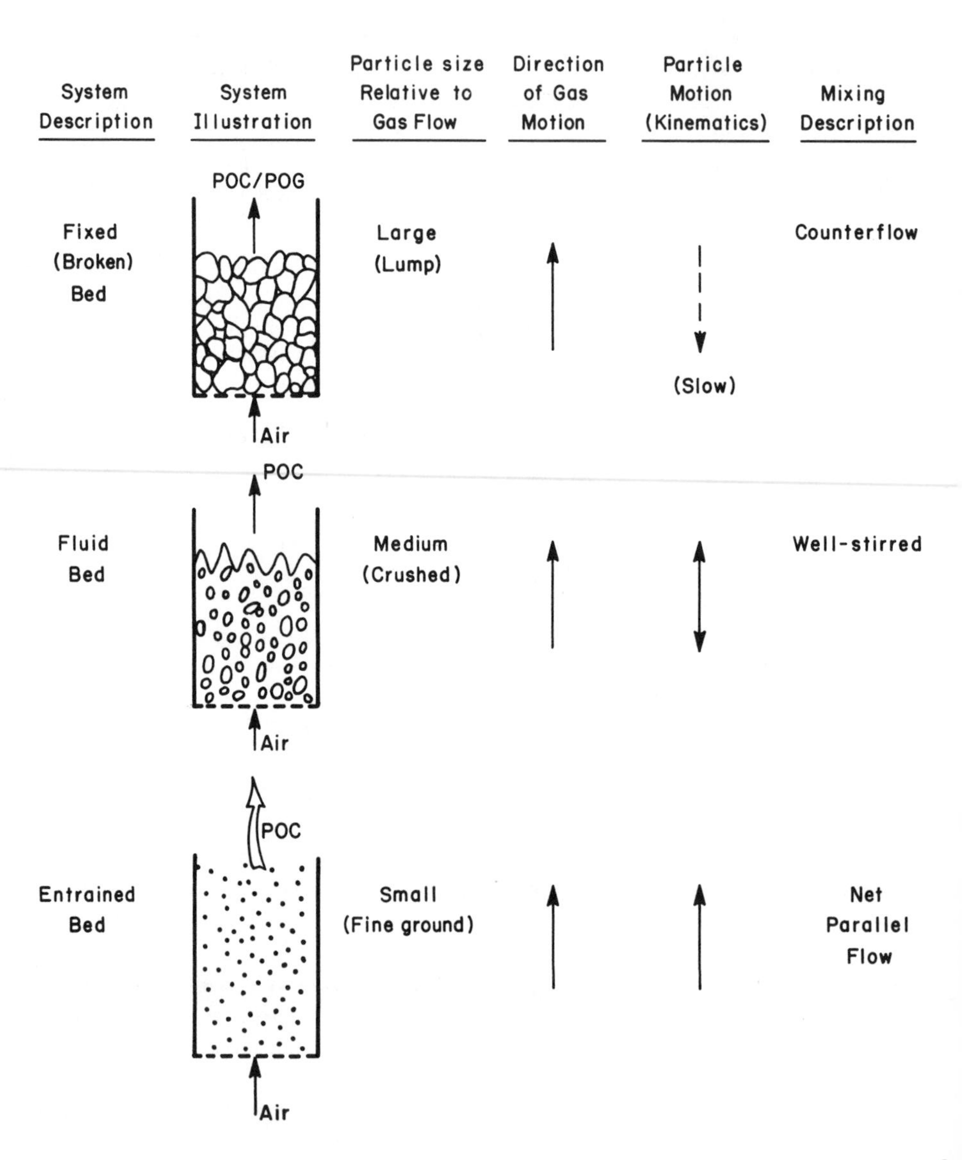

Figure 2.1. Schematic comparison of solid particle kinematics in heterogeneous reacting systems. (Source: Ref. 1.)

ISBN 0-201-08300-0

The classification of Table 2.1 and Figure 2.1 implies that particle size is also a significant descriptor of the different systems. This is not so in principle, but it is in practice. In principle even the largest particle sizes can be fluidized or entrained if gas velocities are high enough. There are engineering and/or cost limitations, however, on equipment size, fan power, and similar design requirements that roughly limit the range of particle size that is appropriate to the different systems.

1. For best performance lump coal fired or gasified on grates or in shafts should be graded, with particular attention given to removing the fines. One-half in. (1–1.5 cm) is frequently regarded as optimum, usually with a preferred top size of 2 in. (5 cm). Firing ungraded coal with all sizes from 2 or 3 in. (5 to 7.5 cm) down to fines is quite common, however.

2. The preferred range of particle sizes for fluid bed use is more limited. It is approximately 1/16–1/4 in. (roughly 1–5 mm), although Dainton [2] has described experiments using 1/2–1 in. coal floated on fine but dense ballast to reduce the fluidizing velocity. Particle sizes below 1 mm are generally present because of natural attrition in handling, and smaller size cuts have been used in experimental systems. When the particles are too small, however, they tend to be quickly entrained and lost as unburned combustibles at the gas flow rates designed to yield reasonable heat release ratings; fly ash reinjection has been attempted to control combustible loss, but it does not appear to be too effective. The preferred solution at present is a special burn-up cell [3].

3. In entrained beds most particles are between 1 and 100 μ with number peaks usually in the range of 10–25 μ (distributions are approximately log-normal).

The three mixing descriptions of the solid particle kinematics outlined in Figure 2.1 are therefore matched by the three size cuts described above. It is then a matter of chance that the factors that dominate the reaction mechanism happen to change at about 100 μ, so that emphasis on kinetic behavior and reaction mechanisms changes between the entrained beds, on the one hand, and the fluid and fixed beds, on the other. Particle size is a parameter that runs through all discussions of coal firing.

2.3. Applications

The principal current use of coal is in boilers, and the largest fraction of that use is as pulverized coal (p.c.) in utility plants. There are also a limited number of process applications, notably p.c. firing of rotary cement kilns (dating from the 1890s).

At one time, of course, all boilers of all types were fired by coal (and coke), displacing wood and charcoal, and always on fixed grates until the advent of p.c.

ISBN 0-201-08300-0

firing and low Btu shaft gasifiers in about the 1860s. Pulverized coal firing of boilers was particularly successful (from the 1920s onward) in the boiler size range above about 350,000 lb steam/hr, which was about the limit of mechanical stokers. Pulverized coal was also used widely for firing a number of metallurgical furnaces (both ferrous and nonferrous) until displaced by oil and natural gas. Aluminum remelt furnaces were fired with traveling grate stokers up to the end of World War II, and fixed grate firing of many refractory furnaces and brick kilns seems to have persisted even longer. Other furnaces were fired with coal-derived fuels – producer gas, water gas, coal gas (coke oven gas).

Future prospects stimulated by the forced revival of interest in coal are taking several forms. A few shaft gasifiers are being built or renovated. There are experiments on p.c. firing of process furnaces, for example, glass tanks, aluminum remelt furnaces, and refractory kilns. There may also be prospects for reviving or enlarging interest in direct power production (diesel [4] and gas turbine). The strongest interest, however, at this time of writing (1979) seems to be in coal-oil mixtures and fluid bed combustion. Emphasis so far has been mainly on boiler applications, but process furnace applications are now being considered. Application of the fluid bed to process furnaces would not take advantage of its unique characteristic of very high heat transfer rates in the bed. The fluid bed would act essentially as a hot gas generator with some entrained particulates, but, with the ballasted bed using limestone to trap SO_2, the pollution characteristics could be acceptable. The gas temperatures, however, might then be so low that furnace applications would be severely restricted. Bed temperatures, and thus gas temperatures, could be raised, but perhaps at the expense of the pollution characteristics. For "direct" firing of high temperature furnaces there would seem to be scope for the development of self-cleaning p.c. burners or some other plant-site processing combustor such as the Cohogg [5].

3. EQUIPMENT SCHEMATICS

3.1. Furnaces and Boilers

Detailed descriptions of different boilers and furnaces are out of place here, but some appreciation of design and function is necessary as context for the discussions and analyses that follow. Figure 3.1–3.5 illustrate some specific examples. Fuller details are available in standard texts [6–16] and data sources [17].

Figures 3.1 and 3.2 represent, respectively, a revolving-drum cement kiln and a simple reverbatory furnace for melting metals.

ISBN 0-201-08300-0

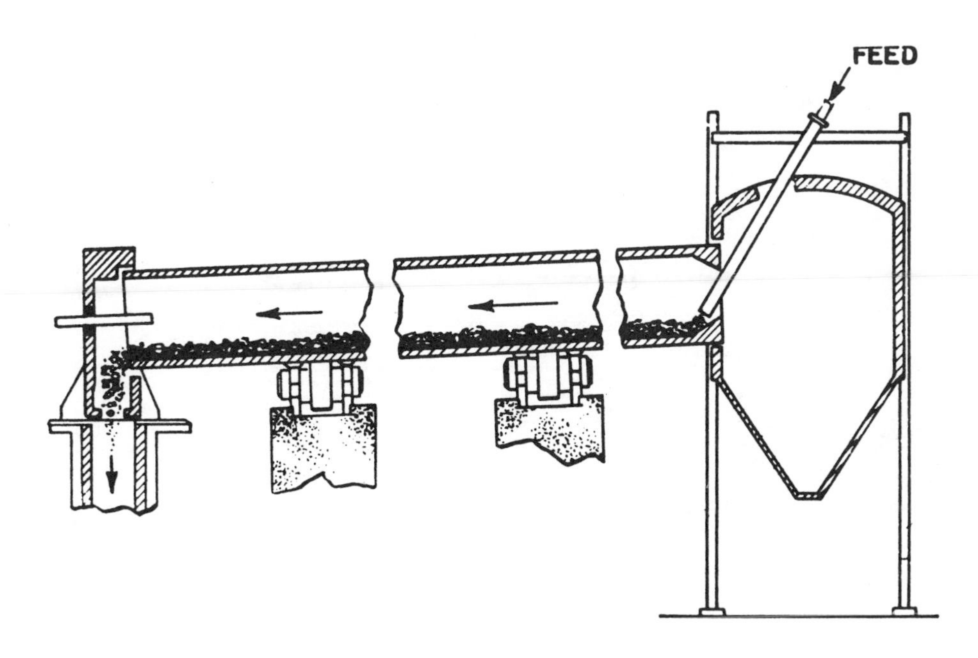

Figure 3.1. Revolving drum cement kiln.

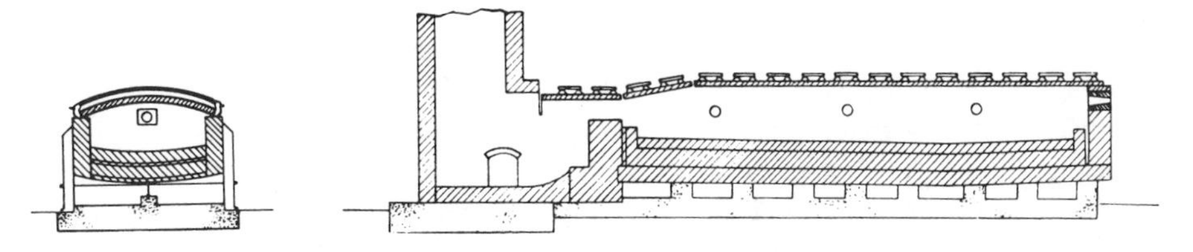

Figure 3.2. Reverberatory furnace for malleable iron melting.

ISBN 0-201-08300-0

Figure 3.3 is a horizontal, single-return packaged boiler of the fire tube or shell type. The firing tube and the flue passes are surrounded by water contained in the cylindrical shell, and they provide heat exchange surface.

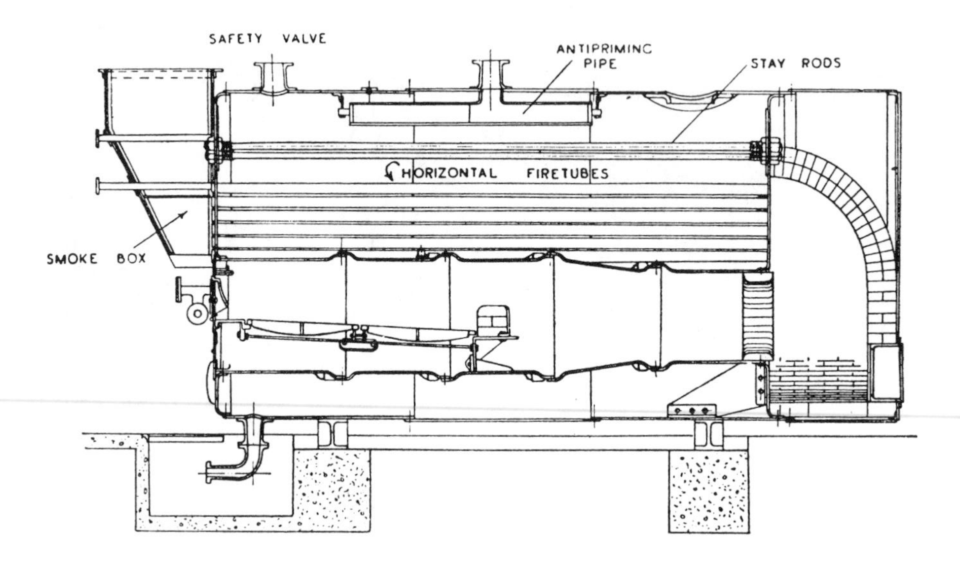

Figure 3.3. Horizontal return tube (HRT) shell boiler.

The design is reversed with larger boilers and/or higher pressures: water or steam is contained in tubes of small inner diameter, with the hot gases on the outside. The tubes may be spaced apart to provide convective heat exchange sections, or they may be effectively touching to form combustion chamber walls. These are the water tube or water wall boilers. Smaller units are usually shop assembled or "packaged" (i.e., transportable as assembled units on flat cars). Larger units are field erected, the smaller (and generally older) units having brick settings to help to form the combustion chamber. Figure 3.4 illustrates a brick-set bent tube boiler. Such units are generally fitted with mechanical grates. With either bent tube or straight tube designs the tubes provide convective sections above the main combustion chamber, which may have brick walls or water walls.

ISBN 0-201-08300-0

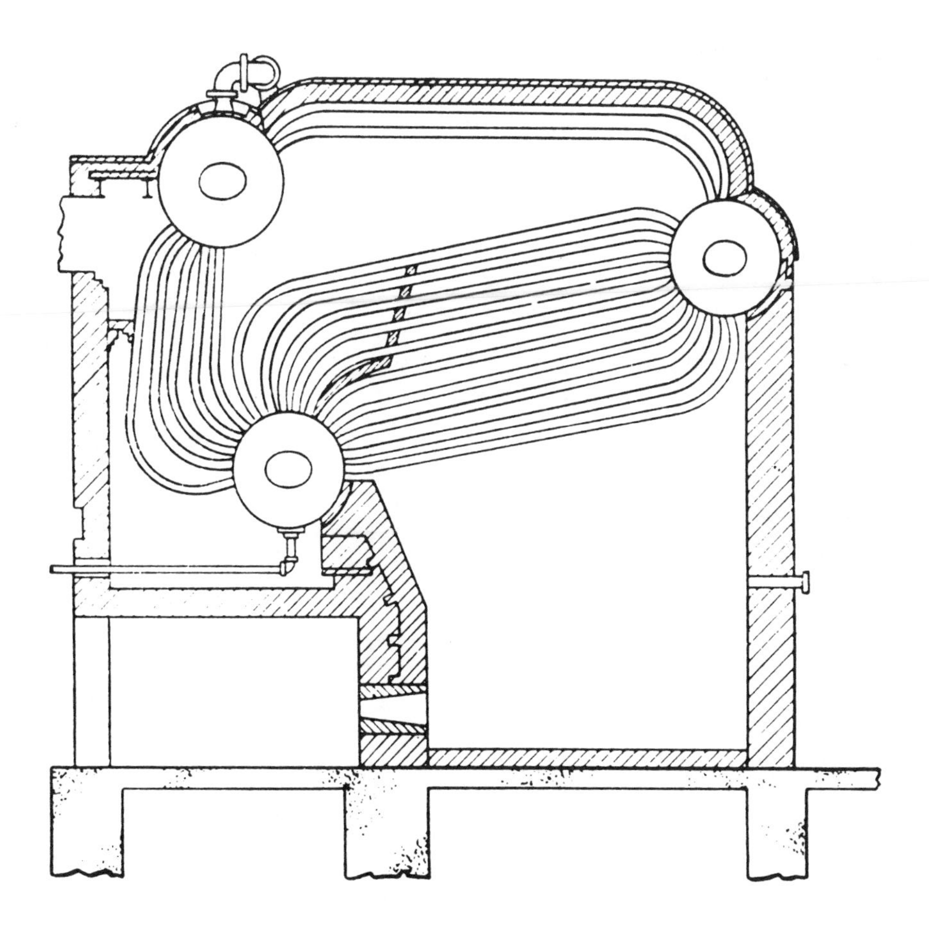

Figure 3.4. Bent water tube type boiler.

The largest boiler or furnace systems of all are the p.c.-fired, radiant water wall boilers with radiant/convective superheaters. Again, layouts differ, often in considerable detail from plant to plant, but Figure 3.5 is typical of many. This is a vertical box, up to 150 ft high in some instances, usually of rectangular cross section, 30 to 40 ft on the side, with coal burners (of different possible designs) in

ISBN 0-201-08300-0

the corners or on the front and/or back walls in the lower third of the chamber. The largest boilers may be double (back-to-back) units.

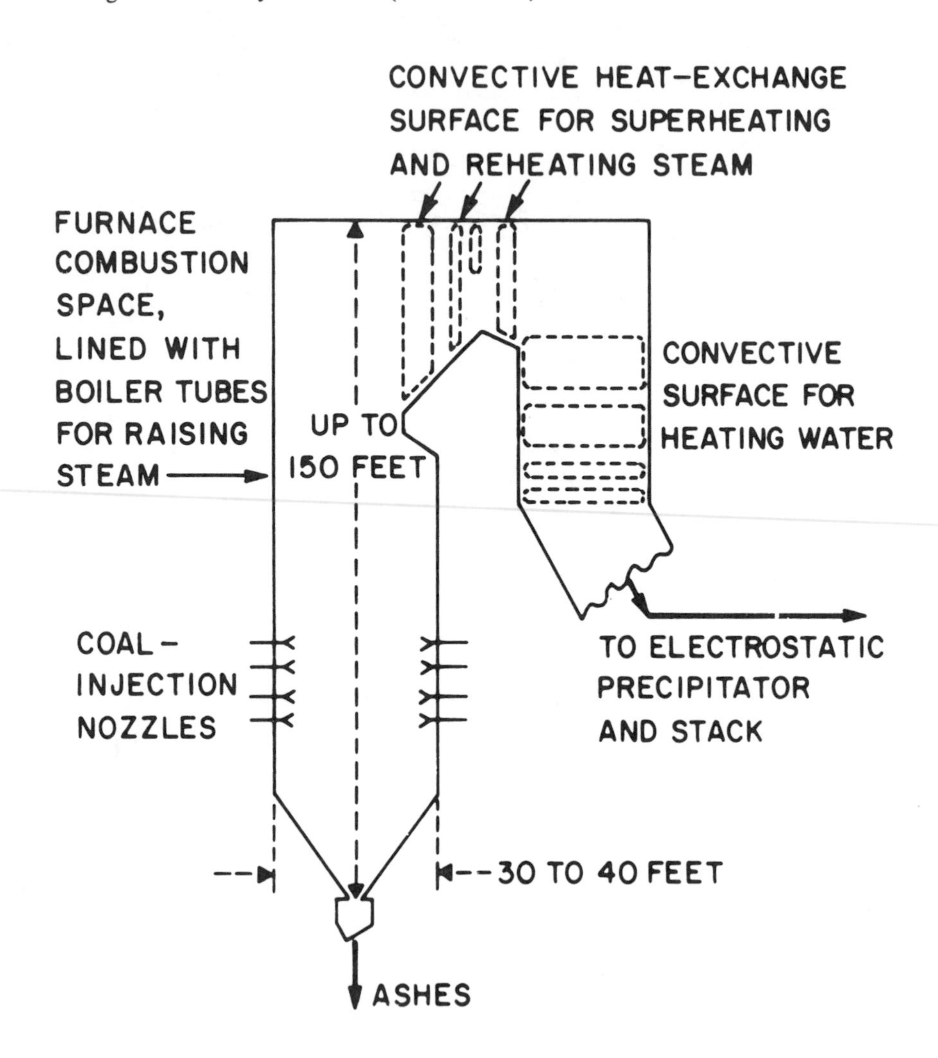

Figure 3.5. Schematic of radiant water-wall utility boiler.

3.2. Grate Firing

In grate firing, coal is heaped on a perforated support (the grate) and air is blown through the bed from underneath (underfire air), as illustrated schematically in Figure 3.6.

ISBN 0-201-08300-0

The reactions in the bed are mainly pyrolysis (Section 5.3) (with drying) and gasification (Section 7.2). The gases leaving the top of the bed are then burned up in additional overfire air (Section 5.4). The various stoker designs and principal applications listed in Tables 2.1 and 2.2 then represent different approaches to solving the problems of fuel supply and combustion and of ash removal.

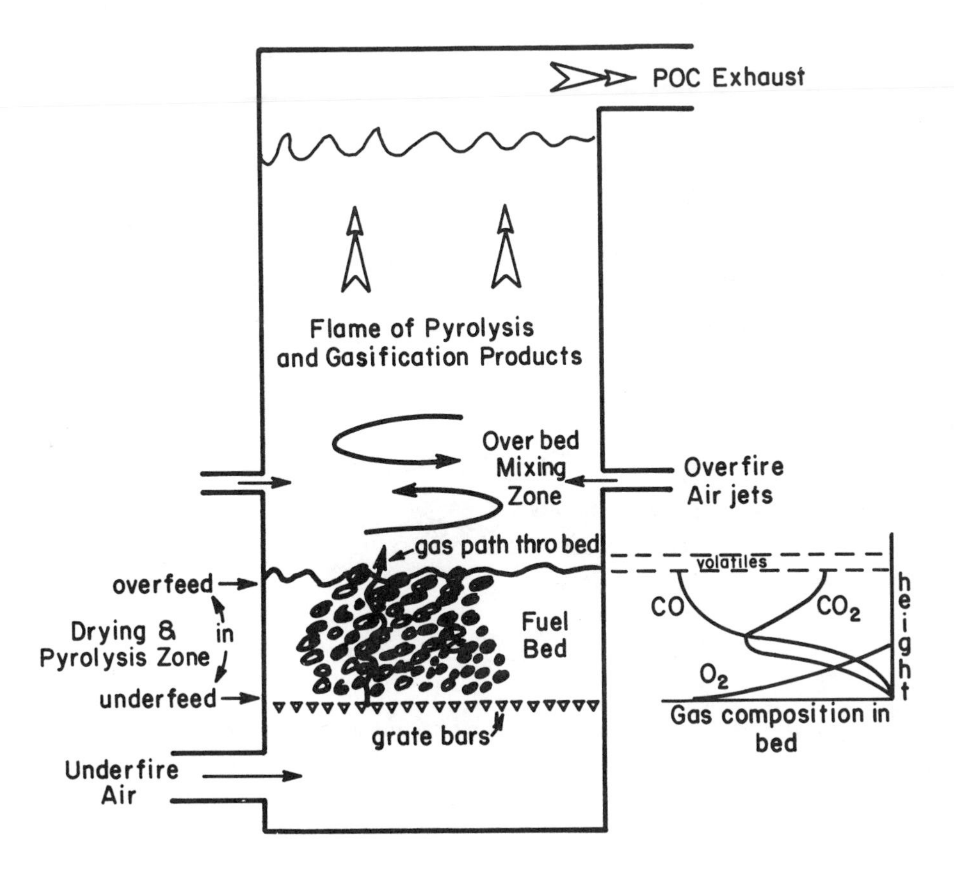

Figure 3.6. Schematic of burning fuel bed, illustrating gas flow in bed, gas composition profiles in bed, overbed mixing, and final burn-out. Variation of local gas composition (oxygen – O_2; carbon dioxide – CO_2; carbon monoxide – CO) with height in bed. Top section of bed is a drying-pyrolysis (devolatilization) zone.

A. SPREADER STOKER

In considering the schematic fuel bed of Figure 3.6 it is obvious that fresh fuel can be supplied in only three ways: from on top, from the side, and from underneath. Top feed or overfeed from a spreader stoker is illustrated in Figure 3.7.

ISBN 0-201-08300-0

With good design, feed is respectably uniform or can be heaped toward one side as required for ash removal. The schematic shows a fixed grate that has to be raked or vibrated in some way to remove the ash that falls into the refuse pit. An alternative design uses a link belt or chain grate similar to the traveling grate, with ash removal under the hopper wall or under the back wall.

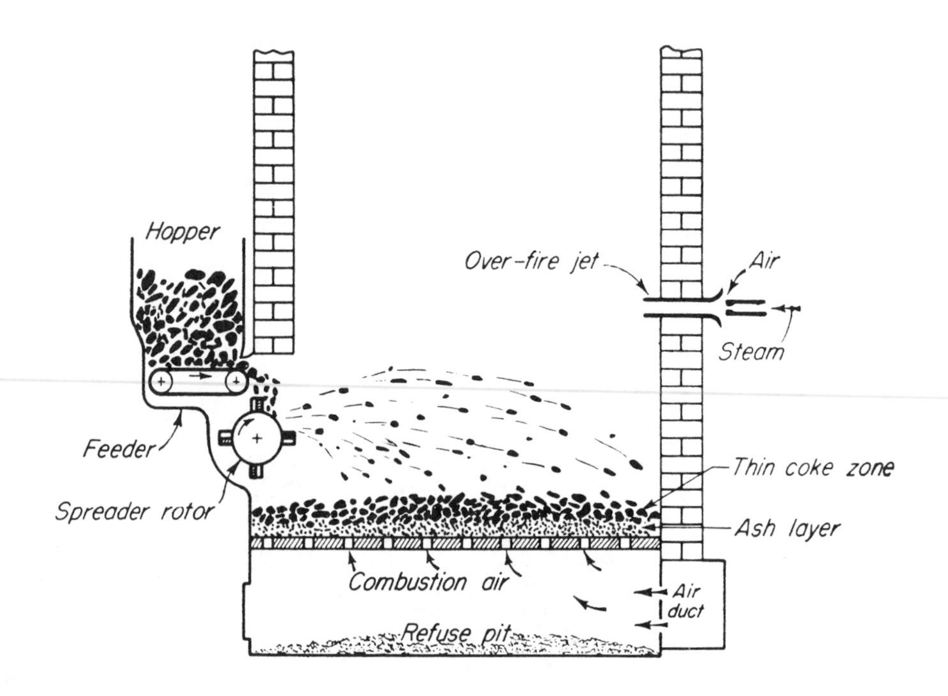

Figure 3.7. Diagram of a spreader stoker and an overfire steam-air jet.

B. TRAVELING GRATE

Figure 3.8 illustrates the traveling grate stoker. Coal is fed out of the hopper onto the chain grate, which then carries the coal across the chamber, delivering ash to the refuse pit under the back wall. Since the bed gets thinner because of combustion on its way across the chamber, the resistance to airflow decreases. The underfire air therefore has to be supplied through a sectioned windbox, with the air quantity approximately matched to the thickness of the fuel above the windbox section. There is also concern about ensuring fast heating of the coal as it enters from the left, and preventing too rapid cooling of the ash as it leaves on the right, as this can affect final burnout and result in too much combustible in the ash. To achieve both these objectives, a variety of different refractory arch shapes have been developed empirically with various claims of effectiveness. Above the feed end

ISBN 0-201-08300-0

the refractory is often referred to as the "ignition" arch. It is assumed to reflect heat from the rest of the firebed or from the overbed flame onto the incoming green coal.

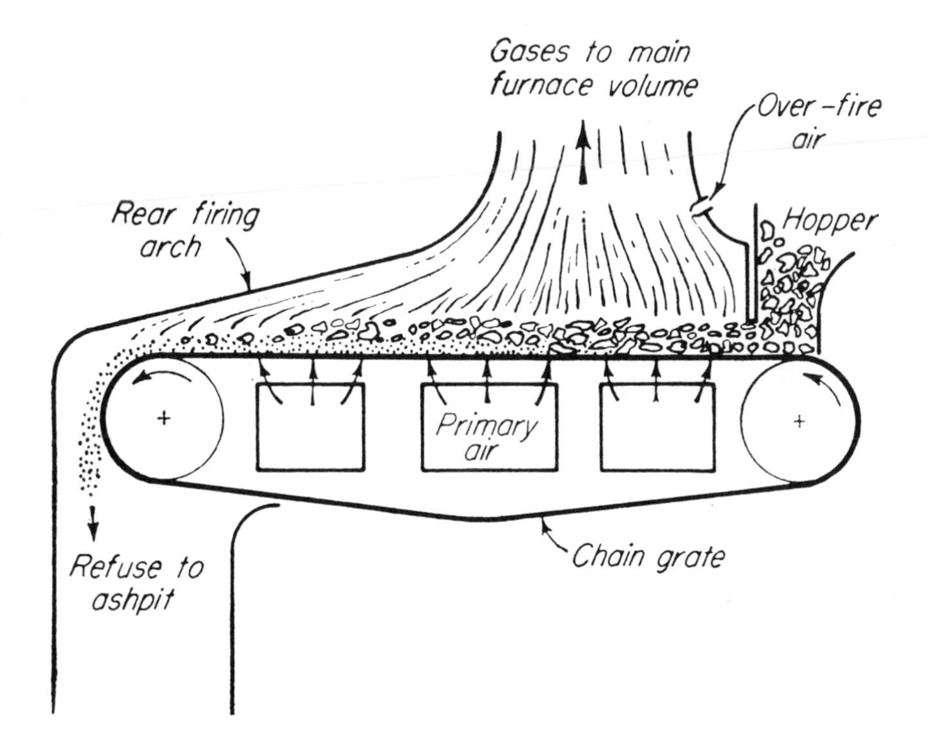

Figure 3.8. Traveling grate stoker with rear firing arch.

C. OVERBED COMBUSTION

The gaseous products from the beds of traveling grate and spreader stokers are a mixture of volatiles and gasification products (CO, H_2) that burn overbed in conjunction with additional overfire air (see Section 5.4). If overbed mixing is poor, however, the unit may smoke. If the units are designed with long ignition and rear arches, the gases rise into the main overbed furnace volume under the water tubes through quite a narrow throat. This can aid mixing to cut down smoke; mixing can also be assisted by proper design and sizing of the overfire air ports. In particular, jet penetration can be achieved by the use of steam assist that may also serve as a substitute for a blower.

ISBN 0-201-08300-0

D. UNDERFEED STOKER

An alternative method of smoke control is to burn up the smoke in the coke bed formed as the coal burns. This requires the coal to be fed from underneath the bed so that the reaction front is traveling down and the volatiles are carried up into the bed. Figure 3.9(a) illustrates the single retort (underfeed) stoker with the coal being forced upward in the channel until it spills over onto downward-sloping grate bars on either side of the channel. Air is supplied through the tuyeres as indicated. A burning/gasifying coke zone develops on the upper levels of the bed with a distillation zone below that (and a drying zone below again). As feed continues, the burning pile rises in height until its profile reaches the natural angle of response and the fuel particles spill over, down the grate and onto the ash bars.

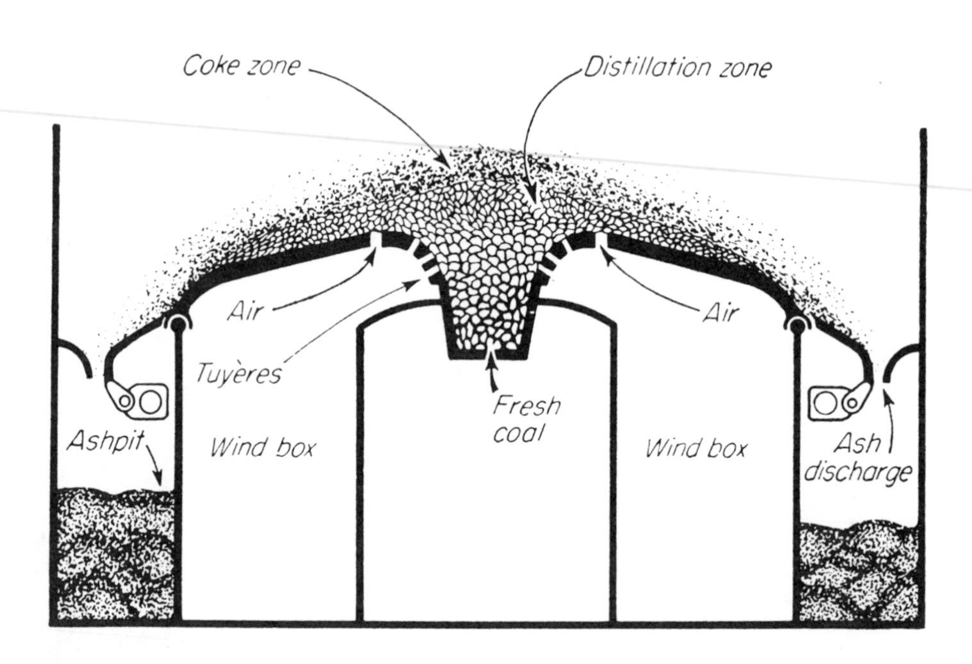

Figure 3.9(a). Cross section of the fuel bed of a single-retort underfeed stoker.

With a single retort the possible grate size is limited by the natural angle of repose. If the grate is too wide, the bed at the center becomes too deep and combustion control becomes impossible. It was then found to be possible to increase

ISBN 0-201-08300-0

boiler size (controlled by grate size) in two ways. The first was to use a moving grate that maintained fuel motion to the ash bars while permitting a bed depth that was more nearly uniform from side to side. The second method was to increase the number of retorts, as illustrated in Figure 3.9(b). With several retorts in parallel, the ash bars between the retorts vanished and the whole retort array was tilted so that the ash traveled to the end of the retorts instead of to the side. Multiple rams then became necessary.

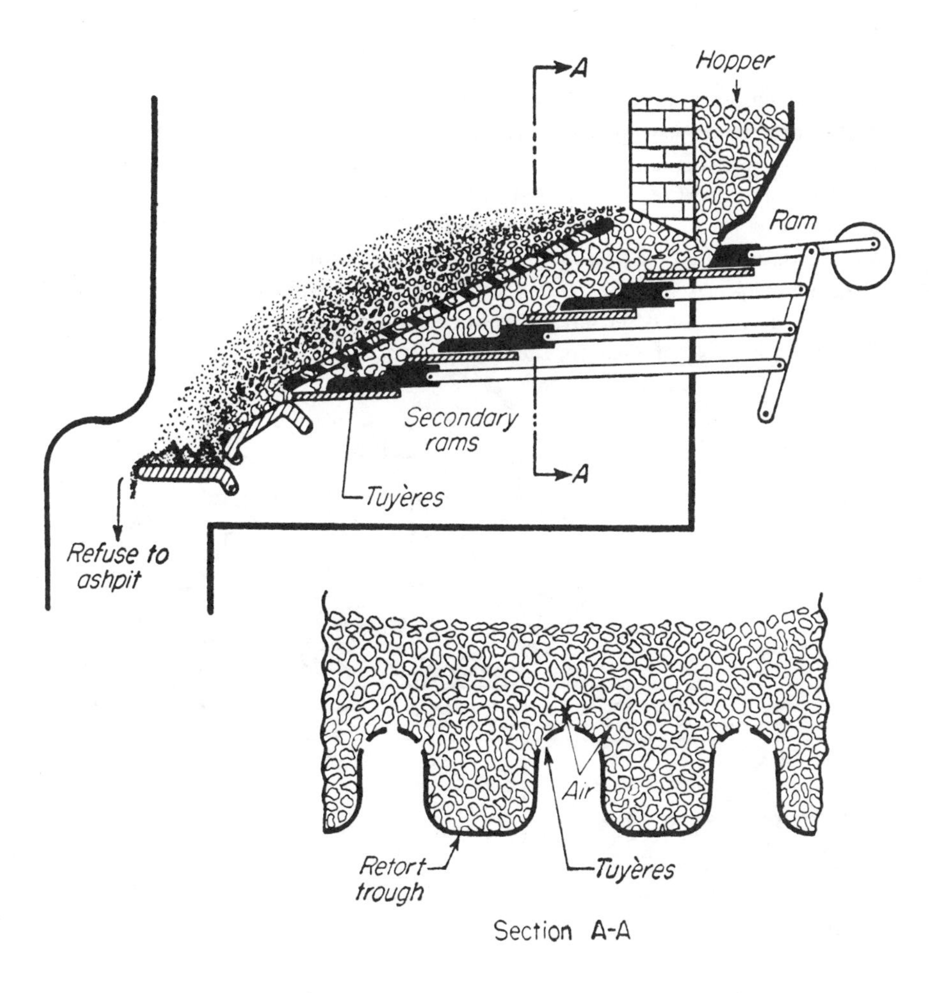

Figure 3.9(b). A multiple-retort underfeed stoker.

ISBN 0-201-08300-0

E. PROCESS FURNACES

The auger or ram feed used for underfeed stokers has also been used for the grate firing of process furnaces (Figure 3.10). In the simplest case the ram or auger merely replaces hand firing. As ultimately developed, however, with forced draft underfire and overfire air supply, such firing systems have been built with complete automatic control. The control is derived from continuous oxygen analysis of the stack gas, and the fuel feed is operated from the controller on an on/off cycle so that the stack gas oxygen hunts between two preset upper and lower limits. This method of control has the advantage of controlling overall air, including in-leakage, a feature which is beneficial for thermal efficiency.

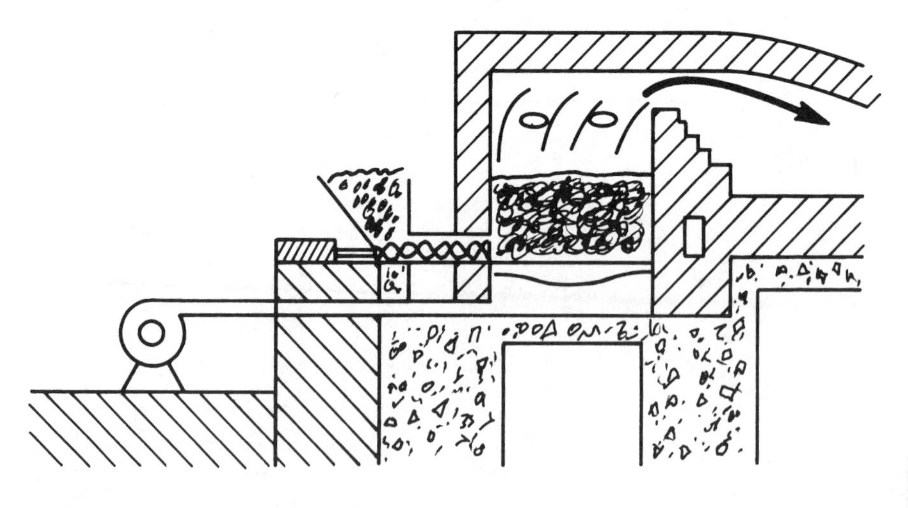

Figure 3.10. Schematic of auger fed, grate-fired process furnace.

ISBN 0-201-08300-0

3.3. Fluid Beds

A. DESCRIPTION

In fluid bed firing the particle size is small enough for the combustion air rising through the bed to "float" the particles. More accurately, the particles move up and down in groups, giving the appearance of convective cells. This vertical "jiggling" motion, as illustrated in Figure 2.1, results in good vertical mixing and even distribution of particles, thus substantially justifying the description of the bed as "well stirred" with respect to the particles. The gas, by comparison, is essentially in plug flow with very limited backmix.

A minimum velocity or volume flow rate is required to fluidize the bed. Volume flow rates in excess of the minimum result in bubbles forming in the bed to carry the excess. The bubbles start forming just above the grate or distributor plate, and they tend to grow as they rise. A wake forms below the rising bubble which entrains particles and assists their upward motion. The bubbles burst when they reach the top of the bed, spraying the particles upward and sideways, and the downward-moving convection cells carry the particles back into the interior. The spraying action can thus be a significant factor in establishing horizontal uniformity in the bed.

If the bed is deep with respect to its diameter, the bubbles can grow to the bed width, and slugging occurs. If most of the bed remains close to the incipient fluidization point and the bubbles concentrate in a few particular paths up through the bed (chaneling), the bed is described as spouting. This can be caused by a bad distributor or a grate design with too few entry points for the air or gas supply.

Of the particles sprayed into the freeboard space by the bursting bubbles, most fall back into the bed but the finest are entrained and removed. There can be appreciable combustible in the fly ash, which can be recovered to some extent by reinjection or in a special burn-up cell [3]. Although most particles fall back into the bed, the height in the freeboard space invaded by the particle spray increases

ISBN 0-201-08300-0

with the velocity excess over the fluidization velocity, and this can amount to multiples of the bed depth. It can be reduced by placing baffles above the bed. The limit is reached when the velocity exceeds the entrainment velocity for all particles in the bed and the bed particles are then blown right out of the reactor chamber; if they are separated from the gas stream and recycled, this is known as the recycle or "fast" fluid bed. Some authors divide the phases as follows: quiescent, incipient or onset, bubbling, turbulent, and fast [cf. Ref. 18].

The beds normally rest on distributor plates of various designs or rely on high velocity at the base of an expanding cone. Exceptions are the Ignifluid boiler [19] (Figure 3.11), which uses a modified chain grate, and the Szikla-Rozinek [20] combustor (Figure 3.12), which has a rotating, cylindrical slotted grate.

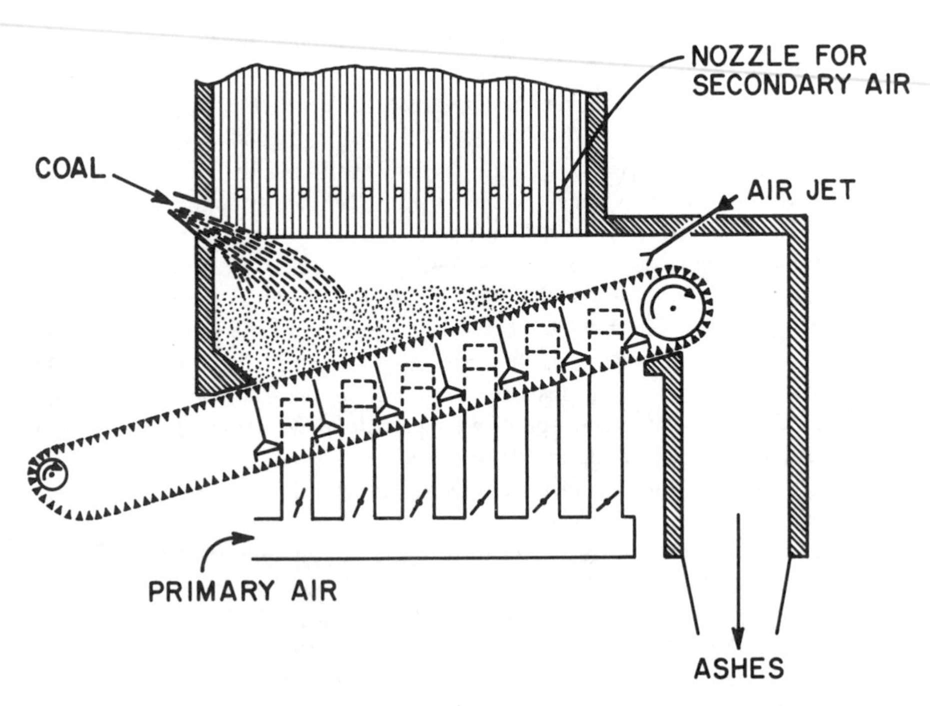

Figure 3.11. Cross-sectional view of lower part of Godel's Ignifluid boiler.

ISBN 0-201-08300-0

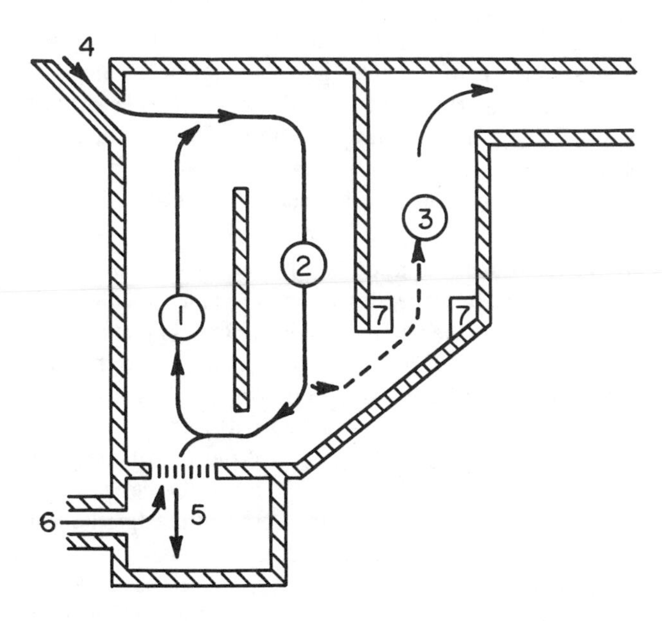

Figure 3.12. Schematic arrangement of Szikla-Rozinek combustor (fast fluid bed). *1* – Gasification chamber; *2* – devolatilization chamber; *3* – combustion chamber; *4* – raw coal feed; *5* – slag tap; *6* – primary air; *7* – secondary air. (Source: Ref. 20b.)

B. MODES OF OPERATION

The engineering aspects of fluid bed design reflect the same objectives that the other firing systems do and governed by the same constraints of heat transfer,

ISBN 0-201-08300-0

ash removal, fuel feed, and so on. The fluid bed potentially provides a range of solutions to these problems. In practice the focus has been on the extreme limiting cases, running the beds quite hot or quite cool. This is achieved by different methods of controlling the heat transfer, and it results in different methods of ash removal. Bed temperature is controlled by bed depth (see Section 4.3B), by the presence or absence of heat exchanger tubes in the bed, and by the use of ballast (i.e., inerts) in the bed.

(i) *Unballasted bed:* Consider a thin fluidized bed of crushed coal in refractory surroundings. The bed operates essentially adiabatically except for radiation loss from the top. If the bed is thin enough for complete combustion, the bed temperature will be close to the maximum adiabatic (about 2000°C or 3600°F). As bed depth increases, however, as in grate firing, endothermic gasification occurs in the upper regions, locally cooling the bed particles. The local cooling is rapidly transmitted to the levels below by the vertical motion of the particles. Therefore, by suitably adjusting the bed depth, either by initial design or during operation, temperatures can be adjusted; for example, even in the Ignifluid process they may be as low as 1000°C (about 1800°F). All coal ashes will slag somewhere in this temperature range (1000–2000°C), and by suitable choice of bed depth for a given coal the optimum temperature for sintering or slagging a given ash may be attained. Ash removal then relies on the phenomenon of agglomeration, by which molten ash or slag particles will wet each other (but they do not wet carbon) and consequently grow in size, or agglomerate. This phenomenon was apparently observed and made use of independently by Szikla and Rozinek [20a], by Thring and coworkers [21], and by Godel [19] for the Szikla-Rozinek combustor, the Downjet furnace, and the Ignifluid boiler, respectively. In the Ignifluid system (Figure 3.11) the slag particles grow in size until they are barely fluidizing; coming to rest on the grate, they are carried upward, out of the bed, to be projected into the ashpit.

The Szikla-Rozinek combustor (Figure 3.12) utilizes the same agglomerating phenomemon, this time in conjunction with a high velocity air supply, the so-called fast fluid bed. The reacting particles are carried out of the reaction chamber (*1,* Figure 3.12), separated from the carrier gas in the secondary chamber (*2*), and fed back into the primary chamber. Agglomerating slag particles above a certain size do not leave the primary chamber and, continuing to increase in size, finally fall onto a cylindrical grate of special design, where they are removed.

Both of the fluid beds so far described are essentially gasifiers, as are the lump beds on grates, and for heat transfer there is reliance on final combustion of the gasification products with secondary air, followed by passage through convective passes of conventional design developed for stokers.

(ii) *Ballasted bed:* In the low temperature fluid beds, by comparison, the focus is on making use of the high heat transfer coefficients obtainable in a fluid bed, which are in the range 50–100 Btu/ft^2 hr °F (250–500 W/m^2 K), compared with 1–5 Btu/ft^2 hr °F (5–25 W/m^2 K) for gas flowing over (boiler) tubes.

ISBN 0-201-08300-0

To take advantage of such high heat transfer coefficients, all combustion should preferably be in the bed, with the water and steam tubes also immersed in the bed. With pure coal, however, the bed depth that would yield combustion without gasification would be so small that the tubes would not be immersed. It would also be so hot, even with cooling from the tubes, that the ash would slag and block the bed. The practice was therefore adopted of using a "ballasted" bed, where the bulk of the bed being fluidized is inert material that provides the necessary heat transfer medium to the tubes. The addition of substantial quantities of inert bed material to a bed of coal so thin that there is total combustion thus has the effective result of greatly diluting the coal in the bed as it is actually operated. The combustible material in the bed is therefore very low, usually of the order of 1% of the total solids, and sometimes less.

The inert particles provide the heat transfer medium, and by adjustment of the bed depth in relation to the heat transfer surface on the immersed tubes, this is used to control the bed temperature (see Section 4.3B). To prevent slagging, bed temperatures must be below about 1800°F (about 1000°C) in the worst cases, although Goldberger is cited by Ehrlich et al. [3] as having reported that particle agglomeration starts as low as 1400°F. The most commonly used bed temperature is about 1600°F, and slagging has not, in fact, been reported as a problem. This temperature has been found to be quite adequate for sufficiently rapid burnout of the coal particles; at the same time it has aided in the control of thermal NO_x. This has not totally removed the NO_x problem, however, since there may still be some production from the fuel nitrogen.

C. DESIGNS

No commercial ballasted fluid bed boiler is operated, so it is not possible to describe commercial designs. Pilot plant and demonstration unit designs vary considerably in detail, but there would seem to be increasing consensus regarding schematic designs of the most likely commercial units. A typical schematic is illustrated in Figure 3.13. The elements of the design are a fluid bed of some inert or ballasted particles on a distributor plate through which all the combustion air passes. Coal is fed through a pipe into the bed, shown schematically in Figure 3.13 as a feed point to the top of the distributor plate. In practice, there are likely to be a number of feed points with design details differing from unit to unit. Cooling coils carrying the water to be heated are immersed in the bed.

Three different tube designs have been tested in the past: first, a design in which the walls of the fluid bed were the pipes; second, a design with vertical tubes in the bed; and, third, a design with horizontal tubes, as illustrated in Figure 3.13. It appears that the horizontal design is currently the most favored. This is the design adopted for the 30 MW(e) Rivesville demonstration unit presently under construction. This is a four-cell multimodule intended for scale-up in due course to 200 MW; four multimodules would be used for an 800 MW(e) power station [22].

ISBN 0-201-08300-0

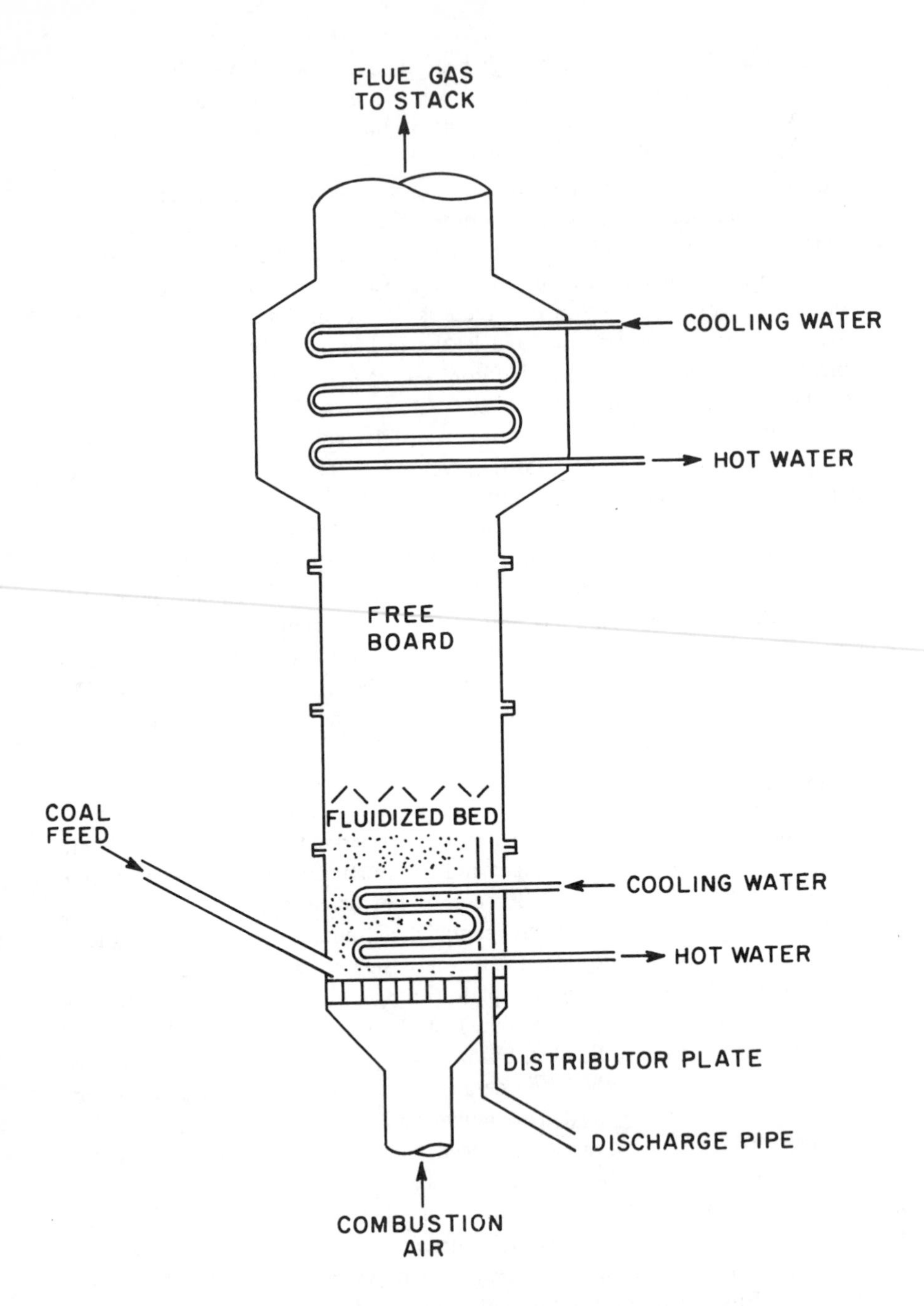

Figure 3.13. Schematic of typical fluid bed boiler designs.

ISBN 0-201-08300-0

As in Figure 3.13, the plant also incorporates additional convective tube banks serving as economizers in the freeboard above the bed. Also indicated schematically in Figure 3.13 is a discharge pipe, which may or may not be present, depending on the precise design, and which may or may not reach nearly to the top of the bed. The function of the pipe or of some alternative device is to control the bed depth since the coal contains ash and limestone supplied to adsorb sulfur gases (generally as SO_2) in the bed, and the bed will grow unless its depth is controlled. Bed material must be removed to maintain bed depth and to remove spent limestone for regeneration or elimination as waste. The preferred bed dimensions for the pilot and demonstration designs seem to be tending toward being thin rather than thick (ratio of depth to width, H/D). In the Rivesville plant, for example, the bed depth will be approximately 4 ft during operation, having expanded from about 2 ft when static. The cell is rectangular in cross section, 12 ft in width and 38 ft in length, that is, an H/D ratio of about 0.3 on the width down to about 0.1 on the length.

3.4 Entrained (Pulverized Coal) Firing Systems

A. GENERAL DESCRIPTION

In pulverized coal firing the particles must be fine enough to be entrained by the combustion air or some fraction of it. In practice, this means that the majority of particles must be less than 100 μm, and pipe diameters must be chosen so that conveying velocities exceed about 60 ft/sec (some sources indicate a minimum of 80 ft/sec for vertical conveying). It is also the usual, though not the invariable, practice to carry the coal in some fraction of the combustion air, known as the conveying or primary air. The rest of the combustion air is supplied separately, the simplest arrangement being to feed the primary air-with-coal and the secondary air simultaneously through two concentric pipes into the boiler or furnace. Most commonly, but again not invariably, the primary air-with-coal is supplied through the center pipe, with the secondary air in the annulus formed between the inner and outer tubes. This "straight shot" configuration was the earliest arrangement, adopted in commercial practice in the late 1800s. It produces a long flame that is particularly appropriate to certain types of industrial furnace, notably the cement kiln, as illustrated in Figure 3.1. It was also widely adopted for firing many other industrial furnaces, such as the reverberatory furnace illustrated in Figure 3.2.

This simple straight shot burner system is illustrated in Figure 3.14. Two standard methods have been used for preparing the fuel before firing. Figure 3.14(a) shows the "unit" system, by which the raw coal is fed to an air-swept mill with a classifier, and the coal swept out of the mill on the primary airstream is fed directly to the furnace. With separate secondary air fans this arrangement is conventionally used for large utility boilers today. Figure 3.14(b) illustrates the bin-

ISBN 0-201-08300-0

and-feeder system, in which the ground coal is removed from the air-swept mill, separated from the conveying air, and stored in a bin for feed to the furnace as required. This arrangement was common until equipment became sufficiently reliable for the unit system to be trusted.

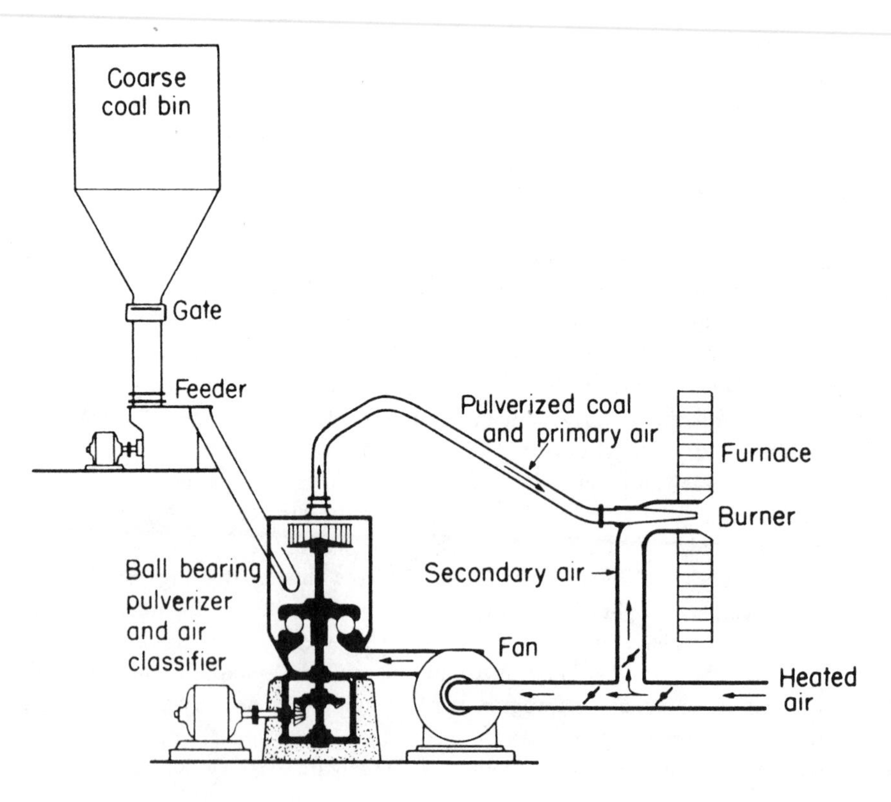

Figure 3.14(a). Unit system of pulverized coal firing.

ISBN 0-201-08300-0

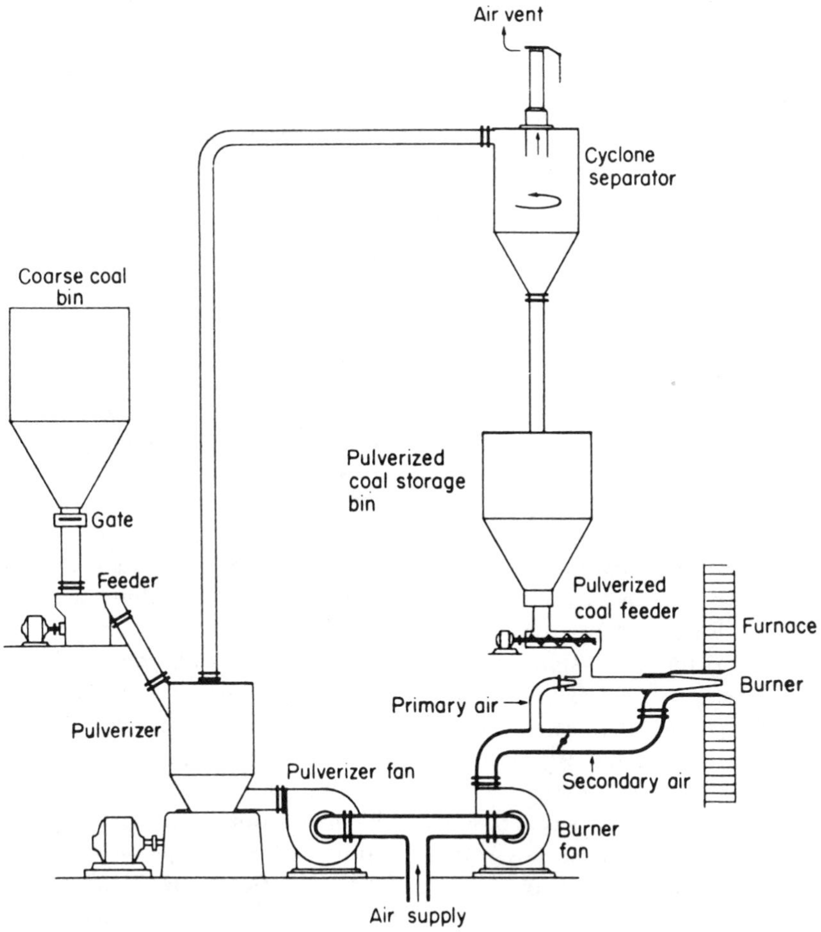

Figure 3.14(b). Storage system of pulverized coal firing.

ISBN 0-201-08300-0

B. SHELL AND FIRE TUBE BOILERS

For firing boilers the long, narrow flame of the straight shot burner is inappropriate. In the case of shell boilers the problem was simply one of scaling down; at the firing rates necessary to raise steam as required, the flame was still narrow but was too long, and there was continued combustion in the smoke tubes. Of the numerous attempts made over several decades to develop a suitable burner for shell boilers the most successful would appear to be the grid and multijet burners. The multijet burner, illustrated in Figure 3.15, can be thought of, in essence, as a matrix array of the original but scaled-down straight shot burners, so that the flame width is increased without significantly increasing the flame length. The multijet burner is similar to the grid burner with the difference that the horizontal burner nozzles of the grid are joined as one. A particularly interesting aspect of this burner design is the careful attention to minimizing turbulence in the streams leaving the burner block since turbulence was not considered by the designers to be important. What they were aiming at was to break up the primary air-with-coal stream into a number of individual streams and to interweave these with a number of secondary airstreams [23].

C. U-FLAME BOILERS

Another solution to the problem of long flame length with a narrow flame was to construct a boiler in which the fuel was fired vertically down from the roof, offset to one side. As the flame developed, it was lifted by buoyancy, forming a U shape, with the uprising tail of the U terminating just under the exhaust stack. From the flame shape this boiler design was given the name "U-flame" boiler. Two such boilers back to back with the dividing wall removed provided a means of increasing boiler capacity and formed the W-flame boiler. All fuels (bituminous and anthracite) were fired in those early boilers. In the course of time, however, they were reserved for low volatile fuels, particularly anthracite. With low volatile fuels it was the common belief for several decades that the primary stream had to be rich, using perhaps no more than 20% of the total combustion air; the balance of the air was injected through the front wall. It is now known, however, that this requirement is unnecessary, and existing anthracite boilers have been successfully fired with primary air as high as 80% of stoichiometric, with improved combustion, lower carbon in the ash, and less wear on the front wall. Few of the U-flame boilers are left. They may be necessary, however, for the combustion of low volatile, low reactivity chars from gasification and liquefaction plants where steam raising from a fraction of the char is required. The best alternative could be a fluid bed boiler.

ISBN 0-201-08300-0

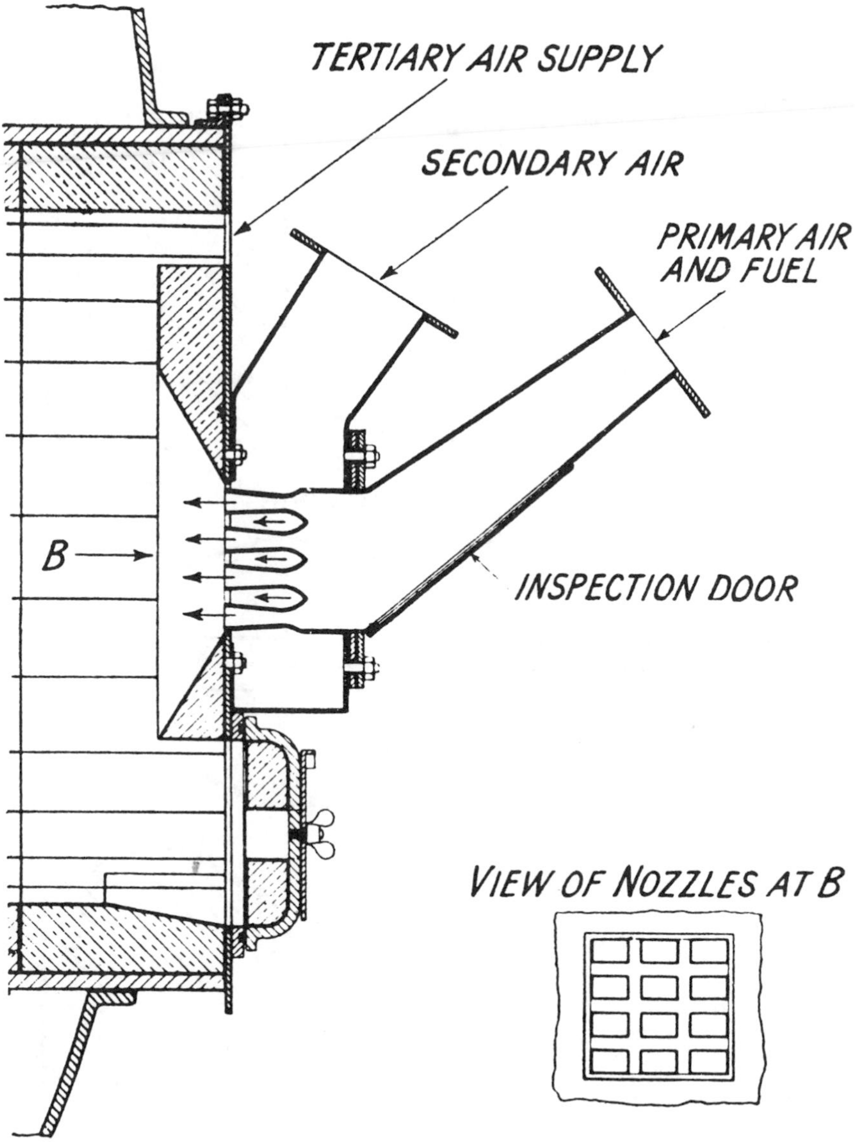

TERTIARY AIR SUPPLY
SECONDARY AIR
PRIMARY AIR AND FUEL
B
INSPECTION DOOR
VIEW OF NOZZLES AT B

Figure 3.15. Multijet burner for shell boilers.

ISBN 0-201-08300-0

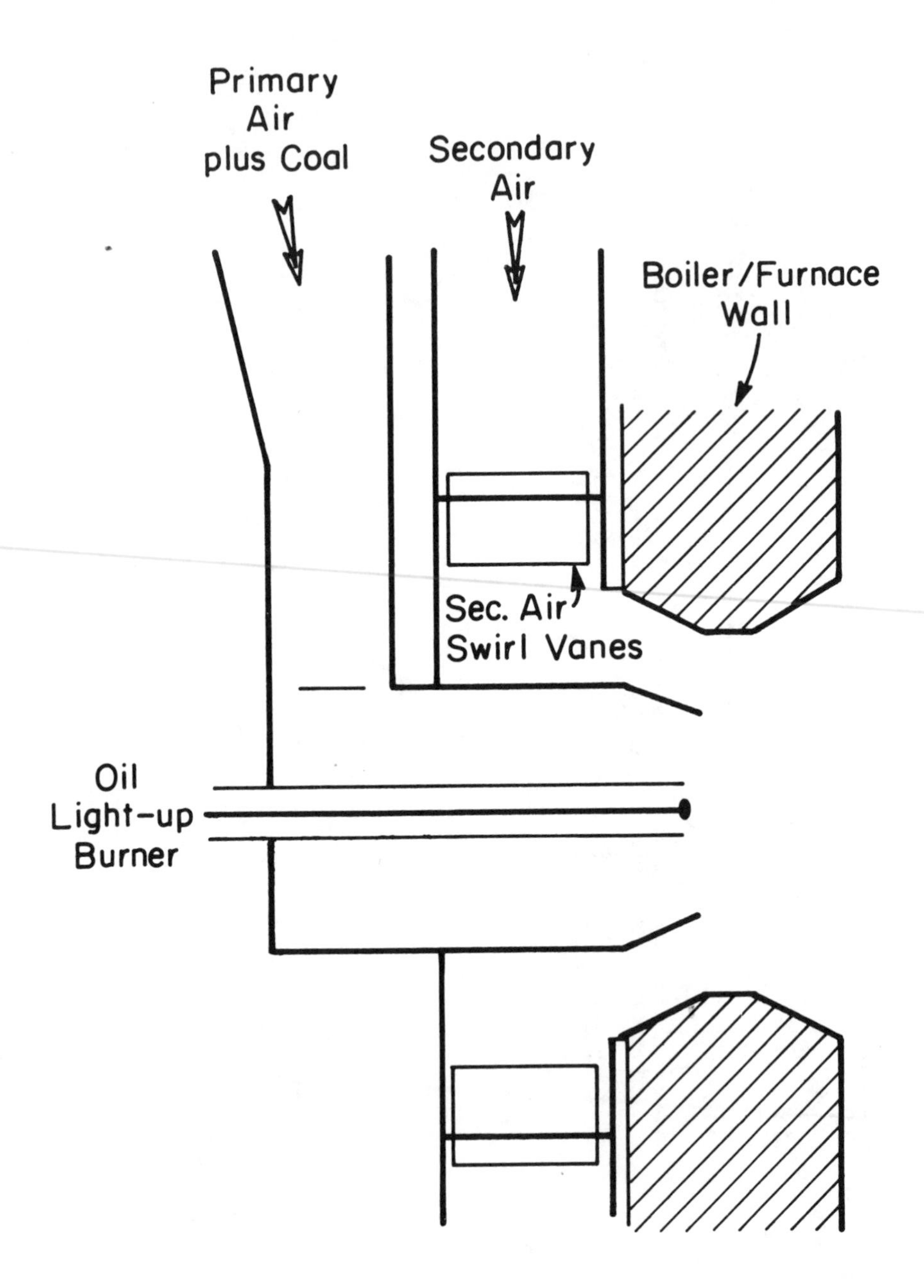

Figure 3.16(a). Flare or "self-turbulating" burner design.

ISBN 0-201-08300-0

D. FLARE BURNERS

An alternative development that might also be regarded as a variant of the original straight shot burner is the so-called flare or "self-turbulating" burner, illustrated schematically in Figure 3.16(a). This is a swirl burner producing a more "bushy" flame. There are many detailed variations, but the essence is, again, a basic design of two concentric tubes with the primary air-with-coal supplied through the central tube and the secondary air through the surrounding annulus, as with the straight shot burner, but with the secondary air directed through an array of swirl vanes so that the stream emerging from the burner array is rotating. The rotation opens out the jet angle and tends to produce a shorter, fatter flame. As a result of the rotation and in conjunction with proper design of the burner aperture in the wall, the flame generated stabilizes in a zone or "ball" just clear of the burner. Stabilization is identified with an aerodynamic backmix flow caused by the swirl, by which gases are brought back from the region of largely complete combustion toward the burner to supply up to 80 or 90% of the heat required for ignition (Figure 3.16(b)). The flame ball can be regarded as a backmix or stirred reactor region, and conditions for stability are discussed in Section 6.3. In a large boiler such as is illustrated in Figure 3.5, there is a matrix array of 8, 16, or more such burners on the front wall of the boiler, and some very large furnaces have an equal number on the back wall. Each can be thought of as having its own stable flame ball in which something like two thirds of the combustion is achieved, with the remaining one third of the combustion taking place in the combining gas streams as they flow to the center of the boiler and up to the top. The self-stabilizing aspects of this burner were attributed at one time to the turbulence caused by the rotational flow, and this was the origin of the description of the burner as "self-turbulating."

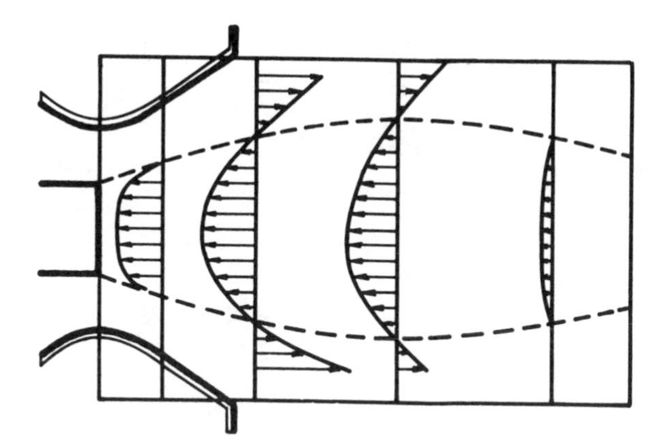

Figure 3.16(b). Characteristic backmix flow pattern with swirled flow.

ISBN 0-201-08300-0

E. TANGENTIAL TILTING BURNERS

An alternative approach to large boilers is to regard the complete boiler as the "burner." In this case, jet streams are injected into the boiler to mix in the center and to produce a single stabilizing "flame ball" in the lower third of the boiler. The practical application is the so-called tangential corner-fired boiler system illustrated in Figure 3.17. At each corner there is a compound burner, with three to five compound burners at different firing levels, all aimed tangentially at an imaginary circle in the center of the boiler. Each compound burner can be thought of as a stack of single-shot burners with rectangular primary pipes and a common secondary air shroud. The primary pipes also have tips that can be moved up and down for between 10 and 30 deg maximum. This moves the stabilizing flame ball in the center of the boiler up or down and provides a means of assisting superheat control, either as a boiler progressively slags up or on low loads.

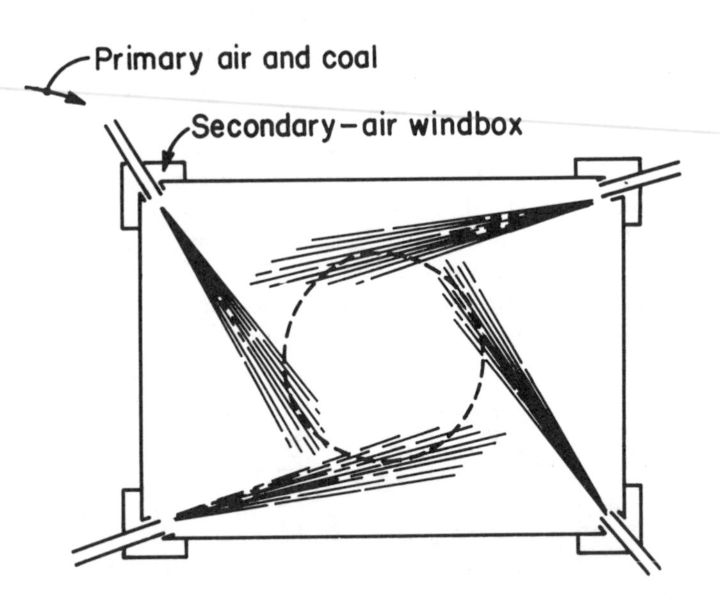

Figure 3.17. Schematic of cross section of corner-fired boiler with tangential injection. Burner tips are also tilting with numerous burner levels (see text).

F. CYCLONE

A burner of entirely different design, intended only for large utility boilers, is the cyclone burner illustrated in Figure 3.18. The figure shows the burner firing into the lower half of an otherwise reasonably conventional radiant boiler. In this

ISBN 0-201-08300-0

system the coal is only crushed, not ground, at least in the original design. The crushed coal and air are supplied to the primary burner pipe, leading into the cyclone by a tangential fitting. The secondary air is likewise introduced tangentially with rotation in the same sense. The fuel-air mixture thus produces rotation of the gases inside the cyclone, and the theoretical concept is that the crushed coal

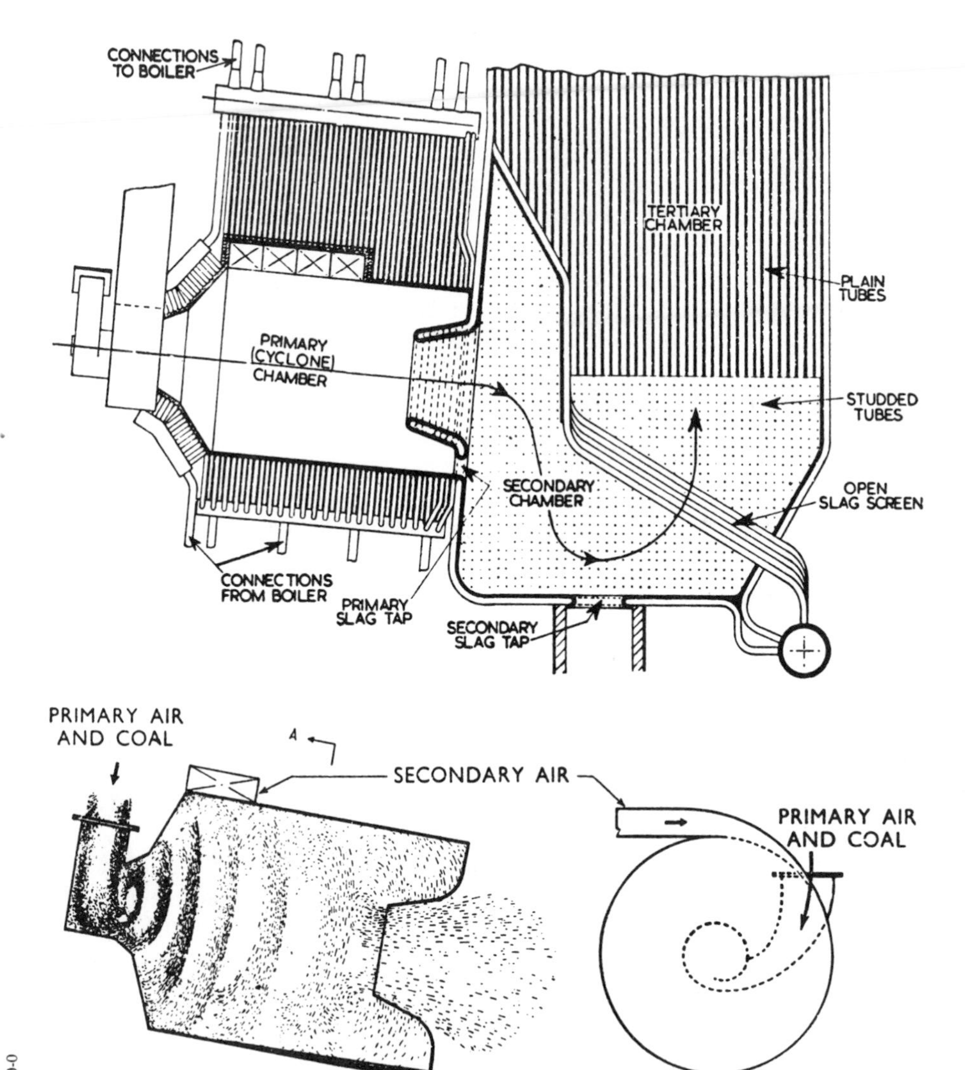

Figure 3.18. Cyclone combustor.

ISBN 0-201-08300-0

particles are thrown by centrifugal force to the walls of the chamber. Combustion is so intense that flame temperatures are high and ash in the coal slags. The slag particles are likewise thrown to the walls of the chamber, where they freeze on contact with bare tubes cooled with water. A slag layer builds up, providing a natural refractory covering for the water tubes; and, as the thickness increases, the surface temperature rises until it reaches liquid temperature and the slag starts to flow. The cyclone chamber is angled slightly downward to allow the surplus slag to flow out through a slag tap for removal. When the coal particles are thrown to the walls, they stick in the molten slag and, being steadily swept by the flame gases, are burned out in due course.

In an alternative cyclone design pulverized coal is used in place of crushed coal and is injected tangentially into the cyclone near the downstream end. One objective of the original axial firing design, using crushed coal, was to eliminate the energy penalty of the fine grinders. With crushed coal the residence times had to be increased substantially, and the idea of embedding in the slag layer at the walls was a means of achieving this. It would appear that the success of the system is due, in part at least, to the phenomenon described above, namely, that slag does not wet carbon. Otherwise we could anticipate that the particles would be engulfed by the slag and carried out of the chamber to give an unacceptably high combustible loss in the ash.

4. MAGNITUDES: HEAT RATES AND FIRING DENSITIES

4.1. Comparative Heat Rates

The heat rate of a furnace is the rate of heat supply: the product of the firing rate, J_f, and the heat of combustion of the fuel, h_f. Table 4.1 shows a range of about 5 orders of magnitude in the heat rates for different systems. The rates are given in three different units: Btu/hr, MW (thermal, t), and boiler HP. The last, an old method of specifying boiler sizes, is still used by some manufacturers, though never for the largest units (10 MW and above). One boiler HP $\simeq$ 10 kW(t) and requires 10 ft^2/BHP, or about 1 ft^2/kW(t) [approx. 0.1 m^2/kW(t)].

Heat rates for individual units range from about 10^5 to 10^{10} Btu/hr (approx. 0.03 to 3000 MW). The smallest single industrial gas or oil burners are about the same size as domestic furnace (boiler) burners, rated at about 100,000 Btu/hr (0.03 MW). The use of coal at this rating is essentially confined to occasional domestic use, and the units are almost invariably hand stoked.

Boiler sizes cover the entire five-decade range. Utility power station boilers, however, are confined for all practical purposes to the size range 10^8 Btu/hr [30 MW(t)] and above. In contrast, most industrial furnaces and industrial (process steam) boilers are in the size range below 10^8 Btu/hr. Boilers larger than this (10^8 Btu/hr) are to be found mostly in paper mills and chemical plants, or in power stations of large integrated plants generating electricity for the plants. Process furnaces, however, rarely exceed 10^8 Btu/hr heat input.

ISBN 0-201-08300-0

Table 4.1

Comparative Heat Rates and Sizes of Furnaces and Boilers

Heat Rate					Fuel Rate, J_f [c]			POC/Air Rate,[d]
Btu/hr	MW(t)[a]	Boiler[b] HP	Application	Coal Firing Method	Coal, lb/hr	Oil, gal/hr	Gas, ft^3/hr	lb/hr
10^5	0.03	3	Domestic furnace (boiler), small industrial furnace	Hand stoked	8	2/3	10^2	10^2
10^6	0.3	30	Commercial/small industrial boiler Auto: 60 mph at 10 mpg		80	6.5	10^3	10^3
10^7	3	300	Industrial furnaces and boilers[e]	Mechanical stokers (boilers)[e] Fluid beds[e]	800	65	10^4	10^4
					tons/hr			tons/hr
10^8	30	–	10 MW(e) power station		4	650	10^5	50
10^9	300	–	100 MW(e) power station	Pulverized coal[e]	40	6,500	10^6	500
10^{10}	3000	–	1000 MW(e) power station		400	65,000	10^7	5000

[a] SI units.

[b] One boiler HP is an evaporation rate of 3350 Btu/ft^2 hr (34.5 lb water "from and at" 212°F) over a heat transfer surface area of 10 ft^2; 1 BHP ≅ 10 kW(t). Heat exchange surface is about 1 kW(t)/ft^2.

[c] For these approximate values heats of combustion adopted were as follows: coal, 12,000 Btu/lb; oil, 150,000 Btu/gal; gas, 1000 Btu/ft^3. Values of actual fuel samples, particularly coal, are variable.

[d] The products of combustion (POC) and combustion air masses differ by about 10% or less.

[e] Bracket indicates range of application. Dotted lines represent: possible range but not commonly used in pulverized coal firing; and projected range for fluid beds.

For firing this range of furnaces, pulverized coal (p.c.) can be used, and has been in the past, down to commercial boilers and small industrial furnaces of about 10^6 Btu/hr [0.3 MW(t)] using such burners as the grid and multijet (Figure 3.15) in shell boilers. Today, as mentioned earlier, p.c. firing is used mainly in large utility boilers (Figure 3.5) and for selected industrial applications such as rotary cement and lime kilns (Figure 3.1).

In contrast to p.c. firing, mechanical stokers have an upper size limit of about 10^8 Btu/hr [30 MW(t)], and applications are confined exclusively, for all practical purposes, to industrial boilers and to small power stations (including on-site industrial or commercial electricity production). The upper size limit is essentially a mechanical constraint imposed by the size of the grate.

The first applications of ballasted fluid bed firing are being aimed at utility boilers, with proposals for single modules (as described above) at the 200 MW size, and integrated systems at the 800 MW size (Section 3.3C). In principle, the system is an alternative to both p.c. and stoker firing at all sizes. Whether it will prove to be competitive remains to be seen. A basis for being so could be the inherent prospects

ISBN 0-201-08300-0

for SO_2 control by direct absorption in the bed, as outlined in Section 3.3 (see also Section 8).

The unballasted Ignifluid and fast recycle Szikla-Rozinek units are at about the 10 MW size. Possible objections to further development of either of these could be lack of SO_2 control and high temperatures generating high NO_x, although the high temperatures occur in the bed where conditions are fuel rich which itself tends to reduce NO_x formation.

Table 4.1 also compares the firing rates of different fuels, given in their customary units. The final column in the table gives the approximate mass flow rate of combustion air, or the products of combustion (POC). It should be noted that the greatest weight of material moving through a furnace is very often the nitrogen in the combustion air; in a boiler the steam rate and the POC or air rate are approximately the same. For a given heat rate the combustion air required is essentially independent of the (fossil) fuel used. A common rule of thumb is as follows: 1 ft^3 cold air/100 Btu fired. On a mass basis this is about 1 lb air/1000 Btu fired, so the heat in Btu/hr and the air rate in lb/hr differ by about 10^3. (More accurate estimates yield values of 1200 and should be checked for individual fuels when accuracy is required.) The value of this result is that it is possible to monitor the firing rate by setting the air rate and adjusting the fuel rate to yield a predetermined level of O_2 in the stack gas. The O_2 level determines the excess air and is very insensitive to fuel type (hydrogen/carbon ratio of the fuel). This is particularly valuable when firing a fuel such as coal that is difficult to meter, or when combining fuels, such as gas over coal, or coal and oil.

Part of the reason for the statistical insensitivity of air requirements to fuel type derives from the (mostly) high values of the air/fuel ratio on a mass basis. For coal this is 10–12:1; for oil, 16–18:1; and for (natural) gas, about 20:1. Since the POC weight is the weight of air plus fuel, the POC weight exceeds the combustion air weight by about 10% (in the case of coal), or less for other fuels and/or for increasing excess air.

These factors all bear on the prospects for retrofit for coal. Retrofit will probably require a redistribution of the air supply, but it should not call for an increased air rate beyond that required for a possible increase in excess air. There should be no (significant) derating of any furnace or boiler system on retrofit since the mass flowthrough and (adiabatic) flame temperatures are largely independent of fuel type until the poorer, low Btu gases are reached (less than 5000 Btu/lb calorific value, or about 10 MJ/kg).

ISBN 0-201-08300-0

4.2. Firing Densities

Firing densities are expressed in two ways: as the volumetric combustion intensity, I_v, or as the area firing intensity, I_A. Both can be expressed either on a Btu basis or on a mass (gallons or cubic feet) basis. The Btu basis is common for the volumetric intensity; both mass and Btu bases are used for the area density. The parameters represent average-practice values that are targets for conventional design and can be exceeded only with careful attention to special design.

If the fuel rate is J_f (fuel units per unit time) of fuel of calorific value h_f (energy units per fuel unit) into a combustion chamber of volume V_c at pressure P (atm), the volumetric combustion intensity is defined as

$$I_v = \frac{J_f h_f}{V_c P} \tag{4.1}$$

In the principal systems of concern here, P is nearly always unity (the pressurized fluid bed is the exception). If the combustion chamber has some reference cross-sectional area A_c, the area firing density is

$$I_A = \frac{J_f h_f}{A_c} \tag{4.2}$$

The reference area, A_c, would be the cross section of a fluid bed distributor plate or the grate area of a mechanical stoker. In p.c. firing the term is not commonly used, but it could refer to the principal boiler cross section.

The mass firing density is I/h_f, and it also corresponds in the volumetric case to the mean volumetric reaction rate, $\overline{R}_v$.

If the overall combustion volume has an effective height H_c for an area A_c, then

$$H_c = \frac{I_A}{I_v} \tag{4.3}$$

In p.c. (and oil and gas) firing, where the fuel travels with the combustion air, the transit time through the chamber, t_r, can be related to the air (and fuel) supply

ISBN 0-201-08300-0

rates, as well as to the volumetric combustion intensity, by the Rosin equation [24]. If the stoichiometric air/fuel ratio on a mass basis is G_s, and the equivalence ratio is ϕ [= 1/(1 + $E\%$/100), where $E\%$ is the overall excess air], then [25]

$$I_v = \frac{K(T_0/\overline{T}_f)}{t_r} \tag{4.4}$$

where $\overline{T}_f$ is an "average" flame temperature, T_0 is ambient temperature, and K is a constant given by

$$K = \frac{\rho_0 h_f}{1 + G_s/\phi} \tag{4.5}$$

where ρ_0 is the POC density at standard temperature and pressure.

Invoking the FPS unit approximations summarized in Section 4.1, that G_s/ϕ is generally 10 or greater and that $h_f/100 \simeq G_s/\rho_0$ (representing 1 ft^3 cold air or $1/\rho_0$ lb cold air/100 Btu released), we have $K \simeq 100\phi$. Since the adiabatic flame temperatures of most carbonaceous fuels are about 3600°F (roughly 2000°C), representing an upper limit, and actual flame temperatures lie mostly in the range 2200–3200°F (roughly 1200–1800°C), the ratio $\overline{T}_f/T_0$ can be taken as approximately 6.5 ± 1.5, with the exception of the ballasted fluid bed, where $\overline{T}_f$ is about 1600°F (almost 900°C) and the ratio is about 4. Excess air values range from close to zero as a sometime target for fluid bed operation up to 30–50% for grate firing, so $(1/\phi)$ can be taken at 1.2 ± 0.2 as covering most of the range. This gives us (multiplying by 3600 to convert hours to seconds) approximately, in rounded-off values,

$$I_v \simeq \frac{(50 \pm 15) \times 10^3}{t_r} \text{ Btu/ft}^3 \text{ hr} \tag{4.6a}$$

$$\simeq \frac{500 \pm 150}{t_r} \text{ kW/m}^3 \tag{4.6b}$$

for t_r in *seconds.*

For p.c. boilers a residence time of about 2 sec yields a combustion intensity of about 25,000 Btu/ft^3 hr (250 kW/m^3), or a mass firing density of about 2 lb/hr ft^3 (approx. 30 kg/hr m^3). These are values typically found in practice. Actual combustion times are usually somewhat less, being closer to about 1 sec, with the flame filling one half to two thirds of the combustion volumes of larger boilers. The additional space is required for heat transfer surface rather than combustion.

Equation 4.6 is a general equation not limited to coal, as discussed elsewhere [25]. It brings out most clearly that, when combustion time, t_c, is about equal to

ISBN 0-201-08300-0

residence time, t_r, volumetric combustion intensity depends primarily on combustion time. In the range of fuel systems from pulverized coal to detonating gases, combustion times range over 5 orders of magnitude, from 1 sec for the first to about 10^{-2} *msec* for the last, with oil drops between 1 and 100 msec. Research reactors fired with gas have volumetric combustion intensities in the region of 10^8–10^9 Btu/ft^3 hr (atm) (10^3–10^4 MW/m^3); industrial gas burners have volumetric intensities, defined on *flame* volume, of about 10^7 Btu/ft^{3}hr (10^2 MW/m^3).

In comparison, intensities in many gas- or oil-fired industrial furnaces are about 10^4 Btu/ft^3 hr (100 kW/m^3), which is of the order of the boiler combustion intensity for p.c. firing. Comparing the values for oil and gas in slightly greater detail, the discrepancy between the intensity values defined on flame volume and the values defined on furnace volume leads to the conclusion that oil flames typically occupy 10% or less of the furnace volume, and that for premixed gas flames the occupancy is 1% or less. Even in a cross-fired glass tank, with turbulent diffusion gas flames adjusted to a combustion time of about 1 sec, the flame gases still occupy only one fifth to one tenth of the freeboard volume (with mean residence times, based on the total freeboard volume, of 5–10 sec). The implications of these figures is that, in retrofit of industrial furnaces by p.c. firing, combustion space to accommodate the flame should not normally be a particular problem. It could be a serious problem, however, in retrofitting boilers that were originally designed for oil or gas so as to take advantage of the higher flame-volume combustion intensities of these fuels. In such cases there will generally be insufficient space unless the boiler is significantly derated. (Even if there is space in the combustion chamber, the tube spacings may be too close if the boiler was designed for oil or gas, and the tube banks will foul.)

In applying the volumetric combustion intensity analysis to grate firing, the total combustion volume includes the space between the bed and the convective tube sections. This space was insufficient in the early boilers, and to eliminate tube burnouts the space was steadily increased until the overall combustion intensity dropped to about 75,000 Btu/ft^3 hr (750 kW/m^3) as an upper limit. Values for existing installations lie between 25,000 and 75,000. Since combustion is thought to be mainly carbon monoxide with some gaseous volatiles, these combustion intensity figures seem to be on the low side. The problem may be one of poor overbed mixing rather than of combustion, and in spite of past efforts further attention to improved overbed mixing might be worthwhile.

In the fluid bed the volumetric combustion intensity equations would apply only to the POC volume created for a given feed and combustion rate. The values, which have little meaning, are in the range 25,000–50,000 Btu/ft^3 hr (250–500 kW/m^3). More to the point are the values based on the bed volume, but no simple expression is available to predict these. Experimental values are in the region of 200,000 to 10^6 Btu/ft^3 hr (2 to 10 MW/m^3). Similar values are obtained in the cyclone combustor (Figure 3.18) and in the solid bed of a grate for essentially the same reason: the particles are trapped in a small volume as the combustion gases are swept through. Although such high values have relatively little meaning scientif-

ISBN 0-201-08300-0

ically, they are highly significant for design as they determine the physical scale of the equipment. The ratio of intensities between the fluid bed and the p.c. boiler is about 10, implying that the fluid bed combustion volume would be about 10% of a p.c. boiler for the same duty. Size reductions should also reduce costs.

Area firing densities, I_A, are the more common means of specifying grate heat release rates. Depending on coal type, grate design, and method of operation, area densities range up to 6 x 10^5 Btu/ft^2 hr (nearly 2 MW/m^2), corresponding to 50 lb/ft^2 hr (about 245 kg/m^2 hr). In normal pressure shaft gasifiers, which may be as much as 10–12 ft (3–3.5 m) deep, gasification rates are similar. In locomotives with forced draft firing, densities of 100 lb/ft^2 hr (nearly 500 kg/m^2 hr) were achieved, so there seems to be no fundamental reason for a 50 lb/ft^2 hr limit.

Area firing densities in the fluid bed are projected to be comparable, with expected ranges of 10–100 lb/ft^2 hr (50–500 kg/m^2 hr). Combustion intensities are higher, reflecting the smaller volume per square foot on account of faster reaction from smaller particles.

Area densities for p.c. firing, based on boiler cross section, are double the fluid bed or greater, but for this system the figure has little physical meaning.

4.3. Elementary Design Equations and Constraints

Table 4.2 summarizes some of the foregoing data. Many can also be summarized in simple equations that provide an initial basis for design. In all cases both fuel and air are metered into the combustion chamber at (nominally) controlled rates. If combustion is stable and reasonably complete, and the system is steady state, the airflow rate, J_a, is directly proportional to the fuel flow rate, J_f, by the equation defining the equivalence ratio, ϕ:

$$J_a = \left(\frac{G_s}{\phi}\right) J_f \tag{4.7}$$

where G_s is the previously defined air/fuel stoichiometric ratio. The equation applies either to the combustor as a whole or to unit cross-sectional values of flow rate. In particular, eliminating J_f by Equation 4.2 and introducing the POC density ρ, we obtain for a mean velocity, v, of combustion products in the combustion chamber:

$$v = \left(\frac{G_s}{\rho_0 h_f}\right)\left(\frac{\overline{T}_f}{T_0}\right)\left(\frac{1}{\phi}\right) I_A \tag{4.8a}$$

$$\simeq \left(\frac{1}{100\phi}\right)\left(\frac{\overline{T}_f}{T_0}\right) I_A \text{ ft/hr} \tag{4.8b}$$

ISBN 0-201-08300-0

Table 4.2

Comparative Combustion Intensities and Related Quantities

Note on units. SI conversions from FPS values have been rounded off.

Parameter	Grate	Fluid Bed	Pulverized Coal
Heat rate		(Projected)	
Btu/hr	$(10^5)10^6-10^8$	Up to 10^8	$(10^6-)10^8-10^{10}$
MW(t)	(0.03)0.3–30	Up to 30	(0.3)30–3000
Volumetric combustion intensity, I_v, Btu/ft³ hr	25,000–75,000 (based on bed and freeboard volume)	(i) Up to 200,000 (based on bed volume) (ii) Up to 50,000 (based on bed and freeboard volume)	15,000–25,000
kW/m³	250–750	(i) Up to 2000 (ii) Up to 500	150–250
Area combustion intensity, I_A			
Btu/ft² hr	$10^5-6 \times 10^5$	Up to 10^6	(Up to 2.5×10^6)
kW/m²	300–1800	Up to 3	(Up to 7.5)
Effective reactor height, $H_c = I_A/I_v$			
ft	5–25	(i) Up to 5 (ii) Up to 20	Up to 150
m	1.5–7.5	(i) Up to 1.5 (ii) Up to 6	45
Coal firing density, $J_{f,V}$			
lb/ft³ hr	2–6	(∼ 15)	1–2
kg/m³ hr	30–100	(∼ 250)	15–30

ISBN 0-201-08300-0

Table 4.2 (Continued)

Parameter	Grate	Fluid Bed	Pulverized Coal
Area firing density, $J_{f,A}$			
lb/ft^2 hr	8–50	Up to 100	(Up to 200)
kg/m^2 hr	40–250	Up to 500	(Up to 1000)
POC velocity (hot)			
ft/sec	Up to 10	Up to 12 (15)	Up to 60 (70)
m/sec	Up to 3	Up to 3.5 (4.5)	Up to 20
POC (air) velocity (cold)			
ft/sec	Up to 1.5	Up to 3 (4)	Up to 10
m/sec	Up to 0.5	Up to 1	Up to 3
Combustion time, sec	Up to 5000	100–500	~ 1
Particle heating rate, $^{\circ}C$/sec	<1	10^3–10^4	10^4
Boilers only			
Heat transfer coefficients, h			
Btu/ft^2 hr $^{\circ}F$	2	50–100	10–100
W/m^2 K	10	250–500	50–500 (variable over surface)
Heat transfer fluxes to heat exchange surfaces			
Btu/ft^2 hr	>3500	35,000	$(15–150) \times 10^3$
kW/m^2	>10	100	50–500 (variable over surface)
Heat exchange surface area per unit cross-sectional area of combustion chamber, Φ	~ 200	~ 30	25–250

ISBN 0-201-08300-0

Table 4.2 (Continued)

Conversion Factors (Reference – *Metric Practice Guide:* ASTM 380-72[ε]; ANS Z 210.1-1973)

1 Btu	= 1054 J = W–sec
1 Btu/hr	= 0.293 W
1 Btu/ft^2 hr	= 3.152 W/m^2
1 Btu/ft^3 hr (atm)	= 10.35 W/m^3
1 lb/ft^2 hr	= 4.89 kg/m^2 hr
1 lb/ft^3 hr	= 16 kg/m^3 hr
1 ft	= 0.305 m
1 ft/sec	= 0.305 m/sec
1 Btu/ft^2 hr ^oF	= 5.674 W/m^2 K
1 Btu/ft^2 hr	= 3.152 W/m^2
1 kcal	= 4.18 kJ

By adopting suitable values for the temperature ratio (4 for fluid beds and 6.5 for grate and p.c. firing) and taking $1/\phi = 1.2$ (20% excess air), hot and cold velocities corresponding to the previously cited area combustion intensities have been calculated and are given, with other data, in Table 4.2. The POC velocities for the grate and the fluid bed are very similar, reflecting the comparable area firing densities at the upper ranges. Velocities in the p.c. boiler are a 0.5–1 order of magnitude higher. At 60 or 70 ft/sec (approx. 20 m/sec) the transit time in a 150 ft (45 m) boiler is about 2 sec. Combustion times are generally less (see table). In the fluid bed and the grate, POC residence times in the beds are mostly in the range of 0.1–1 sec in a 4 ft (1–1/4 m) deep fluid bed, or a 1 ft (1/3 m) deep solid bed on a grate, but depending also on the fraction of air fired as underfire air in the latter case. Combustion times, however, are very different, ranging from 100 to 500 sec in the fluid bed up to possibly 5000 sec on the grate.

Heat transfer comparisons show a different ordering (Table 4.2, last three entries). The boiler HP rating yields the lowest mean heat transfer flux in the region of 3500 (strictly 3350) Btu/ft^2 hr (11 kW/m^2) of heat transfer surface, yielding a heat transfer coefficient of about 2 Btu/ft^2 hr ^oF (approx. 10 W/m^2 K). This is appropriate for shell and for some water tube packaged boilers. For large utility boilers, flux values range [26] from 15,000 (nearly 50 kW/m^2) to 150,000 Btu/

ISBN 0-201-08300-0

ft^2 hr (nearly 500 kW/m^2), with these higher values due to radiation. Black-body fluxes are about 20,000 (63 kW/m^2) at 1400°F, 100,000 (315 kW/m^2) at 2300°F, and 150,000 Btu/ft^2 hr (nearly 500 kW/m^2) at 2600°F; corresponding heat transfer coefficients are approximately 15, 40, and 60 Btu/ft^2 hr °F (85, 230, and 340 W/m^2 K), respectively, the strong increase with temperature following the usual radiation T^3 law. The high peak flux values present design problems in heat dissipation. By contrast the fluid bed has heat transfer coefficients of 50–100 Btu/ft^2 hr °F (approx. 280–560 W/m^2 K), yielding fluxes to the heat exchange surface of about 35,000 Btu/ft^2 hr (over 100 kW/m^2). The heat transfer coefficients are thus comparable to the higher radiation values, the fluxes are an order of magnitude larger than those for grate-fired boilers, and the uniform values at all points avoid the hot-spot dissipation problem.

A final comparative parameter in Table 4.2 is the ratio of the heat transfer surface area to the combustion chamber cross section. The low value for the fluid bed reflects again the potential for size reduction as compared with other firing methods.

As the numbers tabulated indicate, certain system constraints exist quite independently of the kinetics of the fuel combustion. Practice also seems to have identified some limits that may not, in fact, be real limitations. Some of these limitations are briefly outlined below.

A. GRATE FIRING

The first limit on grate firing is that of scale. A practical engineering limit seems to be reached when the length and width of the grate are about 30 ft (9 m), with grate areas about 1000 ft^2 (80 m^2). At 600,000 Btu/ft^2 hr (nearly 2 MW/m^2), the steam capacity at 85% efficiency would be about 5 x 10^8 Btu/hr (nearly 150 MW; 500,000 lb steam/hr, or 227,000 kg/hr). In practice stokers have rarely exceeded 300,000 lb steam/hr (135,000 kg/hr). The limitation is therefore partly grate area and partly firing density. Equation 4.7 indicates that the area firing density, $J_{f,A}$, is determined by the underfire air rate, which is normally unlimited. A limitation may exist, however, where there is a reaction plane moving into a bed of fresh fuel; if the rate of movement of the reaction plane cannot match the (opposed) rate of fuel flow, a classical flame "blow-off" will occur and the bed will go out. In simplest terms, if the rate of advance of the reaction front is R ft/sec, the fuel flow rate perpendicular to the reaction front, per square foot of front, is $R \cdot \rho_f = J_{f,A}$, where ρ_f is the fuel bulk density, if the reaction front is stationary and perpendicular to the direction of feed. In a deep fuel bed where there is gasification in a gasifier or on a grate, the overall system is essentially fuel-rich and we have

$$R = \frac{J_{f,A}}{\rho_f} = \frac{J_{a,A}\,\phi}{G_s \rho_f} = \frac{v_a^0 \rho_a^0 \phi}{G_s \rho_f} \qquad (4.9)$$

where v_a^0 and ρ_a^0 are the velocity and density of the cold underfire air. The equiva-

ISBN 0-201-08300-0

lence ratio, ϕ, is given by $\phi = 2(\text{RCS})$, where RCS is the Thring Relative Carbon Saturation factor [27, 14], which must lie between 0.5 (yielding all CO_2) and 1.0 (yielding all CO), with 0.75 as an average value. Taking $\rho_f = 40$ lb/ft^3 (bulk density), $G_s = 12$, and $\rho_a^0 = 0.08$ lb/ft^3, we find that Equation 4.9 yields

$$R = 2.5 \times 10^{-4} v_a^0 \text{ ft/sec} \quad (\text{for } v_a^0 \text{ in ft/sec}) \tag{4.10a}$$

$$= 7.6 \times 10^{-5} v_a^0 \text{ m/sec} \quad (\text{for } v_a^0 \text{ in m/sec}) \tag{4.10b}$$

From Table 4.2 (cold) air velocities are up to 1.5 ft sec (0.45 m/sec). Inserting this value in Equation 4.10 yields a burning rate of about 1 ft/hr (1/3 m/hr), so a bed of 1 ft (1/3 m) depth, burning downward, would require a burning time of about 1 hr (3600 sec), which is about what is found for grates (Table 4.2).

To add anything to the limitations on firing density requires examination of the factors controlling burning times and flame propagation through a fuel bed (Section 6.2).

B. FLUID BED

Since the fluid bed is a (particulate) backmix system, flame stabilization is assured so long as the thermal balance conditions for stirred reactor stability are satisfied (Section 6.3) and so long as the bed is not blown bodily out of the chamber (fast fluid bed). Experiments have established that 1300°F (about 750°C) is a temperature below which reaction is generally incomplete and the RCS factor is rising unacceptably. At 1800–2000°F (1000–1100°C) the lower fusion-point ashes begin to sinter or slag. This sets temperature limits, unless an ash agglomerating unit is to be built. Within these constraints, design then focuses on heat transfer requirements.

A simple heat transfer analysis may be carried out utilizing the parameter of the heat exchanger area per unit cross-sectional area of the combustion chamber, Φ. Values are given in Table 4.2. If the area combustion intensity is $J_{f,A} h_f$, with adiabatic combustion the POC temperature would be T_{ad}, given approximately by

$$J_{f,A} h_f = J_{f,A} \left(1 + \frac{G_s}{\phi}\right) \bar{C}_p (T_{ad} - T_0) \tag{4.11}$$

where $\bar{C}_p$ is the mean POC specific heat between T_0 and T_{ad}. If there is heat exchanger surface in the bed at temperature T_s the same amount of input heat will be shared between the POC, now assumed to be emerging at T_b, the bed temperature, and the heat exchanger; thus

$$J_{f,A} h_f = J_{f,A} \left(1 + \frac{G_s}{\phi}\right) \bar{C}_p (T_b - T_0) + \Phi h_T (T_b - T_s) \tag{4.12}$$

ISBN 0-201-08300-0

where h_T is the heat exchange coefficient. Rearranging, utilizing Equation 4.7 and the approximation $(G_s/\phi) >> 1$, yields

$$\Phi = \left[\frac{\bar{C}_p\,(T_{\text{ad}} - T_b)}{(T_b - T_w)\,h_T}\right] J_{a,A} \tag{4.13}$$

Adopting approximate values as follows: $T_{\text{ad}} = 3600^{\text{o}}$F, $T_b = 1600^{\text{o}}$F, $T_w = 1000^{\text{o}}$F, $\bar{C}_p = 0.25$ Btu/lb ^{o}F, $h_T = 50$ Btu/ft^2 hr ^{o}F, and $J_{a,A} = 1200$ lb/ft^2 hr, we find that $\Phi = 20$. This compares well with the value of 30 in Table 4.2, as the latter may also include the freeboard gas-convective heat exchange surface. The calculation broadly validates Equation 4.13, which therefore indicates the prime factors of concern in design and also provides a basis for setting design limits on required values of Φ as a function of operating temperature ranges and of fuel or air supply rates. Alternatively, at a fixed Φ, the equation indicates the sensitivity of T_b to varying air rates at a fixed air/fuel ratio and fixed T_s. The expression obtained, inserting the values used above, is

$$\frac{T_b}{1000} = \frac{1 + 9 \times 10^{-4} J_{a,A}}{1 + 2.5 \times 10^{-4} J_{a,A}} \tag{4.14}$$

for $J_{a,A}$ in lb/ft^2 hr. The bed temperature, on these values, will range from about 1300 to 1900°F (750 to 1050°C) if the airflow rate changes from 500 to 2000 lb/ft^2 hr (2500 to 10,000 kg/m^2 hr), which yields hot gas velocities of 7–30 ft/sec (2–10 m/sec).

The calculations indicate orders of magnitude and also show the extent to which design is possible – given stable and reasonably complete combustion – without recourse to kinetic data.

C. ENTRAINED BED

The limit on p.c. firing is the point at which the particle combustion time, t_c, equals the residence time in the combustion chamber, t_r. A simple design equation can be constructed from Equations 4.1 and 4.6. The possible firing density, $\bar{R}_v$ (in lb/ft^3 or kg/m^3) is given approximately by

$$\bar{R}_v = \frac{J_f}{V_c} = \frac{I_v}{h_f} \simeq \frac{4}{t_c} \text{ lb/ft}^3 \tag{4.15a}$$

$$= \frac{64}{t_c} \text{ kg/m}^3 \tag{4.15b}$$

writing $t_r = t_c$, and for t_c in *seconds.* This is the one case where combustion

ISBN 0-201-08300-0

kinetics, indicated by t_c as an integral of a rate equation, enters these approximate design evaluations directly.

In summary, further studies must focus on the general mechanisms of coal reaction, the mechanisms of flame stabilization in different mixing configurations, and application of these principles to the different combustion systems.

5. COAL BEHAVIOR IN COMBUSTION

5.1. Scope of the Problem

Elaboration of the engineering analyses outlined in the preceding sections requires, first, an understanding of how coal burns, and, second, a quantification of the models described. There are two stages in this quantification: (a) determination of the numerical values of relative quantities – burning rates, rates of heating, and so forth; and (b) generalization of the quantitative data by conversion of the physical models into validated mathematical models. Execution of this program starts with a summary of the general behavior of coal in combustion.

5.2. Qualitative Behavior

The physical and chemical characteristics of coals vary widely. Their qualitative behavior on heating and in combustion, however, is common, although it should be noted that virtually all statements must carry the qualification "governed by coal type, and/or experimental conditions such as particle size, heating rate, particulate density, and reaction temperature." On heating, coals are reasonably stable up to a "pyrolysis" temperature (which varies with coal type), at which point thermal degradation becomes substantial. In many cases pyrolysis is accompanied by swelling; rates of pyrolysis increase with temperature up to a maximum; and pyrolysis is essentially complete at about 1000°C for most coals if the reaction time is on the order of seconds or minutes, depending on particle size. Analysis of the chars shows, however, that all still contain some "volatile" material, consisting principally of about 1% each of hydrogen and nitrogen [28]. These volatile traces evidently require temperatures up to 2000°C for their removal.

As many volatile loss experiments are carried out in practice, the weight losses range from 5 to 40 or 45%, increasing as coal rank falls. Weight losses of these magnitudes are usually obtained by fairly slow heating of dense packings of particles. If the packing density is substantially reduced – for example, by generating a dilute phase dispersion in a gas – rapid heating rates of the order of 10^4 °C/sec can be obtained, and the weight loss or "volatile matter content" is often found to be increased, and may even be doubled. There are differences of opinion as to whether the increased weight loss is due to the rapid heating or to the dilute phase dispersion; the latter reduces chances for secondary cracking of volatiles and

ISBN 0-201-08300-0

pick-up of carbon on the char particles being formed. The author believes that the present balance of opinion is tentatively in favor of the second explanation (Section 5.4).

The char formed by pyrolysis from the organic component, after sufficient heating time at 1000°C, is mainly carbon, and the particles are generally openwork structures of substantial internal porosity and surface area [29]. Mineral matter in the coal is converted into ash. If temperatures go high enough, the ash can then slag, the slagging temperature being dependent on the ash composition but for most coals lying roughly in the range 1200–1800°C (2200–3200°F). The precise nature of the ash or slag, and the slag softening temperatures and viscosities, also depend on the atmosphere – oxidizing, neutral, or reducing – under which pyrolysis occurred.

If the ambient gas during pyrolysis is air (or oxygen), combustion of the carbon char can occur simultaneously with pyrolysis or in sequence after it. Studies of single particles [30–37] show that the volatiles tend to come off in concentrated and randomly distributed jets at different points on the particle surface, with the larger jets continuing right through the evolution of volatiles but the smaller ones starting and finishing in a fraction of that time. If the air surrounding the particle is hot enough, the volatiles will ignite and the volatile jets will become jets of flame. Considerable pressures can build up in the particles from the pyrolysis (one report [38] suggesting values as high as 1000 atm); but studies of fracturing in p.c. flames have been negative [39, 40], and the occurrence with larger (1 mm) captive particles at lower heating rates was quite small [30]. During pyrolysis the volatiles evolution is probably able to sweep the particle surface clear of oxygen, at least for the larger particles, and the char residue being formed cannot be attacked until volatiles evolution dies down. There is reason to believe, however, that this screening process is not effective for particles in the smallest p.c. range (roughly below 25 μm [41]). With the largest particles it is also possible that the jets of volatiles could draw oxygen in to the surface.

The characteristic jetting effect has been observed with both millimeter-sized captive particles [30], and with particles in the p.c. size range injected into electric furnaces [42]. With the smaller sizes the effect is to propel the particle rapidly and erratically through the ambient gas. With the larger particles (say, 100 μm and above) the jetting effect seems to be invariable; with the smaller particles, however, there is some reason to believe that the behavior is the exception rather than the rule. It has also been observed [30] that the jetting effect can be eliminated by precarbonization of the particles at about 500°C.

When the solid particle or its resulting residue is hot enough, and the oxygen has access to the surface, the particle then starts to react by gas-solid reactions. The char burnout is, of course, very much slower, typically taking about 10 times as long as the pyrolysis evolution and combustion [30–35].

If particles are being heated on a grate, the rates of heating are quite slow, on the order of a few degrees per second or per minute. As particle size is reduced, the particles can heat faster, depending also on the circumstances of contact between

ISBN 0-201-08300-0

the particles and the heating source. In the fluid bed and the pulverized coal flame the particles are in effect plunged, in a stream, into hot or very hot environments. With millimeter-sized particles the rates of heating in the fluid bed may approach 10^3 $^oC/sec$, and in p.c. flames the rates of heating generally exceed 10^4 $^oC/sec$. In explosion flames the heating rates are probably of the order of 10^5–10^6 $^oC/sec$, and heating rates of this order and range are apparently achievable in stirred reactors ([43], see Table 2.2, entry 4(i)). One effect of increasing the heating rate is to delay the onset of significant pyrolysis [44] – reasonably described as increasing the pyrolysis temperature. At the heating rates found in p.c. flames pyrolysis may not start until the temperatures reach 1000 or 1200°C. The solid particles, nevertheless, are then hot enough to start reacting, and there is evidence that in such circumstances ignition may start on the solid particles, before the onset of pyrolysis [41, 45].

Pyrolysis produces a range of products of widely varying molecular weights, from free hydrogen gas at the one extreme to heavy tars or pitch (when cold) at the other. Many coals then become plastic or fluid, with measurable viscosities, before the material thermosets to produce coke or char [46]. This can be important in firing coal on grates (or reacting in shaft gasifiers) on account of coke formation that makes the bed impermeable (so that combustion becomes very inefficient), but in the fluid bed, and particularly in p.c. firing, caking or coking is not normally regarded as a problem (but see Ref. 47).

When the particle size and cloud dispersion density are reduced (as in p.c. firing), the particles can heat more rapidly and the volatile matter tends to be expelled from the particles much faster. There is reduced opportunity for secondary cracking in the particle with the consequence that a greater fraction of the coal molecule is gasified. Since the increased weight loss must be due mainly to additional loss of carbon, the mean molecular weight of the products must increase, as compared with the volatile products produced during slow heating. In slow heating (degrees per minute), however, the tars produced at the lower temperatures are often removed at about the same temperatures, so that the longer chain hydrocarbons that cool to pitch are likely to remain intact. In rapid heating, by contrast, if the volatile matter produced is subjected to higher temperatures or greater thermal stress, the longer molecular chains are more likely to be broken up and the range in molecular weights will be reduced, although the products are richer in carbon overall. These differences in behavior could affect combustion in circumstances such that the volatiles production is dense, as at the exit of a fluid bed feed pipe, and the composition and secondary cracking behavior could be critical to burn-up inside the bed (see Section 5.4). When the volatiles react as they emerge from the particle, the details of composition are probably of secondary importance.

Models of coal pyrolysis are influenced by views of coal composition. Some investigators assumed that coal is so significantly homogeneous that it could be treated as a material that degrades thermally, in one step, into volatile matter and carbon residue [e.g., 48]. At the other extreme it was viewed as an effective assembly of many varied constituents that decompose independently, with pyrolysis

ISBN 0-201-08300-0

products then undergoing further decomposition, either independently or in interaction with each other, and the overall pyrolysis process had to be treated essentially as a statistical assembly of reactions [e.g., 49, 50]. More recent investigators, however, now appear to take an essentially intermediate view. It is universally agreed that the single-component, one-step reaction is altogether too simple. In the next stage in complication there are assumed to be two components in the coal, pyrolyzing in parallel and independently of each other, and this is probably the most favored operational model at present [51–53]. Other models advanced recently [e.g., 38] are essentially modifications of the two-component, one-step model (excluding secondary reactions). These are pursued in Sections 5.3. and 5.4.

5.3. A View of Coal Constitution

Models of coal constitution are required for two purposes in combustion: first, as a basis for mechanistic models of behavior, particularly pyrolysis; and, second, as a basis for generalizing from the limited experimental data to untested experimental conditions, as well as to other coals. The two-component hypothesis (Section 5.2) is the simplest model with some claim to reality, particularly with regard to explaining combustion behavior [54].

The two-component hypothesis of coal constitution [55] is an operational model, first proposed in 1913, that was subsequently given a more mechanistic basis by Horton's two-stage theory of coalification [56] and some direct evidence in its favor from X-ray studies [57]. Coal is viewed as having two alternative possible structures, as illustrated in Figure 5.1 [29]: an "open structure" with C% $<$ 90%, and an "anthracite" structure with C% $>$ 90%. Coalification is the process of transforming the open into the anthracitic structure. The intermediate "liquid" structure, at about 90% C, is in the transition from Stage I to Stage II.

Whole coal analyses and coal densities reflect the two-stage coalification. If we write y as either the H/C ratio or the s/C ratio, where s is the specific volume per unit carbon in the coal, we find [54] that

$$y = \frac{A(\mathrm{O/C})}{1 + B(\mathrm{O/C})} \tag{5.1}$$

where A and B are empirical constants with values of 80 and 90, respectively, for y = H/C, and with values of 1000 and 900 for $y = s$/C. For the first case Equation 5.1 is the expression for the so-called coal band on the plot of H/C versus O/C.

ISBN 0-201-08300-0

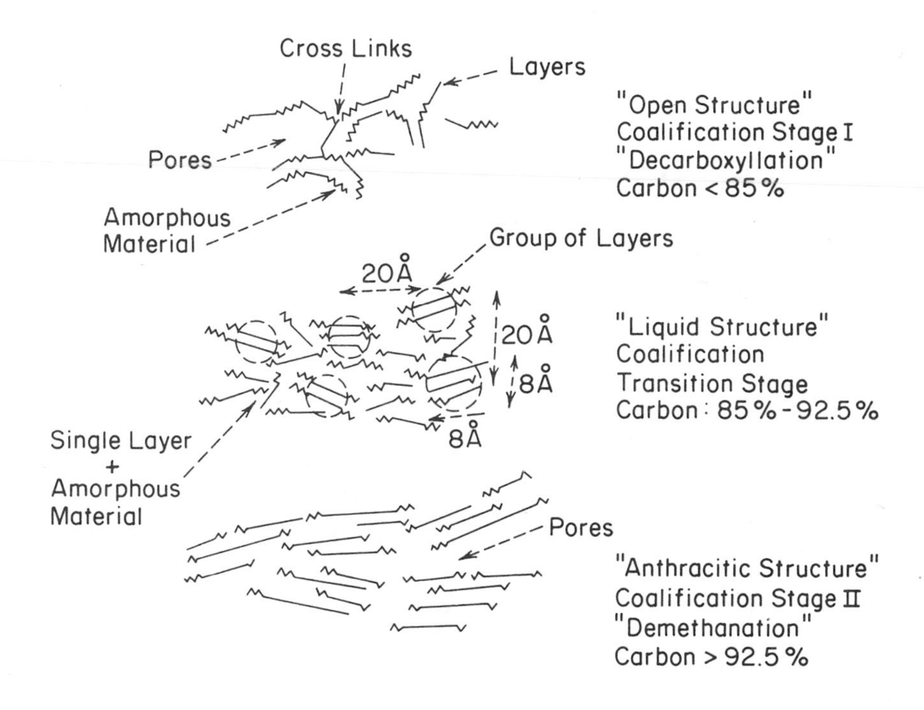

Figure 5.1. Schematic illustrating coal structure during coalification Stage I or decarboxylation *(top)* – "open structure"; transition stage *(center)* – "liquid structure"; and coalification Stage II or demethanation *(bottom)* – "anthracitic structure." (After Hirsch [57] and Mahajan and Walker [29].)

The band has two regions. At high values of O/C (C% < 90), both the H/C and *s*/C ratios are almost constant as O/C falls (coalification by "decarboxylation"). Above 90% C the H/C and *s*/C ratios fall almost linearly with O/C (coalification by "demethanation"). The transition between the two regions is roughly 85–92.5% C. This range contains the minimum coal density, the most highly swelling coals, and a minimum micropore (CO_2) and a maximum macropore (N_2) internal surface (see Figure 5.2) [29].

The two-component constitution model can be used to explain the coalification changes by supposing that coalification Stage I is a process by which some of

ISBN 0-201-08300-0

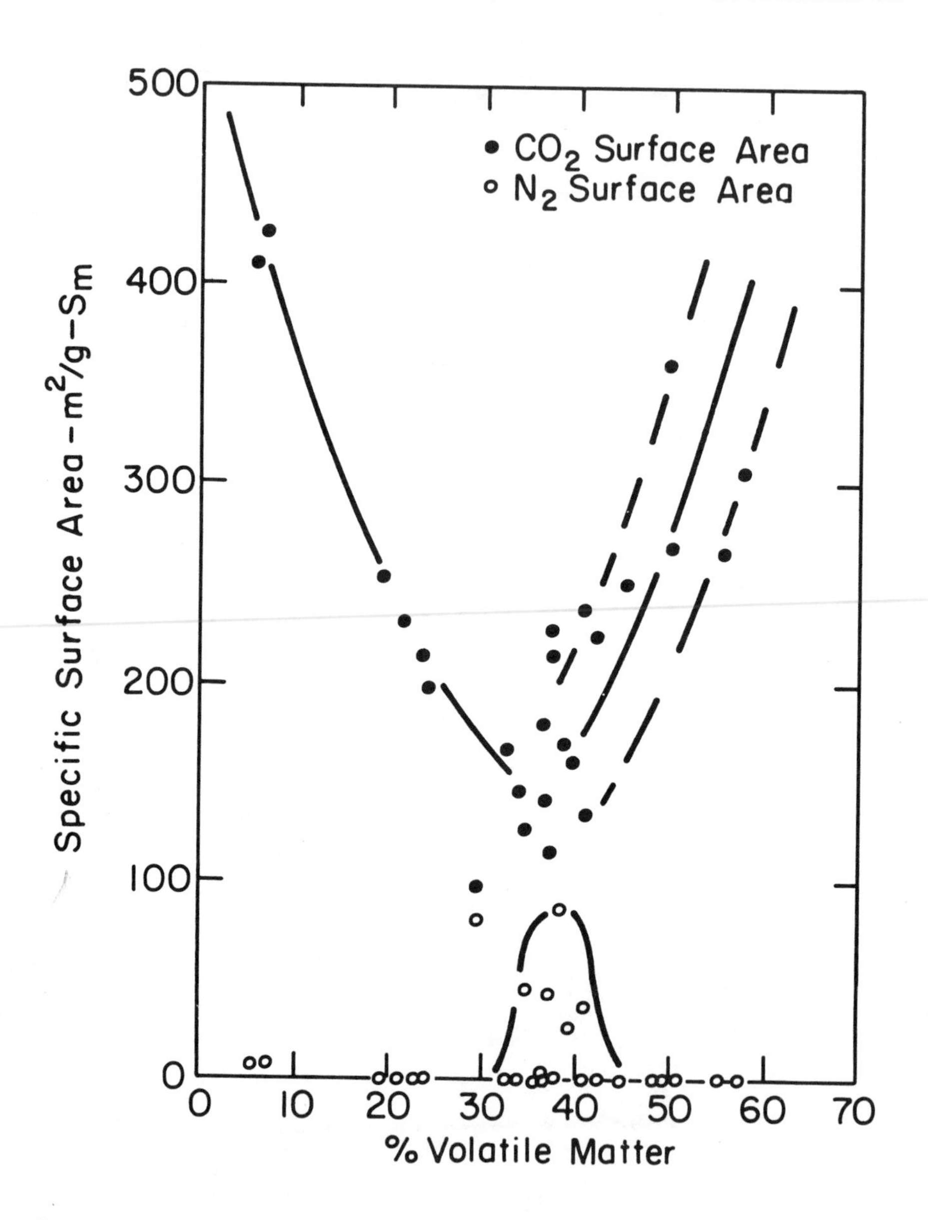

Figure 5.2. Variation of internal specific surface area (m^2/g) with percent volatile matter as rank index. Data source: Ref. 29.

coal Component I is transformed into Component II. Coalification Stage II is then a process of Component II consolidation or condensation [54, 58, 59]. X-ray analysis supports the model by showing the existence of two constituents: first, ordered C-H ring structures stacked in lamellae or crystallites; and, second, X-ray

ISBN 0-201-08300-0

amorphous material, probably providing cross-links between the lamellae. At low carbon, in the lignites, the material is highly porous, and the number of rings or lamellae per crystallite is low, frequently one, with the lamellae cross-linked and randomly ordered with respect to each other. As the carbon percentage rises, the ordering improves, the fraction of carbon in the ordered rings increases (lamellae being created), the volume per unit mass of carbon (s/C) and the density fall, and the pores become increasingly isolated from each other. At approximately 90% carbon, the crystallite packing (now of two to four layers and one to two rings per layer) reaches an optimum; the density is at a minimum, but a large fraction of the internal pores must be isolated from each other so that the internal surface accessible from the exterior is also a minimum (Figure 5.2). Above 90% C the lamellae condense: there is an increase in the number of rings per layer and in the number of layers per crystallite, and a greater fraction of the total carbon is found ordered in the lamellae [57].

Table 5.1

Coalification, Constitution, and Pyrolysis of Coals or the Two-Component Hypothesis[a]

(a) Coalification (Coal Formation)

	Range	Description
Stage 1:	up to 92.5% C	Coalification reaction – transformation of Component I (parent material) into Componet II (lamellae).
Transition:	85.0% C to 92.5% C	Stage 1 goes to completion – the natural end point of this is Hirsch's perfect "liquid" structure at 90% C (minimum porosity).
Stage 2:	85.0% to 100% C	Graphitization – condensation and coalescing of lamellae (Component II); end point, graphite. The "liquid" structure breaks up, and the intermediate structures formed are less perfect.

(b) Constitution and Pyrolysis

	Material Decomposing	Temperature Range, °C	Product
	Material liquated out	Up to decomposition, T_r	Moisture; absorbed and occluded gases; hydrocarbons
Coal	Component I	Ia: T_r to 550	Gases, vapors, tars; smoke forming volatiles generally, containing the volatile carbon
		Ib: 550 to 650	
	Component II	IIa: 650 to 1000	Mainly H_2 gas
		IIb: above 1000	Fixed carbon

Pyrolysis Model Comparisons: Simplified Model

$$\text{Coal} \begin{cases} \text{I} \longrightarrow \text{VM}_1 \ (+\text{char}_1) \\ \text{II} \longrightarrow \text{VM}_2 + \text{char} \end{cases}$$

is equivalent to above model with VM_1 = Ia + Ib and $char_1 = 0$. Corresponds to model of Stickler et al. [51] if $char_1 \neq 0$.

[a] Sources: Refs. 59 and 88.

ISBN 0-201-08300-0

In an earlier summary of these views [59] the X-ray amorphous material was identified as Component I and the ordered lamellae as Component II. This identification is consistent with the expectation that the C-H ring structures (lamellae) would be proportionately more stable to heating than the cross-linking Component I material and, on pyrolysis, would produce mainly hydrogen and char, as is observed experimentally above 700°C. It is also qualitatively consistent with a number of quantitative combustion results [54].

These views are summarized in Table 5.1: part (a) illustrates the two-stage coalification; part (b) indicates the pyrolytic behavior of the two-component structure, noting that subcomponents are also identified here (based on 30) as a first step beyond the simple division.

5.4. Pyrolysis

Pyrolysis is thermal degradation. Experimentally the simplest and most common method of measuring the extent of the degradation is to determine the weight loss. Some investigators have also or alternatively made measurements of volatile yields and their compositions. Methods of experiment have varied quite widely, and both the methods and their results are described in greater detail elsewhere in this book. For the purposes of this chapter, the relevance of pyrolysis to combustion, what follows is a summary of essential methods and results.

A. METHODS OF EXPERIMENT

These show wide variations in broad design and in detail, but the most informative methods appear to have been the following.

1. Captive particle experiments using either single large particles, on the order of 1 mm [30–36], or fine ground particles (50–70 μ) trapped in the folds of a fine wire mesh [50, 38].

2. Dense bed experiments with fine crushed particles (about 100 μm to 1 mm) in a crucible heated in a tube furnace. The specification of 1 g of −60 mesh (250 μ), heated at 950 ± 10°C for 7 min, constitutes the ASTM test for volatile matter. In this experiment the evolving gases so stir up the particles that they fluidize.

3. Fluid bed experiments in which packets of fine crushed particles are injected into preheated sand beds [49].

4. Entrained particle experiments with fine ground particles (less than 100 μ) carried into a reactor on a stream of hot gas, usually but not always nitrogen [e.g., 48, 53]. Most reactors are tubes with the carrier gas and coal in once-through or plug flow. In some recent experiments [43] a backmix or stirred reactor was used.

The results of the experiments depend on the experimental method or the conditions of experiment. The experimental conditions differ primarily in the following ways: (a) in particle size, where dense bed and captive particles are 1–2

ISBN 0-201-08300-0

orders of magnitude larger than those in the entrained particle experiments; (b) in rate of heating, where the fine (< 100 μ) particles can be heated at up to 10^6 ^{0}C/sec, but with the coarse particles (1 mm and above) the heating rate is restricted; (c) in particle bed or stream density, where the crucible beds are dense (on the order of 0.5 g/cm^3) but the entrained particles vary widely in cloud density down to 10^{-5} g/cm^3 (cold); and (d) in different limits on heating time, with the entrained particles generally limited to 1 sec as a maximum but with unlimited heating time with all other experimental methods.

B. PRINCIPAL RESULTS

1. With particles above 100 μm, the time required for pyrolysis is strongly dependent on particle size. Using the volatile burning time of captive particles, t_v, as an estimate of the pyrolysis time, a square-law dependence on *initial* diameter, d_0, was obtained experimentally [30]; thus

$$t_v = K_v d_0^2 \tag{5.2}$$

In cgs units, the proportionality constant, K_v, was found to have a value of about 100 sec/cm^2. Equation 5.2 is entirely empirical. A square-law expression can be obtained on the basis of an internally shrinking liquid-drop model [32], but the model appears to have some internal contradictions. The rate of diffusional escape of volatiles or the rate of heat transfer to the interior, or both, must be important, however, to account for the particle size dependence.

2. Dense bed experiments with fine crushed particles have traditionally been used to determine the so-called volatile matter content, or VM%, under standardized conditions of sample weight, furnace temperature, and pyrolysis time (ASTM, BSS, DIN, and so forth). The experimental conditions had to be carefully standardized since it was known that the weight loss varied with experimental conditions. This is illustrated in Figure 5.3, showing the variation in percentage weight loss with sample weight (bed depth in crucible), furnace temperature, and initial heating rate. The noticable characteristic is the increased weight loss with reduced sample size or with increased heating rate.

3. With reduced particle size and particulate density (number of particles per unit volume), weight loss measurements amplify the observations of the dense bed experiments that the VM yield can increase. Typically reported increases are up to twice the ASTM VM values. The multiplier is generally now referred to as the "*Q* factor" [48]. The increases are also often referred to as "anomalous" VM yields, but it may be preferable to regard the ASTM measurement as anomalous, as explained below.

4. To summarize, rates of volatile release are clearly temperature and time dependent. As a broad generalization, with exceptions, the rates are also particle size dependent above 100 μm and probably particle size independent below 50 μm, with a transition range from 50 to 100 μm.. The total weight loss or VM yield

ISBN 0-201-08300-0

increases as particle size and particulate density drop; the increased yield also correlates with increased heating rate, but this may or may not be the causal factor – in the author's opinion, probably not. Maximum VM yield is also temperature dependent up to about 1000°C (above that, it is still uncertain).

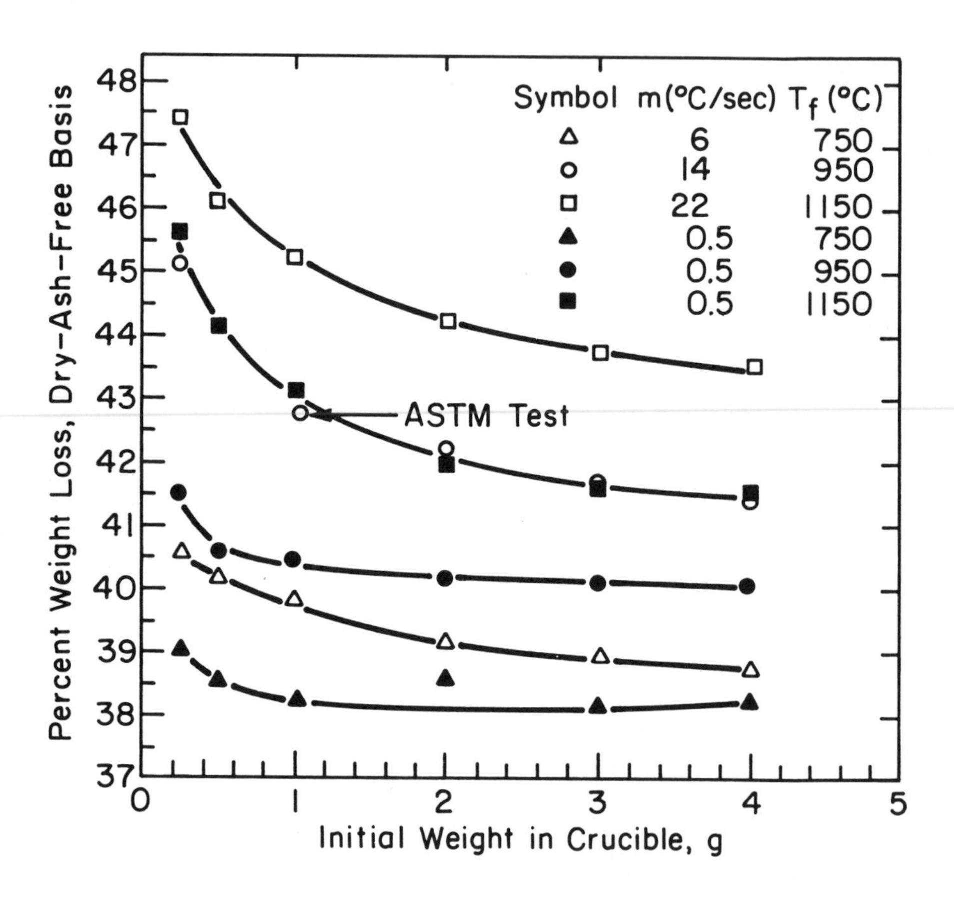

Figure 5.3. Percent weight loss versus initial weight of sample in crucible for different heating rates and different final pyrolysis temperatures for PSOC-296 coal. (Source: Ref. 53.)

5. Particles also swell, although this factor neither appears yet in any quantitative pyrolysis models nor seems to have any particular significance in pyrolysis, though it certainly affects char combustion properties. The swollen particles are now known as "cenospheres" [60] (originally, "carbospheres"), with the degree of swelling and the form of the resulting structure strongly dependent on the nature of the ambient gas and the maximum temperature reached. As originally defined [60],

ISBN 0-201-08300-0

true cenosphere formation occurred only in neutral and reducing conditions with maximum swelling at about 700°C (completion of pyrolytic removal of Component I on the Table 5.1(b) interpretation). Swelling was reduced at both lower and higher temperatures and/or in an oxidizing atmosphere. The degree of swelling is generally said to be rank dependent on the basis of crucible swelling tests, and this can be crucial in coking ovens. The swelling factors of single burning particles of a set of 10 coals, however, displayed much scatter but an almost constant average value (of 1.5), except for anthracite, although crucible tests showed the coals to range from noncaking to highly swelling [33].

C. INTERPRETATION

A reasonable, self-consistent interpretation of these varied results, which is partly quantitative, may now be constructed. What follows is at variance with some views that can be found in the literature, and there are still many uncertain points of detail. What is presented, however, is a working construct, still open to debate, but nevertheless satisfying currently available experimental information.

Coal is viewed here, as described in Section 5.3, as a solid behaving *as if* it consisted principally of two separate components. These pyrolyze as indicated in Table 5.1(b), with Component I yielding only volatiles V_I and Component II yielding coke C_{II} and volatiles V_{II}. Some analyses [51, 52, 61] assume a first-stage formation of coke C_I; but the evidence is not clear, and the two model variants cannot be differentiated experimentally at present. (Some models [e.g., 62, 63] also include a prior "activation" step producing an intermediate "metaplast," but the need for this may be based on a false interpretation; the models are also based on selected experimental results and are then shown to be in agreement with the results they are based on.) The volatiles can then crack in secondary reactions, depositing carbon in the particle interior as the volatiles escape or on the surface of other adjacent particles [52]. Internal deposition is encouraged by large particle size. External deposition is encouraged by dense particulate beds or dispersions. These two factors account qualitatively for the effects of reduced yield with increased particle size or particulate density.

D. TRUE VOLATILE CONTENT: DEFINITION

The true or "intrinsic" VM yield may, therefore, be defined as the percentage weight loss obtained under the following conditions: (1) particle size small enough to prevent internal carbon capture on cracking (probably less than 50 μm); (2) particulate densities low enough to prevent external carbon capture (cloud densities less than 10^{-5} g/cm^3); (3) "instantaneous" heating (exceeding at least 10^4 °C/sec); and (4) isothermal pyrolysis at some temperature above which there is no further significant weight loss (excluding total dissociation and/or vaporization or sublimation of the residue). This may be 1000–1200°C, with reservations. These conditions are approximately met in some wire mesh and entrained particle experiments.

ISBN 0-201-08300-0

E. CAPTURE FACTOR

With respect to these "true yield" conditions ASTM measurements are anomalous because of cracking and capture, but consistently so. This can be analyzed as follows [64]. Assume that the true or intrinsic VM yield is V_0. If isothermal experiments are carried out at different temperatures less than the upper limit value, only a weight loss $V < V_0$ will be obtained. The balance of the volatiles, $V_0 - V$, can be expelled by conventional ASTM VM analysis, but the actual yield will be only a fraction of the balance because of capture. If the fraction captured is α, the recorded weight loss, R, by ASTM analysis, referred to the weight of the original coal, is

$$R = (1 - \alpha)(V_0 - V) \tag{5.3}$$

and

$$R_0 = (1 - \alpha) V_0 = (\mathrm{VM})_0 \tag{5.4}$$

where R_0 is the maximum ASTM weight loss when treating the raw coal ($V = 0$), so R_0 is the conventional ASTM VM for the raw coal, $(VM)_0$. Rearranging we have

$$V = \frac{(VM)_0 - R}{1 - \alpha} = Q[(VM)_0 - R] \tag{5.5}$$

Figure 5.4 is a typical plot of V against $[(VM)_0 - R]$. Such plots were first obtained by Badzioch and Hawksley [48] but have also been reported by others [e.g., 53]. The slope of the plot is the multiplying "Q factor" described above, representing the increased VM yield over the ASTM determination. On this analysis the reciprocal of the Q factor is a divisor, identified with the capture factor, α, that represents the *reduced* VM yield in the ASTM determination compared with the true or intrinsic value. On this interpretation a Q factor of 2 yields a capture factor of 0.5.

An important characteristic of Figure 5.4 is its linearity; the implication is that the capture factor is constant. Another characteristic is the nonzero (but small) intercept on the $[(VM)_0 - R]$ axis. This has been interpreted [48] as due to experimental error in the collection of residue, but could represent a different capture factor between coal and char [64]; however, the point does not seem to be significant.

Applying the above analysis to the data of Figure 5.3 illustrates some additional problems. For an ASTM value of 42.7% (1 g) and on the assumption of a Q factor of 1.5–2, the true VM value, V_0, would be 64–85.4%. We may then see the capture factor as possibly being made up of three parts. First, there is the particle size effect, where the ASTM requirement is 250 μm (or less), permitting capture in

ISBN 0-201-08300-0

the particle pores; second, there is the crucible depth effect, permitting surface pick-up; and, third, there is a probable rate-of-heating effect, whereby the relatively *slow* rates of heating (0.5–20 ^{o}C/sec) could allow time for a fundamental reordering or crystallization of the coal structure, thus binding material that would otherwise be lost by pyrolysis at more rapid heating rates. This interpretation suggests that, if there are three Q factors that can be summed, the particulate density effect could be responsible for 0.1 or 0.2 in the overall factor of 1.5–2.0.

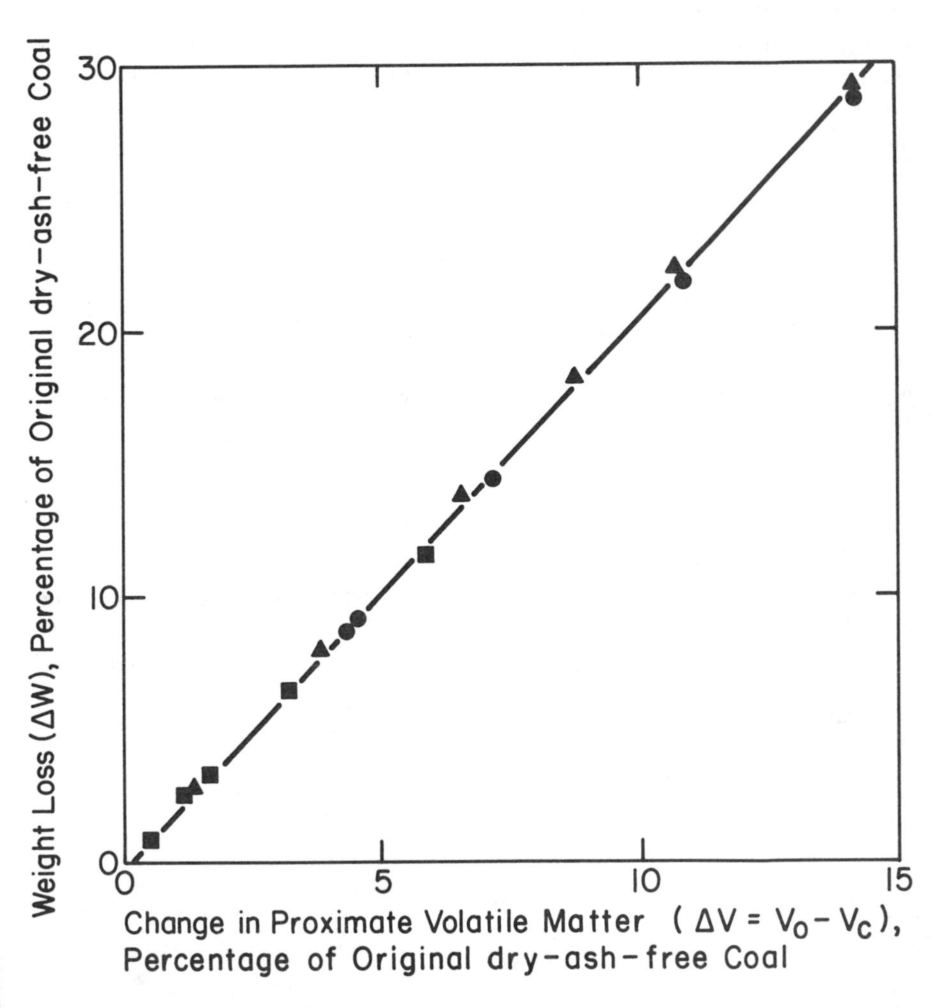

Figure 5.4. Variation of weight loss with change in ASTM volatile matter of chars for Montana lignite (PSOC-90, 43.50% VM). (Source: Ref. 53.)

ISBN 0-201-08300-0

Other evidence of particulate density effects is scarce, but it has been reported by Ubhayakar et al. [52] in entrained flow experiments. The details are complex and contained mostly in computer programs, but the conclusions are clear: volatiles gasification decreases with particulate density because of capture on the surfaces of the particles.

F. KINETIC ANALYSIS

This is based on the two-component constitution discussed in Section 5.3, and the result parallels some kinetic theories of liquefaction [e.g., 65]. If components I and II at concentrations C_1 and C_2 decompose by independent, first-order reactions (with velocity constants k_1 and k_2), then, with C_1^0 and C_2^0 for the initial concentrations of the components, the appropriate differential equations integrate to yield [53]

$$C_1 = C_1^0 e^{-k_1 t} \tag{5.6a}$$

$$C_2 = C_2^0 e^{-k_2 t} \tag{5.6b}$$

The total weight loss per 100 parts of initial weight, V, is $(G_1 + G_2)$, where $G_1 = (C_1^0 - C_1)$ and $G_2 = [(C_2^0 - C_2) - C]$. If χ is a proportioning factor between G_2 and C, given by $\chi G_2 = C$, the fractional weight left of the pyrolyzing sample at time t can be expressed as

$$1 - \frac{V}{V_0} = \left(\frac{C_1^0}{V_0}\right) e^{-k_1 t} + \left(1 - \frac{C_1^0}{V_0}\right) e^{-k_2 t} \tag{5.7}$$

where V_0, the weight loss at infinite time, is obtainable from plots such as Figure 5.4, since $V_0 = Q(VM)_0$.

An equation close to Equation 5.7, with only the first exponential term included, was first obtained empirically by Badzioch and Hawksley [48]. Since this yielded less than complete loss of weight at infinite time, an empirical correction factor, C^*, was included to make the numbers fit. It has been shown [53] that at small t Equation 5.7 can be reduced to the Badzioch and Hawksley expression with

$$C^* = \left(1 - \frac{C_1^0}{V_0}\right)\left(1 - \frac{k_2}{k_1}\right) < 1 \tag{5.8}$$

There are some numerical difficulties with this interpretation [53], but the broad identification appears to be valid.

There are few data available to test Equation 5.7. Figure 5.5 [53] is still about the best example. The data were obtained in an entrained flow experiment, at low particulate density, carrying particles through a tube furnace at constant

ISBN 0-201-08300-0

temperature after heating at 8000 °C/sec. This corresponds to an initial delay time before reaching pyrolysis temperature of 0.1 sec, during which, according to special tests, there was no significant pyrolysis. The figure then illustrates the two-part structure of Equation 5.7 with slopes yielding the kinetic constants k_1 and k_2. Interpretation is restricted at present since the experiments were carried out at only one temperature on account of equipment limitations. The results do indicate, however, that the escape of the Component I pyrolysis products is still influenced by particle size down to 64 μm, and extrapolation would suggest that this will possibly remain true down to 25 μm; but the Component II products could very well be uninfluenced by particle size up to 180 μm and possibly above.

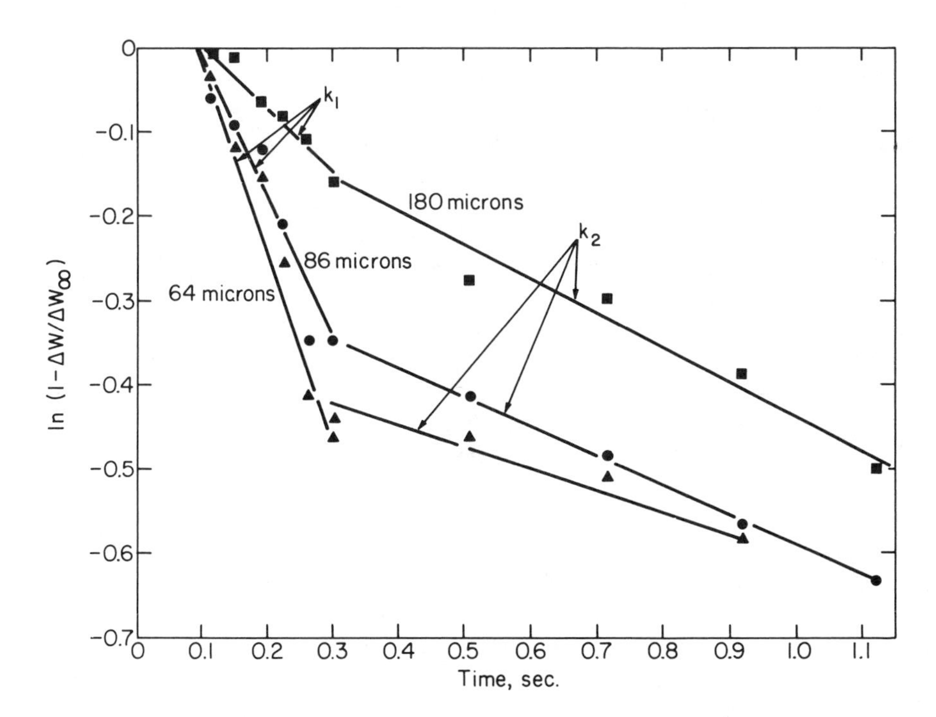

Figure 5.5. Variation in ln $(1 - \Delta W/\Delta W_\infty)$ with cumulative residence time in the pyrolysis furnace for North Dakota lignite (PSOC-246, 47.5% VM). (Source: Ref. 53.)

G. OTHER MODELS AND RESULTS

In other models multiple simultaneous or sequential reactions have been proposed. In a sequence of experiments, Juntgen and associates [44] followed the production of individual constitutents and found that the appearance of the individual products apparently obeyed one-step, first-order rate equations. More recently Suuberg et al. [38] reported the same result on a smaller number of

ISBN 0-201-08300-0

"individual" components: CO_2, CO, H_2O, H_2, CH_4, C_2H_6, "hydrocarbons," "oils," and "tars." Only one particle size of a single coal (Montana lignite) was studied but at a wide range of heating rates and temperatures. The variation of component yield with temperature is illustrated in Figure 5.6. The results are considered indicative: numerical values should vary with coal rank – oxygenated compounds, in particular, should drop with rising rank – but there is substantial background reason now to believe that the pattern for other coals and particle sizes will be generally similar.* When an initial drying period and a problematical jump in CO_2 production above 1100^oC are excluded, there are two principal stages of decomposition, with the first stage possibly split into two substages corresponding very well to the temperature ranges in Table 5.1(b). A particularly interesting additional point is the demonstration [38] that the nine-constituent model closely matches the distribution of activation energies model [49] (Figure 5.7).

The smooth curves of Figure 5.6 are best-fit curves assuming two reactions per product, essentially following Equation 5.7 per component. One ambiguous point remains, however. Examination of equilibria for the different constituents suggests, as the authors point out [38], that what may have been obtained are effective kinetic constants for the equilibration reactions, not for the primary pyrolysis. If this is so, it may be true for all kinetic studies of coal pyrolysis.

The matter of the kinetic constants presents a point of additional interest. Decomposing the velocity constants, k, into frequency factors and activation energies reveals a very wide spread in the values obtained by different investigators. To some degree, however, the rise in frequency factor is offset to such an extent by the rise in activation energy that critical calculations of extent of decomposition show much narrower limits. This is illustrated for a particular set of circumstances in Figure 5.8 [53], which is a plot of activation energy as it varies with frequency factor if some component loses 1%, 5%, or 10% of its mass in 95 msec when heating at 8000^oC/sec. Included on the plot are values of E and k_0 reported by different investigators. Points to the left and right of the lines represent slower and faster reactions, respectively.

In summary, coal decomposition can be approximately described by a two-term rate equation for weight loss, representing independent and parallel pyrolysis of two components as a minimum. The appearance of individual components can also be described by one-step, first-order reactions, with a minimum of two reactions per component, except for hydrogen production. This description is generally consistent at this level of detail with a two-component hypothesis of coal constitution that is consistent, in turn, with a two-stage process of coalification. This reasonably simple picture still lacks much detail; also, in particular, substantiating data are needed on a wide range of coals and conditions. It is, however,

*This expectation is now substantiated by work just reported by P. R. Solomon at the 17th Symposium (International) on Combustion, held in Leeds (England) in August, 1978.

ISBN 0-201-08300-0

reasonably self consistent and is acceptable as a working hypothesis for use in coal combustion models. The model will undoubtedly be modified with time, but present indications are that changes are more likely to take the form of expansion of detail than of fundamental restructuring.

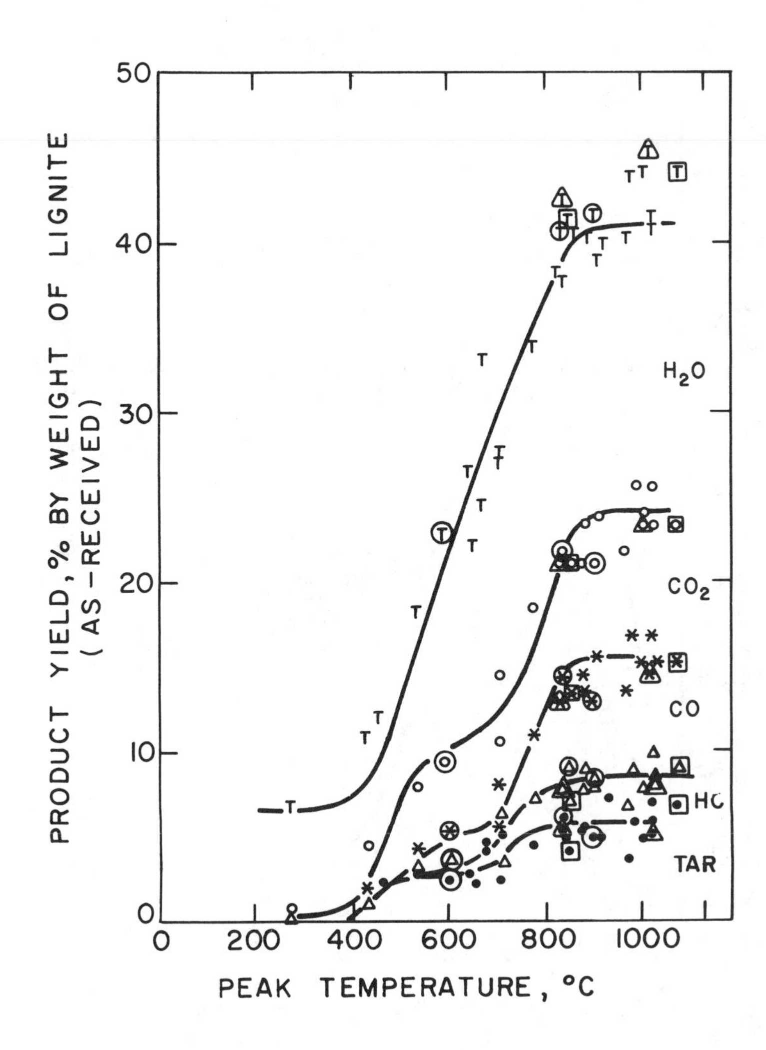

Figure 5.6. Pyrolyis product distributions from lignite heated to different peak temperatures. (•) Tar; (Δ) tar and other hydrocarbons (HC); (*) tar, HC, and CO; (o) tar, HC, CO, and CO_2; (T) total, that is, tar, HC, CO, CO_2, and H_2O. Pressure = 1 atm (He). Heating rate: single points, 1000°C/sec; points inside 0, 7100–10,000°C/sec; points inside Δ, 270–470°C/sec; points inside □, 1000°C/sec but two-step heating; curves, first-order model using parameters in cited source: Ref. 38.

ISBN 0-201-08300-0

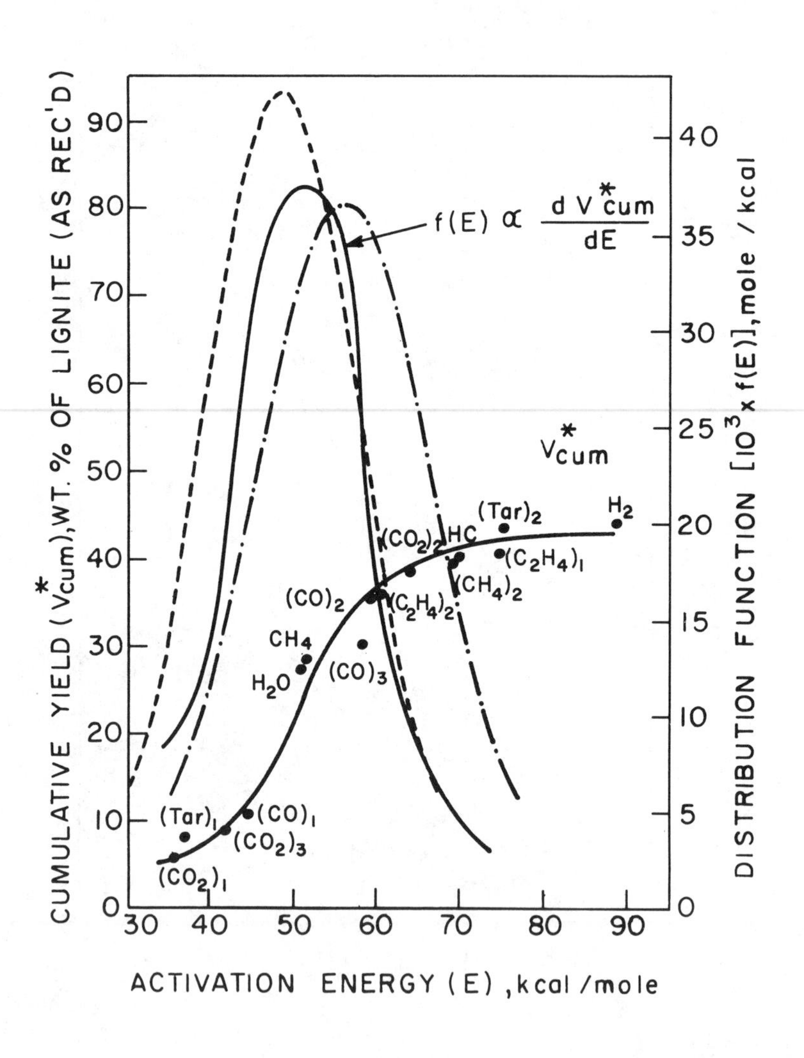

Figure 5.7. Distribution of activation energies of pyrolysis reactions. (●) Cumulative yields, from present base data, of components indicated to left of and including a given point; solid curves, results based on volatiles yields; broken curves, previous results based on weight loss data and two different sets of kinetic parameters. (Source: Ref. 38.)

ISBN 0-201-08300-0

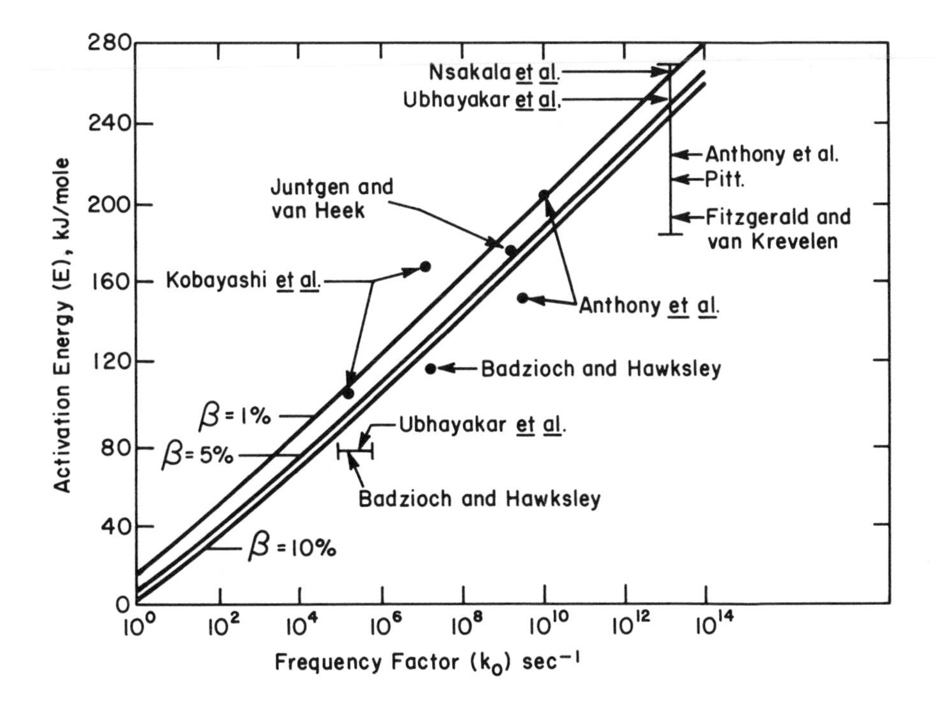

Figure 5.8. Variation of frequency factor with activation energy for 1%, 5%, and 10% loss of volatiles in 95 msec at 8000°C/sec heating rate. (Source: Ref. 53.)

5.5. Volatiles Combustion

Volatiles produced on pyrolysis can burn, as described earlier, either as they leave the particles, or only after traveling some distance if the particulate density is high and the local oxygen concentration is overwhelmed. An extreme example of the latter case is injection of crushed coal into a ballasted fluid bed in too dense a stream. It was reported [66] that a volatiles jet could be formed that would break through to the top of the bed and burn (sometimes with much smoke) in the free-board space (see Figure 3.13).

ISBN 0-201-08300-0

At the two extremes of immediate and delayed combustion the volatiles combustion rate is controlled, in the first case, by the rate of coal pyrolysis (Section 5.3) and, in the second case, by the rate of mixing with additional air. In intermediate cases the volatiles combustion rate may be controlled by the gas phase kinetics of the volatile constituents.

Information on the kinetics of volatiles combustion, however, is almost totally absent. Twenty years ago Underwood and Thring [67] reported that they had been unable to find any reference whatever on the combustion of smoke from coal or any other relevant material. The first experiments were evidently reported by Finch [68] and by Underwood and Thring [67] using Finch's method. A tray carrying a thin layer of coal was pulled steadily through an electric furnace, and the volatiles driven off were mixed with air higher up and burned. In both investigations only experimental information on smoke burn-up was obtained; there was no presentation of results in the context of any kinetic schemes or sets of equations, empirical or otherwise. In a later book review by Field et al. [69] one chapter was devoted to the topic of coal volatiles combustion, but here also there was a lack of kinetic information beyond two approximations. The first approximation borrowed from oil drop combustion theory, where the volatiles were assumed to form a uniform spherical film around the particle and to burn at a rate controlled by boundary layer diffusion. The diffusive model, however, assumed infinitely fast kinetics. An alternative, nondiffusive model was based on the common assumption that hydrocarbons react rapidly to form CO and H_2 with CO burn-up as the final, slow, rate-determining step. A calculation using Hottel et al. 's [70] stirred reactor relation for CO combustion yielded a time of 3 msec for combustion of volatiles at 1800°K (see also Section 7.3C(ii)).

An oil drop model was also used by Howard and Essenhigh [41, 45], but to estimate the screening influence of evolving volatiles rather than to estimate burning times. Burning times were equated to pyrolysis times, and the results indicated that particles above 65 μm were totally screened by the volatile flux and partially screened below that size with substantial simultaneous heterogeneous reaction occurring below 15 μm.

The only other smoke combustion studies appear to be those by Biswas and Essenhigh [71] and by Shieh and Essenhigh [72] in work on cellulose smoke. Kinetic data were obtained by fitting to a second-order expression of the form

$$\frac{dC_s}{dt} = -k_0 C_s C_a e^{-E/RT} \tag{5.9}$$

where C_s and C_a are the mass fraction concentrations of smoke and oxygen, respectively. Activation energies in the two sets of experiments were found to be 16.5 and 20 kcal/mol (70 and 84 kJ/mol) with mass basis velocity constants, k_m^0 [25], of 6×10^6 and 1.6×10^5 sec^{-1}, respectively (see Section 6.3A).

No other sources of data on the kinetics of smoke combustion are known.

ISBN 0-201-08300-0

5.6. Char Combustion

The mechanism of char combustion is, as a whole, better understood than that of either pyrolysis or volatiles combustion. There are still some quantitative problems, but qualitatively the model is quite clear. The model is primarily intended to describe carbon or char combustion, but it can also apply to any porous solid that reacts with a gas, particularly if the reaction products are themselves gases. Conceptually, therefore, the model can also apply to whole coal combustion in circumstances where there is competition between whole coal combustion and pyrolysis, and the first reaction is faster than the second. It has also been used to model SO_2 sorption by lime and limestone.

A. QUALITATIVE PHYSICAL MODEL

This is illustrated schematically in Figure 5.9, which represents part of a porous solid with an adjacent diffusion boundary layer in the gas phase of some thickness δ. The pores are represented schematically as an array of internal channels of progressively diminishing diameter. Parallel branch arrays may be independent of each other or they may cross-communicate. Reaction between oxygen and carbon takes place heterogeneously at all available surfaces, exterior and interior. On account of the reaction an oxygen gradient is established in the boundary layer that then controls the rate of supply of oxygen to the solid surface, and thence into the porous interior. The reaction generates both CO and CO_2, with the CO/CO_2 ratio rising with temperature and CO becoming the principal product at about 1000°C and above [73]. The CO also reacts in the gas phase, to form CO_2, partly in the particle pores and partly in the boundary layer [74]. As the oxygen concentration in the main stream is enriched, there is more CO burn-up inside the solid, and, other things being unchanged, there is a rise in the equilibrium temperature of the solid or of the reacting layer of the solid. This temperature increase can accelerate the reaction and has caused confusion in the past over the interpretation of some experimental results [74, 75].

At sufficiently high temperatures both the C-CO_2 and the C-H_2O reactions also become important and must be included in high temperature flame models. The two reactions have similar levels of reactivity. The reactions are also strongly catalyzed by numerous metals. Platinum has been widely used in many experiments, but salts containing, for example, sodium, potassium, calcium, and iron have also been found [e.g., 76, 77] to have significant effects, a point that could be important on account of the presence of such constituents in typical coal ashes.

The mechanics of the reaction between oxygen and the solid surface are still arguable to some degree. The basic fact is that oxygen forms a chemisorbed layer on the solid surface, and this layer can be removed only by heating to high temperatures in an inert gas or a vacuum unless adsorption occurs at very low (liquid air) temperatures, in which case the oxygen can be recovered as O_2 by heating to temperatures less than −70°C [78]. Above this temperature, chemisorption occurs,

ISBN 0-201-08300-0

implying a low activation energy of adsorption (probably below 5 kcal (20 kJ)). Removal of the chemisorbed layer produces CO and CO_2, as stated above, with an activation energy in the region of 40 kcal (about 170 kJ). Measurement of the activation energies, however, is influenced by sample porosity, with the nominal or apparent activation energy influenced by the depth of penetration of the reaction (see below). Both adsorption and desorption activation energies may be changed by the extent of the adsorbed layer, although they are usually taken as constant. The exact nature of the adsorbed layer is also uncertain. Oxygen can be adsorbed either as the molecule or, if first dissociating, as the atom; adsorption as atoms would require dissociation on approach or dissociation on impact or soon thereafter. Reaction is completed, however, after desorption of the atom or molecule when the underlying carbon atom(s) is removed with the oxygen.

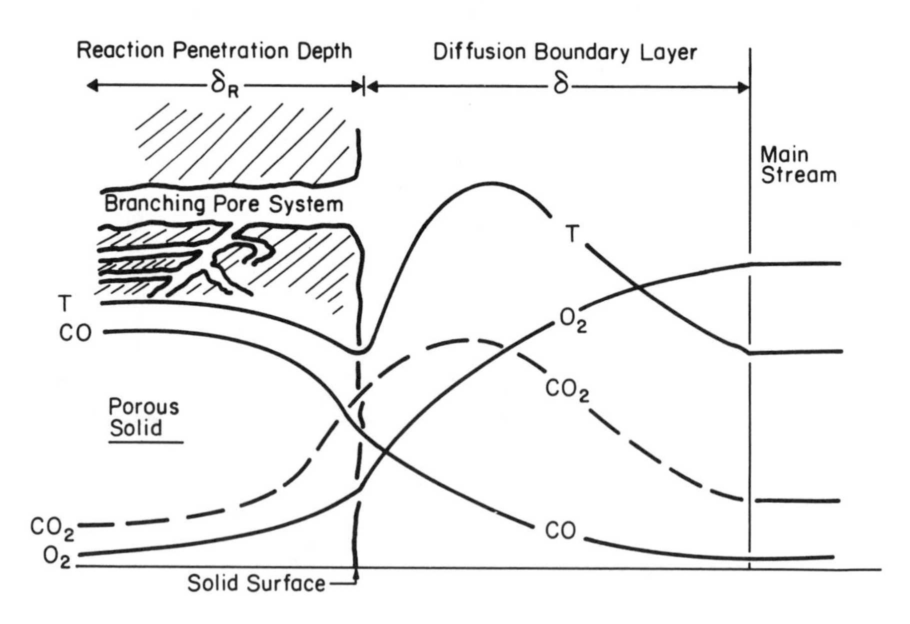

Figure 5.9. Schematic representation of porous solid reacting in air with boundary layer of width δ, and reaction penetration depth of δ_R; showing also temperature and gas concentration profiles.

ISBN 0-201-08300-0

B. METHODS AND DATA SOURCES

Quantitative or mathematical models based essentially on the above qualitative description now exist with roots to be found in work of Langmuir and Nusselt. For comparison with theory a wealth of experimental data also exists on the reactions of numerous carbons and chars with oxygen, CO_2, and steam. Few really definitive and unambiguous experiments have ever been carried out, however, and nearly all comparisons between theory and experiment are open to objections or criticisms of one type or another. A complete review of this complex literature is out of place here; for background, it has been extensively reviewed in the past [64, 69, 79–87]. For our purposes here we may focus on certain more limiting aspects that are most directly relevant to the engineering systems of concern.

The engineering systems of interest can be represented either as tortuous tubes in a solid, schematically describing a fixed bed on a grate or in a shaft, or as single particles representative of those found in fluid or entrained beds. The factor of interest to be entered in the engineering model is the local or specific reaction rate (mass rate of removal per unit area per unit time). Such measurements have been made by many experimental methods but principally on single suspended particles, crushed particles in packings of different thicknesses, electrically heated carbon rods, entrained particles in electric furnaces, and particle flames. Measurements have been made over a wide range of different particle sizes, from 2.5 cm (1 in.) to flame-formed soot particles of 500 Å or less; temperatures have ranged from ambient to 2500°K; total and oxygen partial pressures from ambient to 10^{-4} torr (for references sources, see Ref. 88).

Figure 5.10 [86] is typical of many results. This shows the variation of specific reaction rate (g/cm^2 sec) as a function of temperature on an Arrhenius plot (ln R_T vs. $1/T$) for a number of different carbons and coals reacted under different conditions. Many additional results are now available (for references see Ref. 88), but Figure 5.10 is still characteristic of all additional data in the following respects: (a) the spread over 2–3 orders of magnitude in rate at a given temperature; (b) the generally high slope of the data, corresponding to activation energies mostly between about 20 and 40 kcal (85 and 170 kJ); (c) the evidence in many cases of the rate curve finally peaking at the highest temperature and the rate falling off thereafter; and (d) the possible asymptotic approach of the highest rates to the "diffusion" limit.

ISBN 0-201-08300-0

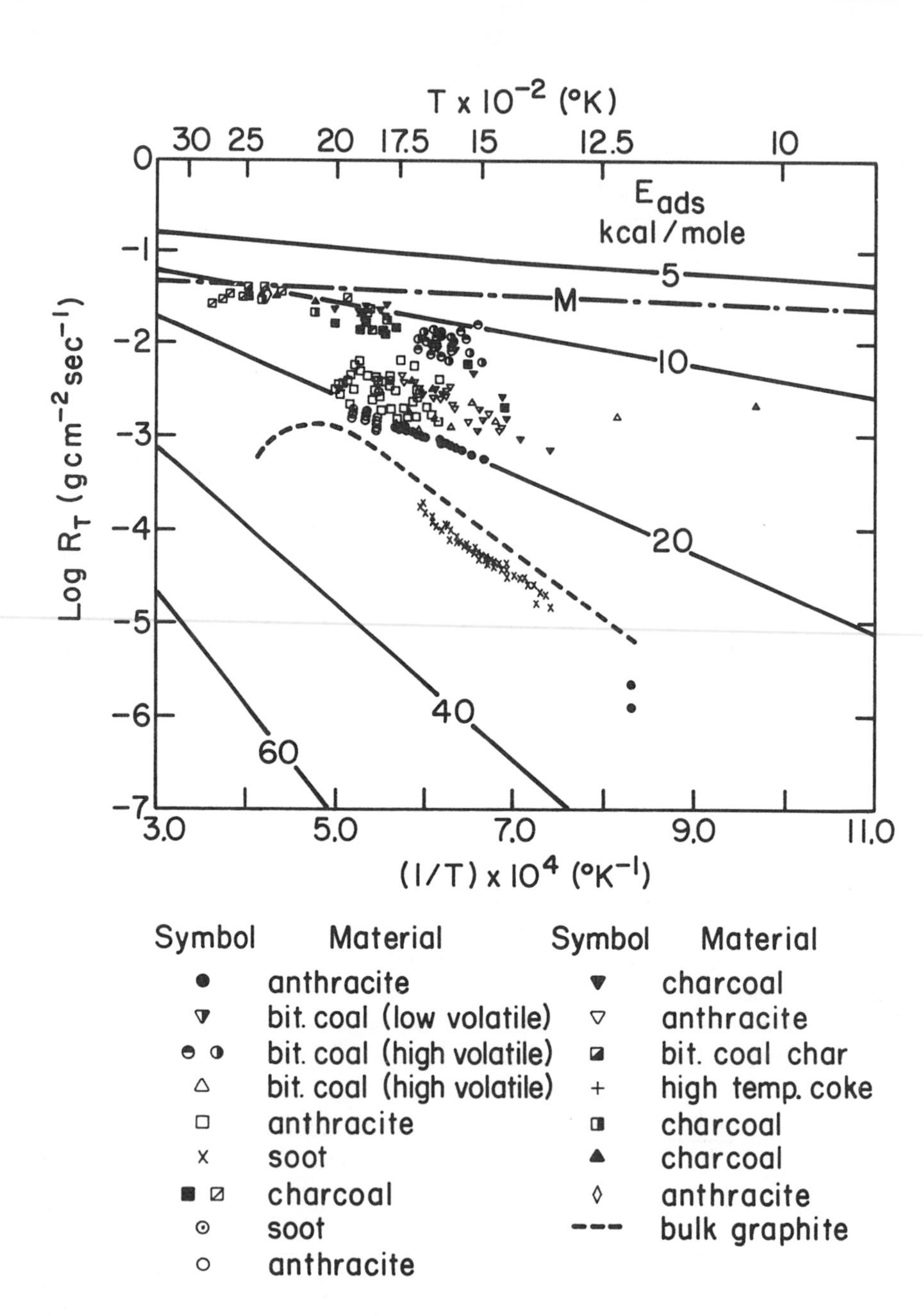

Figure 5.10. Burning rates of particles (calculated to $p_{O_2} = 0.1$ atm). *M*, mass transfer (40 μm particle). (Source: Ref. 86.)

ISBN 0-201-08300-0

C. THEORETICAL ASPECTS

Theories constructed to explain and generalize such results have the following aspects in common. The mass transfer of oxygen from the main stream to the surface across the diffusion boundary layer of thickness δ can be written quite generally as

$$R_T = k_D(p_0 - p_s) \tag{5.10}$$

where p_0 and p_s are the mainstream and surface oxygen percentages, respectively; R_T is the mass rate of carbon removal or specific reaction rate referred to unit area of the superficial carbon surface; and k_D is the velocity constant for mass transfer in appropriate units (converting oxygen moles arriving into carbon mass removed). As a first approximation it is correct in many cases to write $k_D \propto (D/\delta)$, where D is the diffusion coefficient of oxygen through the boundary layer gases with or without inclusion of the Stefan flow and convective flow effects [89]. In the case of a sphere in a quiescent medium, $\delta = a$, where a is the particle radius. If the sphere is large enough for relative velocities to be important (only in the fluid bed and large particles in research reactors), the influence of velocity on k_D can be obtained from Sherwood number correlations (see Sections 7.2 and 7.3). For other geometries other fluid boundary layer treatments are appropriate.

Oxygen arriving at the surface is used up at a rate that determines the surface concentration, p_s. If the particle is so unreactive that p_s is not zero, oxygen will diffuse into the particle interior, as illustrated in Figure 5.9. The oxygen usage at the surface is the combination of superficial surface reaction and diffusion (with internal reaction). If the solid is infinite in extent, or reaction is appropriately fast, reaction is confined to some layer defined by a reaction "penetration depth," δ_R. In solids of finite thickness (e.g., particles) conditions can be found such that the penetration depth is equal to or can exceed the half thickness. There is then a finite oxygen concentration throughout the porous solid.

Since the reaction rate is proportional to the accessible surface area, which is thus determined by the particle thickness or the penetration depth, the reaction rate can be controlled by the ease of access to the interior: clearly, if the solid is very porous, diffusion into the interior is faster; the penetration depth for a given "intrinsic" reactivity is greater; more area is accessible for reaction, which thereby increases in rate with respect to unit superficial surface; and p_s falls until the increased boundary layer diffusion rate comes into balance with the increased internal reaction.

Difference in ease of accessibility to the internal surface, and thus increased reactive area and rate of internal reaction, is most likely a major source of the spread in reactivities at a given temperature, as indicated in Figure 5.10. Indeed,

ISBN 0-201-08300-0

many analysts are increasingly adopting the view that the "intrinsic" reactivity or rate of reaction at a given temperature and surface oxygen concentration, per unit area of accessible surface, is the same for all carbons or chars, and variations in reactivity are then mainly a function of porosity and permeability. This could only be true, if at all, in the absence of catalysts. It is also a difficult concept to test since it requires knowledge of the equations governing intrinsic reactivity, the surface area per unit volume or per unit weight, the penetration depth, and the porosity and permeability or rate of diffusion in and out of the sample – and rarely have all these measurements been made. If "running" reactions are to be modeled, this also requires assumptions regarding pore diameter, diameter distribution, and variation of diameter with reaction progress or burnoff, with provision for pore coalescence and drop in internal surface as reaction proceeds to completion. In spite of these complexities, modeling of all such factors is in progress. Simons and Lewis [90] have recently described one such approach and are claiming such generally good agreement in bounding widely varied experimental data that they indicate some support for a common intrinsic reactivity. Alternative views are that intrinsic reactivity is swamped as a significant parameter by other, mostly transportation factors; or that the model contains sufficient arbitrary constants so that the equations can be made to fit any results. As application of models is extended to a wider and wider range of experimental data, however, they will either survive or fail on the basis of continued agreement or the lack thereof. At this time of writing, such compound models must be considered interim and tentative, but still informative. The problem is that they can demonstrate sufficiency, but rarely also necessity.

D. REACTIONS AND RATE EQUATIONS

It is difficult to believe that intrinsic reactivity can be anything but a key factor. This is the local specific rate of reaction of a surface with oxygen or some other reactive gas. All analyses can reasonably be said to have originated with the Langmuir adsorption isotherm (originally developed for the reaction of oxygen with carbon and tungsten), modified in some instances to take into account variable heats and/or activation energies of adsorption or desorption, thus generating other isotherms (Freundlich, Tempkin, etc.; see, e.g., Ref. 91), and frequently extended on a semiempirical basis to accommodate actual results. For the reaction of carbon with oxygen, or CO_2 or H_2O, producing CO, CO_2, and H_2 by the following reactions:

$$\text{(A)} \quad C + O_2 = C(Ox)_A = CO, CO_2$$

$$\text{(B)} \quad C + CO_2 = C(Ox)_B = CO$$

$$\text{(C)} \quad C + H_2O = C(Ox)_C = CO, CO_2 + H_2$$

these represent potentially reversible reactions between solid carbon and gas, first

ISBN 0-201-08300-0

forming a chemisorbed oxide complex $(Ox)_i$ (i = A, B, or C) that can then decompose, also potentially reversibly, to yield final products. The oxide complex may or may not be of similar or identical form in all cases. The formation or adsorption velocity constant and its reverse are listed as $k_{1,i}$ and $k_{-1,i}$; the decomposition and reverse velocity constants, as $k_{2,i}$ and $k_{-2,i}$. When the above reaction schemes or variants thereof (see Ref. 82) are used, the rate equations, typically quoted with experimental substantiation, have the form [82]

$$\text{(A)} \quad R_{s,O_2} = \frac{k_{1,A}k_{2,A}p_{O_2}}{k_{1,A}p_{O_2} + k_{2,A}(+k_{-1,A})} \tag{5.11a}$$

$$\text{(B)} \quad R_{s,CO_2} = \frac{k_{1,B}k_{2,B}p_{CO_2}}{k_{1,B}p_{CO_2} + k_{-2,B}p_{CO} + k_{2,B}} \tag{5.11b}$$

$$\text{(C)} \quad R_{s,H_2O} = \frac{k_{1,C}k_{2,C}p_{H_2O}}{k_{1,C}p_{H_2O} + k_{-2,C}p_{H_2} + k_{2,C}} \tag{5.11c}$$

where all gas concentrations are values at the superficial surface (p_s in Equation 5.10).

The equations have also been combined in a single rate expression representing the total rate behavior when all three reactions are occurring simultaneously [90]. Implicit assumptions are that the surface is covered uniformly with equally accessible adsorption sites; the heat of adsorption is independent of coverage; and the adsorbing molecule occupies only a single site, that is, it does not dissociate. (If the heat of adsorption falls logarithmically or linearily with coverage, the Freundlich and Temkin isotherms, respectively, are obtained [78].)

Equations 5.11b and 5.11c are identical in form but differ from Equation 5.11a in an important respect. With $k_{-1,A}$ = O we have the classical form of the Langmuir adsorption isotherm. The form given here is that used by Simons and Lewis [90], and it includes the assumption of reverse chemisorption with recovery of O_2 unchanged. This is contrary to evidence from Trapnell [78], quoted above, but the extra term does account for the rate peaking at high temperatures, as shown in Figure 5.10. This writer takes the view that this may be what is needed mathematically, but the theoretical origin is arguable. The three equations are otherwise identical but for the inclusion of product resorption (respectively, CO and H_2) for the CO_2 and H_2O reactions, and the omission of the reverse adsorption term (for reasons not given). One further problem is accounting for both CO and CO_2 as primary products. Typically this is resolved by postulating two different oxide film structures indicated, for example, by C(O) and C(O_2). Numerous hypothetical structures can be thought of; however, direct evidence does not support molecular adsorption. Nevertheless for our purposes here it will suffice that Equations 5.11a, 5.11b, and 5.11c have a reasonable theoretical basis and show good agreement with

ISBN 0-201-08300-0

experiment, noting in passing that distinguishing experimentally between Langmuir, Freundlich, and Temkin isotherms is usually very difficult, and that evidence seems to support dissociative adsorption [88a].

E. REDUCED EQUATIONS AND LIMIT APPROXIMATIONS

Equations 5.11 classically describe reaction rates that, at constant temperature, rise monotonically from zero at zero concentration of the reactant gas up to an asymptotic limit at high reactant gas concentrations. In the limits with the reverse reactions neglected, all the equations reduce to the first-order form at low reactant concentration, p_s:

$$R_s = k_{1,i}p_s = k_{1,i}^0 p_s e^{(-E_{1,i}/RT)} \tag{5.12}$$

and they reduce to the zero-order form at high p_s:

$$R_s = k_{2,i} = k_{2,i}^0 e^{-E_{2,i}/RT} \tag{5.13}$$

Reaction order is therefore a key index for identifying the dominant mechanism (if one is dominant), noting from an earlier argument that the corresponding activation energies in the case of oxidation, $E_{1,A}$ and $E_{2,A}$, can be expected to have values in the regions of 5 kcal or less, and 40 kcal, respectively.

Equations 5.11 are also the expressions describing the intrinsic reactivity of the accessible surface. This would be the superficial surface of a totally dense, smooth solid; the surface area is greatly increased, however, when there is internal reaction, and the rate equations are then modified substantially. The modified expressions have been obtained both phenomenologically and mechanistically, yielding essentially the same result. The expressions are generally obtained for simplifications of the intrinsic reactivity equations, using the approximation

$$R_s \simeq k_{\text{eff}} p_s^n \tag{5.14}$$

where n is the effective reaction order, and k_{eff} is a composite or effective rate constant. Equation 5.14 clearly reduces to the limit expressions 5.12 and 5.13 for $n = 1$ and 0, respectively. With this reduction the specific total reaction rate, R_T, for all internal and external reactions, but referred to unit area of external surface, can be written [82] as

$$R_T = f \cdot k_{\text{eff}} p_s^n + \left(\frac{2}{s}\right) \sqrt{k_{\text{eff}} A_i D_i} \;\; p_s^{(1+n)/2} \tag{5.15}$$

where f is a roughness factor; s is a sample shape factor (= 1 for a plane, 2 for a cylinder, and 3 for a sphere); A_i is the internal surface area per unit volume; and D_i is the internal diffusion coefficient. The first right-hand term is for the external

ISBN 0-201-08300-0

(superficial) surface reaction, and the second one is for the internal, pore-surface reaction. It may be noted that, if $n = 1$ (intrinsic first-order reaction), $(1 + n)/2 = 1$, so the effective reaction is also first order; but if $n = 0$, an intrinsic zero-order reaction appears as an effective half-order reaction. The internal reaction term frequently outweighs the superficial surface term, which is therefore often neglected. The overall reaction is influenced as much by the internal transport rate as by the chemical kinetics. The effect of the square-root term is to halve the activation energy.

The above expressions are valid for the condition that the penetration depth of the reaction is smaller than the half thickness, or radius in the case of a cylinder or sphere. There is then a nonreacting region in the sample center where the oxygen concentration is zero. As penetration increases (for example, with reduced reactivity, due to reduced temperature), the oxygen concentration at the center becomes finite and the internal transport rate no longer influences the reaction rate. In the limit when the oxygen concentration approaches the surface value, p_s, the total rate of reaction for the whole sample, R_w, can be written as

$$R_w = (A_s \cdot f + A_i V) k_{\text{eff}} p_s^n \tag{5.16}$$

where V is the sample volume, and A_s is the sample superficial area. If a is a radius or half thickness, the specific total reaction rate becomes size and shape dependent:

$$R_T = \left[\text{f} + \left(\frac{a}{s} \right) A_i \right] k_{\text{eff}} p_s^n \tag{5.17}$$

and the temperature coefficient of the reaction is determined by the operational activation energy assigned to k_{eff}.

When the reaction is influenced jointly by boundary layer diffusion and by the kinetic behavior, it is necessary to solve Equations 5.10, 5.11, and 5.15 simultaneously (noting that 5.15 already incorporates an approximation of 5.11).

1. If the surface is nonpermeable, Equations 5.10 and 5.11 can be solved simultaneously, yielding a quadratic in R_T [92]. For the condition that k_D is large (fast diffusion), the quadratic reduces to the Langmuir isotherm with $p_s \simeq p_0$. For k_2 very large (fast desorption) the quadratic reduces to what is often known as the "resistance equation": $1/R_T = (1/k_D + 1/k_1) p_0$. For k_1 very large (fast adsorption), however, the residue of the quadratic factorizes, yielding [92, 64]

$$(R_T - k_D p_0)(R_T - k_2) = 0 \tag{5.18}$$

which implies that the reaction is *either* diffusion controlled ($R_T = k_0 p_0$) *or* desorption controlled ($R_T = k_2$), with no transition region in passing from one to the other as temperature rises.

ISBN 0-201-08300-0

2. If the solid is porous, we can solve between Equations 5.10 and 5.15 or 5.16 for specific cases to establish the equation form. For $n = 0$ in Equation 5.16 the reaction rate is desorption controlled and independent of boundary layer diffusion. For $n = 1$ in Equation 5.15 or 5.16 a modified resistance equation is obtained with some compound k in place of k_1. For $n = 1/2$ in Equation 5.16 (zero order, $n = 0$ in Equation 5.15) we obtain another quadratic:

$$R_T^2 + \left(\frac{4}{s^2}\right)\left(\frac{A_i D_i k_{\text{eff}}}{k_D}\right) R_T - \left(\frac{4}{s^2}\right)(A_i D_i k_{\text{eff}}) p_0 = 0 \tag{5.19}$$

Since this also factorizes, there is again the potential for a two-branched rate curve with discontinuous slope between the two branches. Such curves have been reported.

3. If the particle is porous but reaction at the surface is very rapid, that is, k_1 and k_2 are both large, p_s in Equation 5.15 or 5.17 must become very small for any finite R_T, and Equation 5.10 yields

$$R_T = k_D p_0 \tag{5.20}$$

The same result is obtained from Equation 5.19 by dividing by k_{eff} and allowing k_{eff} to become very large. Equation 5.20 specifies the diffusion-limited reaction.

6. PROPAGATION AND STABILIZATION OF COAL FLAMES

6.1. Definition and Objectives

Flames may be generally classified as either stabilized (steady state) or traveling (unsteady state). Flames are usually described as stable if time-averaged values of local temperatures, compositions, gas concentrations, and so forth are invariant with time. There are generally fluctuations about the mean, and these can be either regular or random with low or high amplitude. An increase in the amplitude of the fluctuations, particularly random ones, indicates that the flame is becoming unstable. In a combustion chamber total instability can mean loss of ignition, and the flame extinguishes or is blown off. Unscheduled exposion flames are not unstable in the same sense and are better described as traveling flames. The preceding comments apply generally to all fuels. Coal, however, includes some special cases.

Flames stabilize for one of two reasons or for a mixture of the two. We consider three cases.

1. If flame is propagating through a combustible mixture in a tube, there is a traveling reaction front that, in principle, can be brought to rest by blowing the

ISBN 0-201-08300-0

cloud in the direction opposite to the direction of travel of the flame, and at the same speed. This is velocity balancing. It applies to grate-fired systems and to the plug flow special reactor furnace. It would also apply to a free-jet, unenclosed flame.

2. For all practical purposes all flames in combustion chambers are enclosed jet flames. As mentioned briefly in Section 3.4, the enclosed jet (and, more particularly, the swirled jet) generates a backmix flow of hot combustion products into the issuing jet stream. In the particular case of the U- and W-flame boilers a vertical O-shaped or OO-shaped flow is set up and fresh fuel is injected, in effect, into the downflowing arm of the hot circulating gases. The cold particles heat rapidly after being plunged into the hot gases and ignite some distance downstream from the point of injection. A flame front forms at this location, and a flame speed can therefore be defined; however, this is now a dependent, not an independent, function of conditions, and there is no velocity balancing in the sense described above. This condition applies to U and W flames and to unswirled straight shot flames. It might also apply to the recycled fast fluid bed.

3. When backmix is very intense, the limiting case is the perfectly stirred reactor. The stirred volume may be defined by the physical walls of a combustion chamber, or it can be defined as an aerodynamic volume effectively defining a stabilized "flame ball." In the second case there may be a definable "flame surface," but with perfect stirring there is no definable flame front as a surface effectively separating unburned from ignited material in a definable flow stream. Consequently, without a flame front no flame speed can be defined. With real burners, such as the swirled "self-turbulating" burner illustrated in Figure 3.16, the flame ball is only "well stirred" and a flame front can generally be defined. As in case (2) above, however, the definable flame speed is a dependent function of conditions and has no primary influence on stability. Stability in these circumstances is dependent on two requirements. First, the flame ball must be mechanically stable, which evidently requires that the integral of momentum of all flows over the flame ball surface be zero. Second, the thermal criteria of stabilization must be satisfied (as amplified below). These conditions apply to backmix p.c. flames and to the fluid bed.

6.2. Thermal Propagation of Plane Flames

In a plane flame propagating through a dust cloud or a fixed bed there is an attenuating temperature field ahead of the flame front because of heat transferred from the flame, as illustrated in Figure 6.1. In this representation a finite ignition distance, L_{ig}, is assumed that can also be related to the ignition time, t_{ig}. The mechanism of heat transfer can be (a) conduction, (b) radiation, or (c) turbulent dispersion exchange. Conduction was long regarded as the true mechanism in solid beds [14], but it is now recognized that radiation is as important or more so [14, 93]. Radiation is the dominant heat exchange mode in stabilized p.c. flames; turbulent dispersion dominates in explosion flames.

ISBN 0-201-08300-0

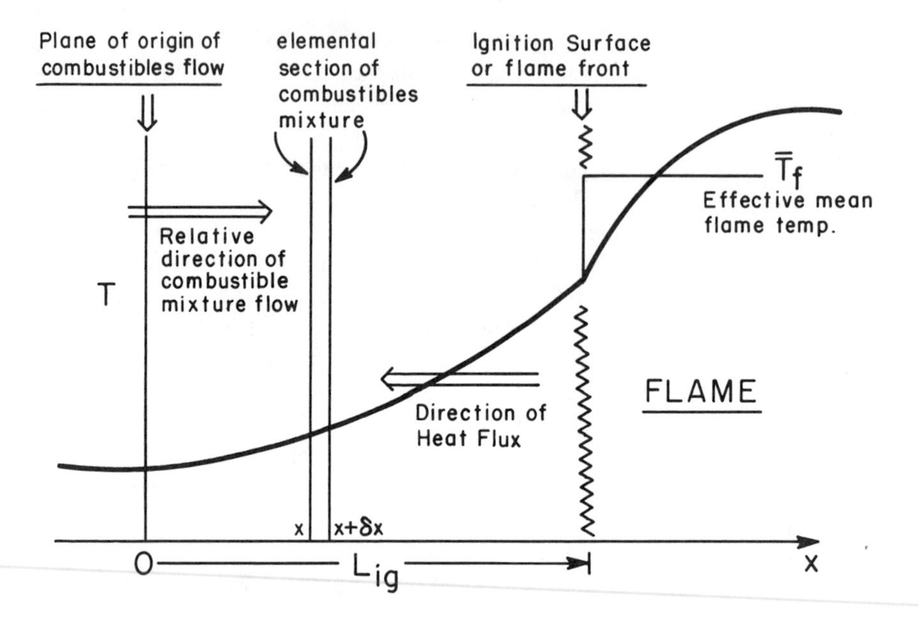

Figure 6.1. Schematic representation of a preheat zone between a combustibles injection plane at $x = 0$ and a flame front or ignition surface at $x = L$ig.

A. ANALYSIS

Analysis of radiating flames is complex, but it can be very substantially simplified. If we assume that there is a thermal flux due to conduction, radiation, or thermal dispersion crossing the flame front to enter the unburned cloud ahead of the flame, this flux is progressively attenuated as it penetrates the cloud. A heat balance on a thin element of the cloud establishes that the fraction of flux adsorbed in each element is a constant, or approximately so within the limits of the radiation assumptions. The solution obtained to the appropriate differential equation describing the system is that the flux attenuates exponentially with distance. If T is the temperature at some location x from the flame front, and if for this argument the assumption of a fixed ignition temperature, T_i, is adopted for the three cases of conduction, radiation, and turbulent exchange, the expression obtained [94–97, 93, 88] for the temperature profile is

$$T - T_0 = (T_i - T_0)e^{(-Kx/d) + (-K_T)} \tag{6.1}$$

where d is some characteristic dimension of the system, usually the burner, or furnace, or explosion tube diameter; K is a dimensionless attenuation constant; and K_T is a term that is a function of temperature and that appears only in the

ISBN 0-201-08300-0

solid bed analysis with radiation [93] (treated as radiative conductivity). If the whole of the flux crossing the flame front, written as $(\text{flux})_i$, is absorbed by the coal cloud or bed, of particulate concentration D_0, this is related to the temperature rise by

$$(\text{Flux})_i = (\rho_0 \bar{C}_p + D_0 \bar{C}_d) S_0 (T_i - T_0) \tag{6.2}$$

where ρ_0 is the (cold) air density; $\bar{C}_p$ and $\bar{C}_d$ are the mean specific heats of air and solid particles, respectively; and S_0 is the flame speed. If the heating zone thickness, δ_i, is defined as the distance from the flame front at which the temperature has risen by 1%, δ_i is given by

$$\delta_i = \frac{4.6d}{K} \tag{6.3}$$

A mean rate of heating can also be defined by averaging linearly through the ignition point and the point $x = d/K$. This yields

$$\dot{T} = \frac{0.63 S_0 (T_i - T_0) K}{d} = \frac{3 S_0 (T_i - T_0)}{\delta_i} \tag{6.4}$$

These expressions can be evaluated only by adopting experimental values for certain coefficients. Nevertheless, the numerical values obtained are informative. Evaluation requires an estimate of the ignition temperature. This is itself a function of conditions, but with high activation energies of reaction it will generally be somewhat insensitive to conditions except when the fuel sample is small or the heating rate is high. In the "crossing-point" method, for example, coal samples are heated slowly at known rates, 15–75°C/min, and the self-ignition point is determined as the temperature where the lagging sample temperature matches the furnace temperature. Ignition temperatures reported range from 300 to 600°C [98]. In a p.c. flame operated by Csaba [99], with values of 10^3 °C/sec for $\dot{T}$, T_i was estimated to be about 300°C. In another p.c. flame, however, heating at 10^4 °C/sec, T_i was estimated as 1000°C [41, 45]. These differences in ignition temperature can be accounted for as differences in ignition mechanism. At the lower heating rates, and particularly in the p.c. flame, ignition can be identified as the onset of pyrolysis. This was first proposed by Faraday and Lyell [100] and is generally known as the Faraday mechanism. Pyrolysis is governed by the equations given in Section 5.3. As the heating rate increases, the temperature of onset of pyrolysis also increases, and at a heating rate of 10^4 °C/sec pyrolysis is delayed until about 1200°C. In Howard and Essenhigh's experiments [41, 45] it was then concluded that ignition was occurring before pyrolysis, heterogeneously, and governed by thermal considerations.

Delayed pyrolysis also appears to have been occurring in a number of small,

ISBN 0-201-08300-0

experimental p.c. flames [101–103, 40]. The flames were proportionately small with cross sections on the order of 10 cm or substantially less. Most of the experimental designs had such a substantial thermal load on the flame that the char combustion was not sustained. The flames were mainly volatile flames and could be stabilized only at low supply velocities, on the order of 10 cm/sec. Computer analysis of the flames [40] then indicated that the particles were pyrolyzing behind the flame front, but the flow speeds were slow enough for the volatiles to diffuse upstream so that the flame formed was a diffusion flame.

B. EVALUATION

We thus have the following flame systems.

(i) *Conduction/diffusion (p.c.) flame.* This is the class of small flames that pyrolyze downstream of the flame front with upstream diffusion (Smoot-Horton-Williams mechanism [40]). For a pure conduction flame

$$K = \frac{S_0 d}{\alpha} \tag{6.5}$$

Here α is the thermal diffusivity of the cloud $= \lambda/\rho_0 c_p$, where ρ_0 is the cold cloud density, and λ is the thermal conductivity.

In the pure conduction case, with pyrolysis ignition (Faraday mechanism), the approximate values of δ_i, T, and S_0, from Equations 6.1–6.5, are then inconsistent with experimental results. Flame speeds would be 10 cm/sec with an ignition distance of 1 mm and a heating rate of 10^5 $^{\circ}$C/sec. The ignition distance is too short, and the heating rate is too high, to permit pyrolysis ignition. At a more realistic ignition distance and heating rate of 1 cm and 10^3 $^{\circ}$C/sec, respectively, the flame speed would be only 1 cm/sec, which is an order of magnitude too slow. The proposed Smoot mechanism satisfies conditions reasonably quantitatively. The result is not amenable, unfortunately, to demonstration by simple calculation. Details are available, however, in the published record [40].

(ii) *Conduction/radiation (solid bed) flame.* This is the class of solid bed reaction systems that includes fuel beds burning on grates, fuel beds in shaft reactors, and porous bed models of *in situ* gasification. Here again the simple conduction model is inadequate. Data are sparse, but an evaluation of burning dowel rods in a simulated porous bed [93] showed that the conduction prediction was too low by a factor of 5. The inclusion of radiation to create the additional K_T term in Equation 6.1 satisfied the numerical requirements. The K_T term is given by

$$K_T \simeq \left(\frac{5\sigma}{k_r \lambda_b}\right) f(T) \tag{6.6a}$$

ISBN 0-201-08300-0

with $$f(T) = 1/3\,(T_i^3 - T^3) + \left(\frac{T_0}{2}\right)(T_i^2 - T^2) + T_0^2\,(T_i - T)$$

$$+ T_0^3 \ln\left(\frac{T_i - T_0}{T - T_0}\right) \qquad (6.6b)$$

where σ is the Stefan-Boltzmann radiation constant, k_r is the radiation attenuation coefficient in the fuel bed, and λ_b is the effective thermal conductivity of the bed. Equation 6.5 still applies. Evaluating K_T for the value of T at δ_i, and adopting a value of k_r of 10 cm^{-1}, which corresponds to a δ_i of about 1 cm, we find that the propagation rate of the burning or reaction front is about 0.01 cm/sec, or roughly 1 ft/hr, in agreement with the estimate of design requirements in Section 4.3.

The limit to fuel feed rate on a grate is provided by the conduction/radiative heat transfer into green fuel. If the feed rate is too high, the net motion of the reaction front will be downstream instead of being stationary or upstream.

This simple analysis ignores the effect of convection due to gas flow. In the case of overfeed beds, this can accelerate the heating of green fuel being fed on top of the bed, but it does so at the expense of removing heat from the lower levels of the bed. If these are overcooled, reaction can be extinguished and this could eventually blow the flame out. Orning [104] has discussed this aspect qualitatively (see also Ref. 105), but a quantitative analysis is still lacking. Increased air supply can raise the fuel bed temperature, and with radiation the flux across the ignition plane will increase by the fourth power. By Equation 6.2 the flame speed will increase likewise, being proportional to the ignition plane flux if other parameters are unchanged.

(iii) *Radiation (p.c.) flames.* For these flames we have $K_T = 0$ and

$$K = k_r d \qquad (6.7)$$

where k_r is the attenuation coefficient of the cloud, given approximately by $3D_0/4a\sigma_d$, where a is the particle radius and σ_d is the particle density. Values of k in p.c. range from 0.04 to 0.4 cm^{-1}, and from Equation 6.3 σ_i is 10–100 cm, rising as the dust concentration, D_0, drops. In this system the quantity $(\text{flux})_i$ can be estimated independently, in principle, as the radiative flux from a flame of appropriate grayness at an effective flame temperature, $\overline{T}_f$. On the assumption that this flux and the ignition temperature were constant, Equation 6.2 was tested, using Csaba's data [99], by plotting $1/S_0$ against D_0. The result is illustrated in Figure 6.2. Evaluating the additional coefficients from the slope and intercept of the line obtained yielded the following results: $T_i = 370°C$, $\dot{T}_i = 615°C/sec$, $\overline{T}_f = 1400°C$. The ignition temperature implies pyrolysis or Faraday ignition. This is evidently consistent with the heating rate $< 10^3$ °C/sec.

In other one-dimensional flame experiments [41, 45, 106] different results were obtained. Ignition temperature and heating rate were found to be much

ISBN 0-201-08300-0

higher, with values of about 1000°C and 20,000°C/sec, respectively. The furnace system used was similar to Csaba's but had a water-cooled tube bank for the flame to stabilize against. With the inlet flow adjusted to give an ignition distance of 5–10 cm, there was some absorption of heat by the tube bank and the flame speed was reduced to 30 cm/sec (see below). These numerical values are none the less consistent with each other when inserted into Equations 6.1–6.4 and 6.7.

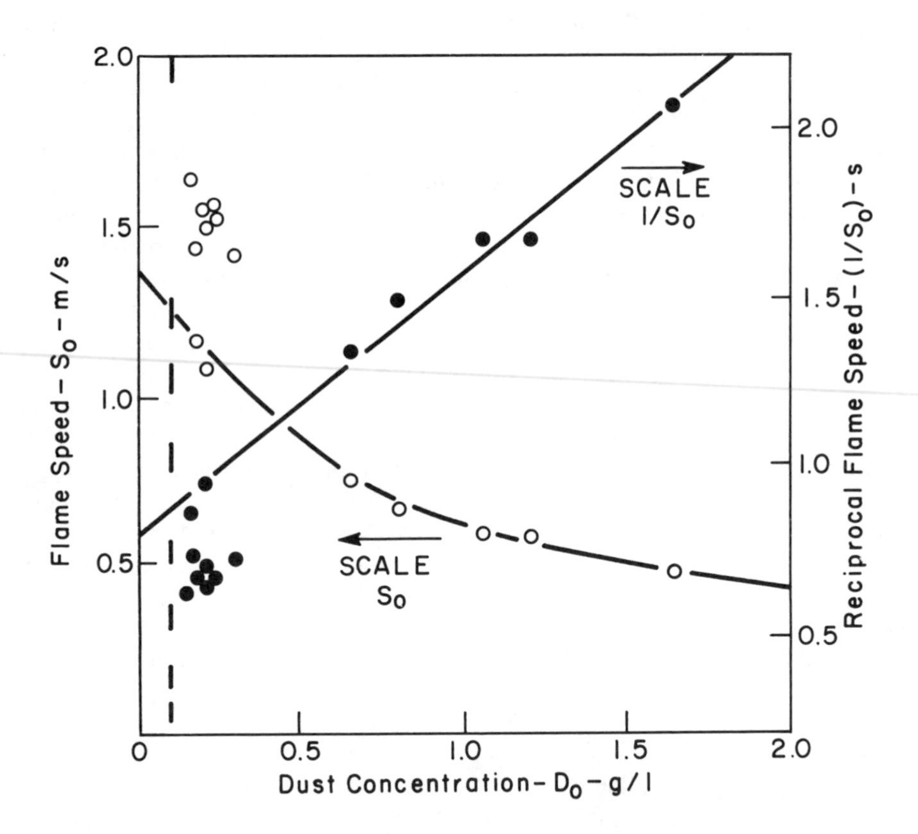

Figure 6.2. Variation of flame speed with concentration in coal dust clouds. (Data source: Csaba [99]; see also Ref. 88.)

More recent experiments [106] have also provided more detailed substantiation of the simple radiation theory with finite ignition distance or time, t_i [97]. With infinite ignition distance there is a maximum or "fundamental" flame speed, S_0, that is a function of the flame conditions given by Equation 6.2. With finite ignition distance the ignition time is related to the cold inlet velocity by [97]

$$V_0 = S_0 (1 - e^{-k_r V_0 t_i}) \tag{6.8}$$

ISBN 0-201-08300-0

Substituting for k_r yields the exponent group $3D_0V_0t_i/4a\sigma_d$. Since this depends on both D_0 and V_0, Equation 6.8 is not a direct relation between V_0 and t_i. As experiments were carried out by Cogoli et al. [106], the fuel feed rate was kept constant and the air rate was changed, thus changing both V_0 and D_0. The product, D_0V_0, however, was constant under such conditions, being equal to the area firing density, J/A, at a value of 5.8 x 10^{-3} g/cm^2 sec (42.6 lb/ft^2 hr, or about 500,000 Btu/ft^2 hr; cf. Table 4.2). Insertion of appropriate values yields

$$t_i = 0.105 \ln\left(1 - \frac{V_0}{S_0}\right) \quad \text{(sec)} \tag{6.9}$$

for the fixed particle sizes and feed rate density used by Cogoli. In those experiments S_0 was 42 cm/sec, and Equation 6.9 is plotted, as illustration, in Figure 6.3, together with the limited experimental data. There is considerable scatter, but these data are the only extant ones on ignition time in such experiments, and they do bracket the calculated curve.

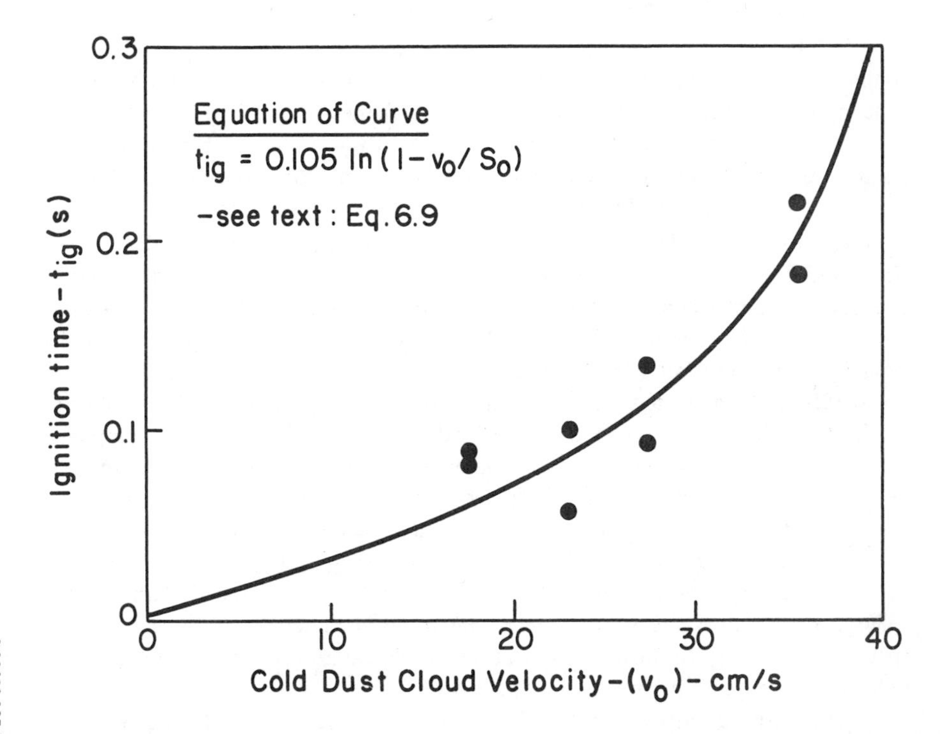

ISBN 0-201-08300-0

Figure 6.3. Variation of ignition time with cold coal-dust cloud velocity in a plug-flow furnace. (Data source: Ref. 106.)

If the fuel feed rate is changed to keep D_0, instead of the firing rate, constant as the air rate is changed, a family of similar curves is obtained, one curve for each (constant) D_0. These curves are discussed in more detail, based on the analysis cited [97], by Field et al. [69]. The curves have the same upward-concave shape. Two other characteristics of the curves noted in the original article [97] are more evident in Field's discussion: ignition time goes to a finite low value, not to zero, as velocity goes to zero; and the limit value increases with decreasing concentration. Ignition time goes to infinity as V_0 goes to S_0, with S_0 again increasing as D_0 decreases. The net result is that the curves cross each other as the concentration is increased.

Ignition time increases in proportion to particle diameter, in this analysis, so extrapolation to other particle sizes is easily accomplished.

These flames evidently differ phenomenologically only in flame thickness from the Conduction/Diffusion flames discussed in B(i) above. All the flame systems are in laminar flow, or effectively so. The maximum Reynolds Number in the Howard [41, 45] and Cogoli [106] experiments was about 3000, just above the tube bank. This is barely outside the laminar flow range. The Reynolds Number through the tube bank, which would also act as a flow straightener, was in the range 300 to 3000, depending on the characteristic dimension adopted. With the subsequent temperature rise and increase in gas viscosity, the Reynolds Number remained below 2000. In the Csaba experiments [99], the flow was all in the laminar region. The mechanistic differences between the flames would appear to depend on the different thermal loads on the flames referred to earlier. As the thermal load is increased, the propagation evidently can change from a radiation-dominated mechanism with flames up to one meter thick or more, to a conduction-diffusion mechanism with flames only a centimeter or two in thickness.

(iv) *Explosion flames.* In explosion flames we are concerned with propagation velocities. Explosion flames tend to be erratic in their reproducibility characteristics even under controlled experimental conditions. Nevertheless, a reasonable body of knowledge is now available [107, 108], and is being expanded. In the typical experimental method a tube or gallery, up to 10 or 15 ft in diameter, and up to 1000 ft long, is used. The dust is initially in thin layers on the gallery floor or on displaceable shelves. It has to be dispersed to start with by a gas-air or gunpower explosion or some other combustible charge at one end of the gallery, which must be closed. On that account thermal expansion due to the flame creates a wind down the gallery that aids dispersion of the dust, creating a fast moving, preformed cloud for the flame to travel through. There are, therefore, two flame speeds to be defined: the flame speed through the cloud, S_0; and the flame speed relative to the gallery, S_g, which is the sum of the flame speed in the cloud, S_0, and the cloud speed, v_0. It is easy to show, however, that S_0 and v_0 are about equal in magnitude in high speed explosion flames, so we have $S_g \simeq 2v_0$.

Flame speeds in explosion galleries can reach high values. Figure 6.4 illustrates the variation with time under different experimental conditions. As the figure shows, such flames tend either to slowly die out or to accelerate dramatically with

ISBN 0-201-08300-0

recorded values as high as 1400 m/sec (the speed of sound in cold, dry air is 331 m/sec). The cloud velocity at even half this speed is high, and the flow ahead of the flame is highly turbulent. This is believed to be the main mechanism of heat exchange between the flame tip and the unburned cloud.

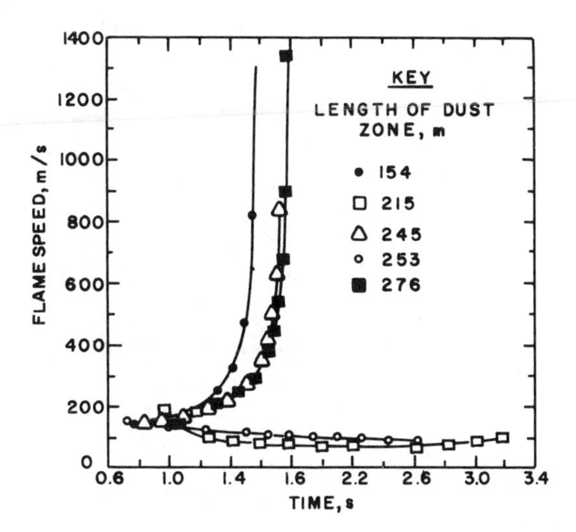

Figure 6.4. "Flame speeds for coal- and rock-dusted zones ignited by flame of a 25 ft zone of 6.5% methane-air." (Data obtained by Edward M. Kawenski, Supervisory Mining Engineer, U.S. Bureau of Mines; source: Ref. 108.)

The basis for an approximate analysis of the system exists in the equations already given. Equation 6.1 still applies (with $K_T = 0$), using turbulent dispersion theory [109], from which we obtain [88]

$$K = \frac{v_0 d}{D_T} = Pe \tag{6.10}$$

where the characteristic dimension, d, is the gallery diameter; v_0 is the *cloud* velocity relative to the gallery but may be approximately equated to the flame speed through the cloud, S_0; D_T is the turbulent dispersion coefficient; and K is thus identified as the cloud Peclet number, Pe. At cloud speeds in the range $0.1S_g$ to S_g (the peak gallery flame speed), the Reynolds numbers mostly exceed 10^5, and in this region Pe tends to a constant value of approximately 5. Inserting this value into Equation 6.3, we find that

$$\delta_i \simeq d \tag{6.11}$$

ISBN 0-201-08300-0

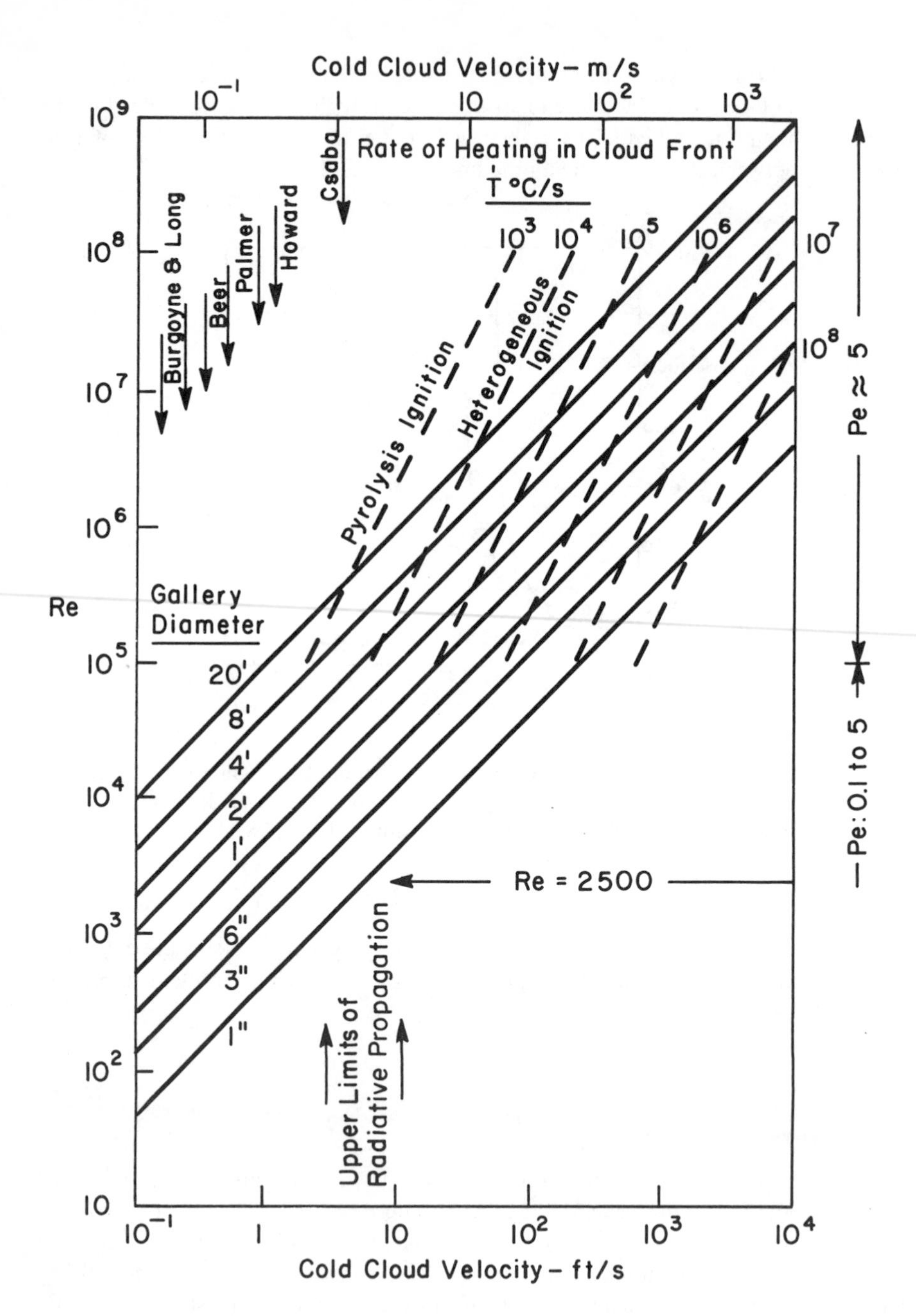

Figure 6.5. Variation of Reynolds number with cold cloud velocity in an explosion gallery for different gallery diameters. (Source: Ref. 88; reference sources: Burgoyne and Long [101], Beer [110], Palmer [102], Howard [41, 45], Csaba [99].)

ISBN 0-201-08300-0

No prediction of flame speed, S_0, is possible from this analysis, only prediction of flame speed ranges. This can be done by equating $S_0 \simeq v_0$, and utilizing Equation 6.4. The method is illustrated in Figure 6.5, which is a plot of two sets of lines. The solid lines are the variations of Reynolds number with cold cloud velocity, v_0, for different tube (gallery) diameters, using the definition of Reynolds' number. The dashed lines are variations of v_0, equated to S_0, with Reynolds' number, using Equations 6.4 and 6.11, at a constant heating rate from 10^3 to 10^8 ^oC/sec. The intersection of a line of constant T with a given gallery diameter is the cold cloud velocity – and hence the approximate flame speed relative to the cloud that would be obtained. Alternatively, a known flame speed and gallery diameter will give the heating rate required by those conditions. At 1000 m/sec, for example, in a 10 or 12 ft diameter gallery, heating rates should be in the region of 10^6 ^oC/sec.

Included on the plot in Figure 6.5, for comparison, are some of the laminar flame speed measurements reported in the past [101, 102, 99, 41, 45, 110, 111].

Table 6.1

Approximate Flame Speeds and Burning Times for Different Ignition Mechanisms at Different Heating Rates[a] and Flame Propagation

Rate of heating deg C/sec.	System	Flame speed (m/sec) Cond/diff	Radiation	Turb. exch.	Mechanism	Burning time (sec) Vols.	Coal Char	Remarks
$10-10^2$	Fixed bed	–	10^{-5}	–	(Faraday & Lyell)	10^2	10^3	Large particles used (run-of-mine or crushed)
$<10^3$	Plane flame	–	0.5–1.5	–	Faraday & Lyell [100]	0.1	1	Ign. T about 370°C at pyrolysis temp. Long flame length ~ meters. Laminar flame
10^4	Plane flame	up to 0.3	–	–	Smoot, Horton, & Williams [40]	0.1	–	Diffusion flame formed by volatiles generated downstream of flame: mainly vols. burning. Short flame length ~ centimeters. Laminar flame with high thermal load
10^4	Plane flame	–	0.3–0.5	–	Howard & Essenhigh [41, 45]	0.1	1	Heterogeneous ignition followed by pyrolysis then by char burn out. Long flame length ~ meters. Laminar flame
$>10^5$	Explosion flame	–	–	$10-10^3$	Wheeler[see 88](?)	0.1 (total coal)		Heterogeneous ignition and combustion at limit of very fast flames. Turbulent flame

[a]Source: Ref. 88.

ISBN 0-201-08300-0

C. SUMMARY OF RESULTS

Table 6.1 summarizes the foregoing results. It compares flame speed for the different systems with rate of heating and propagation mechanism.

6.3. Combustion in Stirred Reactors

The backmix or stirred reactor is more representative of actual burner-combustor combinations as described in Section 6.1. There is essentially no propagation problem; the counterpart, however, might be the mechanical stability aspect. If we have a stirred reactor region or a flame ball created by a bluff body, or aerodynamically (e.g., by swirl) with free surface boundaries, it is important that this region not migrate. If it does, the flame ball can move out of the combustion chamber and extinguish. This is best described as blow-off. The mechanical stability criterion, as remarked in Section 6.1, appears to be a vanishing integral of momentum of all flows over the flame ball surface.

In addition there must be chemical or combustion stability, defined by a match of heat generation and heat loss and satisfaction of what is often known as the Semenov criterion. Analysis is complex because of the involved flow patterns that must be accounted for, except for the limiting case of the perfectly stirred reactor (PSR). Even this can become very complex if several reactions are involved. Here are presented two special cases to illustrate the method and problems involved. First are summarized, in general, the equations for one-step reaction in a PSR.

A. PSR ANALYSIS

In a stirred reactor with reaction between fuel and oxidant, heat is generated that, at steady state, is then lost only by convective flow out of the reactor and by wall loss. If the rate of reaction is proportional to the rth power of the fuel concentration (initial mass fraction value C_f^0), and to the $(n - r)$th power of the oxidant concentration (initial mass fraction value C_{ox}^0), defining the combustion efficiency as $\eta_I = (1 - C_f/C_f^0)$, it can be shown [25] that the combustion efficiency varies with temperature and with residence time, τ_s, in a one-step reaction by

$$\frac{\eta_I}{(1 - \eta_I)^r (1 - \phi\eta_I)^{n-r}} = (C_f^0)^r (C_{ox}^0)^{n-r} [\tau_s(\rho_0^{n-1} k_0)] e^{-E/RT} \qquad (6.12)$$

It may be noted that the group $\rho_0^{n-1} k_0$ has units of reciprocal time and represents a *mass*-based kinetic constant. The variation of η with T is a sigmoid curve rising from zero η at zero T to approach unity for η as $t \rightarrow \infty$. A separate equation in η and T may be written for the heat lost from the system. With convective flow loss

ISBN 0-201-08300-0

and conductive wall loss the expression obtained, writing η_{II} to distinguish between generation and loss, is

$$\eta_{II} = \left(\frac{RT}{E} - \frac{RT^0}{E}\right)\left(\frac{1 + \beta\tau_s^0 k_0}{\omega}\right) \quad (6.13)$$

where ω is a dimensionless initial concentration group, $C_f^0 Rh_f/E\bar{C}_p$; and β is a dimensionless heat exchange coefficient for the wall, $h(A_h/V_c)/\rho_0\bar{C}_p k_0$, where h is the actual heat transfer coefficient, and A_h is the heat transfer surface. Equation 6.13 is the equation of a straight line with a slope controlled by the thermal input, $C_f^0 h_f$, and by the residence time, τ_s (or firing rate). The intersection of Equations 6.12 and 6.13 is the condition for stability, given by

$$\eta_I = \eta_{II} \quad (6.14)$$

Equation 6.14 and the tangency of slopes

$$\frac{d\eta_I}{dT} = \frac{d\eta_{II}}{dT} \quad (6.15)$$

define the boundary between ignition and extinction, or the conditions of criticality. The flame can be extinguished by changing, in particular, the firing rate, affecting τ_s; the fuel/air ratio, affecting C_f^0; the heat transfer coefficient or heat transfer surface, for example, by changing the bed depth in a fluid bed; and possibly the reactivity constants k_0, E, n, and r, by injecting some reaction inhibitor. Further details of relevant equations and their significance may be obtained from appropriate reference sources [25, 112].

B. EVALUATIONS

(i) *PSR Research Reactor with Coal.* In current research [43], using a 200 cm^3 spherical cavity, coal in air is injected through four jets in the same plane aimed at the center. The exhaust is a port perpendicular to the plane of the jets. The reactor operates at about 1200°C, and ASTM VM analysis of char residues indicates that the principal reactions are pyrolysis and (immediate) combustion of most of the pyrolysis products. The VM analysis of the char shows that the VM residue is a function of temperature, and there is evidence of char combustion.

It is concluded that combustion is controlled mainly by the rate of pyrolysis, with combustion occurring instantaneously. Analysis of the VM residue values indicates enhanced VM yield under these conditions, with a Q factor (Section 5.4) of 1.8. The actual solids loss or gasification in the reactor was about 60–70%.

ISBN 0-201-08300-0

Maximum pyrolysis was Q times the ASTM VM value. The difference was checked as Q times the ASTM VM value in the residue extracted. The ratio of loss to maximum possible loss in the reactor defined the pyrolysis (combustion) efficiency. When this was treated initially as a one-step, first-order reaction ($n = r = 1$ in Equation 6.12), a reaction rate group of $\eta/(1 - \eta)\tau_s$, was obtained, and Figure 6.6 is an Arrhenius plot of the resulting data. Scatter is appreciable, but application to the only extant data on coal flames in such circumstances is illustrated. Pyrolysis time is 10–15 msec, which is a value also reported by others [52]. It is also consistent with the kinetic models being developed (Section 5.4). Rates of heating in the reactor lie between 10^5 and 10^6 $^{\circ}$C/sec.

These data have additional interest for the following reason. The prior analysis of the explosion flame (Section 6.2B(iv)) indicated that turbulent dispersion creates a stirred volume of axial length about the width of the explosion gallery. Although explosion flames in galleries can be 100–200 ft in overall length, it now appears that 60–70% of reaction occurs in an axial distance of about one diameter, in a time of 10–20 msec, and reaches temperatures of 1200–1400°C. These times and temperatures are matched quite remarkably closely by the (limited but sufficient) stirred reactor data. Figure 6.6 may, therefore, be indicative of behavior at the tip of an explosion flame.

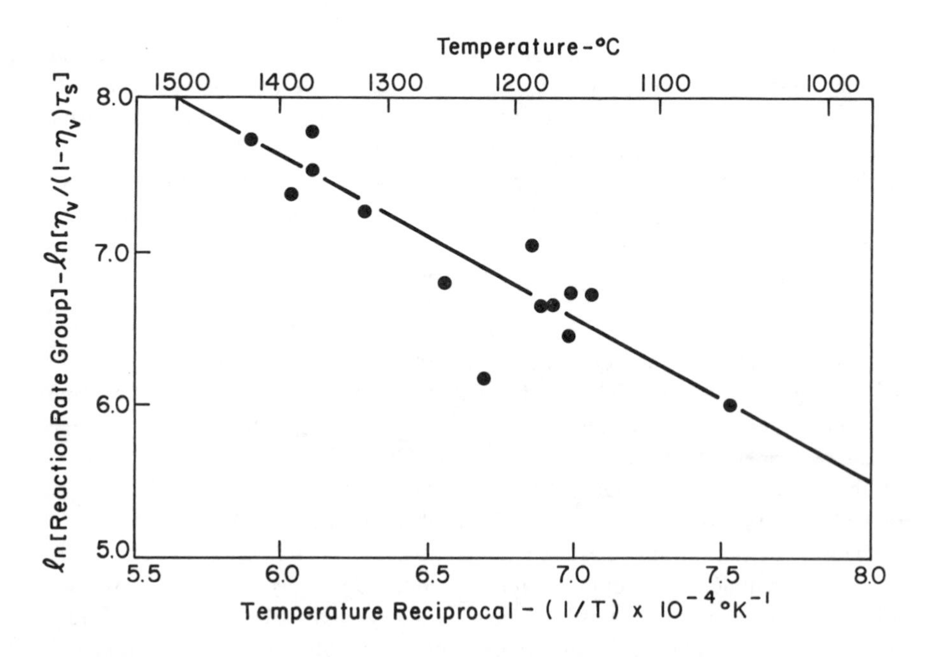

Figure 6.6. Variation of reaction rate, $\eta_v/(1 - \eta_v)\ \tau_s$, with temperature T on an Arrhenius plot for reaction (pyrolysis) of coal in a quadrajet stirred reactor. (Source: Ref. 43.)

ISBN 0-201-08300-0

(ii) *Stirred Reactor Flame Ball Model.* The application here is to a flame ball in a boiler. Boiler dimensions, shapes, and firing equipment were described in Section 3.4. With tangential firing one may approximate the flame volume at the burner levels as that in a single stirred reactor, with a backmix section (see Ref. 26) of dimensions comparable to the boiler width. Multiple flare (front and opposed firing) burners, however, may be viewed as creating a number of much smaller flame balls more comparable in diameter to the burner diameter. Their stability behavior has been analyzed in a simple but informative model [113].

It is assumed for the model that production of volatiles is instantaneous and the balance of heat produced in the stabilized flame ball is from the partial burning of the char particles formed by pyrolysis. The balance of the char combustion takes place in a subsequent plug flow region in the upper section of the boiler. Writing $f_v h_f$ as the fraction of *heat* produced by the combustion of the volatiles, we find that the balance produced by the char combustion is $(1 - m/m_0)h_f$ if m_0 is the initial char particle mass and m is its mass leaving the stirred flame ball region. For a char particle of mass $m = 4/3\ \pi a^3 \sigma_c$, where a and σ_c are its radius and density, respectively, the specific reaction rate, R_T (Section 5.5), can be written as

$$R_T = \sigma_c \left(\frac{da}{dt}\right) + \left(\frac{a}{3}\right)\left(\frac{d\sigma_c}{dt}\right) = k \tag{6.16}$$

General solutions to Equation 6.16 have been discussed by Kuwata et al. [114]; the equation is generally inexact except for the limiting conditions: (a) σ_c = constant, and (b) a = constant. For all other conditions a reaction path, $\sigma_c = f(a)$, must be known, and this requires evaluation of the reactivity equations developed in Section 5.6 applied to the coke particle. Obviously the burning rate may be bracketed by the limiting conditions, where the first condition represents the case of surface reaction and the second condition is the case of completely uniform internal reaction. Integration then yields $m/m_0 = [1 - (3/n)(\tau_s/\tau_c)]^n$, where $n = 3$ for case (a) (σ_c = constant) and $n = 1$ for case (b) (a = constant); τ_s is the residence time in the stirred flame ball; and τ_c is a characteristic reaction time given by $\tau_c = a_0 \sigma_c^0 / k$, where a_0 and σ_c^0 are the initial radius and density, respectively, of the first-formed char particle, and k is the reactivity parameter equated to R_T in Equation 6.16. R_T or k must be evaluated using the equations of Section 5.6, and it is common to use a variant of Equation 5.14 or 5.17 simply writing

$$k = k_{\text{eff}}^0\, p_s^n e^{-E/RT} \tag{6.17}$$

where n may be taken as 0, 1/2, or 1 in selected calculation cases, and k_{eff} may be varied to represent inclusion of the internal surface area term, A_i, of Equation 5.17.

The heat loss equation for the flame ball has the form of Equation 6.13 with

ISBN 0-201-08300-0

β negligible but with the addition of a radiative term based on a mean flame temperature, T_f, a flame ball surface area, A_b, and a view factor, F_b.

The equations obtained for η_I (heat generation) and η_{II} (heat loss) take the forms

$$\eta_I = 1 - \frac{m_0}{M_0}\left[1 - \left(\frac{3}{n}\right)\left(\frac{k\tau_s}{a_0\sigma_c^0}\right)\right]^n \tag{6.18a}$$

$$\eta_{II} = \left(\frac{\bar{C}_p T_0}{C_f^0 h_f}\right)\left(\frac{T}{T_0} - 1\right) + \frac{F_b A_b T_0^4}{JC_f^0}\left[\left(\frac{T}{T_0}\right)^4 - 1\right] \tag{6.18b}$$

where k is specified by Equation 6.17 or some more elaborate variant thereof, as discussed previously, and where m_0/M_0 is the ratio of a char particle mass to its parent coal mass.

The stability and criticality conditions of Equations 6.14 and 6.15 still apply. The equations were evaluated for a number of different conditions: initial particle radius, a_0, = 10, 20, 30 μ; E = 20 kcal/mol (about 85 kJ/mol); constant firing density, Jc_f^0/V_c; pyrolyzed fraction, f_v = 0, 0.1, 0.2, 0.3 (noting that the Q factor effect is not taken into account here); k_0 = 0.2, 0.4, 0.6 g/cm^2 sec.

Typical results of the calculations are illustrated in Figures 6.7(a), 6.7(b), and 6.7(c). All show the typical sigmoid curve of heat generation, η_I, starting from some initial value (0.3 in all the examples quoted) that represents the temperature-independent effect of volatiles pyrolysis and combustion. The curve shape below 1500°F is therefore an artifact of the assumptions, but it has little influence on the stability point, which is the intersection of η_I and η_{II}. In Figures 6.7(a) and 6.7(c) this is shown to be at η = 1, or complete combustion. The stability temperature in each case quoted is seen to be about 2800°F, with a rise to about 3100°F in the case of the 30 ft (about 10 m) diameter flame ball. These values lie inside the typically quoted flame temperatures at the burner level of 2800–3200°F (approx. 1550–1750°C).

In more detail, we should note the following. Figure 6.7(a) shows the relative insensitivity of the difference between surface and internal reaction (n = 3 to n = 1) for the same k_0. The effect of internal surface is taken into account, or partially so, by the values adopted for k_0; these bracket the values quoted by Mulcahy and Smith (Figure 5.10), which represent combustion with internal burning referenced

ISBN 0-201-08300-0

to the external surface. The effect of changing k_0 can be quite substantial, as shown in Figure 6.7(b), where a further drop in k_0 from the lower value would have brought about extinction. Figure 6.7(a) also compares the effects of including and excluding the radiation term (solid and dashed lines, respectively, for η_{II}). The difference in the equilibrium temperature is significant; it is also somewhat overestimated by assuming constant specific heats of the POC, when in actuality they rise above 2500 or 3000°F on account of dissociation to give a stability temperature at η = 1, or an adiabatic flame temperature, of about 3600°F. (It may be noted that the lower intersection points on Figure 6.7(a) are represented by a point of tangency, or condition for spontaneous ignition. Such behavior cannot occur in practice – ignition burners are always required; the effect is another example of the artifact of assumption, but it well illustrates the principle.)

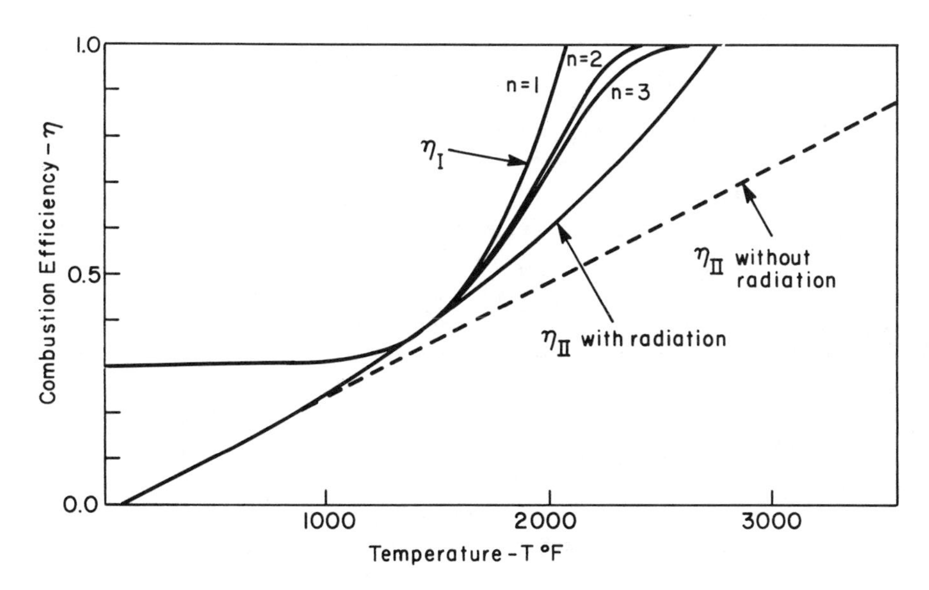

ISBN 0-201-08300-0

Figure 6.7(a). Variation of combustion efficiency of a boiler flame ball with flame ball temperature, *T*: also illustrating the influence of the internal reaction parameter, *n* (= 1 for total internal reaction). (Source: Ref. 113.)

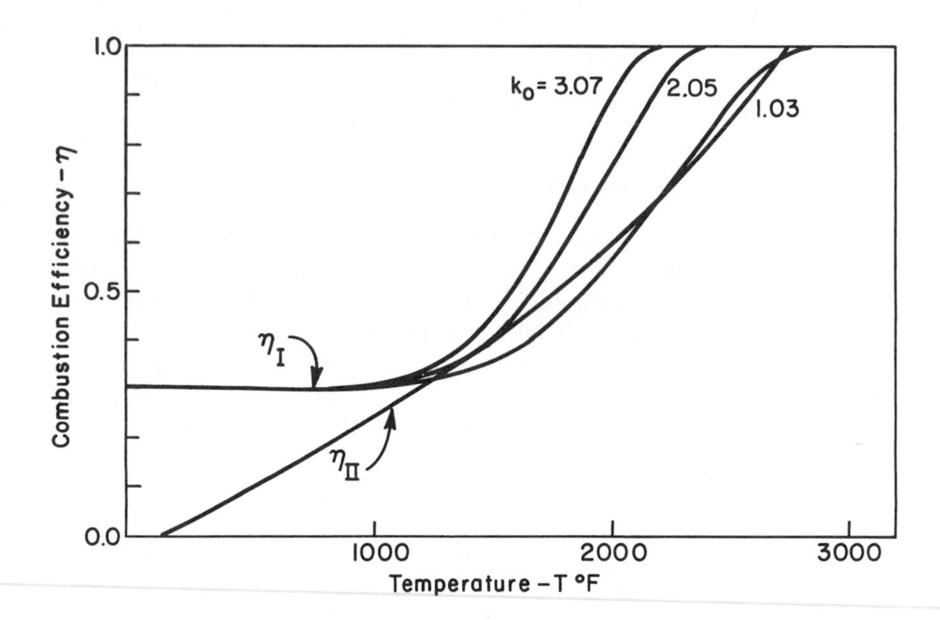

Figure 6.7(b). Combustion efficiency versus temperature, showing effect of variation of k_0, chemical reaction frequency factor. (Source: Ref. 113.)

Finally, Figure 6.7(c) illustrates the effect of reducing the flame ball diameter from 30 ft (boiler dimensions) to 10 ft (burner dimensions). The equilibrium flame temperature drops by about 500°F. In conjunction with change in other parameters, such as k_0, a_0, for example, this reduction can represent the difference between ignition and extinction for single burners under otherwise identical conditions. In a boiler with multiple burners the case is a little different as the flame balls can influence each other: this can be represented by a change in view factor. However, it could be important in scale-down. In this respect it is pertinent that reports of firing low reactivity char in a reduced-scale, cold wall boiler [115] emphasized the need for supplementary gas when the char VM content was below 20%, whereas the same char was used satisfactorily without supplement by Cogoli [39]. An analysis by Cogoli of the discrepancy suggested that it was due to the different radiative loads on the flame in the two cases.

In other results of the calculations, (a) the stability points, as noted, were insensitive to VM content on the assumptions used; and (b) the sigmoid generation curves were sensitive to particle size as expected from Equation 6.18a, although here the more detailed analysis shows, as in Equation 5.17, that to some approximation $k \propto a$, and the particle size influence should vanish in Equation 6.18a, though this was ignored in the calculations. The compactness of the flame or local intensity of combustion is also important. This is specified by the ratio of the

ISBN 0-201-08300-0

stirred reactor volume, V_s, to the total combustor volume, V_c. Varying this ratio from 10 to 50% affected both η_I and η_{II}, with the stability temperature dropping by nearly 1000^oF as the ratio rose. Finally, as expected, initial preheat temperature moved the generation and loss curves apart, conferring greater margin for stability.

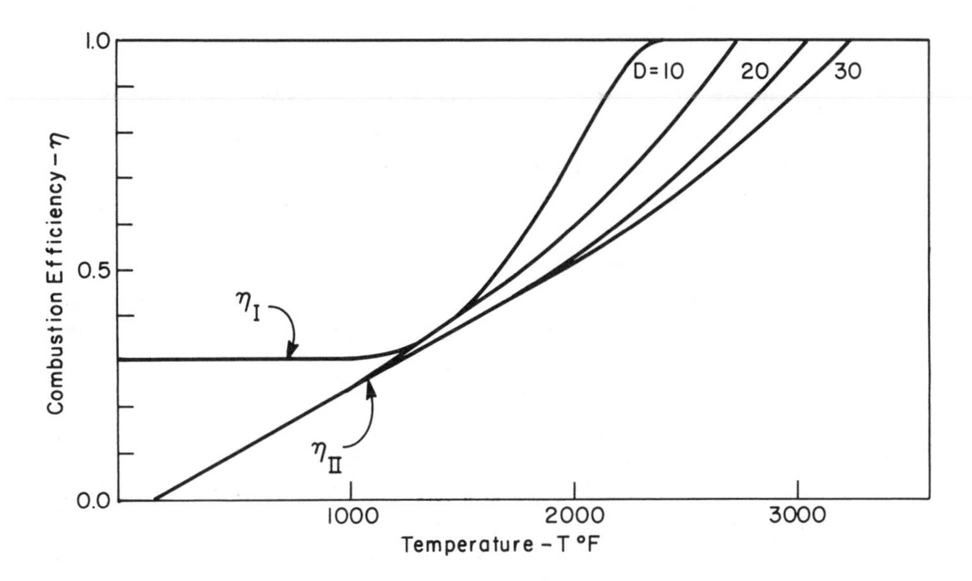

Figure 6.7(c). Combustion efficiency versus temperature, showing effect of variation of *D*, stirred section diameter. (Source: Ref. 113.)

The realism of the results is something to be considered. The range of stable flame temperatures is realistic. The prediction of nearly 100% burn-up in the flame balls is not impossible but a little unlikely, suggesting that the burn-up curves, η_I, rise too rapidly. This is rectified in Figure 6.7(b) reducing k_0 by a factor of 2 (a small correction). It is then seen, however, that there is still 95–98% burn-up and the flame is near extinction. The result may be numerically incorrect but qualitatively realistic. The dislike of utility station superintendents for changing firing conditions is well known; neither is loss of ignition unknown. It may well be that boiler flames are operating closer to extinction than is generally realized. This has critical implications for the use of new sources of solid fuels in the future, particularly gasification and liquefaction chars of potentially low reactivity.

The stirred reactor analysis may also be applicable to a fluid bed, although the application has not yet been made. In the normal fluid bed stability does not seem to be a problem, and analysis would be something of an academic exercise (but possibly informative, nonetheless). This is not the case, however, with the

ISBN 0-201-08300-0

fluid bed burn-up cell to reduce carbon in the ash, where stability is very much a problem. The stability analysis would be appropriate here.

7. COMBUSTION ANALYSIS OF IDEALIZED SYSTEMS

7.1. Scope of the Analyses

In principle, all the varied coal combustion and gasification systems under discussion are amenable to an essentially common theoretical analysis. Predictions of behavior lie in solutions to the simultaneous differential or integral equations for energy, mass, momentum, and so on, with appropriate boundary conditions, and auxiliary and constitutive equations that also define the different systems. In practice such a common analysis is still impossible. At one extreme this has led to the total use of empiricism and past experience. At the other, lack of knowledge has not deterred investigators from attempting to develop solutions to complex sets of equations even in cases where much content, and notably values of physical coefficients, are obtained by guesswork. Here an intermediate approach is adopted. The objective is to sketch out a reasonable theoretical analysis supplemented by sufficient empirical data sources for working solutions to the problems to be obtained. The general target is determination of the burning time or of some related parameter such as combustion zone thickness. This is the type of information that is missing, except on the basis of experiment or past experience, from the approximate design evaluations of Section 4, particularly Section 4.3.

7.2. Solid Bed Combustion

The system to be analyzed is illustrated in Figure 3.6. Present analyses are restricted to beds of coke, so the problems of drying and pyrolysis are generally omitted. One example of their inclusion is to be found in an analysis of incinerator beds [72], but no example is known of such application to coal beds other than as-yet-unverified models of *in situ* gasification [e.g. 116].

In a burning coke bed with air supplied from below, the oxygen is rapidly used up to produce CO and CO_2 in the primary reaction (reaction A in Section 5.6); part of the CO then burns up by further gas phase, homogeneous reaction in the pore spaces between the particles to increase the CO_2 content. The CO_2 can also react with coke, being reduced back to CO by reaction B (Section 5.6). If water is present, it can also react to give more CO and CO_2, and H_2, by reaction C (Section 5.6). The mixture of CO, CO_2, H_2, and H_2O then tends to adjust to suit the value of the water-gas shift equilibrium at the local temperature. The analysis following, however, excludes the water reactions. The gases reacting with the solid are therefore O_2 and CO_2, with CO reacting with O_2 in the gas phase. The consequence of the three competing reactions, however, is illustrated in Figure 7.1. A characteristic is the peaking of the CO_2 curve.

ISBN 0-201-08300-0

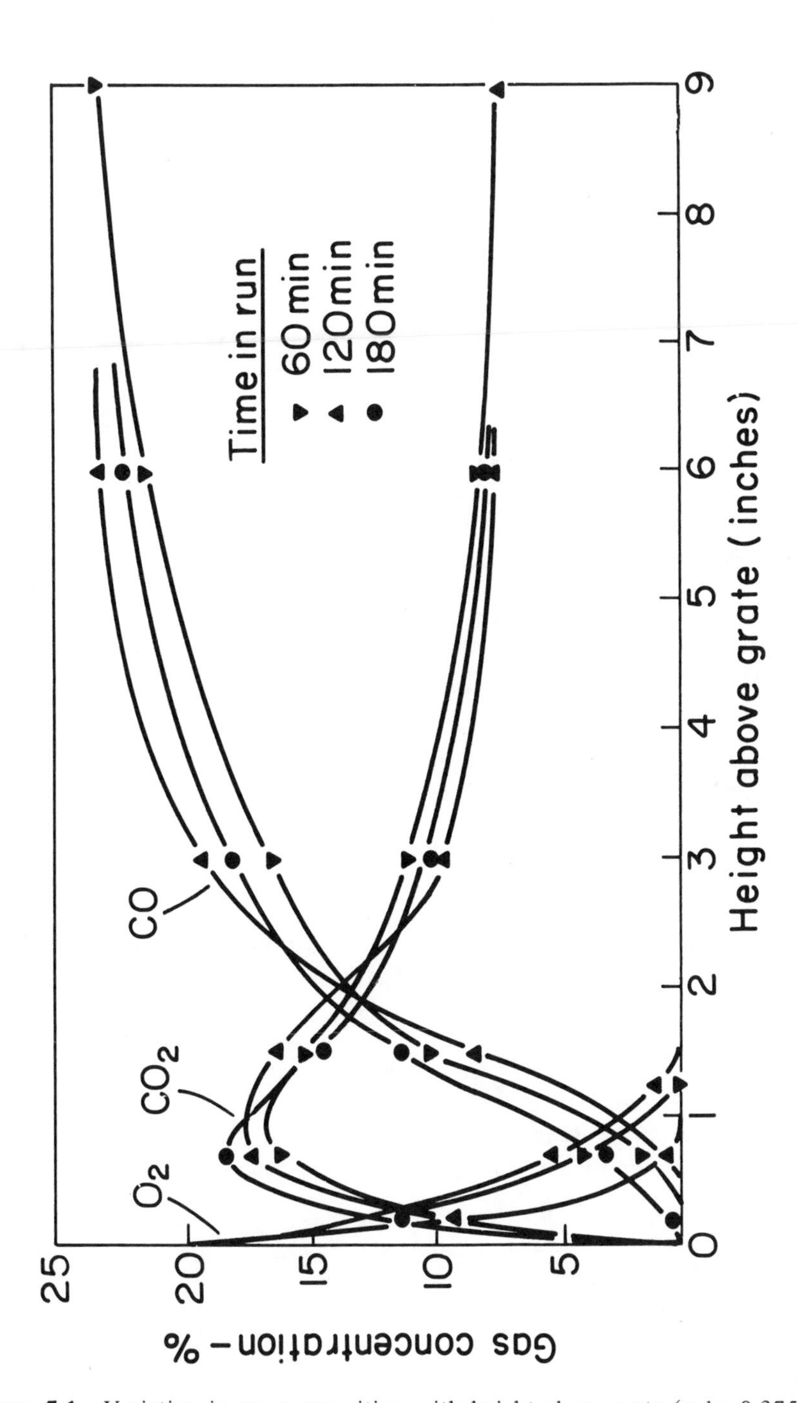

Figure 7.1. Variation in gas composition with height above grate (coke 0.375 in.; air flow 5 scf/min) at different times. (Source: Ref. 117.)

ISBN 0-201-08300-0

If the local mole fraction of the reactive gases *per entering* mole is N_i, using subscripts i = 1 for O_2, 2 for CO_2, and 3 for CO, the differential equations for O_2 and CO_2 reaction with carbon can be written [105, 117] as

$$\frac{dN_1}{dt} = -n_1 N_1 - \frac{1}{2} n_3 N_3 \left(\frac{N_1}{1 + N_3/2} \right)^m \tag{7.1}$$

$$\frac{dN_2}{dt} = -n_2 N_2 + n_3 N_3 \left(\frac{N_1}{1 + N_3/2} \right)^m \tag{7.2}$$

Here n_1 and n_2 are reaction velocity constants for the O_2 and CO_2 reactions, per unit interstitial volume, given by $n_i = (A_s/\epsilon)k_i$, where A_s is the available solid surface for reaction per cubic foot of bed, and k_i are the first-order (assumed) velocity constants for the reaction of O_2 and CO_2 with carbon. The gas phase reaction was treated phenomenologically because of the complexity of the (approximately) known mechanism, with reaction first order in CO and mth order in O_2, as discussed in the evaluation of this reaction by Johnson [118].

Adding Equations 7.1 and 7.2 yields.

$$\left(\frac{d}{dt} \right) \left(N_1 + \frac{N_2}{2} \right) + n_2 \left(N_1 + \frac{N_2}{2} \right) = (n_2 - n_1) N_1 \tag{7.3}$$

which eliminates the problem of knowing the CO kinetics.

The O_2 velocity constant, n_1 (or k_1), is determined primarily by boundary layer diffusion from the mainstream flow in the pores of the bed to the coke surface. Equation 5.10 is applicable with p_s = O. The CO_2 velocity constant, however, is kinetically determined with an activation energy of about 50 kcal (210 kJ). This is at variance with past opinions [e.g., 119], according to which both n_1 and n_2 were diffusionally determined, and thus approximately equal. If $n_1 = n_2$, Equation 7.3 then has an exact analytical solution: written in terms of local mole fraction per *local* mole, Y_i, the solution has the form $(Y_1 + Y_2/2)/(1 - Y_3/2) = Y_1^0 \, e^{-n_1 t}$, where Y_i is the inlet oxygen mole fraction (0.21). For $n_1 \neq n_2$, the solution is much more complex [117], but an approximate solution has the form

$$\frac{Y_1 + Y_2/2}{1 - Y_3/2} \simeq A\, e^{-nt} + B\, e^{-n_2 t} \tag{7.4}$$

where $A = Y_1^0 \, (n_1 - n_2)/(n - n_2)$, $B = Y_1^0 \, (n - n_1)/(n - n_2)$, and n is an arbitrary coefficient $\simeq n_1$ at low t.

Figure 7.2 illustrates the variation of the group $(Y_1 + Y_2/2)/(1 - Y_3/2)$ as a

ISBN 0-201-08300-0

function of time on a log-linear plot for coke burning in a combustion pot. The curvature establishes clearly that $n_1 \neq n_2$ since otherwise the curves would be straight lines if the analysis is valid.

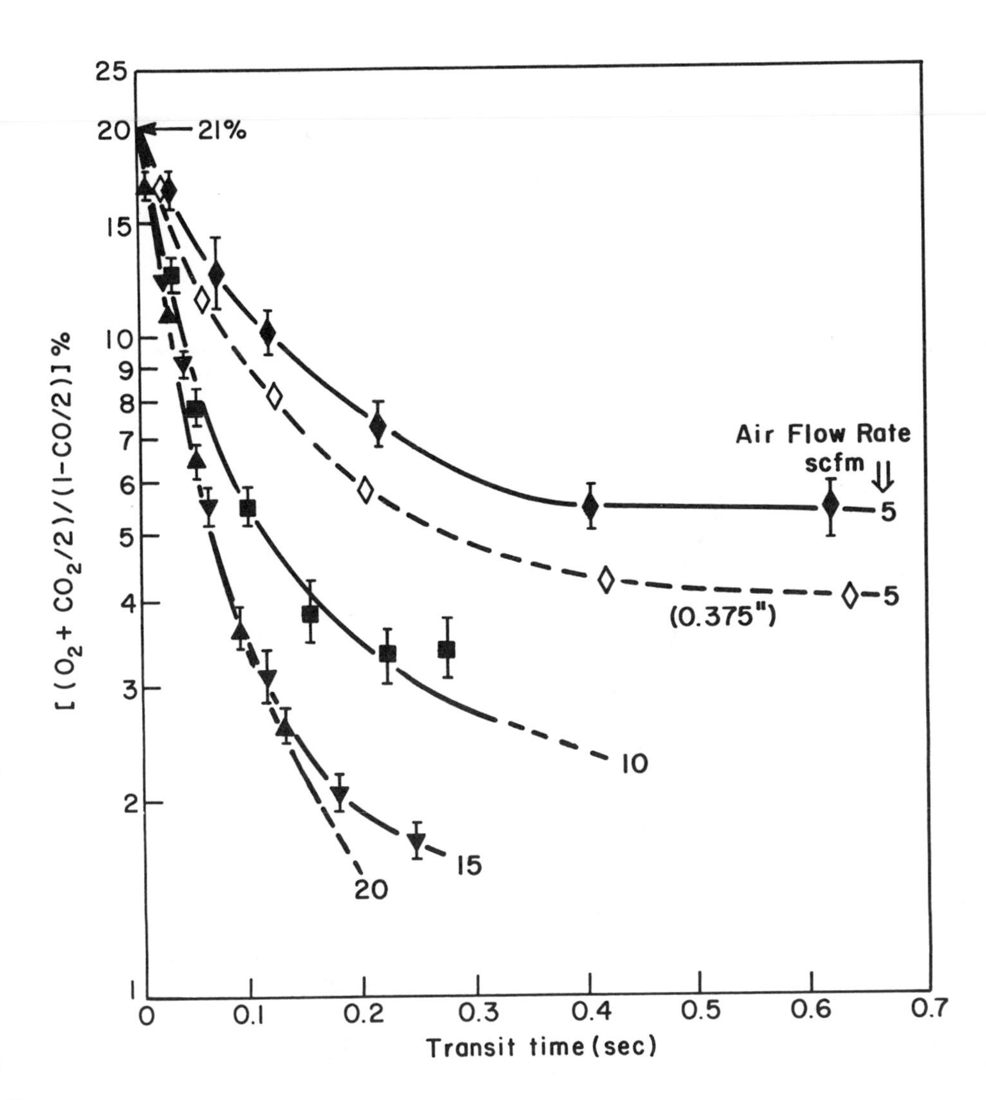

Figure 7.2. Variation of $[(O_2 + CO_2/2)/(1 - CO/2)]$ with time (Equation 7.4) for 0.625 in. coke particles at four airflow rates, and for 0.375 in. coke particles at 5 scf/min airflow rate (dashed line). (Source: Ref. 117.)

ISBN 0-201-08300-0

The slopes of the curves at long t approximate to n_2 by Equation 7.4, and Figure 7.3 is an Arrhenius plot of n_2 for one coal and three different cokes, including Kreisinger et al.'s data of 1916 [120]. The slopes correspond to an activation energy for the CO_2 reaction of about 50 kcal, with this value evidently common to all the fuels tested. In the analysis a first-order reaction was assumed, so the relevant reactivity equation is Equation 5.15 with $n = 1$. Since measurement of particle density indicated no internal reaction, or none detectable, the right-hand term of Equation 5.15 is assumed not to contribute.

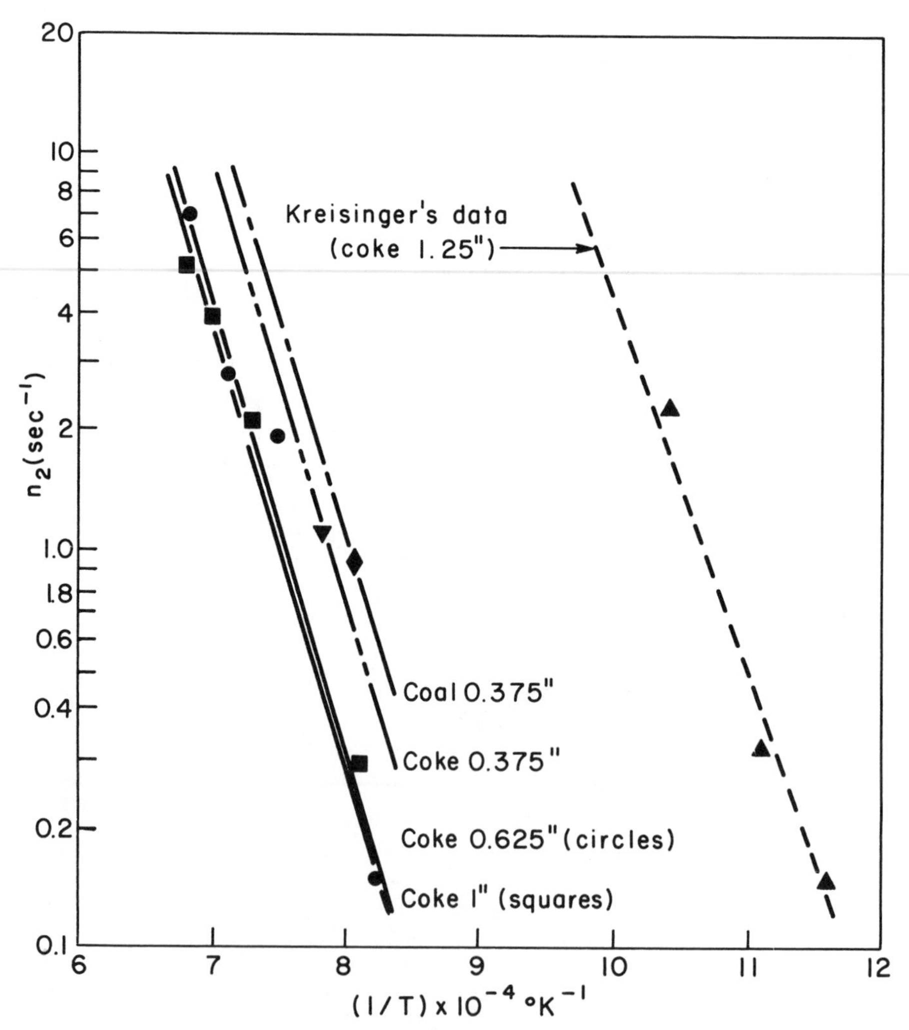

Figure 7.3. Arrhenius graph of velocity constant n_2 with temperature. (Source: Ref. 117.)

ISBN 0-201-08300-0

The tangents to the curves of Figure 7.2 at zero time are the slopes n_1. Being determined by the boundary layer diffusional velocity constant, k_D, of Equation 5.10, n_1 is a function of the gas velocity, V, where the relation between k_D and V can be obtained by Sherwood or Stanton number correlations. The correlations given are all with Reynolds and Prandtl numbers, but the reviewing sources [121–123] are not unanimous on the power index for the Reynolds number. Reduced to the velocity term alone, writing $k_D \propto V^m$, quoted values of m are 0.49, 0.59, 0.7, and 0.83. Figure 7.4 is a log-log plot of n_1 against V, which suggests that $V^{0.7}$ is the closest fit. Further evaluation of the diameter influence suggests that

$$k_D = \text{constant}\left(\frac{V}{d}\right)^{0.7} \tag{7.5}$$

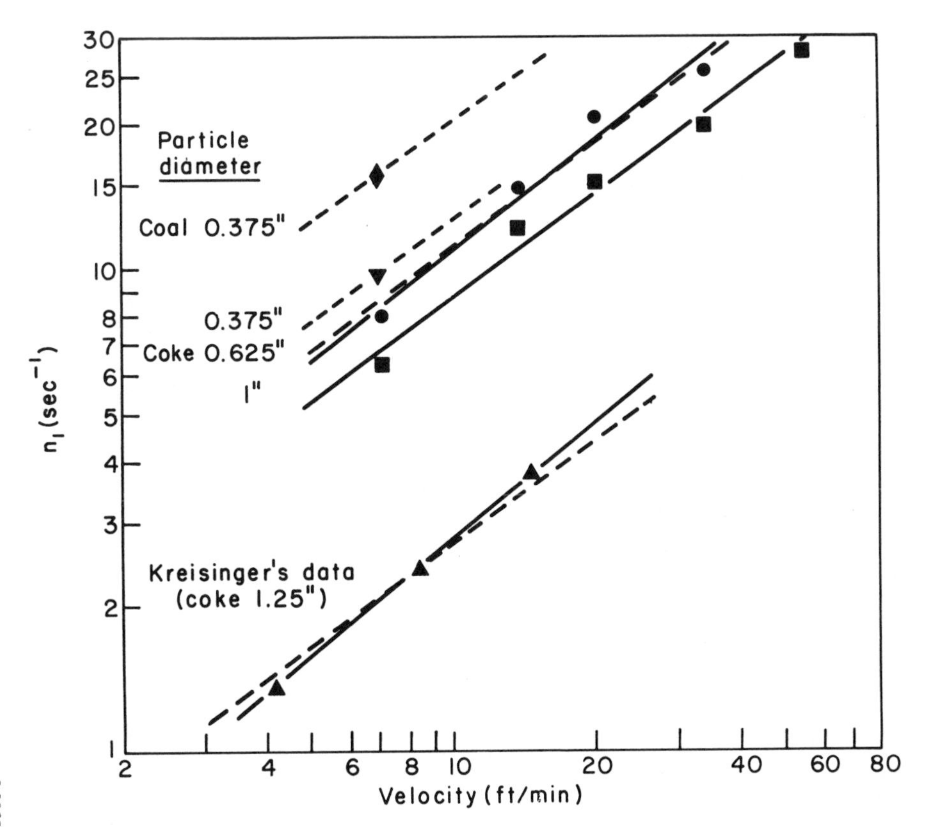

Figure 7.4. Variation of diffusion velocity constant n_1 with velocity. Solid lines are least-squares fit; dashed lines are slope 0.7. (See text.) Data from Kreisinger [120] are also included (bottom line). (Source: Ref. 117.)

ISBN 0-201-08300-0

The model as outlined is incomplete in a number of respects. Just for the carbon reaction an energy balance is also required to account for temperature variations through the bed, and a more formalized analysis sets out the equations to include the following: two continuity equations, one for the gas and one for the solid; likewise two energy conservation equations and one species conservation equation; the equation of state for the gas; and the use of Darcy's law or some variant such as the Blake-Kozeny equation [121] as a substitute for the momentum equation in packed beds. With coal, additional equations must be added to represent drying, pyrolysis, pyrolysis cracking and/or reactions with the solid, and pyrolysis combustion or other gas phase reactions in the pores of the bed. In addition, the physical paths of the volatiles and the gasification products are affected by whether the combustor/gasifier is overfed, underfed, or cross-fed (Section 3.2), and this can affect the reaction history. Models covering all these factors are under investigation by various workers, notably in connection with *in situ* gasification. While such models are necessary for full understanding and prediction of behavior in due course, either in full or in reduced mode, none appears to be sufficiently advanced as yet to be of much greater value than the simple model given here. An example of the more nearly complete approaches to modeling is given by Winslow [116].

The outcome of such analyses with regard to solid bed combustion is twofold. First, the gas-solid kinetics provides the basis for calculating particle lifetimes and, hence, limiting feed rates, bed thickness, required rate of grate travel, and ideal underfire air proportioning. Second, the model predicts the gas composition leaving the top of the bed. In the case of a shaft gasifier or *in situ* gasification the model would then also provide a basis for optimizing the quality of the product gas. In the case of combustion the gas composition should be known to determine the overfire air requirements. Beyond that, however, the combustion is dominated by proper overbed mixing and flame holding, with overbed burnout determined by mixing rates and gas kinetics, as discussed in Section 5.5.

7.3. Fluid Bed Combustion

A. SYSTEM

The system to be analyzed is described in Section 3.3. The bubbling bed consists of two main phases, the bubble phase and the emulsion phase. The bubbles are almost solid-free and are usually treated as such, although they do rise with particles cascading through them. If the particles are hot and combustible and if the

ISBN 0-201-08300-0

oxygen level is high, the particles may flash, causing momentary hot spots. The emulsion phase is correspondingly dense in particles. In the unballasted bed they are mainly coal. In the ballasted (low temperature) bed it is usual to find that more than 99% of the particles are inert.

The physical model to be analyzed, based on the above description, is shown in Figure 7.5. Figure 7.5(a) illustrates a cross section of a fluid bed with rising bubbles of increasing size, based on descriptions and representations by Davidson and Harrison [124]. Figure 7.5(b) illustrates a single bubble with a nearby char particle in the particulate phase, adapted from the illustration by Avedesian and Davidson [125]. The theoretical analysis essentially also follows that of Avedesian and Davidson.

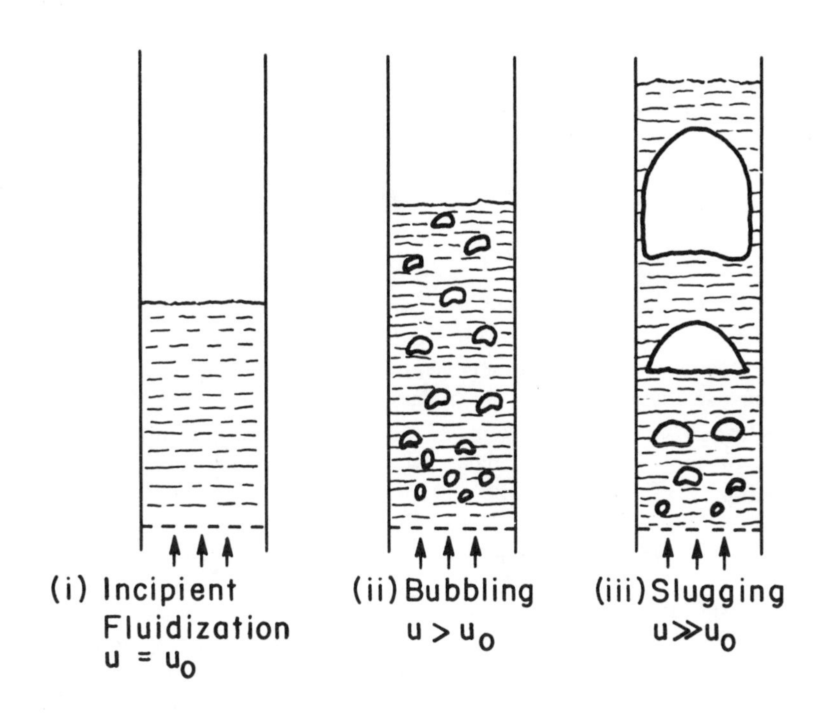

Figure 7.5(a). Schematic representation of fluid bed at (1) incipient fluidization (no bubbles), (2) bubbling (bubbles forming and increasing in size), (3) slugging (bubble diameter increasing to tube diameter). (After Davidson and Harrison [124].)

ISBN 0-201-08300-0

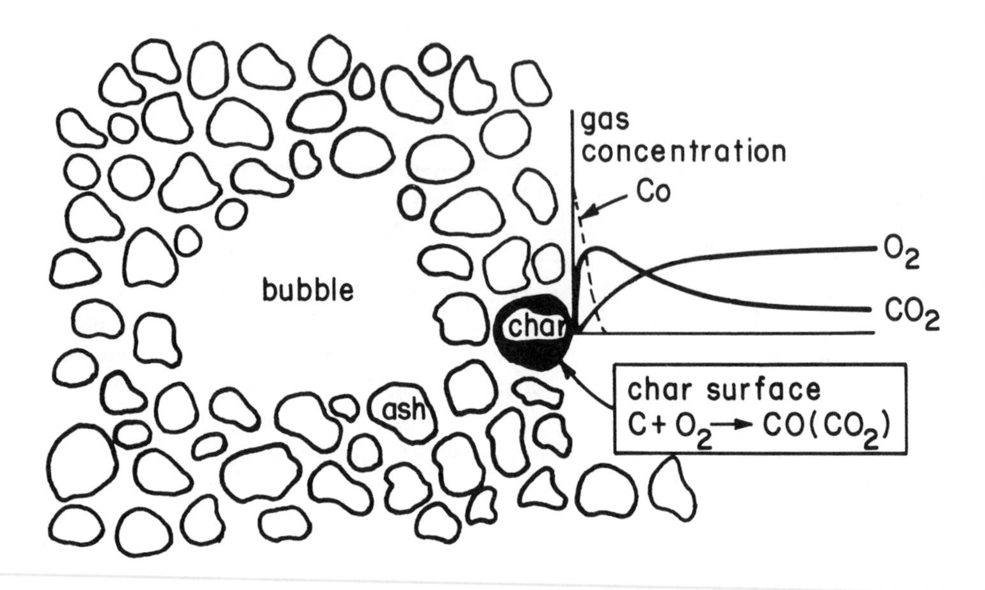

Figure 7.5(b). Schematic representation of a char particle near a bubble in a fluid bed burning in oxygen. A single diffusion film with char surface oxidation and very rapid burnup of CO near the solid surface is assumed. (See text and Ref. 74.) (Adapted from Avedesian and Davidson [125].)

Because of the bypass effect of the largely particle-free gas bubbles and because of oxygen consumption in the emulsion phase, the oxygen concentrations in the two phases can differ and there can be exchange between them. The particle combustion itself is governed by the equations outlined in Section 5.6. Since the initial particle size is generally about 1 mm or larger, combustion tends to be dominated or controlled by boundary layer diffusion, which introduces a major simplification.

B. SINGLE-PARTICLE COMBUSTION

For a single carbon particle burning under diffusion control in an infinite quiescent atmosphere, the oxygen concentration at the superficial surface is near zero, so there is no internal reaction. Combining Equations 5.20 and 6.16 with the condition that $d\sigma_c/dt$ = O (no internal reaction) and $k_D = D/a$ (quiescent atmosphere), we find that differential equation 6.16 takes the form:

$$\frac{da}{dt} = \frac{M_c D C_o/\sigma_c}{a} \tag{7.6}$$

ISBN 0-201-08300-0

if the oxygen partial pressure, p_o, is written as molar concentration, C_o (moles O_2/cm^3), and where M_c is the molecular weight of carbon. This integrates into the well-known Nusselt equation [95]

$$t = K_D(d_0^2 - d^2) \tag{7.7}$$

where d_0 is the initial particle size, and K_D is the diffusional burning constant, given by

$$K_D = \frac{\sigma_c}{8M_c D C_o} \tag{7.8}$$

In particular, when $d = 0$, $t = t_c = K_D d_0^2$, this equation usually being referred to as the Nusselt "square-law" burning time equation (cf. Equation 5.2 for volatiles combustion). Equation 7.6 and the square-law burning time equation have been checked experimentally and found to hold in a number of experiments [32] using 10 different coals, over a range of particle sizes from 300 μ to 4 mm, and over a range of oxygen percentages from 3 to 70%. In the experiments on oxygen enrichment [31] the simple expression $k_D = D/a$ is inadequate on account of the effect of the bulk convective flow or second Stefan flow due to net mass transfer to the solid particle. The molar oxygen concentration $C_o = L_o Y_o$, where L_o is the total molar concentration (moles/cm^3), and Y_o is the mole fraction, must be replaced by L_o $\ln(1 - Y_o)$. The correction to the analysis is standard [89].

The experiments cited were for combustion of coal particles. The volatiles combustion time, t_v, was found to fit an empirical square-law relation, Equation 5.2, as described in Section 5.4. The residue combustion time equation and the burning constant equation also required additional corrections for reduced density, σ_c, on account of loss of mass and swelling. In this last respect, measurements of swelling factors, f, showed them to be about the same for the range of coals examined, with a value of 1.5 [33]. The correction method used was to multiply K_D by $C_f/100f$, where C_f was the fixed carbon percent. With this correction, agreement between theory and experiment was good [32].

The good agreement between theory and experiment implied, significantly, that the effective reaction close to the carbon surface was $C + O_2 \rightarrow CO_2$. This is not what happens in the primary reaction at the temperatures involved, where the primary reaction produces mainly CO. As discussed, however (see Figure 5.9), the CO burns up in the boundary layer; and, if this is fast enough (as seems to have been the case in the quoted experiments), the system behaves *as if* CO_2 were the primary product. If the CO does not burn up rapidly in the boundary layer, the combustion rate can double. On a molar basis we find $k_D = 2D/a$. This halves the burning constant and hence the burning time.

At higher temperatures (above 1000°C) CO_2 starts to react as well, becoming important at about 1200°C. A double film can then form in which CO_2 is reduced

ISBN 0-201-08300-0

to CO at the solid surface and is oxidized back to CO_2 at a surface one radius away from the solid surface. The oxygen concentration becomes zero at a radius $2a$ from the center of the particle over a surface four times the area of the particle. The mass transfer coefficient thus becomes $k_D = 4(D/2a) = 2D/a$, which is identical to the case above when no CO_2 is formed. At a temperature that depends on particle size (the crossover point in Equation 5.18, given by $2DC_o = ak_2$) the double film can become controlling. For 15 mm spheres the controlling temperature was found to be about 1500°C [126]. The onset of single-film diffusion was found to be about 1000°C with a limiting rate about half that for the double film. The CO_2 – and H_2O – reaction rates decline very rapidly below 1200°C, being about 10^{-5} of the C-O_2 reaction rate at 800°C [82].

C. PARTICLE COMBUSTION IN THE FLUID BED

In the fluid bed all the above considerations apply with the additional factor of enhanced mass transfer by forced convection due to the relative velocities between particles and gas. Introducing as a correction for forced convection the Sherwood number, *Sh*, we find that this has a theoretical value of 2 under the (quiescent) conditions already discussed. The burning constant expression can, therefore, be rewritten in the form

$$K_D = \frac{\sigma_c}{4NM_c ShDC_d} \tag{7.9}$$

where C_o can now be written as C_d, the oxygen concentration in the dense particle phase; and N is the number of diffusion films (or takes the value 2 for O_2 to CO production only). With $M_c = 12$ the numerical value of $4NM_c$ is 48 or 96. We should anticipate that $N = 1$ would be appropriate for the low temperature, ballasted beds, and $N = 2$ appropriate for the Ignifluid bed.

Avedesian and Davidson's analysis [125], which is the only extant fluid bed analysis of coal combustion with experimental verification, proceeds with an evaluation to determine the local ambient oxygen concentration, C_d. Following Davidson and Harrison [124], there are two initial steps in the analysis: (i) an oxygen balance on a single bubble; and (ii) an oxygen balance on the dense particulate phase of the bed.

(i) *Bubble Analysis.* In the single-bubble analysis a total exchange factor Q is introduced as the sum of a convective flow, q, in and out of the bubble, and a mass transfer (diffusive) exchange, $k_G A_b$, where k_G is the transfer coefficient and A_b is the bubble surface. For oxygen concentrations in the dense and bubble phases, respectively, of C_d and C_b, the equation $Q(C_d - C_b) = U_A V dC_b/dy$ describes the variation of C_b with height, y, above the distributor plate, where V is

ISBN 0-201-08300-0

the volume of a bubble, and U_A is its vertical velocity. This differential equation can be integrated only if C_d is uniform (perfect mixing of gas in the dense phase) or if C_d is known as a function of y. Davidson and Harrison [124] considered both extremes and showed them to yield closely similar results. As the initially simpler assumption, with C_d uniform and $C_b = C_0$ at $y = 0$, the result obtained, assuming bubbles of constant size, V, is

$$C_b = C_d + (C_0 - C_d)\, e^{-Qy/U_A V} \tag{7.10}$$

In the balance on the particulate phase, on a unit area basis, there is, first, a net rate of oxygen supply between the bottom and the top of the bed, $U_o(C_0 - C_b)$. Second, there is exchange from the dense phase to the bubbles at a rate $N_o QHC_d$, where N_o is the number of bubbles, and H the bed depth; and there is exchange from the bubbles to the dense phase at a rate $N_o Q \int C_b\, dy$, integrated over the bed depth using Equation 7.10. Finally, there is oxygen consumed by the particles at a rate given by $2\pi a Sh D C_d$ moles per particle, or at a rate N_d/A times that in the bed on a unit area basis: this can also be written as $k_v H(1 - N_o V)C_d$. The balance, with $N_o V U_A = U - U_o$, yields

$$\begin{aligned}(C_0 - C_d)\,[U - (U - U_o)\, e^{-QH/U_A V}] &= k_v H_o (1 - N_o V) C_d \\ &= 2\pi \left(\frac{N_d}{A}\right) Sh D a C_d\end{aligned} \tag{7.11}$$

Combining Equation 7.11 with 7.10 at $y = H$ to give the resultant oxygen concentration leaving the bed, C_H, we obtain the expression for the oxygen consumption (reaction efficiency):

$$\eta_{O_2} = (1 - \beta e^{-X}) \left[1 - \frac{1}{1 + k_v'/(1 - \beta e^{-X})}\right] \tag{7.12}$$

where $\beta = (1 - U_o/U)$, $X = QH_o/U_A V$, and $k_v' = k_v H_o/U$.

(ii) *Experimental Results.* Avedesian and Davidson [125] and Campbell and Davidson [127] have investigated Equation 7.11 experimentally in two ways: first, on a batch basis so that the number of particles is known; and, second [127], on a continuous feed basis to represent operational behavior. In the batch system a char sample of mass m_0 and initial diameter d_0 contains N_d particles $= 6m_0/\sigma_c \pi d_0^3$. If the particles burn out uniformly, N_d remains constant. Using Equation 7.6 with modified K_D given by Equation 7.9, we find that the dense phase oxygen concen-

ISBN 0-201-08300-0

tration, C_d, can be eliminated from Equation 7.11, leaving a differential equation in a and t alone. Integrating with diameter d substituted for radius a yields

$$t = f(U, U_0, X)\left(1 - \frac{d^3}{d_0^3}\right) + K_D d_0^2 \left(1 - \frac{d^2}{d_0^2}\right) \tag{7.13a}$$

where

$$f(U, U_0, X) = \frac{m_0 / M_c C_0 A_r}{(U_0 + U - U_0)(1 - e^{-X})} \tag{7.13b}$$

and the burnout time, t_c, obtained at $d = 0$, is given by a modified square-law expression:

$$t_c = f(U, U_0, X) \cdot \left(\frac{m_0}{C_0}\right) + K_D d_0^2 \tag{7.14}$$

where $f(U, U_0, X)$ is a nominal constant for a given bed at constant U.

Experimental test of Equation 7.14 showed generally good agreement, as illustrated in Figures 7.6(a) and 7.6(b). Figure 7.6(a) is the variation of t_c with d_0^2 with a common slope K_D and an intercept governed by U. Figure 7.6(b) is the variation of t_c with m_0 at two different initial diameters, d_0, and three values of U with slopes governed by $f(U, U_0, X)$, and intercepts $K_D d_0^2$. Knowing the function f and the values of U and U_0 gave X from the slope. K_D and hence Sh was obtained from the slope of Figure 7.6(a) and the intercept of Figure 7.6(b). Other details of the model were also satisfactorily checked by functional plots.

Numerical agreement on the constants, notably X and K_D, was also reasonably good with one point of uncertainty. The analysis used by Avedesian and Davidson [125] assumed the double-film model with C-CO_2 at the solid surface, which from reasoning given above should be inapplicable at the temperatures involved. It yielded values for the Sherwood number, however, that were in the range of expectation, with a mean value of 1.4. For single free particles at low velocities $Sh = 2$ and rises with velocity. In packed beds it is expected to be less than 2 on account of blockage by adjacent bed particles, and the value of 1.4 conformed with expectation and independent estimates cited by Avedesian and Davidson [125]. However, when there is an established quiescent diffusional boundary layer round a single particle, a quasi-stationary analysis yield the expression, $p = p_0 (1 - a/r)$, for the oxygen profile, p, with distance, r, from the center of the particle of radius, a. The pressure, p, therefore rises to 99% of the mainstream value, p_0, in a distance of $100a$. The assumption of blockage ignores the possibility of enhanced mixing by entrainment in the wakes of particles as they move in to distances less than $100a$ from the burning particle.

ISBN 0-201-08300-0

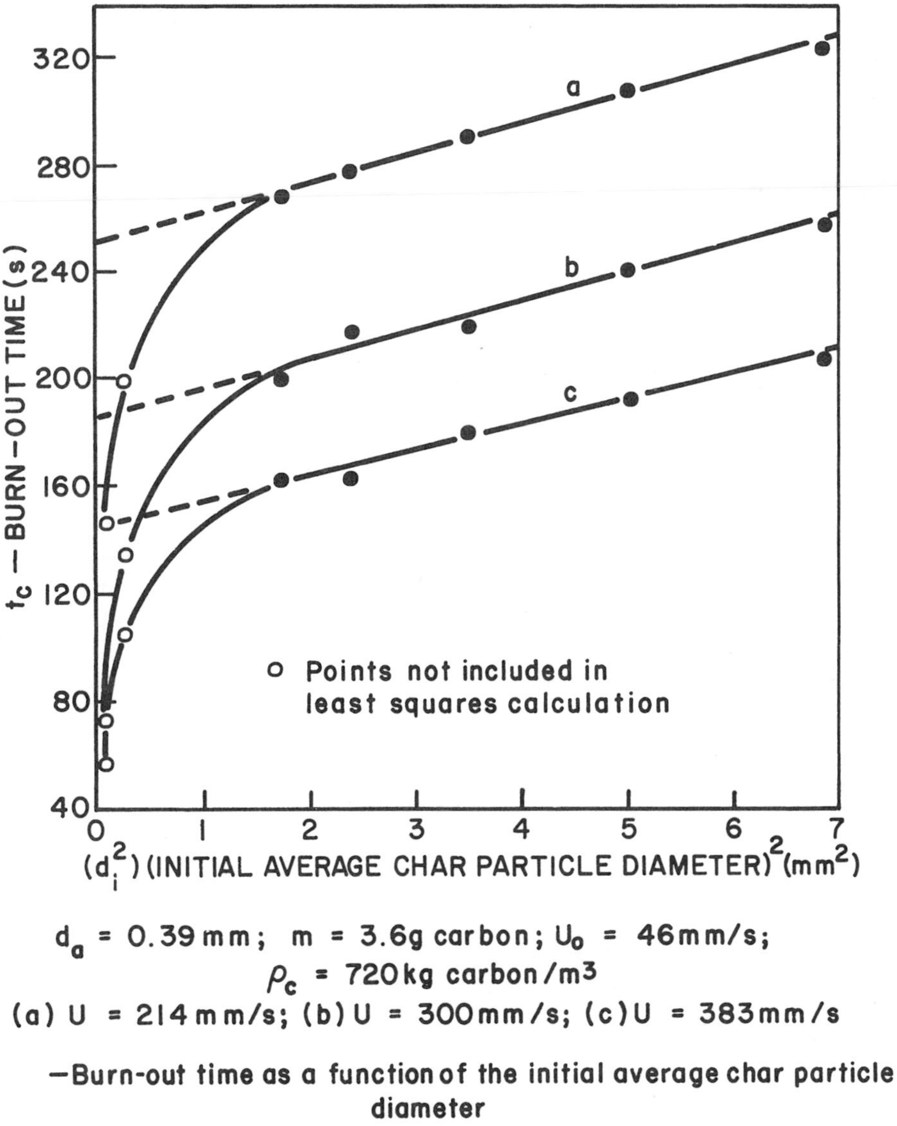

o Points not included in least squares calculation

t_c — BURN-OUT TIME (s)

(d_i^2) (INITIAL AVERAGE CHAR PARTICLE DIAMETER)^2 (mm^2)

d_a = 0.39 mm; m = 3.6g carbon; U_o = 46mm/s;
ρ_c = 720kg carbon/m3
(a) U = 214 mm/s; (b) U = 300mm/s; (c) U = 383mm/s

—Burn-out time as a function of the initial average char particle diameter

Figure 7.6(a). Burnout time as a function of the initial average char particle diameter. (a) U = 214 mm/sec, (b) U = 300 mm/sec, (c) U = 383 mm/sec. d_a = 0.39 mm, m = 3.6 g carbon, U_{O_3} = 46 mm/sec, ρ_c = 720 kg carbon/m^3. (Source: Ref. 125.)

ISBN 0-201-08300-0

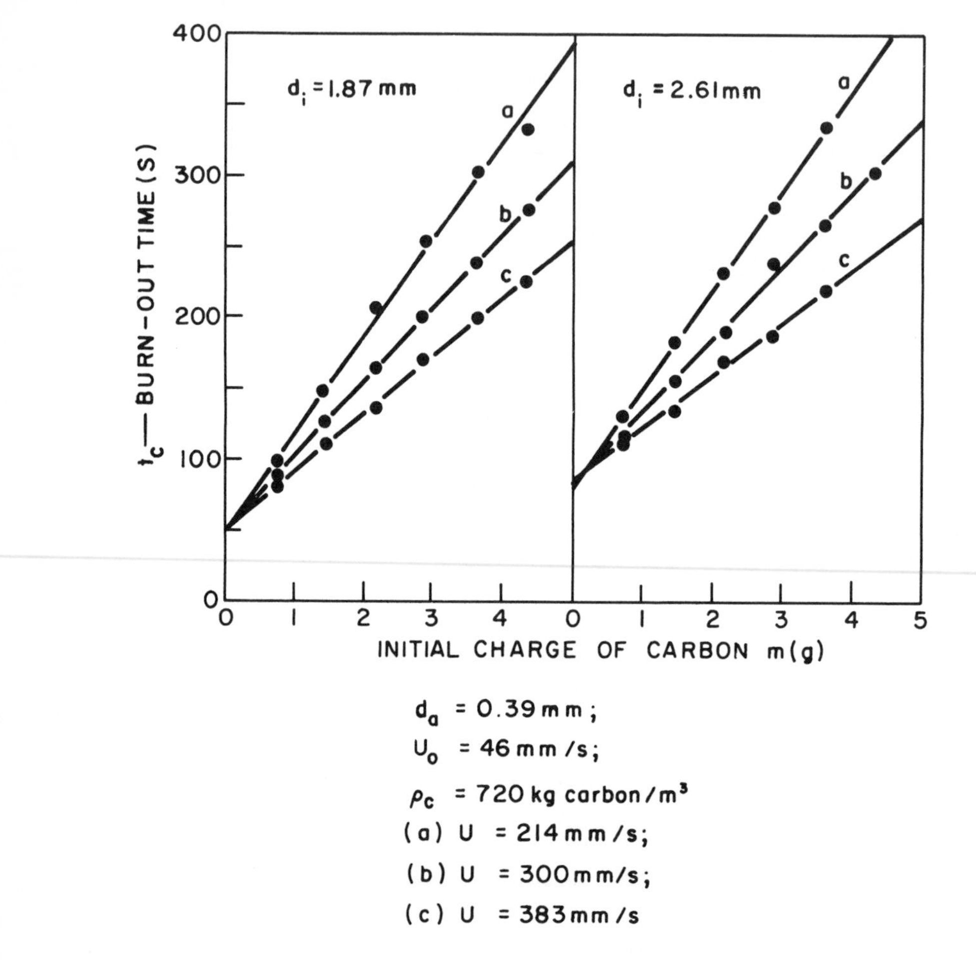

Figure 7.6(b). Burnout time as a function of the initial charge of char. (a) $U = 214$ mm/sec, (b) $U = 300$ mm/sec, (c) $U = 383$ mm/sec. $d_a = 0.39$ mm, $U_o = 46$ mm/sec, $\rho_c = 720$ kg carbon/m^3. (Source: Ref. 125.)

The double-film model is unacceptable on kinetic grounds. It has also been examined by Basu et al. [128], who showed that there was a thermal imbalance if the highly endothermic C-CO_2 reaction predominated at the surface; and it was

ISBN 0-201-08300-0

reexamined by Song and Sarofim [129], as cited by Beer [130], who confirmed that known rate data on the CO_2 reaction were too slow and reanalyzed the diffusional behavior with CO kinetics included in the gas phase. They also showed that CO burn-up fell progressively as excess air was reduced, in reasonable quantitative agreement with experimental data from Skinner [131]. The overall Avedesian and Davidson model, however, is essentially substantiated, with good numerical agreement if it is assumed that the reaction is single-film oxidation with production of CO that is burned in the bed generally and not mainly in the boundary layer. The Song model modification appears to account acceptably for the behavior [129, 130].

In the continuous feed experiments [127] the essence of the model is unchanged: the effect of the particle size distribution has to be included. The particle size distribution is developed, of course, as the result of reaction, and there is a simple relationship in a stirred reactor between the reaction rate dependence on particle size and the particle size distribution function developed in the reactor. Thus if, for particles of radius a, the rate of change of radius, da/dt, due to reaction (combustion or gasification) is governed by some function of radius, $f(a)$, the particle size distribution function, $n(a)$, developed in the reactor at steady state, even with a size distribution in the feed, is related to $f(a)$ by $n(a) \propto 1/f(a)$ [88]. In the particular case of diffusionally controlled single-particle combustion, $f(a) \propto 1/a$ (Equation 7.6), so $n(a) \propto a$. Campbell and Davidson [127] reached, and experimentally confirmed, this conclusion by other means.

In summary, an acceptable model now exists for estimating char reaction times in fluid beds if the bubbles exchange behavior is known. Pereira and Calderbank discuss this problem and outline some solutions [132]. The model also indicates the principal parameters that can be independently varied, and the expected outcomes of these variations. This is important supplementary design data to those indicated in Section 4.3. What is still missing is adequate data on volatiles combustion. This is discussed in Sections 5.4 and 5.5. The best design basis is still Equation 5.2, but this says nothing of the kinetics problem in the rest of the bed if the volatiles are dispersed throughout the bed from the "point" source of an injection pipe. At present the only guide to design would appear to be experience governing the onset of spouting flow by volatiles generated at the top of the injection pipe. The problem has been observed and eliminated by increasing the number of feed points and reducing the feed rate per point.

One other aspect of fluid bed behavior of importance in design is the prediction of heat transfer in the bed. This is a major subject in its own right. Reference here is made only to a recent summary [133] of heat transfer equations by Cole [133], and the plot of Figure 7.7 to illustrate the fit of available data to the best-fit equation developed by Cole [133]. The equations are developed mainly on the basis of the particle "packet" model.

ISBN 0-201-08300-0

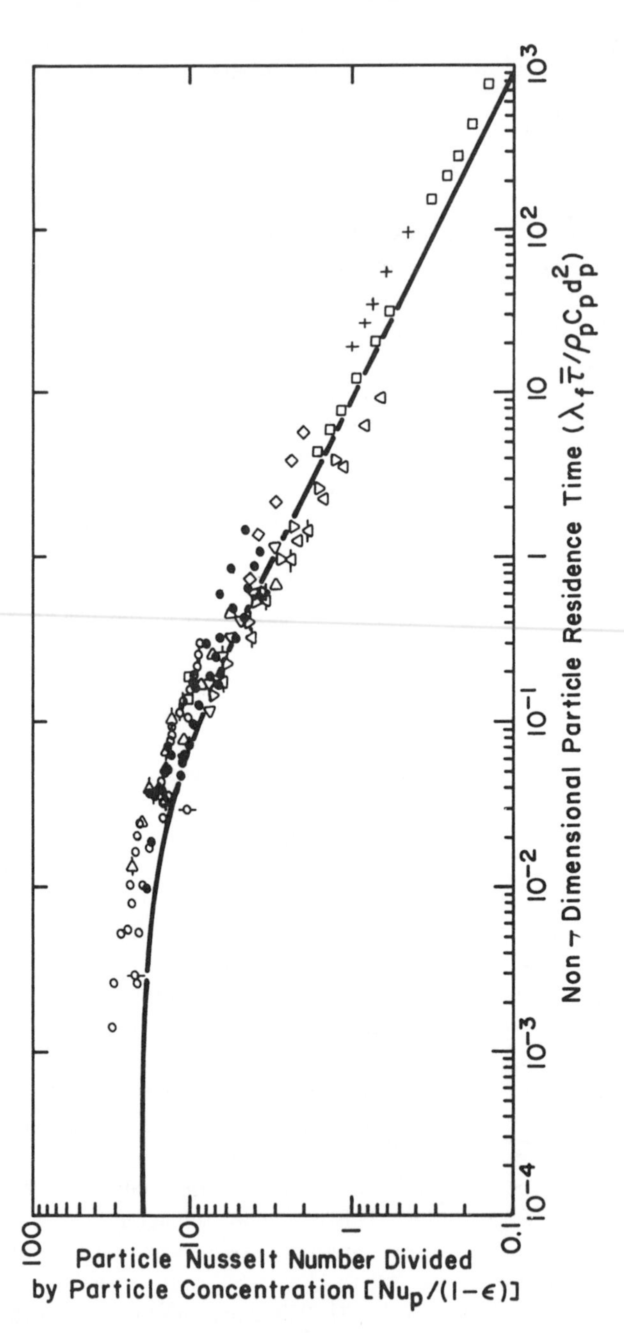

Figure 7.7. Variation of reduced particle Nusselt number, $Nu_p/(1-\epsilon)$ with dimensionless particle residence time ($\lambda_f\, \tau/\rho_p C_p d_p^2$). Curve is modified Ziegler equation. (Source: Ref. 133.)

ISBN 0-201-08300-0

7.4. Pulverized Coal Combustion

A. SYSTEM

The system to be analyzed is described in Section 3.4; the focus of analysis is mainly to determine the combustion time for prediction of firing densities by Equation 4.15.

The combustion time is obtained in principle from integration of Equation 6.16 under suitable boundary conditions. This is difficult because of the general lack of information on the density changes, $d\sigma_c/dt$. In recent years there have been a number of studies (see Ref. 88) of increasing precision aimed at establishing the internal reaction behavior of small particles. These have mostly used injection into electrically heated tube furnaces (as in the pyrolysis experiments, Section 5.4) with reaction time changed by altering the sampling point; heatup is rapid, and reaction is essentially isothermal. One-dimensional flame experiments have also been used [39–41, 45, 99, 101–103, 110, 111]. Relative reactivities have been measured in continuous weighing experiments of small particles (TGA), in air at 1 atm and below 1000°C (and also in CO_2 and steam) (see Ref. 134 and the references cited there). These internal reactivity measurements have developed many data and some important qualitative conclusions. Unification of the data into a coherent quantitative theory has still not been accomplished, however, in spite of some past claims, although valuable detailed modeling investigations are in progress (e.g., 90), as indicated earlier.

B. GENERAL RATE EQUATIONS

Available results and problems can be put most usefully into the framework of exact solutions to limiting cases of combustion in flames. Measurements in flames require one-dimensional plug or piston flow furnace designs for the results to be meaningful. Partly burned coal samples are removed from the flame, and the net mass, or burn-off, is determined, generally using ash as a ratio tracer. If ash is not lost on the walls, then by continuity the ash mass flow is constant through the furnace. The combustible/ash ratio thus gives the burn-off, or the mass flow, of combustible. If the particles are initially monosize, of radius, a_0 and density σ_0, the combustible mass, J, compared with the inlet, J_0, is the mass fraction, m, of a single particle of initial mass m_0; thus

$$\frac{J}{J_0} = \frac{m}{m_0} = \frac{a^3\sigma}{a_0^3\sigma_0} \qquad (7.15)$$

In evaluating the results of flame experiments, it is, then, common practice to draw the curves of m as a function of distance or time, and to differentiate the curves graphically or numerically to obtain dm/dt [110, 135, 136]. The total specific

ISBN 0-201-08300-0

reaction rate per unit area of superficial surface, R_T, is then $(dm/dt)/4\pi a^2$. Substituting m for a, we have

$$R_T = \frac{dm/dt}{4\pi(3m/4\pi\sigma)^{2/3}} \tag{7.16}$$

Expressions for R_T are discussed in Section 5.6. The most commonly adopted expressions in flame evaluations are those for the two limiting cases of diffusional control and chemical control, when $R_T = k_D p_0$, or $= k_1 p_0$ (or $= k_2$). The most common choice for chemical control has been adsorption, k_1, but with a high activation energy 20–40 kcal), although this is contrary to the conclusion of Section 5.6. In selecting these alternative controls, it follows that, for diffusional control, $R_T \propto 1/a \propto (\sigma/m)^{1/3}$, and for chemical control, R_T is independent of m. Collecting terms in m and σ, we find that Equation 7.14 reduces to the general form, with m eliminated from the right-hand side,

$$\left(\frac{1}{m}\right)^n \left(\frac{dm}{dt}\right) = \text{constant} \left(\frac{1}{\sigma}\right)^n p_0^{n'} f(T) \tag{7.17}$$

where p_0 is the local oxygen concentration; $n = 1/3$ in the case of diffusional control and 2/3 in the case of chemical control; and the oxygen power index, n', is 1 for diffusional and adsorption control (k_D or k_1 ruling) and 1/2 or 0 for desorption control, depending on the extent of internal reaction ($k_2 D_i$ ruling; see Equation 5.15). The function $f(T)$ is some power index of T, generally written as $T^{0.75}$ for diffusional control, and $e^{-E/RT}$ for chemical control. It may be noted that Equation 7.17 does *not* include dependence on the initial mass, m_0, or diameter, d_0, although this is often represented to be the case. One further method of representation of interest is the use of the expression with $n = 1$. The left-hand side of Equation 7.17 is then the rate of loss of mass per unit mass. (Another common usage, notably by Walker and associates [134], is $(1/m_0)(dm/dt)$, or the rate of loss of mass per unit of *initial* mass.

C. LIMITING SOLUTIONS FOR IMPERMEABLE SOLID

In using Equation 7.17 it has been common to treat the right-hand side as a function of oxygen and temperature alone. This ignores the density variation, $\sigma(t)$, or treats it as constant. The assumption that it is constant implies only external reaction. It has also been common to assume a first-order reaction, so that the left-hand side of Equation 7.17 is divided by p_0. At constant p_0 and σ Equation 7.17 can be integrated for given values of n. For the alternate cases of diffusional and chemical control ($n = 1/3$ and 2/3, respectively) the burning time, t_c, can be calculated as a function of initial diameter, d_0, with dependencies on the square and on the first power of d_0 for the two respective cases. With both diffusion and first-order chemical reaction (for p_0 and σ constant) the combustion time is found

ISBN 0-201-08300-0

to be a function of the sum of two terms [137]:

$$t_c = K_c d_0 + K_D d_0^2 \tag{7.18}$$

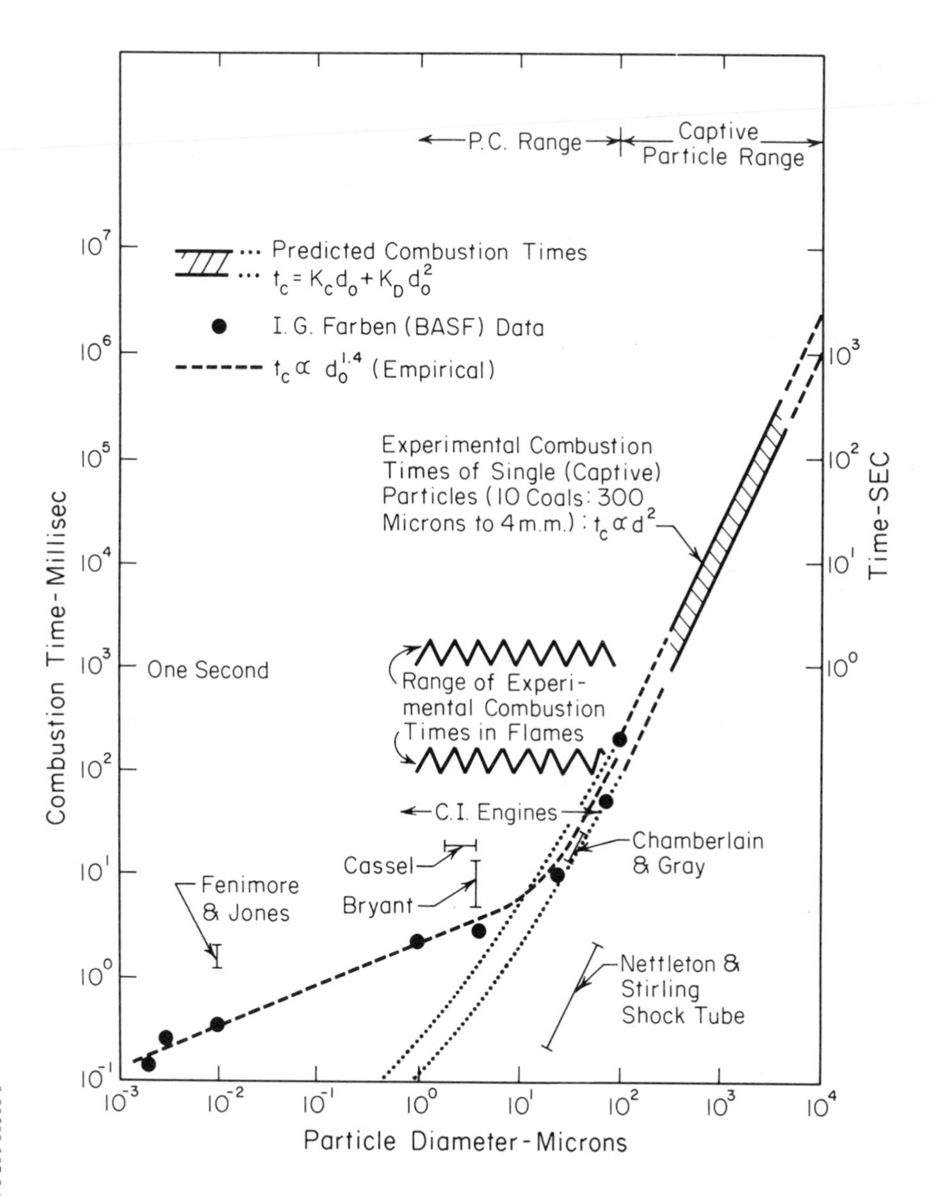

Figure 7.8. Variation of combustion time with particle diameter from various experimental and theoretical sources. (Source: Ref. 88.)

ISBN 0-201-08300-0

where K_c is a chemical burning constant. This follows from integration of the rate equation with R_T obtained from the so-called "resistance" equation (Section 5.6E. It results in the predicted behavior, as illustrated in Figure 7.8. Agreement is excellent above 100 μm, but deviates significantly from experiment below that size.

D. REACTION ORDER AND RATE CONTROL MECHANISM

The source of error is attributed to wrong assumptions about reaction order (1/2 to zero, not unity) and reaction location (internal, not surface). Arrhenius plots of rate data showed that the temperature coefficient of the reaction corresponded to activation energies of 20–40 kcal. On this basis, and for other reasons (see Section 5.6; also discussed more extensively in Refs. 64, 80, 81, 88) this writer concluded some time ago that the reaction control was most probably desorption. This is still a minority point of view but would appear to be being increasingly accepted (e.g., 90). In that case the power index, n', in Equation 7.15 would be 1/2 or zero. Most attempts to check the reaction order directly have resulted in arguable conclusions (see Refs. 87, 88). For example, in one widely quoted set of experiments on combustion of smoke [135], plots of (interpolated) R_T against p_0 at constant T did not go through the origin; indeed, a *second*-order fit was much better, as found by Magnussen [136] in a subsequent analysis of those combustion experiments. Analyzing other experiments, Froberg [75] concluded that the results were in error since the wrong temperature had probably been used – generally the reactor temperature or a mean bed temperature, not the particle surface temperature itself.

In practice the reaction rate is so temperature sensitive and the variation in p_0 is so small (typically varying by only a factor of 2) that it makes little practical difference whether or not the p_0 term is included. If a zero-order reaction is assumed, Figure 7.9 is typical of the results obtained. The original data were obtained by Beer [110], burning anthracite in a plug flow furnace. The points in Figure 7.9 were recalculated from the raw data in the source cited. (The original publication of the data [110] used only four of the runs, as against seven in Figure 7.9, and used a linear plot of R_T against p_0 $(T - 1030)$.) The striking characteristic of Figure 7.9 is that the lines fall into two groups, where runs 1, 2, 5, and 6 have slopes equivalent to 22 kcal and runs 7, 8, and 9 have slopes equivalent to 44 kcal. Run 1 (and possible runs 5 and 9) shows signs of a change of activation energy during the reaction.

The obvious interpretation of these data is to identify the factor of 2 difference in activation energies as due to pore diffusion control (Equation 5.5) in the case of the lower value (22 kcal); the two values are then very well in line with other sources (see Ref. 88). This explanation should be accepted with some reserve, however. If the curves with lower slopes are interpreted as representing reaction with pore diffusion control, the other curves must either represent reaction at the superficial surface only, without penetration, or must represent approximately uniform reaction throughout the particles (or even both extremes in different

ISBN 0-201-08300-0

cases). Since the fuel was anthracite, which has low porosity, reaction at the superficial surface only is not impossible. The analysis and interpretation is then generally acceptable. Where there is partial penetration, however, the evaluation must take into account the change in density, $\sigma = \sigma(t)$, unless the penetration depth of the reaction is slight enough for σ to remain approximately constant over a sufficient fraction of the reaction zone. Even greater reservations must be expressed if reaction is uniform through the particles.

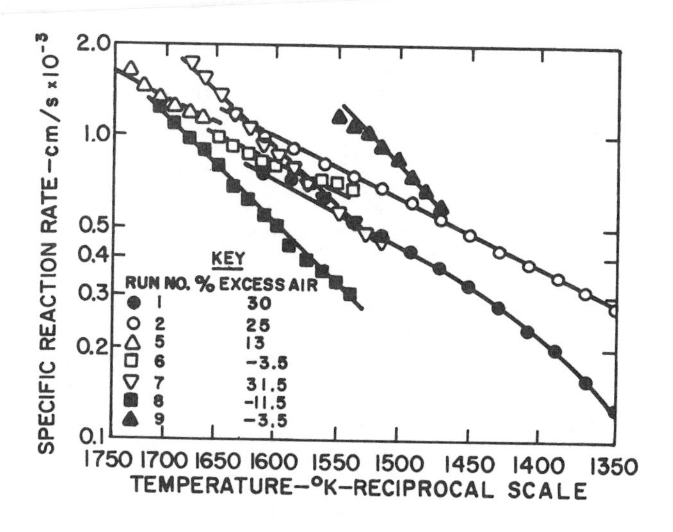

Figure 7.9. Arrhenius plot of reaction rate against temperature. (Oxygen concentration variable through runs.) (Data from J. M. Beer [110]; source of graph: Ref. 87.)

E. INTERNAL REACTION

Attempts to extend the analysis to include internal reaction in a directly testable form have not yet been successful. Theories in computer form do exist containing detailed assumptions about internal reaction behavior, and comparisons with experimental results have shown broad agreements in the gross by suitable choice of somewhat arbitrary rate coefficients. The models are not, therefore, proved, although this degree of agreement may serve to show that high accuracy of assumption is unnecessary. In that event an approximate extension of the above equations serves to illustrate some of the problems to be solved.

When there is internal reaction, Equation 7.16 still applies, but in conjunction with Equation 5.15 or 5.17, and with σ variable. In the extreme case, for which

ISBN 0-201-08300-0

experimental support exists in some instances, the particles may burn at constant radius, a_0, in which case only Equation 5.17 applies, with the first right-hand term zero, and Equation 7.16 can be used in the form $(dm/dt)/4\pi a_0^2$ (also, $m/m_0 = \sigma/\sigma_0$). Equating to Equation 5.17 with the first right-hand term zero, we find that the only mass variable is the internal surface per unit volume, A_i. Internal surface area measurements during burn-off have been made, and wide variations have been reported between different investigations. In general, however, the total surface area of a given sample tends to increase initially with time or burn-off until increasing diameter causes the pores to start to coalesce and surface area falls again. The curves are typically asymmetric, inverted U-shapes. On a unit mass basis, however, there tends to be more regularity. If θ is the surface area per gram, the variation with burn-off, b $(= 1 - m/m_0)$, is a rising flat curve that to a first linear approximation may be expressed as

$$\theta = \theta_0 (1 + \alpha \cdot b) \tag{7.19}$$

At density σ the surface area per unit volume, A_i, is given by

$$A_i = \sigma\theta = \left(\frac{\sigma_0}{m_0}\right) m\theta_0 \left[1 + \alpha\left(1 - \frac{m}{m_0}\right)\right] \tag{7.20}$$

and the rate expression for comparison with Equation 7.17 becomes

$$\frac{dm/dt}{\theta_0 m[(1 + \alpha) - \alpha m/m_0]} = \left(\frac{3}{4\pi a_0^2 S}\right) k_{\mathrm{eff}} p_0^{n'} \tag{7.21}$$

The right-hand side clearly has the same general form in both equations (for σ = constant in 7.17), although with a dependence now on the initial particle size, a_0. In particular, for small α the left-hand sides of both equations compare, with $n = 1$ in Equation 7.17. This corresponds to constant internal surface per unit mass, which was the assumption successfully used by Stewart and Diehl [138] for analyzing char particle gasification in fluid beds.

There is no publication known to this writer describing any test of any equation of the form of Equation 7.21 except for the simple approximations cited (n = 1/3, 2/3, or 1 in Equation 7.17). Some comparison may be made with the thin particle bed data described by Walker et al. [134]. These are TGA data of 40 x 100 ASTM mesh char particles prepared from a range of different coals. Burn-off of the char particles in different atmospheres was determined as a function of time at constant temperature under conditions such that boundary layer diffusion was considered negligible. (Sample size was about 3 mg.) At constant temperature and with the mainstream reactant concentration kept constant, the reactant being O_2, CO_2, or H_2O, Equation 7.21 can be integrated. With $[\theta_0 (1 + \alpha)]$ $(3/4\pi a_0^2 S) k_{\mathrm{eff}} p^{n'}$

ISBN 0-201-08300-0

written as $1/\tau$, where τ is now a characteristic time for the reaction, the integral has the form

$$\text{Burn-off: } b = 1 - \frac{m}{m_0} = \frac{1 - e^{-t/\tau}}{1 + \alpha e^{-t/\tau}} \tag{7.22}$$

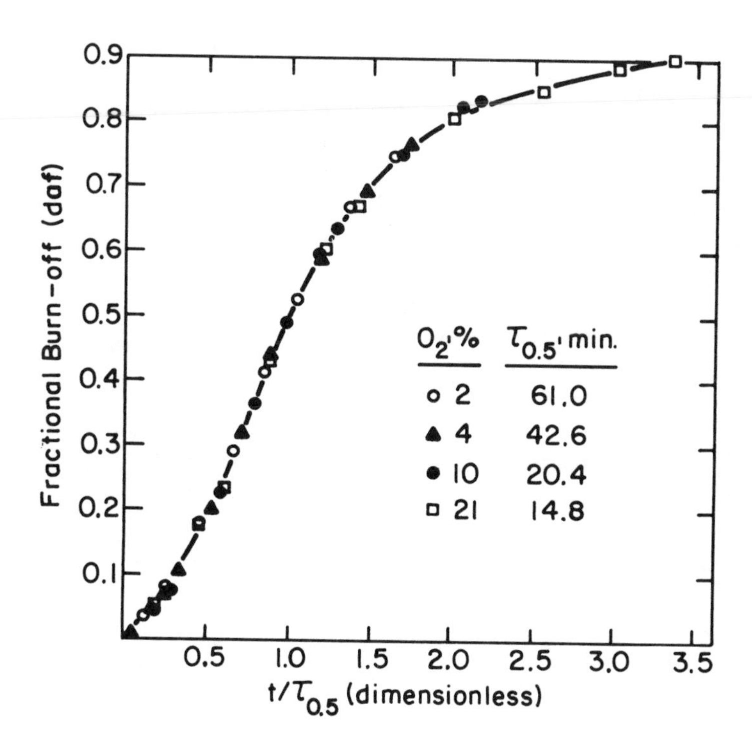

Figure 7.10. Normalized plot for reaction of PSOC-91 char at 405°C in different concentrations of O_2. (Source: Ref. 134.)

This equation qualitatively describes the normalized data obtained by Walker et al. [134], illustrated in Figure 7.10. The quantitative agreement is less satisfactory. The data represented burn-off against time, normalized by a time, $\tau_{0.5}$, chosen at 50% burn-off. From Equation 7.22 it follows that $\tau_{0.5}/\tau = \ln(2 + \alpha)$, which justifies use of $\tau_{0.5}$ as a normalizing factor for time. It also follows that $1/\tau$ or $1/\tau_{0.5}$ should be proportional to $p^{n'}$. From the data provided [134], n' was found to be about 0.5, implying that either pore diffusion is still important with a zero-order primary reaction or that the reaction lies in the transition between zero and first order. Equation 7.22 also predicts the existence of a point of inflection in the curve, as found experimentally. There is not yet numerical agreement, however,

ISBN 0-201-08300-0

between the equation and the data without empirical adjustment of the constants involved. The preceding evaluation nevertheless suggests a possible course of action for further analysis.

In summary, the simplest calculation base for predicting or evaluating burning times of actual particles in flames is inadequate since it does not take into account the internal burning behavior (e.g., Equation 7.17 with $n = 1/3$ or $2/3$). Use of Equation 7.17 with $n = 1$ may be a little more realistic. The problem is that there is no clear idea as yet of how detailed models must be for prediction to be acceptable. In actual flame systems the mixing pattern may become overriding. Figure 7.11(a) shows the effect of different combinations of stirred and plug flow mixing patterns (see Section 6.3) on the burnout of anthracite, and Figure 7.11(b) roughly compares prediction with experiment, showing how experimental data can be approximately bounded by different swirl assumptions [139]. Before it can be said that more detailed reactivity models are necessary, more accurate measurements and more carefully bounded comparisons between theory and experiment are required. If better knowledge of reactivity is needed, however, Equation 7.19 suggests the path to take. What is required is better general knowledge of the variation of internal surface, θ, with burn-off. For operational purposes Equation 7.19 can be regarded as the first two terms of an empirical series expansion to describe θ as a function of b. Further investigations are needed to establish the fundamental basis for the actual curves and the coefficients involved in any pertinent equations.

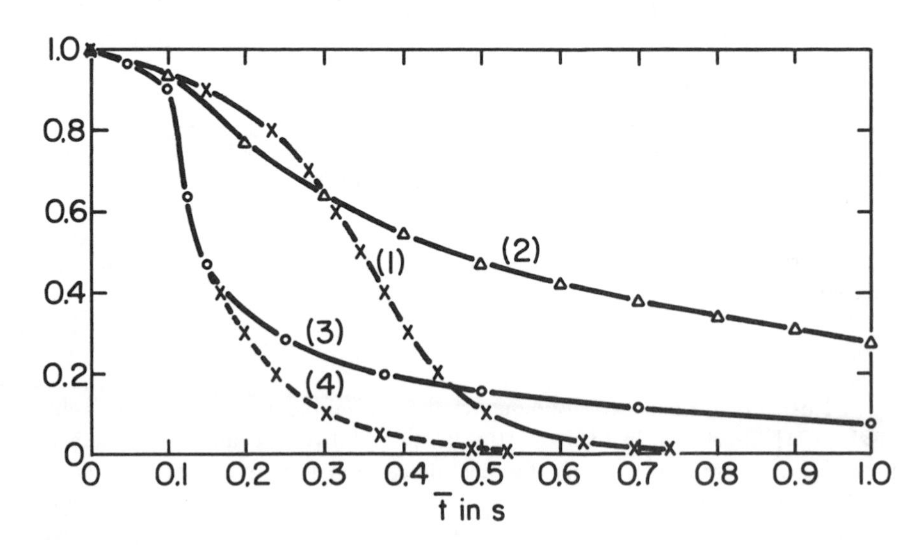

Figure 7.11(a). Fraction of unburned mass of fuel or combustion efficiency as a function of mean residence time, calculated for stirred and plug flow combustors. Curve 1: plug flow combustor; curve 2: stirred combustor with complete "segregation"; curve 3: stirred combustor with complete micromixing; curve 4: stirred and plug flow combustors in series. (Source: Ref. 139.)

ISBN 0-201-08300-0

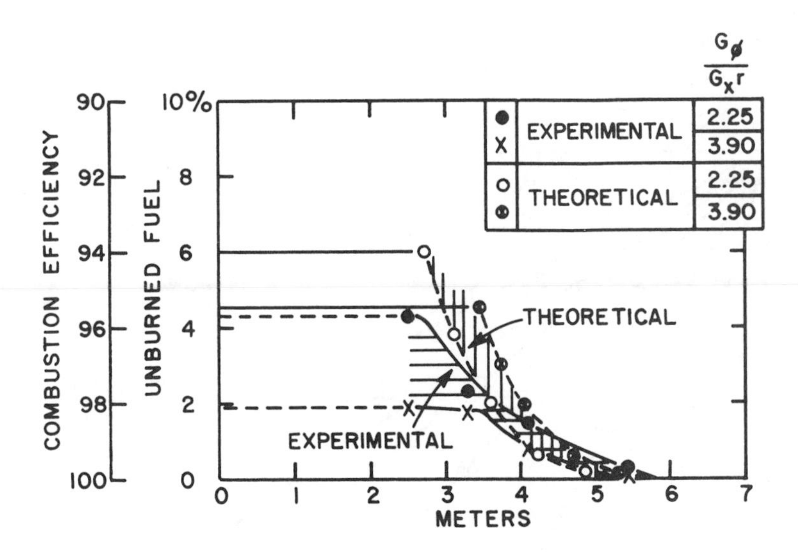

Figure 7.11(b). Comparison of experimental and calculated fractions of unburned fuel, or combustion efficiencies, at the boundary of the stirred and plug flow regions and in the plug flow part of the furnace, plotted as a function of distance from the burner and for two values of the dimensionless degree of swirl. (Source: Ref. 139.)

7.5. Explosion Flames

A. NATURE OF THE PROBLEM

Although the dust explosion hazard was first recognized almost two centuries ago (some would say three; see the review in Ref. 107), with an almost continuous history of investigation for 100–150 years, it has proved to be a particularly intractable problem. The understanding of propagation and extinction requirements remains largely qualitative, with too little that is quantitative or has been successfully analyzed theoretically from the mechanistic point of view. There has been more success with phenomenological models. In an examination of the record of 50 years of experimental explosion research in England, the results were found to be consistent with an overall model of behavior [107, 140]. Section 6.2(iv) briefly describes the typical method of experiment: dust is laid on a gallery floor and dispersed by an igniter. The blast from the igniter that initially disperses the dust is known as the "pioneer" wave. As the intensity of the pioneer wave declines, it is progressively reinforced by combustion of the dispersed coal. The usual method of experiment is to use mixtures of coal dust and rock dust, and to determine the minimum rock dust percentage in the mixture, S_p, at which propagation is stopped.

ISBN 0-201-08300-0

As the coal concentration rises, S_p also rises to an asymptotic limit, as illustrated in Figure 7.12 [141].

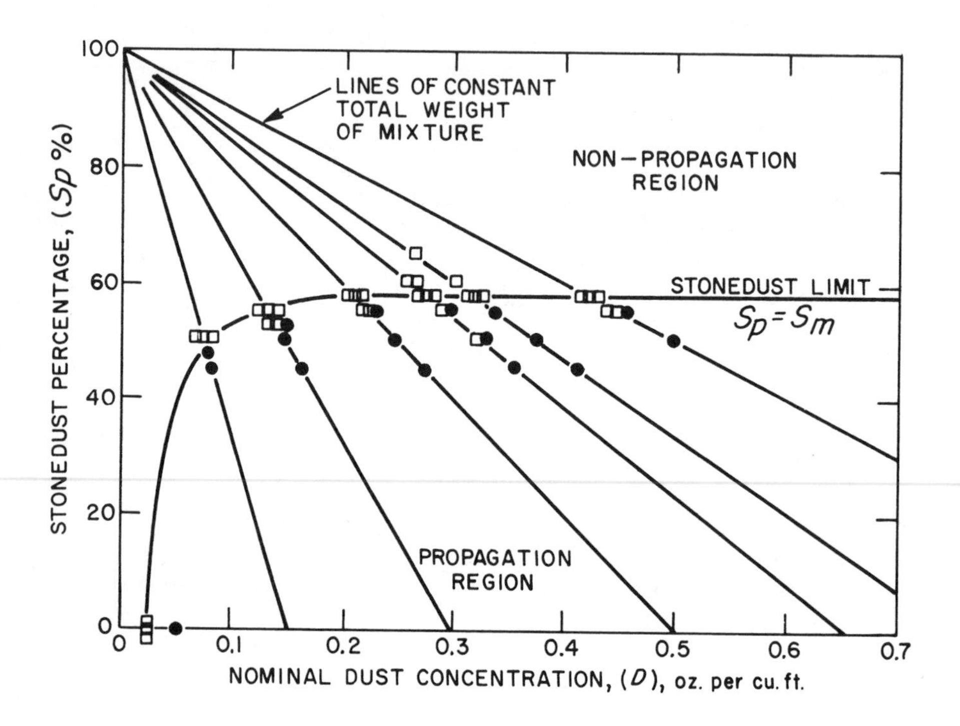

Figure 7.12. Variation of stone dust extinction, % S_p, with nominal dust concentration, D, fit of calculated curve to original data. (Source: Ref. 107.)

B. ROCK DUST EXTINCTION

When the igniter is fired, not all the dust is necessarily dispersed initially; and, as a consequence of more dust being dispersed, or the dust being dispersed more uniformly as the explosion gathers speed (see Figure 6.4), there is more intense combustion as mixing is improved with increased cloud speed down the gallery. This is illustrated schematically in Figure 7.13, which also indicates that the most likely point of propagation failure is at the blast wave intensity minimum. It is further proposed with substantiating arguments [107] that extinction is likely to be a low limit phenomenon in most experiments, even with addition of rock dust to the mixture. If the normal concentration at the failure point without rock dust added is the low limit, this low limit is preserved at greater loading weights of coal dust in the gallery if a larger fraction of dust is dispersed but the fraction in excess of the low limit is all rock dust. This argument was used to develop the following

ISBN 0-201-08300-0

expression for the variation in the stone dust percentage required just to suppress propagation, S_p, and the coal concentration in the gallery, D:

$$\frac{S_p}{100} = \frac{(S_m/100) - e^{-D/D_0}}{1 - e^{-D/D_0}} \tag{7.23}$$

where S_m is an upper limiting value of S_p as D becomes large, and D_0 is a parameter of the dust layer that is governed by its dispersion characteristics. Equation 7.23 describes the behavior of Figure 7.12.

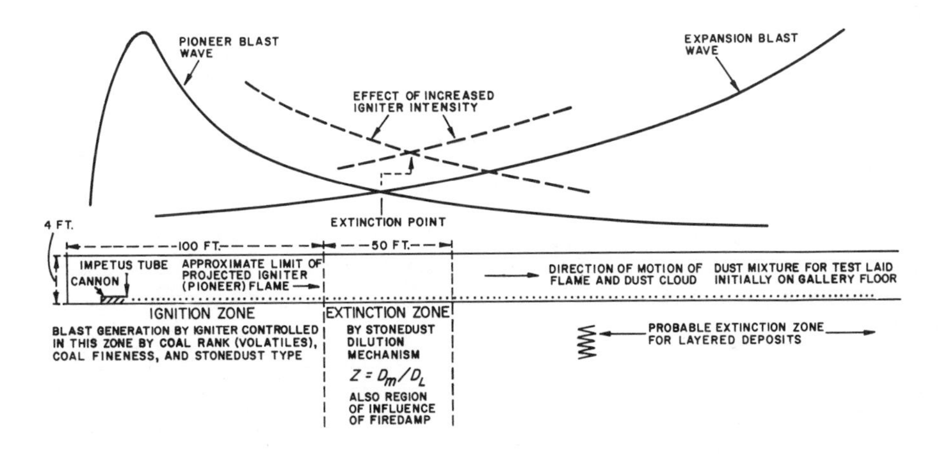

Figure 7.13. Diagramatic representation of explosion gallery (325 ft) with probable regions of ignition and extinction, illustrating also the possible behavior of the pioneer and expansion waves. (Source: Ref. 107.)

C. FLAMMABILITY INDEX

The suggested reason for the asymptotic rise of S_p to the limit S_m, which then becomes insensitive to or independent of D, is that only a maximum quantity of dust is dispersed in the extinction zone (Figure 7.13). If this maximum concentration of the *mixed* dusts is D_m, and the low limit is D_L, a "flammability index" for the gallery can be defined as $Z = D_m/D_L = 100/(100 - S_m)$. In several decades of experiments (as reviewed in Ref. 107; also see Refs. 141, 142) it was established for a given gallery that Z is a simple function of volatile content, V, and fineness, the fineness index, F, being effectively the external surface area per gram. The empirical relations obtained can be combined as follows [140]:

$$Z - Z_0 = K(V - V_0)(F - F_0) \tag{7.24}$$

ISBN 0-201-08300-0

where the empirical constants Z_0, V_0, and F_0 had values of 0.95, 12.5% and 0.15, respectively. These were then independent of coal rank and fineness. The slope factor, K, was evidently dependent on the rock dust type; for a stable dust it has a value of 0.042. The equation implies that coal with VM content less than V_0 (12.5%) and of fineness less than F_0 (0.15) will not explode under the specified gallery conditions. The former conclusion is used in some mining legislation regarding requirements for the use of rock dust in coal mines for safety purposes when rock dust is only required for coals exceeding some minimum VM.

D. ROCK DUST TYPE

The influence of rock dust is of some importance. Both alkali halides and dusts containing water of crystallization have been reported to be more effective than the standard limestone. Based on the details of the experimental procedure, there are arguments [107] that these materials will not influence the dilution behavior in the extinction zone, but they can affect the igniter. On the basis of thermal replacement of the inert heat sink by endothermic decomposition of the materials with water of crystallization (gypsum, $CaSO_4 \cdot 2H_2O$; Epsom salts, $MgSO_4 \cdot 7H_2O$; and borax, $Na_2B_4O_7 \cdot 10H_2O$) the differential effect of S_m between limestone and these listed materials was predicted to be proportional to the endothermic heat of decomposition, with the proportionality constant the product of the specific heats of the materials and the igniter (gunpowder) flame. Available experimental data [142] substantiated the prediction [107] with a reasonable value for the gunpowder flame temperature (2300°C est.). On this basis alkali halides are also presumed to affect the dispersing blast, acting as gas reaction inhibitors of the gunpowder flame.

E. EFFECT OF ADDED METHANE

The model was also used to account for the change in rock dust limit when methane is present in the gallery. Here the effect is presumed to be one-to-one replacement of coal by methane, on a calorific value basis, in the extinction zone, where the coal is presumed to be at its low limit concentration in the absence of methane. The addition of methane at concentration p_m% required an increase in the rock dust limit percent from S_m to S'_m. The thermal replacement argument resulted in the equation

ISBN 0-201-08300-0

$$S'_m = S_m + \left(1 - \frac{S_m}{100}\right)\left(\frac{\rho_m}{D_L}\right)\left(\frac{\Delta H_m}{\Delta H}\right) p_m \qquad (7.25)$$

where ρ_m and ΔH_m are the density and heat of combustion, respectively, of methane; and D_L and ΔH are the low limit concentration and heat of combustion of coal.

The proportionality between S'_m and p_m was established experimentally by Mason and Wheeler [142] for seven different coals. The slopes of the lines for the seven coals were found [142] to be proportional to $(1 - S_m/100)$, and the slope of the line then obtained [107] was in almost exact agreement with the value calculated from the residuals of Equation 7.25.

The model of an explosion flame in a gallery as described above is consistent, then, with broad assumptions and analyses of specific behavior, accounting qualitatively, and in some instances quantitatively, for such behavior as variation of rock dust percent with coal loading in the gallery, rock dust type, and effect of added methane.

Two additional questions emerging from this evaluation concern (a) interpretation of the volatile influence in Equation 7.24; and (b) the value of the inert thermal limit by rock dust. On the first question there is now such a wealth of data to show that flammability in other tests does *not* correlate with VM percent that an alternative explanation would seem to be necessary. The nature of Equation 7.24 and the association with surface area suggest that VM percent might be an index of internal surface except for the apparent trend of Figure 5.2. At present, however, it must be considered that the problem remains open. The other question of the rock dust thermal limit relates to safety regulations. Flame suppression by addition of 60–70% rock dust was long regarded as due to thermal loading. Simple thermal considerations, however, can be used to predict an upper limit of 1.5–2 g/l of coal in air (noting Csaba's measurements of 1.6 g/l in Figure 6.2). Substitution of rock dust for coal would require, on a simple thermal basis, something in excess of 90% stone dust or 80% water to quench the flame. Water has additional effectiveness because it also reduces all concentrations by dilution. In Australian brown coal firing [143], where the coal of 66% moisture was mill dried with hot POC, the coal concentration in the feed was found to be close to the low limit, the POC providing additional dilution. Table 7.1 compares gallery extinction limits with low grade fuel firing conditions [107]. The high loading levels of inerts required to extinguish flames must be recognized.

ISBN 0-201-08300-0

Table 7.1

Comparative Data on Gallery Extinction Limits in Explosion Tests, and Inert Loading Using Low Grade Fuels Employed in Combustion Systems[a]

Reference	Coal or Other Material	VM (daf), %	Principal Inert, %	Total Inerts/ Fuel Ratio
(a) EXPLOSION – Test Gallery Extinction Limits by Inerts (Water and/or Ash)				
Mason and Wheeler	Red Vein	22	Stone dust, 40	0.67
	Silkstone	36	Stone dust, 60	1.5
	Seven foot	44	Stone dust, 70	2.3
Jones and Tideswell	Red Vein	22 (layered deposits)		$\nless 5$
Cybulski	Barbara (Poland)	41	Water, 42.5	1.0
			Ash, 49.0	2.0
	Water was found on average to be 2–2.5 times as effective as ash or stone dust in preventing explosion			
Safety in Mines	Lycopodium	(High, unknown)	Limestone, 80% fully dispersed in vertical tube apparatus	4
(b) COMBUSTION – Economic Operations, Generally at 98–99% Combustion Efficiency, Mostly in Water Tube (Cold Wall) Boilers				
Roberts	Australian brown coals			
	Yallourn	51	Moisture, 66.5	2.0
	Morwell	57.5	Moisture, 62.0	1.7
Bouttes and Vellard	French blended coals and slurries	20–35	Ash, 25–45 Moisture, up to 20	About 1.0
Beer	Hungarian coals			
	Black (in muffle furnace)	27	Ash, 53	1.2
	Nograd (brown)	59	Moisture, 43 Ash, 12.6	1.25

[a] Source: Ref. 107.

8. POLLUTION CONTROL

The problems of pollution control require major separate treatment. This brief summary serves to list the problems, not to discuss their solutions to any great extent (see, e.g., Refs. 144–152).

The majority of the man-made pollutants present in the air today originate from combustion sources, and about 90% of these are accounted for by emissions of only five types: carbon monoxide, hydrocarbons, particulates, nitric oxides, and sulfur oxides. (In connection with a possible greenhouse effect some authors now also include CO_2.) The emissions most particularly associated with coal are the last three, due to ash, fuel nitrogen, and sulfur.

A. COMBUSTIBLE EMISSIONS

Coal has been notorious in past ages for dense hydrocarbon emissions –

ISBN 0-201-08300-0

smoke – due mostly to bad combustion control, but in some instances produced deliberately in early attempts at atmosphere control in steel heat treating furnaces. In modern furnaces and boilers steady smoke emissions usually represent bad furnace design, poor maintenance, or poor operation. When such emission occurs, it should be regarded as curable by improved operational practices, which can include better mixing of air and pyrolysis products, even if this requires refitting with more overfire air ports, higher air pressures, and improved placement [153]. Smoke (and CO) problems are mostly associated with cold wall boilers. In hot wall furnaces mixing has to be very poor to prevent ultimate combustion of all volatiles and nearly all carbon, although this fault is still common with many incinerators.

B. PARTICULATES

Ash is fired mineral matter. Some fraction will remain in a boiler or a furnace as bottom ash; the rest leaves as fly ash. Removal of fly ash by electrostatic and other precipitators, bag houses or other filters, or other wash systems is, for the most part, standard practice [144–152], so that review here is out of place. It may be noted that recent reductions in the SO_2 contents of stack gases have sometimes caused problems with electrostatic precipitators on account of reduced electrical conductivity. One approach has been to bleed SO_3 into the cleaned stack gas just before the precipitator, but this is clearly undesirable. It may also be noted that particulate emissions from fluid bed combustions are proving to be troublesome, and the development of improved control methods is called for in a recent review of the current status of the Fluid Bed Combustor development program [154]. For these units standard practice is not good enough, and improved methods are not yet available.

C. NITROGEN OXIDES

Nitrogen oxides are formed either by reaction of nitrogen with oxygen at high temperatures (in excess of 1500°C) – thermal NO_x – or by radical reactions of fuel nitrogen. Thermal NO_x production from all fuels has now been studied extensively, and necessary practices to alleviate the problem have been reasonably well established. The focus in most cases is to lower the temperature and/or to reduce temperature peaks and hot spots by altering the mixing pattern. In the latter respect it is notable that short flames of the type shown in Figure 3.16 are evidently bigger NO_x producers in most circumstances than are the tangential tilting (corner-fired) burners. The local combustion intensity tends to be higher in the former case. The greater turbulence level may also be an adverse factor here, with a greater probability of forming long lived turbulence vortices that can react almost independently and nearly adiabatically to form exothermic centers that can be prime sources of NO_x [155]. Conventional practices to reduce thermal NO_x

ISBN 0-201-08300-0

were initially developed with a principal focus on oil and gas firing. They include the following:

1. Staging, to fire in two zones, the first being fuel rich and the second having the balance of the combustion air. This helps to reduce flame temperature peaks.
2. Reduced preheat, POC recycle, steam injection, or boiler derating (reduced firing rate), all designed to reduce the overall temperature.
3. Reduced excess air.
4. Altered firing pattern.

The thermal NO_x control measures that work with oil and gas also work with coal. The NO_x balance, which comes from the fuel nitrogen, is more difficult to control and less well understood. It is much less sensitive to reduction in flame temperature but is evidently sensitive to fuel stoichiometry, with reduced production when firing fuel rich. Staged combustion is therefore a possible control measure: if the fuel N becomes N_2 during the fuel-rich gasification, the residual problem is that of controlling thermal NO_x by the practices listed above.

More recently, chemical means of NO_x control have been suggested, based on observations (e.g., 156) that NO can react with nitrogenous compounds such as ammonia to produce N_2. There is evidently an optimum reaction temperature, reported in the region of $1200^\circ C$ in some experiments, below which the destruction is too slow and above which the ammonia oxidation itself may be forming more NO than it is destroying. If materials can be found whose reaction with NO is fast at fluid bed temperatures, this could be a good approach to control. The two general problems likely to be assoicated with this prospective method of control are, first, being able to carry out the reactions at the location where the reaction temperature is optimum, and, second, being able to supply the reactant in necessary quantities.

The control of NO_x is still a problem. Fortunately, only a minor fraction of the fuel nitrogen is converted into NO in most circumstances, but better control is still warranted.

D. SULFUR OXIDES

The most serious pollution problem from coal is generally regarded as SO_x emissions. The oxides derive from both organic and inorganic sulfur (the latter existing mainly as pyrites). Sulfur trioxide, SO_3, is formed by direct oxidation of SO_2 and, like thermal NO, is open to combustion control by reducing the overall oxygen level (excess air). Unfortunately SO_2 is not subject to combustion control, and other strategies are necessary. The principal ones are listed in Table 8.1. The three principal options are removal before combustion, during combustion, or after combustion. The problems involved in these various options are numerous and too specialized to be detailed here. The following summary must suffice.

ISBN 0-201-08300-0

Table 8.1

Sulfur Control Strategies

Strategy	Methods
I. Precombustion Cleaning	
A. Mechanical (physical) (inorganic sulfur only)	(i) Dry methods (ii) Wet methods (flotation, etc.: Trent process, etc.)
B. Chemical (total sulfur)	(i) Hydrodesulfurization (ii) Oxydesulfurization (iii) Solvation or other removal of mineral matter
II. Interstage Cleaning	
A. Gasification strategy: sulfur removed as H_2S in an interstage of the processing	(i) Producer/water (synthesis) gas, including combined cycle operation (ii) Pyrolysis (destructive distillation in coke ovens) (iii) Indirect methane production (HI gas)
B. Liquefaction strategy	(i) Solvation (SRC) (ii) Synthoil production
C. Other	Removal in flame or combustor
III. Postcombustion Cleaning	
A. Stack gas cleaning (FGD)	
[B. Tall stack dispersion (dilution[a])]	

[a] Dispersion is not advocated as a strategy for control of sulfur emissions, but it is advanced as an interim strategy for control of ambient air levels.

(i) *Precombustion cleaning:* Both commercial and experimental methods are available for the mechanical/physical cleaning of coal – gravity separation, froth flotation – some of which are well established and capable of handling very large tonnages (e.g., 1000 tons/hr). They are not necessarily applicable, however, to all coals. Mechanical/physical methods of cleaning, and likewise ash solvation methods, remove only the mineral matter and thus only the inorganic sulfur. The latter is highly variable although it tends to constitute between one third and two

ISBN 0-201-08300-0

thirds of the total sulfur (it is, unfortunately, lower in the low sulfur coals). The mechanical methods also generally require the mineral matter to occur in the coal in typically large lumps, usually on cleavage planes, so that separation from the coal requires only a fine crush or a coarse grind. When the mineral particles are very fine (e.g., micron sized) and quite uniformly distributed through the coal, such methods will usually fail.

Chemical methods are still experimental or represent only paper studies for the most part. Both hydrodesulfurization and oxydesulfurization have been proposed; the first requires temperatures of 700–800^oC [157], but the second requires temperature of only about 400^oC.

(ii) *Cleaning during combustion:* This is a special case of what is listed in Table 8.1 as "Interstage Cleaning." Processes to convert coal to gaseous and liquid fuels include sulfur removal at that time as the strategy for sulfur control. This is outside our interests here other than to note that, in gasification, sulfur usually becomes H_2S, which is generally easier to remove and for which there are numerous established, but expensive, commercial processes. The sulfur compounds are also present in much higher concentration than is SO_x during combustion because of the reduced quantity of nitrogen (e.g., in low Btu gas) or its virtual absence (e.g., in syngas).

The removal of sulfur oxides during combustion requires the injection of some absorber into the flame. This is not a process even suggested for grate firing. It has been used with varying reports of success in p.c. boiler firing; it is one of several claims for the superiority of fluid bed firing over other firing methods. In all cases, however, the central issue can be said to be one of relative and absolute quantities of materials. The problem can be seen more clearly in the context of Table 4.1. Firing 100 tons of coal produces between 1000 and 1500 tons of gaseous products of combustion. A coal of 1% sulfur will therefore produce 2 tons of SO_2. A 1000 MW utility station firing such coal will therefore produce about 70,000 tons/year of SO_2, which is a disconcertingly large quantity. The concentration in the POC, however, is proportionately low. The POC output per year from the same station approximates 45 million tons. The SO_2 concentration in the POC is about 0.15%, which is very dilute seen as a process stream. Furthermore, absorbers are essentially restricted in this instance to limestone or dolomite. If the SO_2 is absorbed stoichiometrically, the required solids/SO_2 ratio is about 3:2 for limestone and nearly 3:1 for dolomite. In practice, efficiencies of absorption in p.c. injection have been rather low because of the short contact time, ranging from 45% down to 20%. The required solids ratio therefore rises to 6:2 up to 15:2 for limestone and 6:1 up to 15:1 for dolomite. This represents the equivalent of firing coal with increases in ash content ranging from 6 to 30 percentage points. If an average of 1/10 ton of absorber per ton of coal is required, applied to all coal fired in the United States, the annual requirement would be of the order of 60 million tons/

ISBN 0-201-08300-0

year, which is about 3 times the present lime industry production. This quantity clearly presents horrendous logistic problems of mining, transportation, storage, and disposal.

In regard to p.c. injection Engdahl [158] recently noted that of five installations using this method (Table 8.2) none has been satisfactory, four have been abandoned or are about to be, and the system is no longer offered by manufacturers. The limestone must be removed from the POC, and the chosen method was wet scrubbing. Some stations had plugging in the boiler tubes; others encountered scaling and plugging problems in the scrubbers.

In fluid beds the high coal/solids ratio is of no significance in the combustion system, although storage and disposal remain a problem. The behavior of different limestones and dolomites in fluid beds has been the subject of intense, ongoing effort. (Even the recent literature is far too extensive for detailed listings. Significant recent sources include Refs. 131, 159–165; see also Ref. 130.) The rate of uptake of SO_2 to convert CaO to $CaSO_4$ roughly follows a (1 – exp) curve with time, and there is a limiting conversion percent which is generally highest with fresh material of up to 60 or 70% in periods of 1–10 hr. The material can then be regenerated, but the maximum uptake falls steadily with the number of regeneration cycles to about half the original maximum. The absorption limit is determined in large part by the accessible internal surface, which tends to become blocked to access as sulfation proceeds. Internal surface measurements are therefore seen as crucial to understanding behavior. Vogel et al. [160] report, for example, that the pore volume below about 1 μ tends to be most affected and reduced as sulfation proceeds. It is also evident that this region of the pore volume is most prone to permanent blockage after regeneration.

The percentage reduction in SO_2 emissions varies with different absorbers, Ca/S mole ratios, temperatures, and other firing conditions. In general, dolomite is reported to be superior to limestone at the same Ca/S ratio, evidently on account of the faster blockage of the micropore structure of the limestone. Percentage reductions in SO_2 emissions then correlate roughly with the percentage conversion for fresh material: 80% for dolomite at unit mole ratio, and about 70% for limestone at a mole ratio of 2. In general, the percentage reduction increases to an asymptotic limit as the mole ratio increases, with a reported 95–98% removal by dolomite at a mole ratio of 2.

Temperature and pressure also have significant effects. Under pressurization the percentage reduction in emissions is evidently higher with dolomite and lower with limestone, again probably reflecting the different mechanical responses of the absorbers to possible closure of the pores under pressure. The effect of temperature on both materials is similar. With rising temperature the sulfur retention reaches a peak, with the peak moving to a higher temperature with increasing pressure.

ISBN 0-201-08300-0

(iii) *Postcombustion control:* Postcombustion control is essentially limited to stack gas scrubbing or flue gas desulfurization (FGD) when the sole objective is SO_2 removal. Dilution by use of tall stacks can reduce local ground level concentrations but without, of course, affecting the total tonnage of SO_2 dispersed into the air. Such dilution is widely practiced in England, in particular, but is not regarded as an acceptable control strategy in the United States.

Of all possible FGD systems that have been proposed, Table 8.2(a) lists 9 in use on 33 coal-fired boilers (10 systems on 38 boilers if limestone injection is included), but the majority of installations are either lime, limestone, or alkali scrubbing [158]. The lime and limestone processes are throwaway; the alkali process can be regenerative (double alkali). Table 8.2(b) lists the regenerable processes [158]. The essence of any system is a contacting reactor (the scrubber), in which the POC gases are forced through the absorbing fluid for scrubbing. Since wet scrubbing is practiced, the cleaned POC must be demisted and (usually) reheated before going to the stack induced draft fans. The scrubber fluid is circulated in a closed loop with a tap-off for spent material and an injection tap for fresh makeup material. Numerous problems have been encountered that can be summarized as mostly damage or blockage. There are two principal sources of damage: *erosion* by flow of solid particles; and *corrosion,* due mostly to high pH. There are also two principal sources of blockage: *plugging,* by a paste of solids; and chemical *scale* formation of sulfites and sulfates. Sulfites are soft scale and can be washed away; sulfates are hard scale and may have to be chipped off. Plugging of the demisters has proved to be particularly troublesome. There are also the two general problems of pH control and waste disposal. Evaluation of future needs has also suggested that lime and limestone availability may become a problem in due course. This could be evident from the quantity estimates given above.

If there is to be large increase in coal use in the future, pollution control must be given even more attention than it has received recently. In the past the problem has been muted to some degree by the prevalence of coal use in large quantities in single units. The problem, although obvious, has been amenable in principle to massive engineering solutions so far as FGD is concerned in postcleaning operations. The problem takes on a different aspect, however, if direct use of coal returns to the industrial market in furnaces averaging 1/100th or 1/1000th the size of a utility boiler (see Table 4.1). The same quantity of coal would then be burned in 100 or 1000 different furnaces. At that point FGD on each unit is likely to be completely impracticable. Equally impracticable would be attempts to treat all POC at a single processing installation. To collect all the gases would require an impossible spaghetti-like array of flue tubes, some carrying very hot gases, and all converging on the ash removal and FGD units. The engineering complications are not encouraging, apart from the problems of draft control, fan requirements to move the gases, and so forth. The solution is likely to lie in the direction of conversion or of precombustion cleaning: Options I and II of Table 8.1.

ISBN 0-201-08300-0

Table 8.2(a)

Full-Scale Demonstration Programs on Boilers in the United States (Start-up by 1976) (All Coal-Fired Except Two Marked *)[a]

Year of Start-up	Facility	Vendor	Size of Facility, MW
Limestone injection-wet scrubbing			
1968	Union Electric - Meramec	Comb. Engr.	140
1968	Kansas P&L - Lawrence	Comb. Engr.	125
1971	Kansas P&L - Lawrence	Comb. Engr.	400
1972	Kansas City P&L - Hawthorn	Comb. Engr.	125
1972	Kansas City P&L - Hawthorn	Comb. Engr.	140
Limestone scrubbing			
1972	Commonwealth Edison - Will County	B&W	165
1973	*City of Key West - Stock Island	Zurn	42
1973	Kansas City P&L - La Cygne	B&W	820
1973	Arizona Public Service - Cholla	Research - Cottrell	125
1974	Southern California Edison - Mohave	UOP	160[b]
1975	Detroit Edison - St. Clair	Peabody	180
1976	Northern States Power - Sherburn County	Comb. Engr.	680
1976	Central Illinois Light - Duck Creek	Riley Stoker/ Environeering	100
1976	Springfield City Utilities - Southwest	UOP	200
1976	Texas Utilities - Martin Lake	Research - Cottrell	793
Lime scrubbing			
1973	Louisville G&E - Paddy's Run	Comb. Engr.	70
1973	Duquesne Light - Phillips	Chemico	387
1974	Southern California Edison - Mohave	Stearns - Roger	170[b]
1975	Ohio Edison - Bruce Mansfield	Chemico	825
1975	Duquesne Light - Elrama	Chemico	510
1975	Kentucky Utilities - Green River	American Air Filter	64
1976	Columbus and Southern Ohio - Conesville	UOP	400
1976	Louisville G&E - Cane Run	American Air Filter	178
1976	Louisville G&E - Cane Run	Not selected	183
1976	Rickenbacker AFB	Research - Cottrell	20[c]

ISBN 0-201-08300-0

Table 8.2(a) (Continued)

Year of Start-up	Facility	Vendor	Size of Facility, MW
Alkali scrubbing without regeneration			
1972	General Motors - St. Louis, Mo.	CEA	15,8[c]
1974	Nevada Power - Reid Gardner	CEA	125
1974	Nevada Power - Reid Gardner	CEA	125
1976	Nevada Power - Reid Gardner	CEA	125
Alkali scrubbing with calcium regeneration			
1974	General Motors - Parma, Ohio	GM/Koch	32[c,d]
1974	Caterpillar Tractor - Joliet, Ill.	Zurn	10,8[c]
1975	Caterpillar Tractor - Mossville, Ill.	FMC	15,8,8[c]
1975	Gulf Power - Scholz	CEA	20
Alkali scrubbing with thermal regeneration			
1976	Northern Indiana Public Service - D. H. Mitchell	Davy Power Gas	115
Magnesium oxide scrubbing			
1972	*Boston Edison - Mystic	Chemico	150[e]
1973	Potomac Electric - Dickerson	Chemico	100
1975	Philadelphia Electric - Eddystone	United Engineers	120
Catalytic oxidation			
1972	Illinois Power - Wood River	Monsanto	110
Dilute acid scrubbing			
1975	Gulf Power - Scholz	Chiyoda	23
Activated carbon			
1975	Gulf Power - Scholz	Foster Wheeler	20

[a] Source: Ref. 158.

[b] Twenty percent of gas flow from 790 MW unit.

[c] Industrial boiler with equivalent MW rating.

[d] Four stoker-fired boilers.

[e] Oil fuel.

ISBN 0-201-08300-0

Table 8.2(b)

Brief Status Summary on Regenerable Processes[a]

Process Name	Year, Installation Site, Vendor, Size, and Type of Boiler	Status
Wellman-Lord	1976, D.H. Mitchell, Northern Indiana Public Service, Davy Power Gas/Allied Chemical, 115 MW, coal	Under construction
MgO scrubbing	1972, Mystic, Boston Edison, Chemico, 150 MW, oil	Shut down since June 1974
	1974, Dickerson, Potomac Electric, Chemico, 100 MW, coal	Shut down since July 1975
	1975, Eddystone, Philadelphia Electric, United Engineers, 120 MW, coal	Shut down for modification
Cat-Ox	1972, Wood River, Illinois Power, Monsanto, 110 MW, coal	Shut down since 1974
Chiyoda	1975, Scholz power plant, Gulf Power, Chiyoda, 23 MW, coal	Operating since June 1975
FW BF	1975, Scholz power plant, Gulf Power, Foster Wheeler, 20 MW, coal	Started commissioning January 1975; many problems
SFGD	1974, Big Bend Station, Tampa Electric, UOP, 0.6 MW slipstream, coal	Tests are in progress
Citrate[b]	1973, Pfizer's Vigo Chemical complex, McKee/Peabody, 1 MW slipstream, coal	Shut down September 1974 after data collection
Phosphate	1974, Norwalk Harbor Station, Connecticut P&L, Stauffer, 0.1 MW, oil	Shut down June 1974 after data collection

ISBN 0-201-08300-0

Table 8.2(b) (Continued)

Process Name	Year, Installation Site, Vendor, Size, and Type of Boiler	Status
Catalytic-IFP	No data in open literature	No data in open literature
Consol-Potassium Salts	1972, Cromby, Philadelphia Electric, Consol/Bechtel, 10 MW, coal	Shut down since 1972 after data collection (a smaller plant, 1000 ft^3/min, was operated until 1975)
Al-Aqueous Carbonate	1971, Mohave, Southern California Edison, Al, 0.5 MW, oil	Shut down 1972 after data collection on open-loop system
Stone and Webster/Ionics	1973, Valley Station, Wisconsin Electric Power, Stone & Webster/Ionics, 0.75 MW slipstream, coal	Shut down June 1974 after data collection
Westvaco	1970, Westvaco Research Center, Westvaco, 0.2 MW, oil	Shut down 1974 after data collection

[a] Installations in the United States only are listed in this table.

[b] Extensive pilot plant studies on a lead smelter gas are being conducted by U.S. Bureau of Mines, Salt Lake City Metallurgy Research Center.

REFERENCES

1. Essenhigh, R. H., Pratt, D. T., Shull, H. E., and Smoot, L. D., "Current Status of Fluid Mechanics in Practical Heterogeneous Combustors," paper presented at the Spring Meeting, Central States Section, Combustion Institute, Cleveland, 1977.
2. Dainton, H. D., "Status of Fluid Bed Combustion of Coal," Coal Characteristics and Coal Conversion Processes Course, Pennsylvania State University, May 1975.

ISBN 0-201-08300-0

3. Ehrlich, S., *Proceedings of the 2nd International Conference on Fluidized-Bed Combustion,* U.S. Environmental Protection Agency, 1971, p. I-3-1. Ehrlich, S., Robinson, E. B., Bishop, J. W., Gordon, J. S., and Hoerl, A., *AIChE Symp. Ser.,* **70**, 397 (1974).
4. Soehngen, E. E., "Development of Coal-Burning Diesel Engines in Germany," *ERDA Rep.* FE/WAPO/3387-1, 1976.
5. Wormser, A., Private communication, 1977.
6. Brame, J. S. S. and King, J. G., *Fuel,* Edward Arnold, 1955.
7. Forrester, C. (Ed.), *Efficient Use of Fuel,* H.M.S.O., London, 1958.
8. Fryling, G. R. (Ed.), *Combustion Engineering,* Combustion Engineering, Inc., New York, 1967.
9. Griswold, J., *Fuels, Combustion and Furnaces,* McGraw-Hill, New York, 1946.
10. Haslam, R. T. and Russell, R. P., *Fuels and Their Combustion,* McGraw-Hill, New York, 1926.
11. Himus, G. W., *Fuel Technology,* Leonard Hill, London, 1958.
12. Schuhmann, R., *Metallurgical Engineering.* Vol. 1: *Engineering Principles,* Addison-Wesley, Reading, Mass., 1952.
13. Smith, M. L. and Stinson, K. W., *Fuels and Combustion,* McGraw-Hill, New York, 1952.
14. Thring, M. W., *The Science of Flames and Furnaces,* John Wiley & Sons, New York, 1962.
15. Trinks, W. and Mawhinney, M., *Industrial Furnaces,* Vols. I and II, John Wiley & Sons, New York, 1955 and 1961.
16. *Steam,* Babcock and Wilcox Co., New York, 1960.
17. *Industrial Combustion Data,* Hauck Manufacturing Co., Lebanon, Pa., 1953.
18. Yerushalmi, J., Turner, D. H., and Squires, A. M., *Ind. Eng. Chem. Process Des. Dev.,* **15**, 47 (1976).
19. Godel, A. A., *Rev. Gen. Therm.,* **5**, 349 (1966).
20a. Szikla, G. and Rozinek, A., *Feuerungstechnik,* **26**, 97 (1938).
20b. Essenhigh, R. H. and Beer, J. M., *J. Inst. Fuel,* **33**, 206 (1960).
21. Arthur, J., Bangham, D. H., and Thring, M. W., *J. Soc. Chem. Ind. (London),* **68**, 1 (1949).
22. Gamble, R. G., *Proceedings of the 4th International Conference on Fluidized-Bed Combustion,* The Mitre Corp., 1976, p. 133.
23. Hurley, T. F., *J. Inst. Fuel,* **4**, 243 (1931). Hurley, T. F. and Cook, R., *J. Inst. Fuel,* **11**, 195 (1938).
24. Rosin, P. O., *Proceedings of the 1st International Conference on Bituminous Coal,* **1**, 838 (1925); *Braunkohle,* **24**, 241 (1925).
25. Essenhigh, R. H., Chapter XIV in *Combustion Technology,* Palmer and Beer, Eds., Academic Press, New York, 1974.
26. Bueters, K. A., Cogoli, J. G., and Habelt, W. W., *Proceedings of the 15th Symposium (International) on Combustion,* Combustion Institute, Pittsburgh, Pa., 1975, pp. 1245–1260.
27. Thring, M. W., *Coal Res.,* September 1940, p. 70.
28. Essenhigh, R. H., *Fuel,* **38**, 543 (1959).

ISBN 0-201-08300-0

29. Mahajan, O. P. and Walker, P. L., Jr., "Porosity of Coals and Coal Products," in *Analytical Methods for Coal and Coal Products,* C. Karr, Ed., Academic Press, New York, **1**, 125 (1978).
30. Essenhigh, R. H. and Thring, M. W., Paper 29 in *Proceedings of the Conference on Science in the Use of Coal,* Institute of Fuel, London, 1958.
31. Beeston, G. and Essenhigh, R. H., *J. Phy. Chem.,* **67**, 1349 (1963).
32. Essenhigh, R. H., *J. Eng. Power,* **85**, 183 (1965).
33. Essenhigh, R. H. and Yorke, G. C., *Fuel,* **44**, 177 (1965).
34. Ivanova, G. P. and Babii, V., *Teploenergetika,* **13**, 54 (1966) (trans.); *Thermal Eng.,* **13**, 70 (1966).
35. Brookes, F. R., *Fuel,* **48**, 139 (1969).
36. Shibaoka, M., *J. Inst. Fuel,* **42**, 59 (1969).
37. Anson, D., Moles, F. D., and Street, P. J., *Combust. Flame,* **16**, 265 (1971).
38. Suuberg, E. M., Peters, W. A., and Howard, J. B., *Preprints, Am. Chem. Soc., Div. Fuel Chem.,* **22** (1), 112 (1977).
39. Cogoli, J. G. and Essenhigh, R. H., *Preprints, Am. Chem. Soc., Div. Fuel Chem.,* **20** (3), 134 (1975).
40. Smoot, L. D., Horton, M. D., and Williams, G. A., *Proceedings of the 16th Symposium (International) on Combustion,* Combustion Institute, Pittsburgh, Pa., 1977, p. 375.
41. Howard, J. B. and Essenhigh, R. H., *Ind. Eng. Chem. Process Des. Dev.,* **6**, 76 (1967).
42. Orning, A. A., *Trans. ASME,* **64**, 497 (1942); *Proceedings of the 1st Pulverized Fuel Conference,* Institute of Fuel, London, 1947, p. 47.
43. Goldberg, P. and Essenhigh, R. H., Proc. at 17th Symposium (International) on Combustion, Leeds (England), 1979 (in press).
44. Juntgen, H., *Erdoel Kohle,* **17**, 180 (1964). Juntgen, J. and Traenchner, K. C., *Brennstoff-Chem.,* **45**, 105 (1964). Peters, W. and Juntgen, H., *Brennstoff-Chem.,* **46**, 175 (1965). Van Heek, K. H., Juntgen, H., and Peters, W., *Brennstoff-Chem.,* **48**, 35 (1967); *Ber. Bunsenges. Phys. Chem.,* **71**, 113 (1967). Hanbaba, P., Juntgen, H., and Peters, W., *Ber Bunsenges. Phys. Chem.,* **72**, 554 (1968). Juntgen, H. and van Heek, N. H., *Fuel,* **47**, 103 (1968); *Fortschritte der Chemischen Forschung,* Vol. 13, Springer-Verlag, Berlin, 1970, pp. 601–699.
45. Howard, J. B. and Essenhigh, R. H., *Proceedings of the 11th Symposium (International) on Combustion,* Combustion Institute, Pittsburgh, Pa., 1967, p. 399.
46. Neavel, R. C., "Coal Plasticity Mechanism," Symposium on Coal Agglomeration and Conversion, West Virginia University, Morgantown, May 1975.
47. Yerushalmi, J., "Fluid Bed Processing of Agglomerating Coal," paper presented at 81st National Meeting, American Institute of Chemical Engineers, April 1976.
48. Badzioch, S. and Hawksley, P. G. W., *Ind. Eng. Chem. Process Des. Dev.,* **9**, 521 (1970).
49. Pitt, G. J., *Fuel,* **41**, 267 (1962).
50. Anthony, D. B., Howard, J. B., Hottel, H. C., and Meissner, H. P., *Proceedings of the 15th Symposium (International) on Combustion,* Combustion Institute, Pittsburgh, Pa., 1975, pp. 1303–1320.

ISBN 0-201-08300-0

51. Stickler, D. B., Gannon, R. E., and Kobayashi, H., "Rapid Devolatization Modelling of Coal," paper presented at Eastern States Section, Combustion Institute, Fall 1974.
52. Ubhayakar, S. K., Stickler, D. B., von Rosenberg, C. W., Jr., and Gannon, R. E., *Proceedings of the 16th Symposium (International) on Combustion,* Combustion Institute, Pittsburgh, Pa., 1977, p. 427.
53. Nsakala, N., Essenhigh, R. H., and Walker, P. L., *Combust. Sci. Technol.,* **16**, 153 (1977).
54. Essenhigh, R. H. and Howard, J. B., "Combustion Phenomena in Coals and the Two-Component Hypothesis of Coal Constitution," *Penn State Stud. Paper* 31, 1971.
55. Clark, A. H. and Wheeler, R. V., *Trans. Chem. Soc.,* **103**, 1754 (1913).
56. Horton, L., *Fuel,* **31**, 341 (1952).
57. Hirsch, P. B., *Proceedings of the Conference on Science in the Use of Coal,* Institute of Fuel, London, 1958, pp. A29–A33; *Phil. Trans. R. Soc.,* **252A**, 68 (1960).
58. Ayre, J. L. and Essenhigh, R. H., *Sheffield Univ. Fuel Soc. J.,* **8**, 44 (1957).
59. Essenhigh, R. H., *Proceedings of the Conference on Science in the Use of Coal,* Institute of Fuel, London, 1958, p. A74.
60. Slater, L. and Sinnatt, F. S., *Fuel,* **1**, 2 (1922). Newall, H. E. and Sinnatt, F. S., *Fuel,* **3**, 4, 424 (1924); **5**, 335 (1926); **6**, 118 (1927). MacCulloch, A., Newall, H. E., and Sinnatt, F. S., *J. Soc. Chem. Ind.,* **46**, 331T (1927).
61. Kobyashi, H., Howard, J. B., and Sarofim, A. F., *Proceedings of the 16th Symposium (International) on Combustion,* Combustion Institute, Pittsburgh, Pa., 1977, p. 411.
62. Reidelbach, H. and Summerfield, M., *Preprints, Am. Chem. Soc., Div. Fuel Chem.,* **20** (1), 161 (1975).
63. Antal, M. J., Plett, E. G., Chung, T. P., and Summerfield, M., *Preprints, Am. Chem. Soc., Div. Fuel Chem.,* **22** (1), 137 (1977).
64. Gray, D., Cogoli, J. G., and Essenhigh, R. H., Paper 6 in *Coal Gasification, Advances in Chemistry Series* 131, American Chemical Society, Washington, D. C., 1976, pp. 72–91.
65. Curran, G. P., Struck, R. T., and Gorin, E., *Ind. Eng. Chem. Process Des. Dev.,* **6**, 166 (1967).
66. Bishop, J. W., Robinson, E. B., Ehrlich, S., Jain, A. K., and Chen, P. M., "Status of the Direct Contact Heat Transferring Fluidized Bed Boiler," *ASME Paper* 68WA/FU4, 1968.
67. Underwood, G. and Thring, M. W., *Proceedings of the Conference on Science in the Use of Coal,* Institute of Fuel, London, 1958, p. D.27.
68. Finch, D., *Fuel,* **35**, 415 (1956).
69. Field, M. A., Gill, D. W., Morgan, B. B., and Hawksley, P. G. W., *Combustion of Pulverized Coal,* BCURA, Leatherhead, 1967.
70. Hottel, H. C., Williams, G. C., Nerheim, N. W., and Schneider, G. R., *Proceedings of the 10th Symposium (International) on Combustion,* Combustion Institute, Pittsburgh, Pa., 1965, pp. 111–121.
71. Biswas, B. K. and Essenhigh, R. H., "Air Pollution and Its Control," *AIChE Symp. Ser.* 126, **68**, 207–215 (1972).

ISBN 0-201-08300-0

72. Shieh, W. S. and Essenhigh, R. H., *Proceedings of the 5th National Incinerator Conference,* American Society of Mechanical Engineers, New York, 1972, pp. 120–134.
73. Arthur, J. R., *Trans. Faraday Soc.,* **47**, 165 (1951).
74. Kurylko, L. and Essenhigh, R. H., *Proceedings of the 14th Symposium (International) on Combustion,* Combustion Institute, Pittsburgh, Pa., 1973, pp. 1375–1386.
75. Froberg, R. W., "The Carbon-Oxygen Reaction: An Experimental Study of the Oxidation of Suspended Carbon Spheres," Ph.D. thesis, Dept. of Fuel Science, Pennsylvania State University, June 1967. Froberg, R. W. and Essenhigh, R. H., *Proceedings of the 17th Symposium (International) on Combustion,* Combustion Institute, Pittsburgh, Pa., 1979 (in press).
76. Long, F. J. and Sykes, K. W., *Proceedings of Symposium "La Combustion du Carbone,"* Nancy, France, 1949, p. 49.
77. Patai, S., Hoffmann, E., and Rajbenbach, L., *J. Appl. Chem.,* **2**, 306–311 (1952).
78. Trapnell, B. M. W., *Chemisorption,* Butterworths, London, 1955, p. 71.
79. Essenhigh, R. H., *Univ. Sheffield Fuel Soc. J.,* **6**, 15, (1955).
80. Essenhigh, R. H. and Perry, M. G., *Proceedings of the Conference on Science in the Use of Coal,* Institute of Fuel, London, 1958, pp. D1–D11.
81. Blayden, H. E., *Proceedings of the Conference on Science in the Use of Coal,* Institute of Fuel, London, 1958, pp. E1–E6.
82. Walker, P. L., Rusinko, F., and Austin, L. G., *Advances in Catalysis,* Vol. 11, Academic Press, New York, p. 134.
83. Thring, M. W. and Essenhigh, R. H., Chapter 17 in *Chemistry of Coal Utilization,* Supple. Vol., John Wiley & Sons, New York, 1963.
84. Sherman, R. A. and Landry, B. A., Chapter 18 in *Chemistry of Coal Utilization,* Supple. Vol., John Wiley & Sons, New York, 1963.
85. Von Fredersdorff, C. G. and Elliott, M. A., Chapter 20 in *Chemistry of Coal Utilization,* Supple. Vol., John Wiley & Sons, New York, 1963.
86. Mulcahy, M. F. R. and Smith, I. W., *Rev. Pure Appl. Chem.,* **19**, 81 (1969).
87. Essenhigh, R. H., "Dominant Mechanisms in the Combustion of Coal," *ASME Paper* 70-WA/Fu-2, 1970.
88. Essenhigh, R. H., *Proceedings of the 16th Symposium (International) on Combustion,* Combustion Institute, Pittsburgh, Pa., 1977, p. 353.
88a. Essenhigh, R. H., Chapter 19 in Chemistry of Coal Utilization Second Suppl. Vol. (to be published).
89. Frank-Kamenetskii, D. A., *Diffusion and Heat Exchange in Chemical Kinetics,* Princeton University Press, Princeton, N.J., 1955 (Russian 1st ed., 1947).
90. Simons, G. A. and Lewis, P. F., "Mass Transport and Heterogeneous Reactions in a Porous Medium," paper presented to the Spring Technical Meeting, Central States Section, Combustion Institute, Cleveland, Ohio, March 1977. Lewis, P. E., Simons, G. A., Wray, K. L., and Finson, M. L., "Modelling of Coal Gasification," Quarterly Reports to ERDA [Contract No. E(04-3)-1254] PSI TR-62 and 72, 1976.
91. Thomas, J. M. and Thomas, W. J., *Introduction to the Principles of Heterogeneous Calatysis,* Academic Press, New York, 1967.
92. Essenhigh, R. H., Froberg, R., and Howard, J. B., *Ind. Eng. Chem.,* **57**, 33 (1965).

ISBN 0-201-08300-0

93. Stumbar, J. P., Kuwata, M., Kuo, T., and Essenhigh, R. H., *Proceedings of the 4th National Incinerator Conference,* American Society of Mechanical Engineers, 1970, pp. 288–303.

94. Mallard, E. and le Chatelier, H. L., *Ann. Mines,* **1** (ser. 8), 1 (1882).

95. Nusselt, W., *Z. V.D.I.,* **68**, 124 (1924).

96. Mayers, M. A., *Trans. ASME,* **59**, 279 (1937).

97. Essenhigh, R. H. and Csaba, J., *Ninth Symposium (International) on Combustion,* Academic Press, New York, 1963, p. 111.

98. Sebastian, J. J. S. and Mayers, M. A., *Ind. Eng. Chem.,* **29**, 1118 (1937). Sherman, R. A., Pilcher, J. M., and Ostborg, H. N., *ASTM Bull.* 112, p. 23, 1941. Orning, A. A., *Ind. Eng. Chem.,* **36**, 813 (1944).

99. Csaba, J., "Flame Propagation in a Fully Dispersed Coal Dust Suspension," Ph.D. thesis, Dept. of Fuel Technology and Chemical Engineering University of Sheffield, England, 1962.

100. Faraday, M. and Lyell, C., "Report to the Home Secretary on the Explosion at the Haswell Colliery on 28 September 1844" (Report of 1845), *Phil. Mag.,* **26**, 16 (1845).

101. Burgoyne, J. H. and Long, V. D., *Proceedings of the Conference on Science in the Use of Coal,* Institute of Fuel, London, 1958, p. D.16.

102. Marshall, W. F., Palmer, H. B., and Seery, D. J., *J. Inst. Fuel,* **37**, 342 (1964).

103. Milne, T. A. and Beachey, J. E., "Laboratory Studies of the Combustion, Inhibition, and Quenching of Coal-Dust/Air Mixtures," Paper WSS/CI-76-1, Spring Meeting, Western States Section, Combustion Institute, 1976.

104. Orning, A. A., "Principles of Combustion," Chapter 2 in *Principles and Practices of Incineration,* R. C. Corey, Ed., Wiley-Interscience, New York, 1969.

105. Kuwata, M., Kuo, T. J., and Essenhigh, R. H., *Proceedings of the 4th National Incinerator Conference,* American Society of Mechanical Engineers, 1970, p. 272.

106. Cogoli, J. G., Gray, D., and Essenhigh, R. H., "Flame Stabilization of Low Volatile Fuels," *Combust. Sci. Technol.,* **16**, 165 (1977).

107. Essenhigh, R. H., *Colliery Eng.,* **38**, 534 (1961); **39**, 23, 65, 103 (1962).

108. Grumer, J., *Proceedings of the 15th Symposium (International) on Combustion,* Combustion Institute, Pittsburgh, Pa., 1975, p. 103.

109. Levenspiel, O., Chapter 9 in *Chemical Reaction Engineering,* John Wiley & Sons, New York, 1962.

110. Beer, J. M., "Investigation into the Combustion Mechanism of Pulverized Fuel Flames with Special Respect to the Mixing of Flames with Secondary Air," Ph.D. thesis, Dept. of Fuel Technology and Chemical Engineering, University of Sheffield, England, 1960. Thring, M. W. and Beer, J. M., "Proceedings of the Anthracite Conference, *Mineral Industries Experiment Station Bull.* 75, Penn State University, University Park, Pa., 1961, p. 25.

111. Beer, J. M., Lee, K. B., Marsden, D., and Thring, M. W., *Fifth Journeés des Combustibles et de l'Energie,* Paris, May 19–23, 1964, p. 141.

112. Vulis, L. A., *Thermal Regimes of Combustion,* McGraw-Hill, New York, 1961.

ISBN 0-201-08300-0

113. Waibel, R. T. and Essenhigh, R. H., "Requirements for the Characterization of Coal Chars for Combustion," Seminar on Characterization and Characteristics of U.S. Coals for Practical Use, Pennsylvania State University, October 1971.

114. Kuwata, M., Stumbar, J. P., and Essenhigh, R. H., *Proceedings of the 12th Symposium (International) on Combustion,* Combustion Institute, Pittsburgh, Pa., 1969, pp. 663–675.

115. McCann, C. R., Demeter, J. J., Orning, A. A., and Bienstock, D., *Preprints, Am. Chem. Soc., Div. Fuel Chem.,* **15**, 96 (1971). Demeter, J. J., McCann, C. R., and Bienstock, D., *ASME Paper* 73-WA/Fu-2, 1973.

116. Winslow, A. M., *Proceedings of the 16th Symposium (International) on Combustion,* Combustion Institute, Pittsburgh, Pa., 1977, p. 503.

117. Eapen, T., Blackadar, R., and Essenhigh, R. H., *Proceedings of the 16th Symposium (International) on Combustion,* Combustion Institute, Pittsburgh, Pa., 1977, p. 515.

118. Johnson, M. L. M. and Essenhigh, R. H., *AIChE Symp. Ser.* 126, **68**, 311 (1972).

119. Hougen, O. A. and Watson, K. M., *Chemical Process Principles,* John Wiley & Sons, New York, 1947, pp. 1068–1074.

120. Kreisinger, H., Ovitz, F. K., and Augustine, C. E., *U.S. Bur. Mines Tech. Paper* 137, 1916.

121. Bird, R. B., Steward, W. E., and Lightfoot, E. N., *Transport Phenomena,* John Wiley & Sons, New York, 1960.

122. Rohsenow, W. M. and Choi, H. V., *Heat Mass and Momentum Transfer,* Prentice-Hall, Englewood Cliffs, N.J., 1961.

123. Spiers, H. M., *Technical Data on Fuel,* 6th ed., British National Committee World Power Conference, 1961.

124. Davidson, J. F. and Harrison, D., *Fluidized Particles,* Cambridge University Press, 1963.

125. Avedesian, M. M. and Davidson, J. F., *Trans. Inst. Chem. Eng.,* **51**, 121 (1973).

126. Golovina, E. S. and Khaustovich, G. P., "The Interaction of Carbon With Carbon Dioxide and Oxygen at Temperatures up to 3000°K," Paper 84 in *Proceedings of the 8th Symposium (International) on Combustion,* Williams and Wilkens Co., Baltimore, 1962, p. 784.

127. Campbell, E. K. and Davidson, J. F., "The Combustion of Coal in Fluid Beds," Paper A2 in *Proceedings of the Conference on Fluidized Combustion,* Institute of Fuel, London, 1975.

128. Basu, P., Broughton, J., and Elliott, D. E., Paper A3 in *Proceedings of the Conference on Fluidized Combustion,* Institute of Fuel, London, 1975.

129. Song, Y. H. and Sarofim, A. F., Fluidized Bed Head Start Program, Phase 1, M.I.T., Cambridge, Mass., 1975.

130. Beer, J. M., *Proceedings of the 16th Symposium (International) on Combustion,* Combustion Institute, Pittsburgh, Pa., 1977, p. 439.

131. Skinner, D. J., *Fluidized Combustion of Coal,* National Coal Board, London, 1970.

132. Pereira, J. R. and Calderbank, P. H., Paper B2 in *Proceedings of the Conference on Fluidized Combustion,* Institute of Fuel, London, 1971.

ISBN 0-201-08300-0

133. Cole, W. E., "Design Parameters for Fluidized Bed Processors and Natural Gas Combustors," Ph.D. thesis, Dept. of Material Science, Pennsylvania State University, 1975.

134. Walker, P. L., Mahajan, O. P., and Yarzab, R., *Preprints, Am. Chem. Soc., Div. Fuel Chem.*, **22** (1), 7 (1977).

135. Lee, K. B., Thring, M. W., and Beer, J. M., *Combust. Flame,* **6**, 137 (1962).

136. Magnussen, B. F., *Proceedings of the 13th Symposium (International) on Combustion,* Combustion Institute, Pittsburgh, Pa., 1972, pp. 869–877.

137. Essenhigh, R. H., *J. Inst. Fuel,* **34**, 239 (1961).

138. Stewart, J. T. and Diehl, E. K., *Proceedings of the 3rd International Conference on Fluidized-Bed Combustion,* EPA Technology Series EPA-650/2-73-053, 1973, pp. III-1-1 to III-1-18.

139. Beer, J. M. and Lee, K. B., *Proceedings of the 10th Symposium (International) on Combustion,* Combustion Institute, Pittsburgh, Pa., 1965, pp. 1187–1202.

140. Essenhigh, R. H., *Combust. Flame,* **5**, 300 (1961).

141. Gibson, J. W., Grimshaw, H. C., and Woodhead, D. W., Paper 29 in *Proceedings of the 7th International Conference of Directors of Safety in Mines Research,* Buxton, 1952.

142. Mason, T. N. and Wheeler, R. V., *Saf. Mines Res. Board Papers:* 33, 1927; 48, 1928; 64, 1931; 79, 1933; 95, 1936; 96, 1936; H.M.S.O., London. Cybulski, W., Paper 36 in *Proceedings of the 7th International Conference of Directors of Safety in Mines Research,* Buxton, 1952.

143. Roberts, F. H., Paper 15 in *Proceedings of the 2nd Conference on Pulverized Fuel,* Institute of Fuel, London, 1957, p. D1.

144. Cheremisinoff, P. N. and Young, R. A., *Pollution Engineering Practice Handbook,* Ann Arbor Science, Mich., 1975.

145. Danielson, J. A. (Ed.), *Air Pollution Engineering Handbook,* EPA Publ. AP-40, 1973.

146. Edwards, J. B., *Combustion: Formation and Emission of Trace Species,* Ann Arbor Science, Mich., 1974.

147. Hesketh, H. E., *Understanding and Controlling Air Pollution,* Ann Arbor Science, Mich., 1972.

148. Liptak, B. G. (Ed.), *Air Pollution,* Vol. 2: *Environmental Engineers' Handbook,* Chilton Book Co., Radnor, Pa., 1974.

149. Perkins, H. C., *Air Pollution,* McGraw-Hill, New York, 1974.

150. Starkman, E. S. (Ed.), *Combustion Generated Air Pollution,* Plenum Press, New York, 1971.

151. Strauss, W., *Industrial Gas Cleaning,* Pergamon, 1966.

152. Rolke, R. W., Hawthorne, R. D., Garbett, C. R., Slater, E. R., Phillips, T. T., and Towell, G. D., "Afterburner System Study," Rep. S-14121, Shell Development Co., Emeryville, Calif., 1971.

153. Engdahl, R. B., "Application of Overfire Jets," *Tech. Rep.* 7, British Coal Research, Inc., 1944. Engdahl, R. B. and Stang, J. H., "Effect of Overfire Air on the Efficiency of a Small Industrial Boiler," *Natl. Eng.,* May 1947. Engdahl, R. B. and Holton, W. C., *Trans. ASME,* **65**, 74 (1943).

154. Interagency Report, "EPA Fluidized Bed Combustion Program," Status Report for FY1976, Rep. EPA-600/7-77-012, February 1977.

ISBN 0-201-08300-0

155. Kuwata, M. and Essenhigh, R. H., *Combust. Flame,* **20**, 437–439 (1973); *AIAA Paper* 75-1267, 1975.

156. Muzio, L. J., Arand, J. K., and Teixeira, D. P., *Proceedings of the 16th Symposium (International) on Combustion,* Combustion Institute, Pittsburgh, Pa., 1977, p. 199.

157. Yergey, A. L., Lampe, F. W., Vestal, M. L., Day, A. G., Fergusson, G. J., Johnston, W. H., Snyderman, J. S., Essenhigh, R. H., and Hudson, J. E., *Ind. Eng. Chem. Process Des. Dev.,* **13**, 233–240 (1974).

158. Engdahl, R. B., "Air Pollution Control Division News," Hemeon, W. C. L., Ed., *ASME News* 5, April 1977.

159. *Proceedings of the 3rd International Conference on Fluidized-Bed Combustion,* EPA Technology Series EPA-650/2-73-053, 1973.

160. Vogel, G. J., Swift, W. M., Lenc, J. F., Cunningham, P. T., Wilson, W. I., Panek, A. F., Teats, F. G., and Jonke, A. A., "Reduction of Atomspheric Pollution by the Application of Fluidized-Bed Combustion and Regeneration of Sulfur-Containing Additives," EPA Technology Series EPA-650/2-74-104, 1974; "Supportive Studies in Fluidized-Bed Combustion," *Quarterly Prog. Reps.* ANL/ES-CEN-1017 and 1018, Argonne National Laboratory, Illinois, 1976.

161. *Proceedings of the Conference on Fluidized Combustion,* Institute of Fuel, London, 1975.

162. *Proceedings of the 4th International Conference on Fluidized-Bed Combustion,* The Mitre Corp., 1976.

163. Interagency Report, "Studies of the Pressurized Fluidized-Bed Coal Combustion Process," EPA-600/7-76-011, 1976.

164. Interagency Report, "Application of Fluidized-Bed Technology to Industrial Boilers," EPA-600/7-77-011, 1977.

165. Little, J. L., "Sulfur Removal in Fluidized-Bed Combustion," ERDA Morgantown Environmental Research Center, MERC/TPR-77/1, 1977.

ISBN 0-201-08300-0

4. Coal Gasification for High and Low Btu Fuels

L. G. Massey

1. INTRODUCTION

To Jean Baptiste van Helmont we are indebted for the term "gas" and its association with coal [72]. Van Helmont, born in Brussels in 1577, studied medicine but early turned his attention to alchemy. He was a firm believer in the powers of the philosopher's stone and claimed to have succeeded several times in turning metals into gold. One day while working in his crude laboratory he made a startling discovery. In his own words he found that coal in a heated crucible "did belch forth a wild spirit or breath. This spirit, up to the present time unknown, but not susceptible of being confined in vessels, nor capable of being reduced to visible body, I call by the new name of gas." No doubt van Helmont was thoroughly frightened when he saw or felt the presence of this spirit, and it was quite natural for him to refer to it as a *Geest.* The words *Geest* and *Geist* in the Dutch and German languages mean "ghost." Some students, however, think that the word "gas" is derived from the Greek word *Chaos,* meaning "without form."

We are inclined to think of coal as a substance that has been used by mankind for many hundreds of years. In truth, coal has been recognized as a commercial commodity for only about 700 years; its first mention in recorded history was dated AD 852 in the "Saxon Chronicle" of the Abbey of Petersborough [72]. In 1272 coal was first used in London; however, a royal proclamation issued in 1316 forbade its use in that city because of the "noisome smell." In 1580 Queen Elizabeth prohibited the use of coal in London while Parliament was in session because "the health of the Knights of the Shires might suffer during their abode in the Metropolis." As unpopular as the use of coal appears to have been, its employment in the population centers apparently continued unabated, as indicated by the "health tax" levied by King Charles II in 1662, raising the then huge sum of 200,000 pounds through this tax on fireplaces. In 1679 Father Hennepin, a Jesuit missionary, mentioned a "coal mine" near Ottawa, Illinois; and in 1749 coal was

C. Y. Wen and E. Stanley Lee, Editors, Coal Conversion Technology, ISBN 0-201-08300-0

ISBN 0-201-08300-0

mined in Richmond Basin, Virginia. Shortly thereafter (in 1755) coal was discovered in Ohio.

While direct combustion of coal can certainly supply adequate heat for many purposes, it is awkward for some applications and impossible for others. Examples are found in the firing of brick and in the manufacture of glass, where gases are more readily controlled and do not deposit particulate matter into the batch of material being processed. It is not difficult, therefore, to see the reasons for early emphasis on the manufacture of gases from coal for lighting and for certain manufacturing uses.

Table 1.1

Chronology of Coal To Gas[a]

Year A.D.	Event
852	Coal was first mentioned in England, in the "Saxon Chronicle" of the Abbey of Petersborough.
1000	Coal began to replace wood and charcoal.
1180	Coal was first mined systematically in England.
1250	Coal became a commercial commodity.
1259	King Henry III granted charter to mine coal in Newcastle.
1272	Coal was first used in London.
1316	A royal proclamation was issued forbidding the use of coal in London on account of the "noisome smell."
1509	Box stoves were made of cast iron in Ilsenberg.
1580	Use of coal was prohibited by Queen Elizabeth in London while Parliament was in session, because "the health of the Knights of the Shires might suffer during their abode in the Metropolis."
1609	Van Helmont gave the name "gas" (from *Geist,* meaning "ghost" or "spirit") to the gaseous bodies produced by combustion and by fermentation.
1659	Thomas Shirley investigated a natural well in Lancashire, England, and wrote the first description of experiments with natural gas, published in the *Transactions of the Royal Society* for June 1667.
1662	The use of coal had become so extensive that the sum of 200,000 pounds was raised by means of a "health tax" imposed on fireplaces by King Charles II.
1662	Robert Boyle enunciated the statement known as Boyle's law.
1675	Coal was distilled for the production of tar.
1679	A "coal mine" near Ottawa, Illinois, was mentioned by Father Hennepin, a Jesuit missionary.
1700	Coal was exported from England to the continental European countries.
1749	Coal was mined in Richmond Basin, Virginia.

ISBN 0-201-08300-0

Table 1.1 (Continued)

Year A.D.	Event
1754	Dr. Joseph Black discovered carbonic acid gas (carbon dioxide).
1775	Coal was discovered in Ohio.
1781	A patent was granted to the Earl of Dundonald for distilling coal. All the products of distillation were mentioned except gas.
1784	Jean Pierre Minckelers lighted gas distilled from coal as a demonstration to his class in the University of Loubain.
1790	Anthracite was first mined in Pennsylvania.
1796	Manufactured gas was exhibited in a museum in Philadelphia.
1803	Main Street, Richmond, Virginia, was lighted by a huge gas lamp on a 40 foot tower.
1805	Murdock built gas works and lighted the cotton mill of Messrs. Phillips and Lee at Manchester; 900 burners were supplied.
1812	A royal charter was granted to the London and Westminster Gas Light and Coke Co., the first gas company formed.
1817	The Gas Light Company of Baltimore, the first gas company in America, was incorporated (February 5).
1859	Colonel Drake completed his first oil well near Titusville, Pennsylvania, thus giving birth to the American petroleum industry.
1885	The first garbage destroyer, using coal as fuel, was built in the United States by an army officer.
1896	The incandescent gas mantle was applied to street lighting.
1909	The first meeting of the Illuminating Engineering Society, London, was held.

[a] Source: Norman [72].

Probably the simplest coal "gasification" process is simple pyrolysis, or the heating of coal in the absence of air to produce a combustible gas. Large quantities of residual coke, or nearly pure carbon, are left behind in such operations. This may be desirable if coke is wanted as an end product, but may be undesirable if the objective is to maximize the yield of gases produced. Early workers in this field soon discovered that when a mixture of air and steam was blown through an incandescent fuel bed the oxygen of the air would combine with part of the carbon to produce carbon monoxide and some carbon dioxide, liberating enough heat to cause the steam to combine with the carbon to form additional carbon monoxide and hydrogen gas. The product of this operation contains all of the nitrogen supplied with the air and produces a fuel gas having heating values in the range from 130 to 170 Btu/scf.* This gas has been commonly known as producer gas, and is

ISBN 0-201-08300-0

* Btu/scf: British thermal units per standard cubic foot, normally measured at 14.73 psia and 60°F; may contain to 7 lb. of water per million standard cubic feet.

used close to the scene where it is generated to minimize deposition of tars and other condensable substances. Producer gas also contains some sulfur compounds, the amount depending upon the nature of the coal being fed to the process, making the fuel gas suitable only for applications where the presence of sulfur is not detrimental. Producer gas is admirably adapted to many heating operations, particularly where a long, soft, luminous flame is desired. It has a relatively low adiabatic flame temperature of approximately 3100°F, making it especially useful for burning brick, where hot spots must be avoided and where it is often desirable that the entire furnace be filled with a luminous and sometimes slightly reducing flame. The same long, highly radiant characteristics are also valuable in glass melting furnaces. The heating value of producer gas is dependent upon the nature of the fuel used for gas manufacture, as shown in Table 1.2, which, regrettably, shows for producer gas no combustible constituents heavier than carbon monoxide.

Table 1.2

Typical Manufactured Gas Compositions, Mol Percent

	Producer Gas						
Component	Bituminous Coal	Anthracite	Coke	Coke Oven Gas	Blue Water Gas	Carbureted Water Gas	Natural Pipeline Gas
CO	25.0	27.1	31.0	6.3	38.5	34.0	
H_2	14.5	16.6	9.3	46.5	50.5	40.5	
CH_4	3.1	0.9	0.7	32.1	1.0	10.2	92.6
ILL[a]				4.0		8.0	0.31[b]
CO_2	4.7	5.0	3.6	2.2	6.0	3.0	0.90
N_2	52.7	50.4	55.4	8.1	3.5	2.9	0.95
O_2				0.8		0.5	
H_2S					0.5		
C_2H_6							4.27
C_3H_8							0.97
GHV[c]	167	151	137	584	300	550	1051
Reference	23	23	23	23	23	23	85

[a] ILL = illuminants, ethylene and heavier, = C_2+.
[b] Natural gas, 0.31% is heavier than C_3H_8.
[c] GHV = gross heating value, Btu/scf.

ISBN 0-201-08300-0

Blue water gas gets its name from the characteristic pale blue flame when it is burned in air. The process for making water gas was first applied by Ibbetson in

1834 by passing steam through a mass of incandescent coke, or carbon (no volatile matter is present at these temperatures). Within 35 years over 60 patents for water-gas processes were issued. In general, these consisted of blowing with air until the bed of carbon had reached white heat, after which air was shut off and steam was blown through the bed of incandescent carbon. Steam reacts with carbon at these temperatures to form carbon monoxide and hydrogen in approximately equimolal quantities. Formation of water gas is a highly endothermic reaction, however, causing rapid temperature drop and decreased gasification rate within the bed of carbon. The steam is then shut off and air allowed to flow again to reheat the bed to incandescence. By thus cycling the flow of air and of steam, a cyclic production of water gas was carried out.

Professor Thaddeus S. C. Lowe (1832–1913) is the father of the carbureted water gas process, having secured a patent in 1872 and again in 1873 for processes making water gas and coal gas.* Carbureted water gas was made intermittently, the process involving production of two kinds of gases, blue gas (water gas) and oil gas to carburet or "enrich" the blue gas. The blue gas was made as described above and then conducted through a hot brick work or checker work over which was sprayed a gas oil to be cracked thermally. The volatile products of oil decomposition were mixed with and became a part of the total product gas. Carbureted water gas thus consists of blue water gas with added volatile products of pyrolysis from an oil, giving rise to substantially enhanced heating values as shown in Table 1.2. The oil pyrolysis products are referred to as "illuminants" and are commonly thought to be mostly olefinic in chemical structure. These components produce the soft yellow flame commonly attributed to gas lighting without a gas mantle, a flame characteristic similar to that of a candle. Production of carbureted water gas by gas utilities in the United States amounted to 306 million mcf† in 1947, requiring an annual consumption of approximately 5 million tons of bituminous coal. No carbureted water gas is produced by a utility now operating in the United States; any such gas used is generated at the site by the manufacturer employing it in his process.

During the first quarter of the twentieth century in the United States coal was king; during the second quarter substantial inroads into the energy market were made by petroleum and natural gas; during the third quarter cheap petroleum and natural gas reduced the use of coal to little more than electric power generation. The higher efficiency of the internal combustion engine caused U.S. railroads to abandon the steam locomotive completely during the 1950s in favor of the more convenient, cleaner handling, and cleaner burning distillate fuels. For the same reasons natural gas displaced coal and coal-derived fuels in residential, commercial, and industrial applications. By the time the United States was entering the fourth

---

* Lowe first began making gas for balloons in 1862 during the Civil War for the Army of the Potomac, and was said to be planning the construction of a huge airship like the Zeppelin, just before he died in 1913.

† mcf: one thousand standard cubic feet.

ISBN 0-201-08300-0

quarter of the twentieth century it had seriously reduced its petroleum and natural gas reserves, and had become heavily dependent upon foreign sources of petroleum. The nation was threatened with a severe shortage of domestic natural gas and petroleum reserves in the face of a rapidly increasing demand.

When natural gas became available beginning in the 1930s, its clean handling and burning characteristics, combined with increasing availability through pipeline construction, started the inexorable displacement of manufactured gas. To the time of World War II only minor construction of gas transmission pipelines had been accomplished; when the war ended, a strong growth in pipeline construction began. Long distance transmission lines began to reach from the south and southwestern parts of the United States northward and eastward to the heavy industrial areas of the country, delivering natural gas into the Chicago area for about 35¢/mcf (including all transmission expenses) as recently as 1965. Until 1968 discovery of new reserves of natural gas consistently exceeded the rates of production of gas from wells, leading to a continually increasing proved reserve of natural gas. Beginning in 1968, however, discovery rates began to fall behind production rates, and a growing shortage of this nearly ideal gaseous fuel resulted.

Fossil fuel shortages have been forecast for the United States since early in the twentieth century, but during the decade of the 1960s the need for domestic supplementary sources and forms of energy became alarmingly clear. Now there is an intense national interest in various forms of energy, including geothermal, solar, wind, tides, municipal wastes, agricultural wastes, coal, tar sands, oil shale, and, of course, nuclear power. Many suggestions have been offered in these fields, in addition to such newer concepts as growing plant materials on land and in the oceans for conversion of these cellulosic materials to desired forms of energy. Of all the alternatives available, coal appears to hold the best promise for early development to meet the nation's energy requirements. It is believed by many that coal will serve to fill the interim period pending suitable development of nuclear power, geopower, and solar energy. All of these are believed to be relatively long term projects involving an interval before significant commercial realization is likely to develop.

It is clear that if the natural gas industry is to survive it will be required to produce natural gas synthetically from carbonaceous materials, abundantly available in the form of coal.

2. GENERAL CHARACTER OF COALS

2.1. Structure; Behavior on Heating

Coals are highly complex mixtures, with mineral matter more or less adventitiously and nonuniformly dispersed in a hydrogen-deficient hydrocarbon of probably polynuclear aromatic structure. Age has much to do with coal character, as does the geologic processing to which it has been subjected during its long period at rest. Young coals contain considerably more moisture, as well as hydrogen and

ISBN 0-201-08300-0

oxygen, than do older coals. The types of coal, ranked in order of increasing age, are peat, lignite, subbituminous, bituminous, and anthracite. While all of these coals possess substantial macro- and microporosity, the degree of porosity decreases, as do the hydrogen and oxygen contents, as the age of the coal and its rank increase. Anthracite displays the lowest moisture content, the least porosity, and the lowest hydrogen/carbon atomic ratio.

It is interesting to note the rough correlation that exists between the normal state of a hydrocarbon and its atom ratio of hydrogen to carbon, as shown in Table 2.1. As a broad generalization one might say that, as hydrogen is added to pure carbon by chemical combination, the physical structure is softened almost continuously from the solid state through the liquid and into the gaseous state.

Table 2.1

Hydrogen/Carbon Atom Ratios in Hydrocarbons

Hydrocarbons	H/C Ratio	Normal State
Paraffins		
Methane	4.0	G
Ethane	3.0	G
Propane	2.67	G
Butane	2.5	G
Isooctane	2.25	L
Decane	2.2	L
Octadecane	2.1	S
Olefins	2.0	G, L
Acetylene	1.0	G
Aromatics		
Benzene	1.0	L
Toluene	1.143	L
Naphthalene	0.80	S
Anthracene	0.71	S
Hexacene	0.62	S
Gasoline	1.78	L
Fuel oil	1.56	L
Coals (dry, ash-free)		
Lignite	0.87	S
Subbituminous	0.87	S
Bituminous	0.75	S
Anthracite	0.34	S

ISBN 0-201-08300-0

Table 2.1 helps to emphasize the severe hydrogen deficiency of coal substances that must be overcome if liquid or gaseous hydrocarbon fuels are to be made from these materials.

Upon heating, all coals respond similarly but to a degree dependent upon rank. For example, heating lignite in the absence of air and at constant pressure first permits dehydration of the coal to take place; this is followed at higher temperature levels by decomposition (pyrolysis) to yield a vapor of relatively high hydrogen content and a residue deficient in hydrogen. The ultimate residue (char) consists mostly of mineral matter (ash-former) and nearly pure carbon. On the other hand, if a high grade bituminous coal is subjected to heating in the absence of air at constant pressure, relatively less moisture will be driven off at low temperature levels; as temperature is increased, pyrolysis reactions also set in to produce volatile matter relatively rich in hydrogen and a residue deficient in hydrogen. However, as temperature is increased, the hydrocarbon of the bituminous coal also tends to melt. The combination of melting and decomposition tends to cause swelling as the generation of volatile matter opens up the pore structure. Since decomposition leaves behind a more hydrogen-deficient residue than the parent material from which the volatile matter was evolved, the melting point of the residue is increased as pyrolysis proceeds.

Thus at a particular temperature level decomposition will continue until the melting point of the residue reaches at least the temperature level at which the process is controlled. Further decomposition can continue at the fixed temperature level, but at a decreasing rate, until further decomposition ceases. If then the temperature of the process is again elevated to a higher level, the competition between melting and decomposition is again encountered and the process repeats itself until it becomes stabilized at the new temperature level. If the heating of bituminous coal is carried on at a rapid rate, melting and coalescence (agglomeration) of particles will occur before decomposition is complete; the final product will be a porous, monolithic block of coke. However, if the bituminous coal particles are suspended free while subjected to rapid heating, it becomes possible for the particle to melt and be blown (like a balloon) into a bubble (cenosphere) by evolution of volatile matter internally, while the external surfaces of the sphere are in a plastic condition. Further extremes of rapid heating can cause such rapid evolution of volatile matter internally in the pore structure of the coal as to cause rupturing of the particle to take place with the production of fine, solid particles. High speed motion pictures have verified the explosive rupturing of bituminous coal particles through rapid heating.

The preceding description of coal pyrolysis is obviously oversimplified; the actual mechanisms of pyrolysis are complex and not well understood. Rapid pyrolysis is known to produce greater yields of volatile matter than are obtained in slow pyrolysis. For a thorough discussion of this topic the reader is referred to Chapter 2, "Rates of Coal Pyrolysis and Gasification Reaction." Further useful discussion is given by Mentser et al. [67], by Wen and co-workers [99], by Coates et al. [12], and by Nelson et al. [71]. For purposes of coal gasification the

ISBN 0-201-08300-0

tendency of a given coal to agglomerate as described above is of primary importance because of the potential for agglomeration and bogging of fluidized beds of solids, and for sticking and channeling in fixed bed gasifier systems.

2.2. Sulfur in Coal

To a significant extent the need for conversion of coal to gas is dictated by the desire for relief from noxious sulfur oxide emissions accompanying direct coal combustion. The case for conversion to gas is further supported by the need for abatement of particulate solids emissions to the atmosphere.

All coals contain some sulfur in all three forms: combined with the organic coal substance ("organic sulfur"), combined with iron as pyrite or marcasite ("pyritic sulfur"), and combined as calcium and iron sulfates ("sulfate sulfur"). Organic sulfur is distributed throughout the coal substance as part of the "hydrocarbon" molecular structure and therefore is not removable by conventional mechanical cleaning or separation processes. Generally, organic sulfur predominates in the low sulfur coals. As the total sulfur content of coals increases, both pyritic and organic forms tend to increase, although there is no direct relationship between the two.

Pyritic sulfur in coal is present as the minerals pyrite and marcasite, both having the same chemical composition (FeS_2) but differing in physical structure. Pyrites are distributed in coal in many ways, occurring in lenses and bands, joints or cleats, ball or nodules, and as finely disseminated particles. Since this material occurs in an adventitious manner, the size and distribution of these particles play a major role in the mechanical separability of this form of sulfur.

Sulfate sulfur in fresh coals generally is less than 0.05%. Since sulfur occurring as pyrite or marcasite is easily oxidized under moist conditions, the presence of sulfate sulfur in excess of about 0.05% indicates that the coal probably has been weathered. Generally, sulfate sulfur is not an important factor in coal utilization.

Because all coals contain some sulfur in addition to mineral ash forming constituents, any gas making process based upon coal as feed stock will necessarily be faced with the question of removal of sulfur and fine particulate material. Total sulfur content varies markedly from one kind of coal to another and, indeed, within any given coal seam. Table 2.2 is presented only to indicate the relationship of sulfur content to character of the coal. It appears that high sulfur content is encountered preferentially in bituminous coals, and that low sulfur content is common among the low ranked subbituminous coals and lignites. Nearly two thirds of the total U.S. coal reserves contain no more than 1% sulfur by weight; by allowing maximum 4% sulfur this figure is raised to better than 90% of all reserves.

The U.S. Bureau of Mines determined total sulfur and sulfur forms for coals in 283 counties in 29 states and 2 fields of the state of Alaska. Organic, pyritic, and sulfate sulfur were measured for approximately 2900 samples, which include most of the coal beds in the United States. Walker and Hartner [96] have tabulated these

ISBN 0-201-08300-0

data, together with total moisture and ash content, to provide a useful reference for this information.

Table 2.2

Sulfur Content of U.S. Coal Reserves[a]

Type of Coal	Total 1965 Reserves, MM tons[b]	Maximum % S	% of Total Reserve	MM tons[b]
Anthracite	15,180	0.7	96.5	14,652
Bituminous	724,680	4.0	85.7	620,992
Subbituminous	388,665	1.0	99.6	387,202
Lignite	447,641	1.0	90.7	406,012
Total	1,576,166	4.0	90.6	1,428,858
Anthracite	15,180	1.0	97.2	14,748
Bituminous	724,680	1.0	29.8	215,671
Subbituminous	388,665	1.0	99.6	387,202
Lignite	447,641	1.0	90.7	406,012
Total	1,576,166	1.0	64.9	1,023,633

[a] Source: Perry and DeCarlo [76].

[b] Millions of tons.

3. GASIFICATION PRINCIPLES

3.1. General Equilibrium Considerations

Most combustible substances can be gasified by one or more of the following means:

1. Evaporation by heating.
2. Pyrolysis.
3. Combustion in air or oxygen.
4. Reaction with steam.
5. Reaction with hydrogen.
6. Reaction with carbon dioxide.

Undoubtedly there are additional reaction schemes by which combustible material could be gasified, but they, together with items 1 and 2 above, are of much less

ISBN 0-201-08300-0

importance than items 3–6. Since combustible materials normally are found in contact with air, sometimes with steam, and usually with minor amounts of water vapor and carbon dioxide, it is clear that reaction rates are extremely slow at ambient conditions but are substantially increased as temperature is raised. Because the rate of reaction is really the rate of approach to chemical equilibrium at specified conditions, it will be highly instructive to examine what that state of equilibrium may be.

Although we recognized in earlier paragraphs that coal is a complex substance composed primarily of a mixture of hydrocarbonaceous and mineral matters, it will be convenient to regard coal as identical with graphite for purposes of the following discussion. In a way this convenience is forced upon us for lack of detailed thermodynamic information and knowledge of complex coal substances; in another way it is desirable to make this assumption because of the readily available thermodynamic information on graphite and because of its striking similarity in structure to coal substance.

Because steam is the most commonly used substance for gasification of carbonaceous materials, a study of the equilibrium composition of a system composed of equimolal solid graphite and steam was made at 1 atm to temperatures as high as 8000°K. The high temperature permits inclusion of the plasma arc, for example, in these considerations. The interesting results at 1 atm are shown in Figure 3.1 and in Table 3.1. It should be observed that Figure 3.1 is a semilogarithmic plot of the mole fractions in the gas phase (5 log cycles) versus the system temperature in degrees Kelvin on the uniform scale. It illustrates dramatically the rapid disappearance with rising temperature of such species as CH_4 and CO_2, and the nearly complete disappearance of H_2O at temperatures above 3000°K. At temperature levels as low as 2000°K hydrogen atoms begin to appear in increasing quantity as the temperature is raised, reaching levels of over 60% at 5000°K. Atomic carbon gas and atomic oxygen gas begin to appear in appreciable quantities at temperatures of 4000°K and higher. Molecular hydrogen concentration decreases rapidly at temperatures above 3000°K, leaving as the most stable species carbon monoxide, atomic carbon gas, atomic oxygen, and atomic hydrogen as temperatures are raised beyond 3000°K. At temperatures of the plasma arc, which can range as high as 10,000°K [87, 56], carbon monoxide decomposes to yield more atomic carbon gas and atomic oxygen. At 10,000°K the existing species would be primarily atomic carbon, oxygen, and hydrogen; little would remain of any molecular species.

Such high temperatures go far beyond the range of normal gasification systems, which commonly operate in the range 800–1500°K (980–2240°F). Many gasifiers operate in the relatively narrow range of 1400–1750°F (1033–1228°K). Because containing vessels would be inordinately large at atmospheric pressure and because reaction kinetics can be promoted by higher pressures, real gasifiers generally operate at elevated pressure. This may range from as low as a few pounds per square inch gauge to as high as 1000–1500 psi (66.7–100 atm), but it is uncommon for gasifiers to operate above 1000 psi.

ISBN 0-201-08300-0

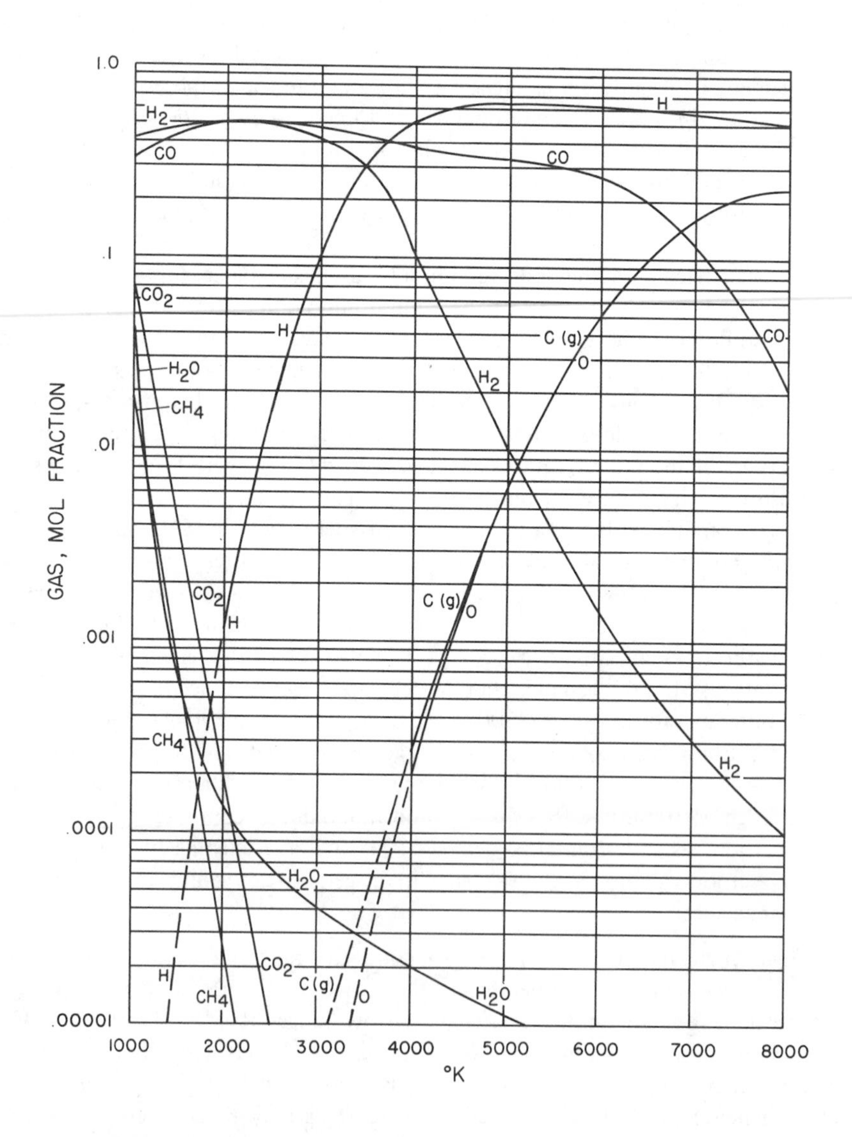

Figure 3.1. Equilibrium in the system $C(s) + H_2O(g)$ at 1 atm.

ISBN 0-201-08300-0

Table 3.1

High Temperature Equilibrium Composition in the System Graphite-Steam at 1 Atm[a]

$C(s) + H_2O(g)$

Component	1000°K	2000°K	3000°K	4000°K	5000°K	6000°K	7000°K[b]	8000°K[b]
C(s)	0.10930							
C				0.00027	0.00601	0.04965	0.16231	0.23378
CH				.00001	.00004	.00008	.00008	.00005
CH_4	.01927	0.00003						
CO	.32301	.49965	0.47440	.36911	.32870	.26735	.11648	.02134
CO_2	.06769	.00002	.00001	.00001	.00001	.00001		
C_2					.00002	.00011	.00022	.00013
C_2H			.00001	.00001				
C_2H_2		.00004	.00003					
H		.00114	.10219	.52186	.64882	.63133	.55736	.51052
HCO				.00001				
H_2	.41985	.49902	.42330	.10843	.01030	.00151	.00031	.00010
H_2O	.06088	.00010	.00006	.00002				
O				.00020	.00591	.04972	.16263	.23398
OH				.00007	.00017	.00022	.00017	.00008
O_2						.00001	.00002	.00001
Sum	1.00000	1.00000	1.00000	1.00000	0.99998	0.99999	0.99958	0.99999

[a] Computed from *JANAF Thermochemical Tables* [91].
[b] Extrapolated.

The interesting region from 34 to 68 atm and from 923 to 1223°K is explored in Tables 3.2 and 3.3 to reveal the equilibrium effect of pressure, temperature, and steam/graphite mole ratio. It is clear from these tables that solid carbon will remain under all given conditions unless the mole ratio R [= $H_2O(g)/C(s)$] exceeds 1.0. Raising the temperature tends to gasify more carbon when $R = 1.0$, but raising the system pressure tends to gasify *less* carbon when $R = 1.0$. Even though decrease of pressure should gasify more carbon, it should be observed in

ISBN 0-201-08300-0

Table 3.1 that, for $R = 1.0$ and at pressures lowered to 1 atm, solid carbon persists at 1000°K to the extent of nearly 11 mol % of the equilibrium mixture. Further increase of temperature gasifies more carbon.

Table 3.2

Equilibrium Mole Fraction in the System[a]

$H_2O(g) + C(s)$

Component		923°K	1023°K	1123°K	1223°K
$P = 34$ atm:	C(s)	0.30004	0.25414	0.17893	0.09312
	CH_4	.13049	.10945	.08473	.06042
	CO	.03279	.09533	.20951	.34459
	CO_2	.16717	.15053	.11156	.06228
	H_2	.10614	.17748	.26315	.34829
	H_2O	.26337	.21307	.15212	.09128
	Sum	1.00000	1.00000	1.00000	0.99998
$P = 68$ atm:	C(s)	0.30609	0.27057	0.21038	0.13032
	CH_4	.14312	.12792	.10777	.08523
	CO	.02332	.06885	.15787	.28243
	CO_2	.17059	.16058	.13174	.08725
	H_2	.07826	.13417	.20579	.28644
	H_2O	.27862	.23791	.18643	.12832
	Sum	1.00000	1.00000	0.99998	0.99999

[a] Data from *JANAF Thermochemical Tables* [91].

Tables 3.2 and 3.3 help us deduce the following: (1) if the objective of gasification is to promote the formation of methane, the gasifier should be operated at high pressure, at high steam/carbon ratio, and at the lowest practicable temperature; or (2) if maximum yield of carbon monoxide and hydrogen is desired, the gasifier should be operated at low pressure for complete carbon gasification, and at low steam/carbon ratio and at the highest practicable temperature to minimize CO_2 in the product. Thus the maximum equilibrium yield of methane (19.5%) occurs at

ISBN 0-201-08300-0

923°K, 68 atm pressure, and a steam/graphite ratio = 2.0 (Table 3.3). The maximum ($CO + H_2$) yield occurs at $R = 1.0$, $T = 1223°K$, and 34 atm (Table 3.2), but the reaction is penalized by incomplete conversion of carbon. By further lowering pressure and increasing temperature to the conditions of Table 3.1 (1 atm, steam/graphite ratio = 1.0), the equilibrium yield of ($CO + H_2$) can be raised to nearly 99.9% at 2000°K (3140°F), or to 74.3% at 1000°K (1340°F).

Table 3.3

Equilibrium Mole Fraction in the System[a]

$2H_2O(g) + C(s)$

	Component	923°K	1023°K	1123°K	1223°K
P = 34 atm:	C(s)	–	–	–	–
	CH_4	0.17563	0.11701	0.05560	0.01684
	CO	.04391	.10543	.17706	.22574
	CO_2	.23088	.18889	.13774	.10198
	H_2	.15440	.24920	.34134	.39602
	H_2O	.39518	.33946	.28826	.25942
	Sum	1.00000	0.99999	1.00000	1.00000
P = 68 atm:	C(s)	–	–	–	–
	CH_4	0.19507	0.14786	0.08997	0.03995
	CO	.03145	.07972	.14638	.20570
	CO_2	.23686	.20432	.15696	.11431
	H_2	.11503	.19266	.28034	.35441
	H_2O	.42159	.37544	.32634	.28562
	Sum	1.00000	1.00000	0.99999	0.99999

[a] Data from *JANAF Thermochemical Tables* [91].

Since equilibrium represents an approachable but unattainable goal of conversion for a real process, these considerations are of major importance in selecting appropriate process conditions for the gasification task at hand. To be sure, there are additional factors, many of which will be discussed in the following sections.

ISBN 0-201-08300-0

3.2. Thermochemistry; Thermodynamics

Gasification of carbon by combustion with oxygen is highly exothermic, as shown in Table 3.4. The heat of reaction in forming CO_2 and CO from the elements is remarkably independent of temperature. Although it is interesting to look at these reactions independently, they occur simultaneously and inseparably in real systems, leading to the partitioning of gasified carbon into CO and CO_2 in a ratio that is kinetically controlled until equilibrium is established. Oxidation of carbon is so slow at low temperatures that little if any change can be measured. However, very slow oxidation does occur, and with the release of heat as shown in Table 3.4. If enough interfacial area could be provided and if heat conduction from the scene of oxidation could be avoided, heat buildup would occur, temperature would rise, and oxidation rate would increase. Such a process is called *autothermic,* and such finely divided material, capable of initiating and completing its own oxidation in ambient air, is *pyrophoric.*

Table 3.4

Graphite-Oxygen System: Heat of Reaction[a]

(1) $C(s) + O_2(g) \rightarrow CO_2(g)$

(2) $C(s) + 1/2\ O_2(g) \rightarrow CO(g)$

T, $^\circ$K	$(\Delta H^\circ_T)_1$, kcal	$(\Delta H^\circ_T)_2$, kcal
298.16	−94.052	−26.416
400	−94.069	−26.317
500	−94.091	−26.295
600	−94.123	−26.330
800	−94.217	−26.512
1000	−94.319	−26.768
1500	−94.555	−27.546

[a] Data from Hougen et al. [42].

Steam gasification of carbon is highly endothermic, as shown in Table 3.5, requiring about 32 kcal of energy input for each gram-mole of carbon gasified. It is worth noting that this is about 21% more heat than can be obtained by partial combustion of 1 g-mol of carbon to form CO. If one could cause these two

ISBN 0-201-08300-0

reactions to behave independently and ideally, transferring the exotherm of combustion to the endotherm of steam gasification, one would have

$$(1)\ 1.2C(s) + \frac{1.2}{2} O_2(g) \rightarrow 1.2\ CO(g); \qquad \Delta H \doteq (1.2)(-26.7) = -32.0 \text{ kcal}$$

$$(2)\ C(s) + H_2O(g) \rightarrow CO(g) + H_2(g); \qquad \Delta H \doteq +32.0 \text{ kcal}$$

$$\text{Sum: } (3)\ 2.2C(s) + \frac{1.2}{2} O_2(g) + H_2O(g) \rightarrow 2.2CO(g) + H_2(g); \qquad \Delta H \doteq 0$$

Table 3.5

Graphite-Steam System: Free Energy and Heat of Reaction[a]

$C(s) + H_2O(g) \rightarrow CO(g) + H_2(g)$

T, °K	ΔG^o_T, kcal	$Log_{10}\ K$	K	ΔH^o_T, kcal
298.16	21.827	−15.998	1.005×10^{-16}	+31.382
400	18.510	−10.113	7.709×10^{-11}	+31.723
500	15.176	− 6.633	2.328×10^{-7}	+31.978
600	11.796	− 4.296	5.058×10^{-5}	+32.164
800	4.966	− 1.356	4.406×10^{-2}	+32.371
1000	− 1.912	+ 0.418	2.617×10^{0}	+32.445
1500	−19.107	+ 2.784	$6.081 \times 10^{+2}$	+32.295

[a] Data from Hougen et al. [42].

If all carbon in real systems were of uniform character and reactivity and if the kinetics of the two reactions were identical, thermally balanced reactions like these could be carried out. Actual gasifiers approach this condition fairly well, but the real behavior of the system entails heat loss to surroundings, use of excess steam, and formation of significant amounts of carbon dioxide, sometimes accompanied by methane formation as well. The real situation is more complex than is indicated

ISBN 0-201-08300-0

for reactions 1–3 above.

Free energy change for the equimolar graphite-steam system (Table 3.5) permits calculation of equilibrium conversion at chosen conditions of temperature and pressure, a powerful tool for selection of process conditions to accomplish the desired result. The calculation quantifies and confirms the adverse effect of pressure expected when the products have a greater volume than the reactants.

Illustration 1. Calculate the equilibrium conversion for the reaction

$$C(s) + H_2O(g) \rightarrow CO(g) + H_2(g)$$

at 1, 10, and 34 atm for each of these temperatures: $800^\circ K$, $1000^\circ K$, $1500^\circ K$.

Solution:

From Table 3.5, $K_{800} = 0.04406, K_{1000} = 2.6170, K_{1500} = 608.1.$

Basis = 1 mol of steam feed
Pressure = π atmospheres
Moles steam converted = x = fractional conversion
At equilibrium:
Steam = $1 - x$ mols
CO = x
H_2 = x

TOTAL = $1 + x$ mols

Activities of gaseous components are assumed equal to partial pressures, and solid graphite activity is unity; thus

$$a_{CO} = \frac{x\pi}{1+x}, \quad a_{H_2} = \frac{x\pi}{1+x}, \quad a_{H_2O} = \frac{(1-x)\pi}{1+x}$$

and

$$K = \frac{a_{CO}a_{H_2}}{a_{H_2O}} = \left(\frac{\pi x}{1+x}\right)^2 \frac{1+x}{(1-x)\pi} = \frac{\pi x^2}{1-x^2}$$

from whence $x = \sqrt{K/(K+\pi)}$. Substitution of values for K and total pressure π give the following values of fractional conversion of steam (x):

ISBN 0-201-08300-0

π, atm	800°K $K = 0.04406$	1000°K $K = 2.617$	1500°K $K = 608.1$
1	0.2054	0.8506	0.9992
10	0.0662	0.4554	0.9919
34	0.0360	0.2673	0.9732

These results illustrate well the trade-off between temperature and pressure effects, since increased temperature permits a higher operating pressure while attaining the same conversion. It can be important to operate at elevated pressure in order to decrease the size (cost) of reactor vessels. In this system the adverse effect of pressure can be compensated for by operation at a higher temperature.

Gasification of carbon by carbon dioxide is somewhat analogous to steam gasification in that 2 mols of products are made from 1 mol of reactant gas, leading qualitatively to the same conclusions about the effect of temperature and pressure:

$$C(s) + CO_2(g) \rightarrow 2CO(g)$$

Table 3.6 shows this reaction to be considerably more endothermic (about 9 more kcal) than that for steam gasification. Furthermore, the equilibrium constant responds much more strongly to temperature here than for steam gasification, becoming significantly large at about 1000°K and above.

Although CO_2 gasification is not considered for commercial gas making purposes, it can have subtle importance in some gasification processes. For example, if hot circulating char is used to supply endothermic heat to a steam gasifier, and if the char is heated for this purpose by exposure to hot flue gases from a fuel combustor, the opportunity exists for some carbon gasification during char heating. If the fuel is carbon, CO_2 gasification of char can occur; but if the fuel is a hydrocarbon, both CO_2 and steam gasification can occur. Since either of these is highly endothermic, such loss by gasification of process carbon into the flue gas represents a process penalty that can be severe. Reference to Tables 3.5 and 3.6 shows that either reaction develops the most unfavorable equilibrium constant in the very temperature range such processes would employ (1000–1500°K). These large equilibrium constants may be regarded as a measure of the driving force responsible for the rate of gasification. Carbon so lost from the process and appearing as CO in the flue gas can have its fuel value recovered by secondary combustion with energy recovery. The capital and operating cost of equipment to accomplish this, however, may outweigh the value of energy recovered. These carbon losses are also capable of unfavorable effects on the process utilities balance when fuel values are recovered, for example, by raising steam.

ISBN 0-201-08300-0

Table 3.6

Graphite-Carbon Dioxide System: Free Energy and Heat of Reaction[a]

$C(s) + CO_2(g) \rightarrow 2CO(g)$

T, °K	ΔG^o_T, kcal	$\text{Log}_{10} K$	K	ΔH^o_T, kcal
298.16	+28.644	−20.9940	1.014×10^{-21}	+41.220
400	24.311	−13.2819	5.225×10^{-14}	41.434
500	20.024	− 8.7519	1.7706×10^{-9}	41.501
600	15.727	− 5.7282	1.870×10^{-6}	41.463
800	7.185	− 1.9626	1.090×10^{-2}	41.194
1000	− 1.275	+ 0.2787	1.900	40.782
1500	−22.034	+ 3.2101	1.622×10^{3}	39.447

[a] Data from Hougen et al. [42].

Table 3.7

Hydrogasification of Carbon: Free Energy and Heat of Reaction[a]

$C(s) + 2H_2(g) \rightarrow CH_4(g)$

T, °K	ΔG^o_T, kcal	$\text{Log}_{10} K$	K	ΔH^o_T, kcal
298.16	−12.140	8.8977	7.902×10^{8}	−17.889
400	−10.723	5.8584	7.218×10^{5}	−18.629
500	− 7.839	3.4261	2.668×10^{3}	−19.302
600	− 5.491	2.000	1.000×10^{2}	−19.895
800	− 0.547	0.1494	1.4107	−20.822
1000	+ 4.610	−1.007	0.0983	−21.434
1500	+17.794	−2.5924	0.00256	−22.059

[a] Data from Hougen et al. [42].

Hydrogasification, the direct formation of methane from carbon and hydrogen, is a strongly exothermic reaction. Table 3.7 shows this, and also depicts the instability of methane as temperature is increased. To some extent this effect

ISBN 0-201-08300-0

can be compensated for by an increase of pressure. For example, equilibrium hydrogen conversion to methane at 800°K is increased from 60% at 1 atm to 93% at 34 atm as computed from the equilibrium constant in Table 3.7.

Illustration 2: Compute the equilibrium conversion of hydrogen to methane at 800°K and at pressures of 1 atm and 34 atm in accordance with the reaction

$$C(s) + 2H_2(g) \rightarrow CH_4(g)$$

Solution: Basis = 2 mols H_2

Moles hydrogen converted = x, and fractional conversion = $x/2$

At equilibrium:

Hydrogen = $2 - x$ mols
Methane = $x/2$
TOTAL = $2 - x/2$ mols

Activities are assumed equal to partial pressures, and activity of solid graphite is unity; then

$$a_{CH_4} = \frac{\pi x/2}{2 - x/2}, \quad a_{H_2} = \frac{(2-x)\,\pi}{2 - x/2}$$

where π = pressure on the system (atm); then

$$K = \frac{a_{CH_4}}{a^2_{H_2}} = \frac{\pi x}{(4 - x)} \; \frac{(4 - x)^2}{4\,(2 - x)^2 \pi^2}$$

from whence

$$x = 2 \pm \frac{2}{\sqrt{1 + 4\pi K}}$$

and fractional conversion of H_2 is

$$\frac{x}{2} = 1 \pm \frac{1}{\sqrt{1 + 4\pi K}}$$

From Table 3.7, K_{800} = 1.4107, and substitution for π gives the following:

$$\text{at } \pi = 1.0, \; \frac{x}{2} = 0.612 = 61.2\%$$

$$\text{at } \pi = 34.0, \; \frac{x}{2} = 0.928 = 92.8\%$$

These results confirm the predictably strong influence of pressure in favoring conversion to CH_4.

Only a small increase in temperature causes a severe decrease in conversion. In Illustration 2 the fractional conversions are reduced to 15.3% at 1 atm and

ISBN 0-201-08300-0

73.6% at 34 atm if temperature is raised from 800 to 1000°K. One can thus calculate with good accuracy the *decomposition* of methane to its elements at any temperature and pressure. The strong dependence of the equilibrium constant upon temperature, as shown in Table 3.7, should be borne in mind in considering direct formation of methane in a coal gasifier.

It is tempting to think of steam gasification to obtain hydrogen and carbon monoxide (endothermic), and then to use this hydrogen to hydrogasify (exothermic), thereby reducing the heat input to the gasifier. But the endotherm to make 1 mol of hydrogen is about three times the exotherm of 1 mol of hydrogen consumed in hydrogasification. This leads, correctly, to the conclusion that only relatively small concentrations of methane are likely to be formed in these reactions:

$$C(s) + H_2O(g) \rightarrow CO(g) + H_2(g); \quad \Delta H^{\circ} \doteq +32.0 \text{ kcal}$$

$$\tfrac{1}{2}C(s) + H_2(g) \rightarrow \tfrac{1}{2}CH_4(g); \quad \Delta H^{\circ} \doteq -10.0 \text{ kcal}$$

Simplistic approaches like those above help greatly in understanding complex phenomena, but they lack the power to account in a proper way for the simultaneity of multiple interactions. Reference is recommended once again to Tables 3.2 and 3.3 to observe the effects of temperature, pressure, and steam/graphite ratio on the distribution of equilibrium products. These tables take all simultaneous component interactions into account by minimization of the free energy of the system at equilibrium. Special attention should be paid to the strong methane sensitivity to temperature and the relatively slight increase in methane concentration when system pressure is doubled. Thus at 1023°K (Table 3.3) the methane mole fraction is increased from 0.117 at 34 atm (500 psi) only to 0.148 at 68 atm (1000 psi). The gain of 3 mol % is purchased at the cost of heavier vessels and piping, combined with higher operating and maintenance costs.

Since gasification atmospheres contain carbon oxides and hydrogen, it is worthwhile to examine the thermodynamic potential for interaction between these components. These are called "methanation" reactions and appear to be similar:

$$CO(g) + 3H_2(g) \rightarrow CH_4(g) + H_2O(g)$$

$$CO_2(g) + 4H_2(g) \rightarrow CH_4(g) + 2H_2O(g)$$

They differ strongly in hydrogen consumption and consequent water production; they are similar in their strong exothermicity and are somewhat alike in thermodynamic behavior. Examination of Tables 3.8 and 3.9 will show that CO methanation is considerably (about 8–10 kcal) more exothermic than CO_2 methanation, and that the equilibrium constant for CO is much more strongly affected by temperature change than is that for CO_2. Figure 3.2 is a plot of the values in Tables 3.8 and 3.9, and serves to illustrate the difference in equilibrium response to temperature.

ISBN 0-201-08300-0

Table 3.8

Methanation Reaction: Free Energy and Heat of Reaction[a]

$CO(g) + 3H_2(g) \rightarrow CH_4(g) + H_2O(g)$

T, °K	$\Delta G^o{}_T$, kcal	$\text{Log}_{10} K$	K	$\Delta H^o{}_T$, kcal
298.16	−33.968	+24.896	7.870×10^{24}	−49.271
400	−28.574	+15.611	4.083×10^{15}	−50.352
500	−23.015	+10.059	1.145×10^{10}	−51.280
600	−17.287	+ 6.296	1.977×10^{6}	−52.059
800	− 5.513	+ 1.506	3.206×10^{1}	−53.194
1000	+ 6.522	− 1.425	3.758×10^{-2}	−53.878
1500	+36.902	− 5.376	4.207×10^{-6}	−54.355

[a] Data from Hougen et al. [42].

Table 3.9

Carbon Dioxide Methanation: Free Energy and Heat of Reaction[a]

$CO_2(g) + 4H_2(g) \rightarrow CH_4(g) + 2H_2O(g)$

T, °K	$\Delta G^o{}_T$, kcal	$\text{Log}_{10} K$	K	$\Delta H^o{}_T$, kcal
298.16	−27.151	5.9334	8.578×10^{5}	−39.435
400	−22.774	4.9768	9.481×10^{4}	−40.640
500	−18.167	3.9700	9.333×10^{3}	−41.757
600	−13.356	2.9186	8.291×10^{2}	−42.760
800	− 3.294	0.7198	5.246	−44.371
1000	+ 7.159	−1.5644	2.727×10^{-2}	−45.542
1500	+33.975	−7.4246	3.762×10^{-8}	−47.187

[a] Data from Hougen et al. [42].

ISBN 0-201-08300-0

In spite of the generally large difference in equilibrium constants (Figure 3.2), there is surprisingly little difference in the temperatures required for identical conversion of CO and CO_2 to methane *in the range from 10 to 90% conversion at*

34 atms. Figure 3.3 shows equal conversions at approximately equal temperatures from 800 to 1500°K, with CO_2 requiring at most 40°K higher temperature than CO at about 1200°K and about 50% conversion. This figure was developed via the approach used in Illustrations 1 and 2, followed by forming the K_{CO_2}/K_{CO} ratio from the equilibrium conversion expression for each. Cancellation of terms can be done by imposing equal conversions of CO and CO_2, leading to the requirement that temperatures be determined to satisfy the ratio of equilibrium constants.

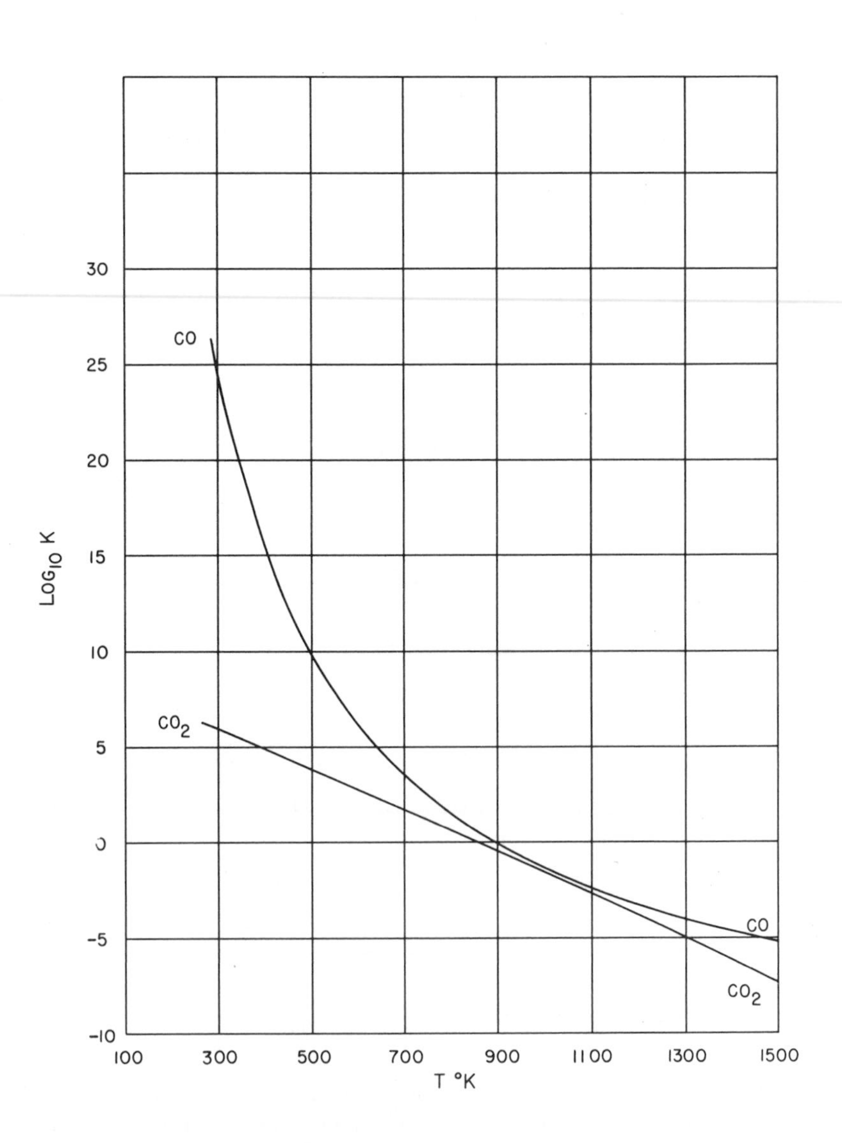

Figure 3.2. Equilibrium methanation of carbon oxides.

ISBN 0-201-08300-0

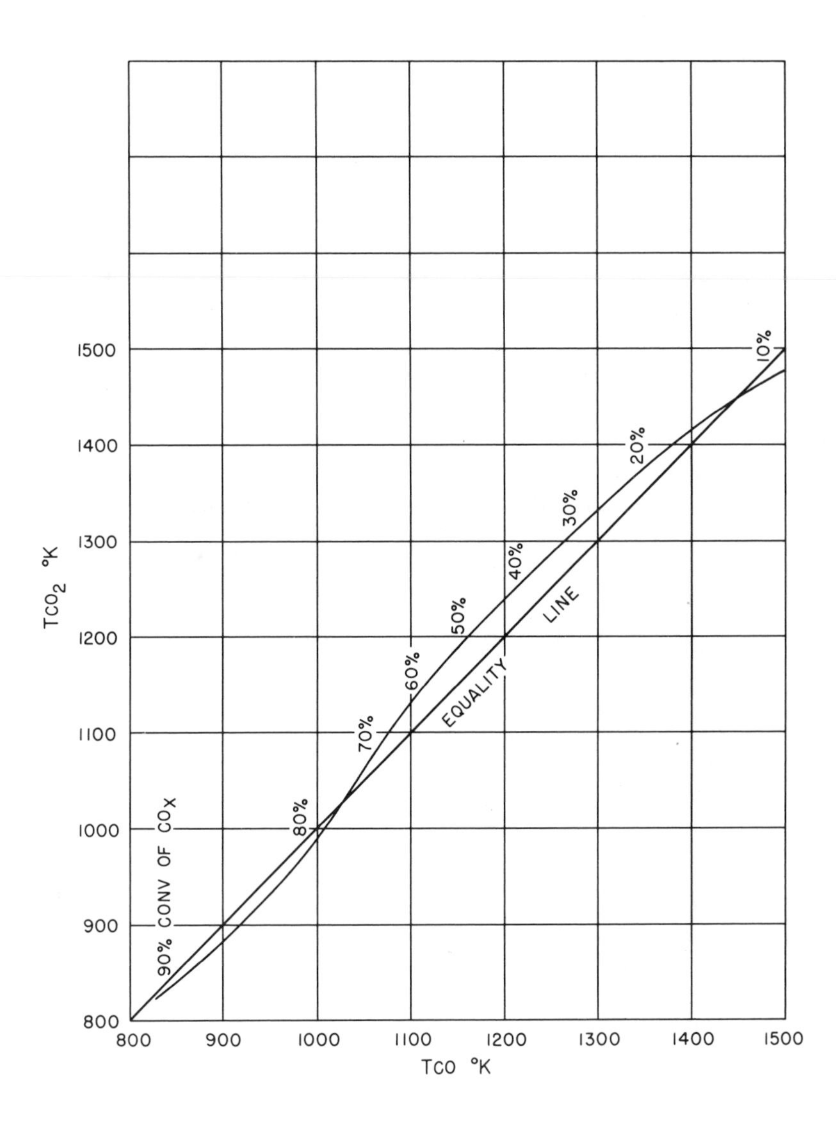

Figure 3.3. Temperatures for equal conversions of CO_x at 34 atm.

It should be noted that both reactions produce the same decrease in gas volume, and that the ratio of equilibrium constants is independent of pressure. However, absolute pressure level strongly affects conversion and thus the temperature required to obtain a given conversion. In turn, this fixes the abscissa range of Figure 3.2 which applies, thus giving rise to unique relationships like that of Figure 3.3 for each selected pressure. The addition of repeated evaluations at varied

ISBN 0-201-08300-0

pressure levels to Figure 3.3 would produce an isoconversion map of equivalent temperatures and pressures for methanation of CO and CO_2. The reader is cautioned that the similarity displayed in Figure 3.3 may be fortuitous; the isoconversion map probably would show significant effects of pressure.

In the synthesis of methane from carbon monoxide and hydrogen, it is necessary to operate the reactor or reactors in a manner that avoids deposition of carbon on catalyst surfaces and to produce as nearly as possible complete conversion of CO to CH_4. Since methanator feed gas compositions can vary considerably because of recycle gas or employment of diluents for temperature control, it is desirable to be able to estimate the effects of feed gas composition in addition to the effects of temperature and pressure.

Calculation of equilibrium composition for a given feed gas composition, temperature, and pressure is not conceptually difficult; it is only more tedious to do this for a range of compositions, temperatures, and pressures. With modern computing equipment the tedium is transferred from the calculational effort to the sensible use of a large volume of computed results. Gruber [34] developed a Cartesian coordinate system of plotting to limit the number of cases to be computed and to make application of the results easy and fast.

The following reactions are sufficient to define the system:

$$(1)\quad CO + 3H_2 \rightleftharpoons CH_4 + H_2O$$

$$(2)\quad CO + H_2O \rightleftharpoons H_2 + CO_2$$

$$(3)\quad 2CO \rightleftharpoons CO_2 + C$$

and the following dependent reaction is also of interest:

$$(4)\quad CH_4 \rightleftharpoons C + 2H_2$$

The corresponding equilibrium constants are as follows:

$$K_1 = \frac{a_{CH_4} a_{H_2O}}{a_{CO} a^3_{H_2}}, \quad K_2 = \frac{a_{H_2} a_{CO_2}}{a_{CO} a_{H_2O}}, \quad K_3 = \frac{a_{CO_2}}{a^2_{CO}}, \quad K_4 = \frac{a^2_{H_2}}{a_{CH_4}}$$

Carbon deposition as graphite will occur when $a_{CO_2}/a^2_{CO} < K_3$, and deposition is incipient when this ratio equals K_3. Thus carbon deposition is avoided by maintaining $a_{CO_2}/a^2_{CO} > K_3$. Similarly, if $a^2_{H_2}/a_{CH_4} > K_4$, carbon deposition is avoided. This is accomplished by adding sufficient steam to the system (reaction 2) to maintain high activity of hydrogen simultaneously with lowered activity of CO and elevated CO_2 activity. Thus the ratios of activities above can be maintained at levels high enough to prevent carbon deposition and consequent catalyst deactivation, and possibly reactor plugging.

Determining whether carbon deposition can occur requires evaluation of the activity ratios, a procedure described below. If a_{CO_2}/a^2_{CO} exceeds the value of

ISBN 0-201-08300-0

K_3, addition of carbon will permit CO_2 to react with it until the ratio is reduced to equal K_3, when equilibrium with solid carbon is established.

Computation of conditions to avoid carbon deposition can be very tedious, as the following example shows. Where possible, a computer should be used, but calculation can be done with a manual calculator.

Illustration 3: Feed to a methanator contains 25 mol % CO and 75 mol % H_2. The methanator operates at 600°C and 30 atm. Compute the equilibrium composition of the reactor effluent, and comment upon the deposition of carbon on the catalyst.

Solution:

Basis = 1 mol CO and 3 mols H_2

Activities are assumed equal to partial pressures. Values of K_1, K_2, and K_4 are obtained by logarithmic interpolation in Tables 3.8, 3.10, and 3.7, respectively:

$$K_1 = 2.729,\ K_2 = 2.210,\ K_4 = 2.13$$

Let x = moles of CO converted in reaction 1.
Let y = moles of H_2O converted in reaction 2.

Then at equilibrium:

$$\begin{aligned} CO &= 1 - x - y \text{ mols} \\ H_2 &= 3 - 3x + y \\ CH_4 &= x \\ H_2O &= x - y \\ CO_2 &= y \\ \text{TOTAL} &= 4 - 2x \text{ mols} \end{aligned}$$

Thus

$$K_1 = \frac{\left[\dfrac{(x)(30)}{4-2x}\right]\left[\dfrac{(x-y)(30)}{4-2x}\right]}{\left[\dfrac{(1-x-y)(30)}{4-2x}\right]\left[\dfrac{(3-3x+y)(30)}{4-2x}\right]^3} = \frac{x\,(x-y)\,(4-2x)^2}{(1-x-y)\,(3-3x+y)^3\,(900)}$$

And, similarly,

$$K_2 = \frac{(3-3x+y)\,y}{(1-x-y)\,(x-y)}$$

ISBN 0-201-08300-0

It is assumed that no carbon is deposited, and the equations for K_1 and K_2 are solved simultaneously for x and y. These results are then used to compute the activity ratios to test for carbon deposition.

The equations for K_1 and K_2 are generally too complex to allow analytic solutions to be found. They are solved by trial and error or by graphical methods. In this example values of x were assigned and the resulting values of y calculated as y_1 and y_2, also by trial and error. These results were as follows:

$$x = 0.90 \quad y_1 = 0.071285 \quad y_2 = 0.082527$$
$$x = 0.80 \quad y_1 = 0.197775 \quad y_2 = 0.152352$$

Then y_1 and y_2 were each fitted with a linear function of x, and the pair solved simultaneously to find a value of $x = 0.88$, closer to the correct solution. This value was used to get a third pair of values of y_1 and y_2 as above.

By using all three points for y_1, and similarly for y_2, parabolic functions of x were fitted and the pair of resulting equations solved simultaneously for $x = 0.8816$ and $y_1 = y_2 = .09764$.

These results were then substituted into the original equations to compute check values for K_1 and K_2:

$$K_1 = 2.719;\ \text{check value} = 2.9134$$
$$K_2 = 2.210;\ \text{check value} = 3.4382$$

It is clear that further improvement can be obtained by continuing the trial and error search for a best value of x and y. However, using the results thus far obtained gives a fair approximation to the ultimate equilibrium composition in mole fractions:

CO	0.00708
H_2	0.19662
CH_4	0.39815
H_2O	0.35430
CO_2	0.04385
	1.00000

and no carbon deposition occurs:

$$\frac{a^2 H_2}{a_{CH_4}} = 5.826 >> 2.13 = K_4$$

The gas phase considered contains five molecular species but only three atomic species. Thus the overall composition can be reduced to atomic fractions of carbon, hydrogen, and oxygen. The three equilibrium equations used to define the

ISBN 0-201-08300-0

system can be reduced to terms of C-H-O for analysis of the carbon forming potential. This is especially convenient in methanation analysis because the C-H-O does not change throughout the reactor. By inspection of the most severe operating conditions in the reactor the thermodynamic potential for carbon deposition can be appraised. The computed equilibria are summarized as atomic fractions of the gas at the carbon deposition boundary and plotted on triangular coordinates. Each curve is at a fixed temperature and pressure (see Figure 7.2). This figure shows a curve for "nonideal carbon," frequently referred to as "Dent carbon," discovered and reported by J. F. Dent [16, 15]. If allowance is made for carbon deposition in a form other than graphite, the equilibrium constants of reactions 3 and 4 must reflect the difference from pure graphite. Dent found that the observed equilibrium constant of reaction 3 was less than the theoretical equilibrium constant for deposition of graphite at 600–1200°K. Maximum departure was observed at 600°K; observed and theoretical equilibrium constants approached equality with rising temperature, becoming substantially the same at or above 1100–1200°K. Pure CH_4 and mixtures of CO and H_2 were decomposed over a nickel catalyst to determine the difference in free energy between graphite and the actual form of carbon deposited. The results confirmed the measurements made by decomposition of pure CO. It is speculated that the anomalous free energy of the actual carbon deposit can be attributed to the formation of carbides and their possible solid solutions. Pursley et al. [77] later confirmed qualitatively the formation of Dent carbon.

An excellent up-to-date survey of methanation chemistry has been presented by Seglin et al. [86]; and Gruber [34] and White et al. [101] give thorough reviews of equilibrium considerations in methane synthesis, particularly with respect to carbon deposition on the catalyst. Many other useful articles dealing with methanation processes, chemistry, and catalysis appear in Seglin's book [86].

Blue water gas, principally a mixture of hydrogen and CO, is the main raw gas product of all processes gasifying coal with steam. If the final product is to be methane (high Btu gas), it is sometimes necessary to adjust the H_2/CO ratio to a bit over 3:1 for methanation. This is done by means of the "water-gas shift" reaction, in which steam and CO are reacted catalytically at low temperature to produce more hydrogen from water at the expense of CO conversion to CO_2. Table 3.10 shows this reaction to be mildly exothermic with a pronounced tendency for increased conversion to hydrogen as temperature is lowered. The water-gas shift has long been practiced in the manufacture of hydrogen, principally operating on gases produced by the steam reforming of methane.

For coal gasification processes aimed at methane as principal product and deriving hydrogen from water, the reaction

$$2C(s) + 2H_2O(g) \rightarrow CH_4(g) + CO_2(g)$$

describes the overall chemical change taking place with almost no endothermic heat of reaction. It is important to emphasize that half the reacting carbon must leave the operation as CO_2 with no fuel value. But the process requires that energy be

ISBN 0-201-08300-0

added to the system as shown in Table 3.11. To be sure, the reaction is only mildly endothermic, requiring only about 1.7% additional carbon to be burned for process energy in a system at 100% process thermal efficiency. The mild endothermicity of this reaction has led many an investigator to speculate upon catalyzing it to take advantage of the implied reduction of capital equipment for heat exchange and recovery.

Table 3.10

Water-Gas Shift: Free Energy and Heat of Reaction[a]

$CO(g) + H_2O(g) \rightarrow CO_2(g) + H_2(g)$

T, °K	ΔG°_T, kcal	$\text{Log}_{10} K$	K	ΔH°_T, kcal
298.16	−6.816	4.9958	9.903×10^4	−9.838
400	−5.800	3.1320	1.355×10^3	−9.712
500	−4.848	2.1190	1.315×10^2	−9.523
600	−3.932	1.4319	2.704×10^1	−9.299
800	−2.219	0.6061	4.038	−8.822
1000	−0.637	0.1391	1.378	−8.337
1500	+2.927	−0.4264	3.746×10^{-1}	−7.168

[a] Data from Hougen et al. [42].

Equilibrium for this reaction, however, is less favorable in the real situation than is implied by the figures in Table 3.11. If the reaction as shown in Table 3.11 were the only one to consider, at 34 atm and 1023°K we would compute the equilibrium mole fraction of CH_4 and CO_2 to be 0.27; but the real problem is complicated by the potential for various other reactions to occur simultaneously:

$$
\begin{aligned}
&CH_4(g) \rightarrow C(s) + 2H_2(g)\\
&C(s) + H_2O(g) \rightarrow CO(g) + H_2(g)\\
&CH_4(g) + H_2O(g) \rightarrow CO(g) + 3H_2(g)\\
&CO(g) + H_2O(g) \rightarrow CO_2(g) + H_2(g)\\
&2CO(g) \rightarrow C(s) + CO_2(g)\\
&CO_2(g) \rightarrow CO(g) + \tfrac{1}{2}O_2(g)\\
&H_2O(g) \rightarrow H_2(g) + \tfrac{1}{2}O_2(g)\\
&2CH_4(g) \rightarrow C_2H_6(g) + H_2(g)
\end{aligned}
$$

ISBN 0-201-08300-0

Table 3.11

Direct Methane Formation: Free Energy and Heat of Reaction[a]

$$2C(s) + 2H_2O(g) \rightarrow CH_4(g) + CO_2(g)$$

T, °K	ΔG^o_T, kcal	$\text{Log}_{10}\,K$	K	ΔH^o_T, kcal
298.16	2.872	−2.1049	0.00785	+3.655
400	2.646	−1.4458	0.0358	+3.383
500	2.489	−1.0878	0.0817	+3.153
600	2.373	−0.8643	0.1367	+2.969
800	2.200	−0.6010	0.2506	+2.727
1000	2.061	−0.4504	0.3545	+2.674
1500	1.614	−0.2351	0.5819	+3.068

[a] Data from Hougen et al. [42].

When all the possibilities are accounted for, disappointingly small equilibrium mole fractions of CH_4 and CO_2 are found, as is shown in Table 3.2. The disappointing results persist (Table 3.3) even after doubling the initial concentration of steam (reactant) to force the reaction to the right to avoid unconverted carbon in the product mix. The results are summarized below:

		Equilibrium Mole Fraction CH_4	CO_2	$C(s)$
Table 3.11:	$2C(s) + 2H_2O(g) \rightarrow CH_4(g) + CO_2(g)$	0.27	0.27	0
Table 3.2:	$C(s) + H_2O(g)$	0.11	0.15	0.25
Table 3.3:	$C(s) + 2H_2O(g)$	0.12	0.19	0

It is clear from the foregoing that a successful catalyst could at best only approximate the simple production of methane and CO_2 as products, that formation of these products would have to possess very rapid kinetics compared with those of other product species, and that the product composition would be kinetically dependent and thus sensitive to process control variations. The probability of finding a single catalyst to accomplish all this is considered to be remote. Chances of a successful mixture of two or more catalysts are somewhat better, as suggested by Willson and co-workers [103], who have tried alkali carbonate mixed

ISBN 0-201-08300-0

with supported nickel. The former promotes steam gasification of carbon; the latter promotes methanation of carbon oxides and some hydrocracking. However, 20 g of K_2CO_3 and 115 g of supported nickel catalyst (35 wt % Ni) were charged for each 100 g of coal to obtain maximum conversion. Since this represents an active catalyst requirement of at least 60% of the coal feed, it is clear that there remains much room for improvement in catalysis of the direct formation of methane from coal and water.

Thus far the discussion has been confined to the reactions of graphite rather than of coal substance. This was rationalized on the basis of similarity between coal and graphite structures, and upon easy access to thermodynamic data for pure graphite in contrast with the lack of such information for coal. Although coal is far from being a single physical or chemical entity, it is still possible to estimate its behavior for comparison with that of graphite. As an illustration, estimation was based on high volatile bituminous coal cores drilled from the Sewickley seam in Greene County, southwestern Pennsylvania. Analysis of this coal is reported in Table 3.12.

Table 3.12

Coal Inspection, Sewickley Seam, Greene County, Pennsylvania (Dry Basis)

Analysis	Content, wt %
Proximate	
Ash	14.29
Volatile	39.01
Fixed carbon	46.70
Ultimate	
Carbon	69.28
Hydrogen	4.97
Nitrogen	1.16
Chlorine	0.03
Sulfur	3.96
Oxygen	6.31
Ash	14.29
Sulfur	
Pyritic	1.84
Organic	2.08
Sulfate	0.04
Total	3.96

ISBN 0-201-08300-0

Table 3.12 (Continued)

Analysis	Content, wt %
Mineral Content	
P_2O_5	0.14
SiO_2	42.90
Fe_2O_3	18.70
Al_2O_3	22.61
TiO_2	1.10
Na_2O	0.44
K_2O	1.91
CaO	4.09
MgO	2.42
SO_3	4.83
Unidentified	0.67
Ash fusion temperature, $^{\circ}$F	
Initial	2096
$H = W$ softening	2157
$H = 1/2W$ softening	2247
Fluid	2298
Coal Properties	
Grindability index	54–58
Free swelling index	7–8
Gross heating value, Btu/lb	12,713

Gross heating value (12,713 Btu/lb) was first credited to the carbon, net hydrogen, and sulfur contents of the coal after expressing these components in their atomic ratios as $CH_{0.7189}S_{0.0214}$. This standard heat of combustion turned out to be 122,437 cal/g-atom of carbon, and from this the standard heat of formation of "coal" was computed to be $(\Delta H^{\circ}_{f})_{298.16} = +2.310$ kcal/g-atom of carbon. For guidance and comparison, similar calculations were done for benzene liquid, benzene solid, naphthalene, and anthracene. Results are given in Table 3.13, where it is observed that the heat of formation of coal is similar to that of naphthalene and anthracene, is endothermic, and is appreciably different from zero, the value for graphite.

A comparative value of standard free energy of formation was found by estimating the absolute entropy for coal substance from values available for benzene, naphthalene, and anthracene solids. When these were expressed per carbon

ISBN 0-201-08300-0

atom and plotted against the hydrogen/carbon ratio, a smooth curve bracketing the hydrogen/carbon ratio for coal was obtained. Quadratic interpolation produced an estimate of $S^{o}_{298.16}$ = 3.56 cal/g-atom ^{o}K for coal from the quadratic equation fitting three pure aromatic hydrocarbons:

$$S^{o}/C = 4.1918 - 6.4141\,(H/C) + 7.7023\,(H/C)^2$$

where $0.714 \leqslant H/C \leqslant 1.0$. Implicit is the assumption that this is a valid step, in view of the unknown molecular configuration of coal substance. However, using this value to compute the entropy change for the formation of "coal" from its elements ($\Delta S^{o}_{298.16}$ = 9.091 cal/g-atom C ^{o}K) permitted computation of the standard free energy of formation:

$$(\Delta G^{o}{}_f)_{298.16} = \Delta H^{o}_{298.16} - T\,\Delta S^{o}_{298.16} = +5.021 \text{ kcal/g-atom C}$$

This value appears to fit suitably into place with the others shown in Table 3.13.

Table 3.13

Comparison of Thermodynamic Properties of Hydrocarbons with Those of Organic Coal Substance, kcal/g-atom C at 298.16^{o}K

Hydrocarbon or Coal Substance	State	H/C Atom Ratio	Heat of Combustion, ΔH_c	Formation: Enthalpy $\Delta H^{o}{}_f$	Formation: Free Energy, $\Delta G^{o}{}_f$
Benzene	l	1.0	−130.163	+1.953	+4.959
Benzene	s	1.0	−129.772	+1.561	+4.987
Naphthalene	s	0.800	−123.16	+1.781	+4.720
Anthracene	s	0.714	−121.07	+2.620	+5.277
Bituminous coal[a]	s	0.719	−122.44	+2.310	+5.021[b]
Graphite	s	0	− 94.0518	0	0

[a] Sewickley Seam, Greene County, Pennsylvania, moisture- and ash-free.

[b] Estimated.

ISBN 0-201-08300-0

The purpose of all this computation and estimation was to develop a comparison of "coal" behavior and graphite behavior in gasification atmospheres. It will be sufficient to make only one or two consequent comparisons.

1. Steam gasification to produce water gas:

$$CH_{0.7189}(s) + H_2O(g) \rightarrow CO(g) + 1.35945H_2(g)$$

$$\Delta H^o_{298} = +29.072 \text{ kcal/g-atom C}$$

$$\Delta G^o_{298} = +16.806 \text{ kcal/g-atom C}$$

From Table 3.5 the corresponding values based on graphite are as follows:

$$\Delta H^o_{298.16} = +31.382 \text{ kcal}$$

$$\Delta G^o_{298.16} = +21.827 \text{ kcal}$$

Clearly both gasifications are endothermic, but "coal" can produce 36% more hydrogen with 7.35% less energy input *at 298.16°K*. The reader is encouraged to investigate whether this advantage persists to gasifier operating temperature levels.

2. Steam gasification to produce methane and carbon dioxide:

$$CH_{0.7189}(s) + (0.82030)H_2O(g) \rightarrow (0.41015)CO_2(g) + (0.58985)CH_4(g)$$

$$\Delta H^o_{298.16} = -4.026 \text{ kcal/g-atom C}$$

$$\Delta G^o_{298.16} = -6.025 \text{ kcal/g-atom C}$$

Corresponding values from Table 3.11 based on graphite are as follows:

$$\Delta H^o_{298.16} = +1.8275 \text{ kcal/g-atom C}$$

$$\Delta G^o_{298.16} = +1.436 \text{ kcal/g-atom C}$$

This comparison is dramatically different from the preceding one, primarily because of the approximation of the reaction to thermoneutrality. It is interesting to note, however, that "coal" (instead of graphite) shifts the nature of the reaction from somewhat endothermic to mildly exothermic. In a perfect reacting system and with no side reactions nearly twice as much heat would be *evolved* using coal as would be *absorbed* using graphite to make 1 mol of methane. While this is a difficult reaction to realize in practice, the thermodynamic implication is much more favorable for coal than for graphite.

Readers are cautioned that the conclusions above are drawn from the assumption that "coal" behaves as a chemical compound attaining complete conversion to

ISBN 0-201-08300-0

the products shown. At real gasifier conditions (T = 1000+°K) "coal" does not exist; its composition changes because of pyrolysis and partial reaction with the steam-hydrogen atmosphere. The unreacted residue thus changes, as reaction proceeds, from "coal" to "char," a solid refractory form of carbon possessing properties more nearly like those of graphite than like those of the parent "coal." One should not be misled into thinking of coal and its chemical behavior as those of a specific chemical compound.

In general, it appears that process calculations based on graphite reactions are conservative, and that rating process performance against graphite criteria will tend to overstate the efficiency of the process.

3.3. Reaction Kinetics

Thermodynamic equilibrium constants, K, like those shown in Tables 3.5–3.11 indicate the extent to which a given reaction will proceed toward the right under given conditions if enough time is allowed for equilibrium to be established. A small K implies little conversion to products; a large K indicates large equilibrium conversion. Exothermic reactions have a tendency to spontaneity. Thus carbon in an oxygen bearing atmosphere (Table 3.4) should be expected to oxidize spontaneously, as, indeed, it does. But the *rate* of oxidation is normally so slow that we consider it zero at ambient conditions. At higher temperatures the rate of oxidation is increased, and the heat evolved tends to raise the temperature further, thus further increasing the rate of oxidation. Temperature ultimately is stabilized at a value fixed by the rate of heat loss from the oxidizing system and the rate at which oxygen is brought to the carbon for reaction. This autothermic reaction is familiar to all who have used a charcoal barbecue grill.

Endothermic reactions, on the other hand, proceed at a faster rate, as temperature is raised only if the endothermic heat of reaction is supplied from a source exterior to the reaction. An endothermic reaction cannot, on this account, be considered spontaneous and is incapable of an autothermic reaction of the kind described above. A good example is shown in the steam gasification of graphite (Table 3.5), in which the reaction is highly endothermic and shows an equilibrium constant so small at room temperature as to make equilibrium conversion of carbon to carbon monoxide essentially zero at this temperature. However, as temperature is increased, the thermodynamic equilibrium constant increases rapidly, but the endothermic heat of reaction remains nearly unchanged. Conversion of carbon to CO is clearly facilitated by an increase in temperature in at least two ways: greater equilibrium conversion is favored, and the *rate* of conversion is increased with increasing temperature.

This discussion inevitably leads to the conclusion that the thermodynamic equilibrium constant is related to the kinetics of a given reaction. Furthermore, since we are dealing largely with interactions between gaseous and solid phases, diffusion effects must be taken into consideration. Thus, in the case of carbon burning in an oxygen bearing atmosphere, diffusion of oxygen to the carbon inter-

ISBN 0-201-08300-0

face must occur before oxidation of carbon can take place. In the process of diffusing oxygen from air to such an interface, the tendency is to accumulate nitrogen near the interface on account of the depletion of oxygen from the interfacial gas mixture. Thus the oxygen diffusing *to* the interface and the product CO or CO_2 diffusing *from* the interface must find their way through the film of nonreacting and "stagnant" nitrogen molecules. The stationary film offers a resistance to the diffusion of reactants and products which can be reduced by higher relative velocities between gas and solids. In the limit, high relative gas velocity erodes the thickness (hence resistance) of the stagnant film to nearly zero at topographical surfaces, but is unable to affect similar film resistances encountered in the pore structures of reacting solids.

Numerous kinetic investigations have been conducted to study the various reactions involved in the steam and hydrogen gasification of coal and other hydrocarbons. A substantial body of literature has developed; a good bibliography can be compounded from the literature cited in Refs. 52 and 98. Most authors agree that the initial reaction rates in steam and hydrogen gasification of coal are substantially more rapid than the later rates of conversion of residual chars. Johnson [52] postulates that overall gasification occurs in three consecutive stages: (1) devolatilization, (2) rapid rate methane formation, and (3) low rate gasification. The reactions in these stages are independent. He further postulates that carbon occurs in two forms; one is called "base" carbon, and the other is the carbon in volatile matter. Volatile carbon can be evolved only by thermal pyrolysis, independently of the gaseous medium in which the reaction occurs. Base carbon remains in the coal char after devolatilization is complete. This carbon can subsequently be gasified in either the rapid rate methane formation stage or the low rate gasification stage.

When devolatilization occurs in the presence of a gas containing hydrogen at an elevated pressure, coals or coal chars containing volatile matter also exhibit a high (although transient) reactivity for methane formation in addition to thermal pyrolysis reactions. Studies performed with good time resolution indicate that this rapid rate methane formation occurs at a rate which is at least an order of magnitude slower than devolatilization [21, 105]. This can be interpreted to mean that rapid rate methane formation occurs after devolatilization.

The amount of carbon gasified to methane during transient high reactivity increases significantly with increased hydrogen partial pressure [21, 105]. Evidence indicates that at sufficiently high hydrogen partial pressures virtually all of the carbon not evolved during devolatilization can be gasified quickly to methane by this process [68].

Devolatilization reactions begin at about 700°F, and the rate of devolatilization increases continuously with temperature up to 1300°F, at which temperature rates are considered to be essentially instantaneous. These rates are actually determined by the temperature, pressure, and gas composition existing during devolatilization. After the devolatilization and rapid rate methane formation stages are completed, char gasification occurs at a very much reduced rate. The new slow

ISBN 0-201-08300-0

rates for residual chars are determined by temperature, pressure, gas composition, carbon conversion, *and prior history,* particularly with respect to temperature.

Johnson [52] used the data of Zielke and Gorin [106, 107] and Goring et al. [31] for fluid bed gasification of Disco char, as well as the bulk of data obtained in Institute of Gas Technology studies with the high pressure thermobalance (Figure 3.4) and pilot-scale fluid beds. All these data were used to evaluate parameters in a quantitative model developed to describe coal char gasification kinetics over a wide range of conditions in the low rate gasification stage. Three basic reactions were assumed to occur in gases containing steam and hydrogen:

Reaction I: $H_2O + C \rightarrow CO + H_2$

Reaction II: $2H_2 + C \rightarrow CH_4$

Reaction III: $H_2 + H_2O + 2C \rightarrow CO + CH_4$

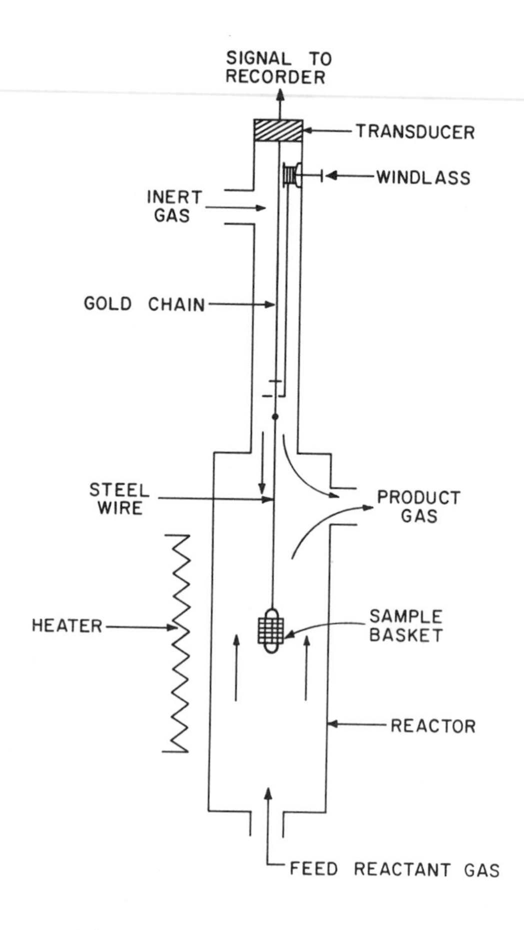

Figure 3.4. Johnson's thermobalance.

ISBN 0-201-08300-0

These three reactions are assumed to behave independently of each other. Reaction I, for example, is assumed to operate in a pure atmosphere of steam; reaction II occurs in a pure atmosphere of hydrogen; reaction III is considered to be a third, independent gasification reaction, arbitrarily assumed to occur to facilitate correlation of experimental data. Reaction I, occurring at elevated temperatures, is affected by thermodynamic reversibility only for relatively high steam conversions. However, the reaction is severely inhibited by the poisoning effects of hydrogen and CO at steam conversions far removed from equilibrium for this reaction.

Johnson's correlation is as follows.

$$dX/dt = f_L k_T (1 - X)^{2/3} e^{-\alpha X^2}$$

Here $X = \dfrac{\text{base carbon gasified}}{\text{base carbon in feed coal char}}$

and $k_T = k_I + k_{II} + k_{III}$

and $f_L = f_o e^{8467/T_o}$

$$k_I = \frac{e^{9.0201 - 31{,}705/T}\left(1 - \dfrac{P_{CO}P_{H_2}}{P_{H_2O}K^E{}_I}\right)}{\left[1 + e^{-22.2160 + 44{,}787/T}\left(\dfrac{1}{P_{H_2O}} + 16.35\dfrac{P_{H_2}}{P_{H_2O}} + 43.5\dfrac{P_{CO}}{P_{H_2O}}\right)\right]^2}$$

$$k_{II} = \frac{P^2{}_{H_2}\, e^{2.6741 - 33{,}076/T}\left(1 - \dfrac{P_{CH_4}}{P^2{}_{H_2}K^E{}_{II}}\right)}{[1 + P_{H_2}\exp(-10.4520 + 19{,}976/T)]}$$

$$k_{III} = \frac{P_{H_2}^{1/2}\, P_{H_2O}\, e^{12.4463 - 44{,}544/T}\left(1 - \dfrac{P_{CH_4}P_{CO}}{P_{H_2}P_{H_2O}K^E{}_{III}}\right)}{\left[1 + e^{-6.6696 + 15{,}198/T}\left((P_{H_2})^{1/2} + 0.85\,P_{CO} + 18.62\dfrac{P_{CH_4}}{P_{H_2}}\right)\right]^2}$$

ISBN 0-201-08300-0

$$\alpha = \frac{52.7P_{H_2}}{1 + 54.3P_{H_2}} + \frac{0.521(P_{H_2})^{1/2}P_{H_2O}}{1 + 0.707P_{H_2O} + 0.50(P_{H_2})^{1/2}P_{H_2O}}$$

where $K^E{}_I, K^E{}_{II}, K^E{}_{III}$ = equilibrium constants for reactions I, II, and III, considering carbon as graphite

T = reaction temperature, ^{0}R

T_0 = maximum temperature to which char has been exposed before gasification, ^{0}R (if $T_0 < T$, a value of $T_0 = T$ is used in Equation 16)

$P_{H_2}, P_{H_2O}, P_{CO}, P_{CH_4}$ = partial pressures of H_2, H_2O, CO, and CH_4, atm

f_0 = relative reactivity factor for low rate gasification, which depends on the particular carbonaceous solid

Much of the work done by Johnson [52] has been carried out on his high pressure thermobalance, diagramed in Figure 3.4. This device permits continuous electronic indication of weight change of the sample, suspended from a gold chain and a steel wire into the reaction atmosphere. It is designed to permit raising the sample with the windlass until it is clear of the reaction zone and is suspended in a flowing inert gas at low temperature. When reaction conditions have been established at the lower part of the instrument, the sample basket is lowered and observations can begin immediately. It should be noted that inert gas is brought in at the upper end of the instrument and flows down toward a gas outlet at the center, that feed reactant gas enters at the bottom (preheated to reactor conditions), and that both inlet gases exit from the system at a point well removed above the position of the sample. The sample is suspended from a highly sensitive transducer and electronic recording device.

Most workers in the field of coal gasification kinetics have noted a wide disparity in the reactivities of various types of coal substance. Johnson defines relative reactivity, f_0, based upon a value of 1.0 for a specific batch of air-pretreated Ireland mine coal char. Various batches of similarly treated coal char from the same mine exhibited some variation in reactivity, with values ranging from about 0.88 to 1.05. Johnson has found that results of tests made with the thermobalance, using a variety of coals and coal chars, indicate that the relative reactivity factor, f_0, generally tends to increase with decreasing rank, although individual exceptions to this trend have been noted. Values have been obtained ranging from 0.3 for a low volatile bituminous coal char to about 10 for a North Dakota lignite. The reactivity of the Disco char used in gasification studies conducted by Consolidation Coal Company [106, 107] is $f_0 = 0.488$. Continued work at the Institute of Gas Technology indicates that the relatively high sodium and calcium contents of the lower rank coals may be largely responsible for the higher reactivities Johnson

ISBN 0-201-08300-0

has noted. It is believed that these metals occur in coal structure in the form of metallic salts of organic acids.

Wen et al. [98] carried out a kinetic study of the reaction of coal char with hydrogen-steam mixtures, using pilot plant data generated at the Institute of Gas Technology before November 1966. These workers also found that the carbon contained in volatile matter is much more reactive than the carbon in the fixed carbon portion of coal, and recommended a model based on a "first-phase reaction" in which highly reactive volatile matter carbon interacts with hydrogen and with steam independently of each other to form methane, carbon monoxide, and hydrogen. When devolatilization has been completed, the residual char proceeds to react directly with hydrogen and with steam. It is believed that the water-gas shift reaction occurs predominantly on the char surface and is catalyzed by the inorganic ash forming material present in char, so that in calculating product gas distribution for moving bed reactors the water-gas shift reaction is assumed to exit at equilibrium. The data indicate a good probability of a diffusion-controlled reaction mechanism.

A detailed discussion on rates of coal pyrolysis and gasification reactions can be found in Chapter 2.

3.4. Catalysis

Thermodynamic equilibrium studies (p. 343) of the direct formation of methane and carbon dioxide from carbon and water lead to the conclusion that a successful catalyst for this reaction could at best only approximate the simple production of methane and CO_2 as products; the formation of these products would have to possess very rapid kinetics, compared with other product species, and the product composition would be kinetically dependent and thus sensitive to process control variations. Discovery of a suitable catalytic agent is improbable, especially if heterogeneous catalysts (e.g., nickel or platinum supported on alumina) are to be sought. It is highly likely that such a catalyst would become poisoned because of the large variety and number of mineral as well as hydrocarbon constituents found in coal. Homogeneous catalysts may be less subject to poisoning difficulties but may offer severe problems in recovery of the catalytic agent from the product stream. This problem is intensified as the cost of the catalytic agent increases.

Haynes et al. [37] carried out an extensive survey of 40 different solid additives in the gasification of coal. Steam-coal gasification tests were carried out in bench-scale units at 850°C and 300 psig with coal containing 5 wt % additive. They found that alkali metal compounds increased carbon gasification most, by 31–66%, and that 20 different metal oxides raised carbon gasification by 20–30%. Inserts coated with Raney nickel were active but lost activity rapidly. Additional tests were carried out in a gasifier at 907–945°C and 40 atm. A 5% admixture of either dolomite or hydrated lime produced significant increases in the amount of carbon gasified and in the amounts of methane, CO, and hydrogen produced.

ISBN 0-201-08300-0

The most active catalytic agents found by these workers were Raney nickel (unactivated spray), potassium chloride, and potassium carbonate. Each of these materials was found to be selectively active for the production of methane, hydrogen, and CO, and for the gasification of carbon as shown below:

	Catalytic Effect, % increase			
Catalyst	CH_4	H_2	CO	C Gasified, %
Raney nickel, unactivated	24[a]	30	8	10
$LiCO_3$	21	55	72	40
KC1	7	105[a]	81	66[a]
K_2CO_3	6	83	91[a]	62

[a]Maximum tested.

It appears from these results that potassium salts are good promoters for the gasification of carbon to form hydrogen and CO, but they are not very effective for the production of methane. The Raney nickel insert was most active in the formation of methane, but it lost activity too rapidly to be considered a suitable catalytic candidate. It is interesting to note that Raney nickel is not particularly effective for carbon gasification or the production of CO. Lithium carbonate was the second most effective agent for the production of methane, and showed a reasonable effectiveness for the production of hydrogen and CO, as well as the gasification of carbon.

Employment of solid agents in simple admixture with particulate coal suggests that catalytic effects were exerted primarily through gas-solid interactions at the catalyst surface and separately at the carbon surfaces, making a sequential type reaction mechanism necessary. The residues after gasification would necessarily be found mixed with ash residues from the coal. Haynes and co-workers tested the catalytic effectiveness of the ash residues with the following results: potassium compounds (K_2CO_3 and KC1) in the residues retained most of their activity in increasing the production of methane, lost part of their capability of increasing hydrogen production, and inhibited CO production.

The pilot plant tests reported by Haynes et al. were carried out in a U.S. Bureau of Mines 4 in. diameter gasifier, employing 5% by weight of dolomite and of hydrated lime. At 914°C and 40 atm, addition of 5% hydrated lime in the coal feed increased the hydrogen yield by approximately 30%, the methane yield by 25%, and the CO yield by 23%. Under the same conditions 5% dolomite in the coal feed increased the hydrogen yield by 17%, the methane yield was not increased, and the CO yield was increased by 26%. At the 2% level dolomite failed to bring any significant increase in the yield of methane, hydrogen, or CO. Dolomite and

ISBN 0-201-08300-0

hydrated lime additives, however, permitted employment of higher peak temperatures in the gasifier without incurring excessive sintering. Local temperatures as high as 1045°C were encountered with no adverse effect on operations, but without additives temperatures in excess of 1000°C invariably caused excessive sintering or slagging of the char ash.

W. G. Willson and co-workers [103] employed two distinct solid catalysts in admixture with coal in a fixed bed bench-scale investigation. The gasification rate and coal conversion were promoted by the alkali carbonate, while the supported nickel catalyst functioned to methanate the carbon oxides and to hydrocrack the liquids produced. About 60% carbon conversion is effected by this system at 650°C and 2 atm pressure. A gaseous product was produced with a CO_2-free heating value of 850 Btu/scf. It was found that 20 wt % potassium carbonate produces optimum gasification results, and that potassium carbonate appears unchanged in the ash product.

Except for the poisoning effect of sulfur compounds from the coal upon the nickel catalytic agent, this work demonstrates the potential for mixed catalysts in the possible direct formation of methane and CO_2 from coal and steam. Willson and co-workers have clearly established the efficacy of each of these catalysts, as shown in Table 3.14. The quantity of nickel catalyst employed is far in excess of that which is capable of being inactivated by the sulfur present in the coal charged to the reactor. If a sulfur-insensitive methanation catalyst of relatively high activity could be found and employed in the manner indicated by Willson and co-workers, it might be possible to effect a very substantial production of methane directly in the gasifier. Such a development could have an effect of major proportions on the production of substitute natural gas from coal.

Studies of effects of potassium carbonate as a catalyst for the steam-coal reaction have been reported by Eakman et al. [18]. These workers have found that 20% potassium carbonate increases the rate of steam gasification, prevents agglomeration when gasifying caking coals, and promotes gas phase equilibrium. They have found it possible to operate at low gasifier temperatures in the range of 1200–1400°F at pressures of about 500 psia. Observed concentrations of methane in the product gases are in excellent agreement with values computed from thermodynamic equilibrium considerations. These authors have studied the effect of separation of CO and hydrogen from the product methane and recycle of the synthesis gas to the gasifier. The only net products from gasification would be methane and CO_2 with small quantities of hydrogen sulfide, and the overall gasification step would remain approximately thermoneutral. From 50 to 75% of the potassium carbonate is recoverable by use of a water wash of the product stream; additional recovery of potassium from insoluble aluminosilicate is being subjected to further study. Such a process has interesting potential because it promises higher thermal efficiency and lower capital costs than those of existing thermal coal gasification processes. Commercial use of such a process would require simple, inexpensive recovery of practically 100% of the catalyst, even at catalyst cost as low as 10¢/lb.

ISBN 0-201-08300-0

Table 3.14

Effect of Mixed Alkali and Nickel Catalysts on Steam Gasification of Coal [103], (620^oC, 32 psia)

Coal, g	100	100	100	100
K_2CO_3, g	0	20	0	20
Nickel catalyst, g	0	0	115	111
Total gas, 1000 scf/ton	12.27	17.32	14.47	19.57
Gas produced, mol %				
H_2	40.2	32.1	17.6	13.0
CO	13.4	31.6	1.7	1.5
CO_2	32.0	22.2	39.9	39.4
CH_4	13.2	12.9	40.8	46.0
C_2H_6	0.7	0.3	0	0
Unsat.	0.5	0.8	0	0
Liquid, bbl/ton	0.3	0.2	0	0
Raw gas, Btu/scf (CO_2-free)	474	454	792	848

Catalyzed hydrogasification of coal chars has been studied by Gardner et al. [26]. These workers deposited catalysts in char particles by evaporation from solution to leave a deposit of approximately 5 wt % metal in char. Catalyst distribution on the char was examined by electron microprobe and scanning electron microscopy. The work was carried on in a high pressure, high temperature thermobalance to obtain kinetic information on the effects of various catalytic agents. Very substantially increased reaction rates were obtained by deposition of potassium bicarbonate, potassium carbonate, and zinc chloride, with catalytic effectiveness ranked in the same order. Potassium bicarbonate, the most effective agent studied, caused gasification rates to be approximately doubled over those for the uncatalyzed reaction (more specifically, the time required to achieve a given gasification fractional conversion was roughly halved by the $KHCO_3$ catalyst).

ISBN 0-201-08300-0

These workers assumed a reaction rate expression employing an activation enthalpy varying linearly with the extent of conversion:

$$\frac{dX}{dt} = kP_{H}^{n}\,(1 - X)\,e^{-\Delta H^{\ddagger}/RT}$$

where X = fractional conversion of char

k = frequency factor

n = order of reaction

P_{H_2} = partial pressure of hydrogen, atm

$\Delta H^{\ddagger}$ = $H^{o} + \alpha X$ = activation enthalpy at conversion X, kcal/g-mole

ΔH^{o} = initial activation enthalpy

α = proportionality factor

R = 1.987 cal/g-mole ^{o}K

T = temperature, ^{o}K

Linear variation of activation enthalpy was assumed in an effort to account for the decreasing reaction rate as conversion proceeds. The expression above was rearranged and integrated:

$$\int_0^X \frac{e^{bX}}{1 - X}\,dX = \int_0^t K\,dt = Kt$$

where $K = kP_{H_2}^{n}\,e^{-\Delta H^{o}/RT}$ and $b = \alpha/RT$. The parameter b was chosen to minimize the sum of the squares of the errors of a least-squares fit to a straight line through the data points relating the conversion integral to time.

The data produced remarkably good straight line fits, leading to the following conclusions: (1) values of b (hence α) are independent of hydrogen pressure;

ISBN 0-201-08300-0

(2) the hydrogasification reaction order is $n = 3/2$; (3) $\Delta H^{\ddagger} = 29.3 + 2.43X$ kcal/mol for an uncatalyzed char; (4) $\Delta H^{\ddagger} = 29.3 - 2.43X$ for a char catalyzed with $KHCO_3$. It should be noted that the catalyst causes activation enthalpy to *decrease* with increasing conversion, resulting in easier reaction as conversion proceeds. This is, of course, directly opposite to the observed behavior of uncatalyzed char conversions.

3.5. Hydrogasification

Hydrogasification of carbon and hydrocarbons is sufficiently different from steam and carbon dioxide gasification to warrant special attention. Reference is made in various places in this chapter to the use of hydrogen with little if any discussion of its special attributes.

Table 3.7 provides thermodynamic information for the reaction of hydrogen on graphite to form methane. From the thermodynamic equilibrium constant and its variation with temperature we can evaluate the tendency to decomposition of methane as temperature is increased. If similar tables were prepared for the hydrogenation of hydrocarbons, additional insights might be gained with respect to coal gasification. Thus it can be shown that aromatic hydrocarbons (benzene, for example) are more stable toward hydrogen than are the corresponding paraffinic hydrocarbons (hexane, for example), and this effect is pronounced at normal gasification temperature levels (1000–1500°K); but the enthalpy change of the reaction (exothermic) is not greatly different when expressed per gram-mole of hydrogen reacted. Hydrogasification of graphite, however, is relatively mildly exothermic when compared with hydrocarbon hydrogasification, where the exotherm is about 50% greater per mole of hydrogen reacted.

The large exothermicity of hydrogen-hydrocarbon reactions approximates 15 kcal/g-mole of hydrogen converted to methane. This is enough to cause large temperature rises and consequently faster reaction rates, leading to autothermic behavior in hydrogen atmospheres of adequate concentration. It is not difficult to make paraffin hydrocarbons "burn" in a hydrogen atmosphere at high pressure; it can be very difficult to control the reaction, however, unless pressure (partial pressure of H_2) can be controlled.

Synthesis gas ($CO + H_2$) can be shifted (water-gas shift) with steam to obtain hydrogen commercially:

$$CO + H_2O \rightarrow CO_2 + H_2$$

and the CO_2 can be removed readily for purification. Free energy and enthalpy changes for this reaction are shown in Table 3.10 and need no elaboration here. Although most discussion in this chapter centers about the use of the water-gas shift to prepare a gas mixture for methanation, it is an important method also for the manufacture of hydrogen, especially in the commercial production of ammonia. It represents one available means for the production of hydrogen for hydrogasifica-

ISBN 0-201-08300-0

tion, the hydrocracking of heavy hydrocarbons, or the hydrotreating of oils for sulfur and nitrogen removal.

Yet another method of making hydrogen for these purposes is based on the reaction of steam on iron and its oxides in a reductive-oxidative cycle:

Reduction (1) $Fe_3O_4 + H_2 \rightarrow 3FeO + H_2O$

(2) $Fe_3O_4 + CO \rightarrow 3FeO + CO_2$

(3) $FeO + H_2 \rightarrow Fe + H_2O$

(4) $FeO + CO \rightarrow Fe + CO_2$

Oxidation (5) $Fe + H_2O \rightarrow FeO + H_2$

(6) $3FeO + H_2O \rightarrow Fe_3O_4 + H_2$

(Reactions 5 and 6 are the reverse of reactions 3 and 1.) These reactions are carried out in the neighborhood of 1500°F to obtain suitable kinetics and equilibria. More specifically, if reduction is carried out at 1750°F and oxidation at 1350°F, near-equilibrium conversions of 80 and 60% can be achieved for the reducing and oxidizing gases, respectively. This requires manipulation of two reaction zone temperatures and hence is technically difficult to achieve because of the large thermal mass represented in the solids. While it is possible to conduct these reactions in a fixed bed of solids with alternate oxidation and reduction gas feeds, a much more attractive alternative is to move the solids continuously from one zone to the other, using fluidized solids technology as described in detail by Tarman [94] and as shown in pages 4c-1 through 4c-6 of Ref. 43.

Other alternatives for hydrogen production employ electrolysis of water (very expensive) and, potentially, thermochemical splitting of water using waste heat from a nuclear reactor [10, 25, 40, 74, 83].

When coal is exposed to hydrogen at high pressure and high temperature, rapid pyrolysis and hydrogenation occur. Evidence to date [79] points to production of liquid products if exposure is brief, and to gaseous products if exposure is prolonged. As severity of treatment is increased and exposure time decreased, more gas and liquid products are made at the expense of decreased char yields. This suggests that coal hydrocarbonaceous substance breaks down first at aliphatic bridges connecting aromatic elements, with broken bonds rapidly satisfied by hydrogenation. There appears to be a tendency, with prolonged exposure, for polynuclear aromatic groups to hydropyrolyze further, yielding additional light hydrocarbons and what appear to be heavier liquids formed by a polymerization or condensation mechanism. Perhaps char is the end product of this mechanism, since char yield appears to decrease as exposure time is shortened. Large yields of benzene-toluene-xylene (BTX) suggest that these compounds are the end products of single aromatic rings stripped of side chains and hydrogenated to stable ring structures before polymerization or condensation could occur.

ISBN 0-201-08300-0

Fair agreement is seen on the effects of prolonged exposure at mild severity when the results of Rosen et al. [79] are compared with those reported by Martin [61] for hydrocarbonization of a similar coal:

	Rosen et al. [79]	Martin [61]
Coal	Wyoming Glenrock	Wyoming Lake de Smet
Pressure, psig	600	600
Temperature, ^{O}F	930	932
Exposure time (vapor), sec	0.15	$\geqslant$ 22
Char yield, lb/ton maf	1000	1140
$CH_4-C_2H_6$ yield, scf/ton maf	2900	3400
Liquids, BTX+, bbl/ton maf	1.4	~ 1.0

One would anticipate the longer residence time for Martin's data to lead to reduced liquid yield and increased yields of gas and char. Not shown in these figures is the decreased volatility of Martin's liquids, possibly a result of prolonged severe treatment and increased opportunity for polymerization or condensation.

Pyrolysis of coal in the presence of hydrogen can be justified only on the basis of hydrogen uptake by the products of pyrolysis. Since this is an exothermic process, the resulting temperature rise tends to increase the velocity of pyrolysis and hydrogenation. If high hydrogen pressure and temperature are available, the process becomes autothermic and tends to get out of control. If, however, temperature or hydrogen pressure is too low, little if any hydrogenation can occur, and the employment of a hydrogen atmosphere is expensively useless and wasteful. This leads to the conclusion that mild hydropyrolysis conditions with extended residence time for products of pyrolysis can lead only to shallow levels of hydrogenation. Under these (noncatalytic) conditions relatively little hydrocracking, hydrodesulfurization, or hydrodenitrogenation can occur, and the heavier products of pyrolysis will remain heavy. Light aliphatic hydrocarbons are more readily hydrocracked; hence their production remains nearly limited to pyrolysis yields in the absence of hydrogen. In short, *without severe conditions* for hydropyrolysis, the products can be only mildly different from those of ordinary pyrolysis in yield and in quality; catalytic hydrotreating may be required to upgrade the products to commercial acceptability.

Prolonged exposure of coal to hot hydrogen at high pressure is another matter entirely, since yields are nearly all gases and are 60–70% char, with liquid yields in the range from 1 to 6 wt % of dry coal feed. Feldman et al. [21] reported yields from their Hydrane process experiments, using free fall of fresh Pittsburgh

ISBN 0-201-08300-0

seam hvAb coal through hot (900°C) hydrogen at 1000–2000 psig. Yields fell between these values, based on moisture- and ash-free coal:

Product	From, wt %	To, wt %
Methane	14.8	25.8
Carbon monoxide	2.3	3.3
Oil	0.5	1.3
Water	2.9	5.1
Char	66.3	70.2

It is interesting to note that the coal used was assayed at 56.7% fixed carbon, indicating that only net volatile matter was hydrogenated to light products, with some of the volatile component apparently degraded to char in the process. This behavior is consistent with the mechanism speculation above concerning the effect of residence time.

Virk et al. [95] studied the decomposition of simple aromatic molecules (Table 3.15) with and without hydrogen atmosphere and reached the following conclusions:

1. The rates of decomposition of simple aromatic molecules are essentially independent of hydrogen partial pressure from near zero to about 100 atm. However, increasing hydrogen concentration does change the dominant decomposition product from solid carbon (coke) to methane gas.

2. The experimentally observed activation energy for the decomposition of an aromatic molecule is linearly related to its delocalization energy as calculated from Dewar's theory.

3. The formation of carbon from benzene at low temperatures (800–1100°C) proceeds through diphenyl as an intermediate and probably does not involve any further benzene-by-benzene addition and dehydrogenation.

4. Equilibrium considerations suggest that the production of a high Btu gas product from benzene without coke formation will require operation at several hundred atmospheres and at relatively low conversions of benzene per pass, using hydrogen partial pressures below those needed for stoichiometric conversion of benzene to methane.

5. In the absence of hydrogen, aromatics are synthesized during pyrolysis of low molecular weight paraffins and olefins; in both cases olefinic intermediates are probably involved. As the ratio of hydrogen to hydrocarbon is increased, the synthesis of aromatics is inhibited; during decomposition of a light paraffinic naphtha at 700°C, no aromatic products were formed when the hydrogen/hydrocarbon mole ratio exceeded 6.

ISBN 0-201-08300-0

Table 3.15

Model Aromatic Molecules[a]

Number of Rings	Name	Structure	Formula	T_b, °F
1	Benzene		C_6H_6	176
2	Naphthalene		$C_{10}H_8$	424
2	Diphenyl		$C_{12}H_{10}$	491
3	Anthracene		$C_{14}H_{10}$	646
3	Phenanthrene		$C_{14}H_{10}$	643
4	Pyrene		$C_{16}H_{10}$	740
4	Chrysene		$C_{18}H_{12}$	827

[a] From Virk et al. [95].

4. GASIFIERS

Direct partial oxidation of carbon to carbon monoxide is strongly exothermic (Table 3.4), leading to unacceptable loss of fuel value to sensible heat in conversion from solid fuel to gaseous fuel. Steam gasification of carbon overcomes this

ISBN 0-201-08300-0

problem by being strongly endothermic (see Table 3.5), requiring additional combustion of solid fuel (or other energy source) to satisfy the reaction energy requirement. The efficient way to accomplish this is to carry out partial oxidation of carbon in the presence of steam so that endothermic and exothermic reactions occur simultaneously in the same volume of space. Thus, to maintain energy balance, about 32/26 or 1.23 g-atoms of carbon must be oxidized to CO for each gram-atom of carbon steam-gasified:

$$1.23C + \frac{1.23}{2} O_2 \rightarrow 1.23CO \quad ; \quad \Delta H = -32.0 \text{ kcal}$$

$$C + H_2O \rightarrow CO + H_2; \quad \Delta H = +32.0 \text{ kcal}$$

$$\text{Sum:} \quad 2.23C + \frac{1.23}{2} O_2 + H_2O \rightarrow 2.23CO + H_2; \quad \Delta H = 0 \text{ kcal}$$

If conversion were complete, as shown, the resulting gas composition would be 31% H_2, 69% CO. Oxygen consumption would be 0.735 lb O_2/lb carbon gasified. Although real gasification systems are more complex than this, they produce raw gas compositions surprisingly close to the above, as reported by Whiteacre and coworkers [102]: 35% H_2, 53% CO, 10% CO_2, 2% H_2S + N_2 and other. Oxygen consumption in real gasifiers of this type ranges from about 1.0 to 1.25 lb O_2/lb carbon gasified. Gross heating values of CO and H_2 are 320.6 and 323.8 Btu/scf, respectively; for the raw gasifier gas above, the value is estimated to be 283 Btu/scf because of dilution with CO_2 and inerts.

If air instead of pure oxygen were used for partial combustion, a "low Btu" fuel gas would be produced because of dilution by nitrogen in the product.

$$1.23C + \frac{1.23}{2} O_2 + \frac{(79)(1.23)}{(21)(2)} N_2 \rightarrow 1.23CO + 2.314N_2;$$
$$\Delta H = -32.0 \text{ kcal}$$

$$C + H_2O \rightarrow CO + H_2;$$
$$\Delta H = +32.0 \text{ kcal}$$

$$\text{Sum:} \quad 2.23C + \frac{1.23}{2} O_2 + \frac{(79)(1.23)}{(21)(2)} N_2 + H_2O \rightarrow 2.23CO + 2.314N_2 + H_2;$$
$$\Delta H = 0 \text{ kcal}$$

The resulting gas composition would be 18% H_2, 40% CO, and 42% N_2; gross heating value would be 187 Btu/scf. A typical Winkler (dry) gasifier product showed 13% H_2, 21% CO, 58% N_2, 7% CO_2, 1% H_2S, CH_4, and so on, with a gross heating value of 118 Btu/scf [8].

ISBN 0-201-08300-0

It is clear from the foregoing that idealized calculations like those above have limited value. They serve to set the limit of performance for real systems without regard to competing side reactions, and therefore are useful for judging real system performance. Thus we can be sure that a real gasifier requires no less than 0.73 lb oxygen/lb carbon gasified, and that a low Btu gas must contain at least 42% nitrogen and will have a gross heating value no higher than 187 Btu/scf. When real performance exceeds these values, it is almost certainly the result of added effects such as pyrolysis or blending in hydrocarbons, or employment of plug flow fixed bed countercurrent gasification, as in the Lurgi gasifier. Reactions like those above can be carried out in one of four types of gasifier: fixed bed, fluidized bed, transport gasifier, liquid medium gasifier. Additionally, the reactions can be carried out directly in the unmined coal deposit (*in situ* gasification).

4.1. Fixed Bed Gasifier

As the name implies, the bed of coal undergoing gasification is held "stationary" while gas flows upward through interstitial spaces. To attain steady-state operation the bed of material slowly slides downward to replace the carbon gasified and the ash withdrawn. Fixed bed gasifiers have been used since the middle 1700s; they work satisfactorily only on noncaking coals of high ash fusion temperature, but progress is being made toward handling caking coals and the slagging of ash.

Operation of a "dry ash" fixed bed gasifier is described below with reference to Figure 4.1, where fresh coal is fed to the top of the bed at E; there it cools the leaving product gas, and inert residue is withdrawn from the bottom at H, where it is cooled by incoming steam-oxygen feed. Thus the preheated steam-oxygen stream enters the oxidation zone to burn out carbon as completely as possible from the ash residues, while the steam and steam-carbon reactions serve to quench the combustion temperature below the ash fusion point. The principal chemical reactions in the oxidation zone are as follows:

$$C + O_2 \rightarrow CO_2$$
$$C + 1/2O_2 \rightarrow CO$$
$$C + CO_2 \rightarrow 2CO$$
$$C + H_2O \rightarrow CO + H_2$$

Hot gases from the oxidation zone enter the reduction zone, giving up heat to the coke and causing steam and CO_2 gasification of carbon:

$$C + H_2O \rightarrow CO + H_2$$
$$C + CO_2 \rightarrow 2CO$$

ISBN 0-201-08300-0

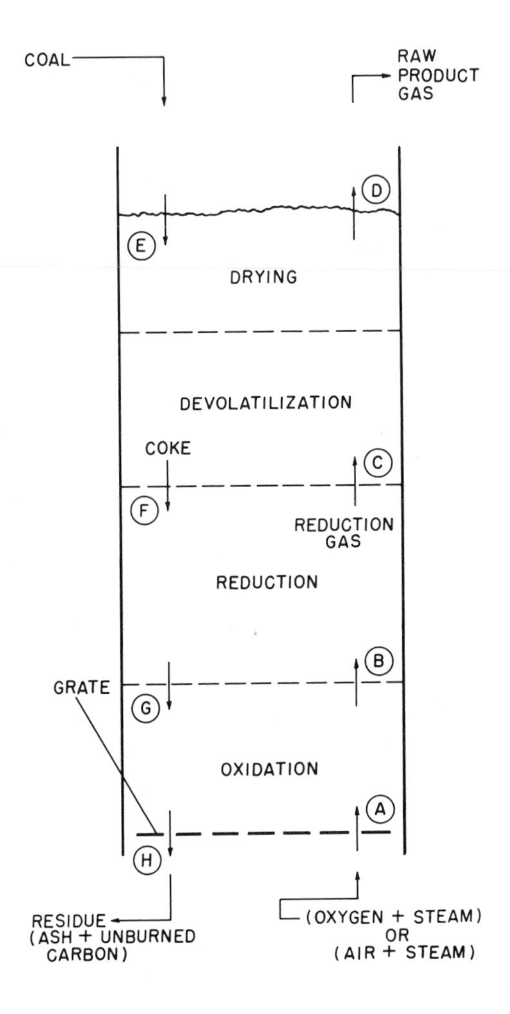

Figure 4.1. Fixed bed coal gasification.

Gases leaving the reduction zone are very substantially cooler because of the endothermicity of these reactions. Gas entering the devolatilization zone then provides sensible heat to reach pyrolysis temperatures and to provide the heat required for pyrolysis reactions. Cooled gas from this exchange, now laden with products of pyrolysis, rises through the incoming fresh coal to preheat and dry it. To some extent, heavy pyrolysis products are believed to condense out in the

ISBN 0-201-08300-0

drying zone to aid in heat transfer. The result is a recycle of heavier hydrocarbons to devolatilization, there to be revolatilized and partially pyrolyzed. Cooled product gases leave the bed (at *D* in Figure 4.1) at the dew point temperature (highly variable, but nominally about 900–1000°F).

Solid coal passes successively downward through the zones, being heated all the way, first for drying, next devolatilization followed by char reduction, and then oxidation; finally, the residues are cooled by incoming steam-oxygen before discharge of ash via the grate.

One should be aware of certain inherent characteristics of the fixed bed gasifier:

1. Fines (particles smaller than about 1/8–1/4 in., and preponderantly dust particles) are elutriated from the bed and carried overhead with dew point product gas. The least bit of condensation tends to form solid deposits with the finer solids on vessel or pipe walls and on heat exchange surfaces. Fines in product condensates are difficult to settle or filter out. Since coal mining, transportation, and handling can produce as much as 50 wt % fines in some coals, it is clear that such coals are poor candidates for fixed bed gasification.
2. Feed coal lumps must be large enough not to be transported out by the gas stream velocities in interstices, yet small enough to form a bed with minimal channeling of gases. The typical particle size range for a Lurgi fixed bed gasifier is controlled at from 1/4 to 1 1/4 in. Any maldistribution of gases causes a severe penalty in lost operating efficiency of the gasifier, and is the result of nonuniform feed sizes, agglomeration, and inadequate rabbling of the bed.
3. The countercurrent nature of the fixed bed gasifier leads to very good thermal efficiency via excellent heat transfer and, in large gasifiers, minimal heat loss to surroundings.

Dry ash fixed bed gasifiers use large amounts of steam to control ash fusion via maximum temperature limitation. This is accomplished largely by the effects of oxygen dilution and partially by employment of the steam-carbon endothermic reaction, resulting in a large economic penalty for steam cost with additional penalties for residual carbon left in the ash and for increased reactor diameter required for passage of unconverted steam along with product gases. Further cost penalties occur downstream of the gasifier, where the excess steam is condensed (heat exchanger requirement) to produce an aqueous effluent requiring costly treatment for removal of noxious substances such as phenols and cyanides.

Operation of a dry ash fixed bed gasifier requires constant attention to the character of the ash residues leaving the gasifier, with frequent and gradual adjustment of the steam-oxygen mixture to maintain oxidation zone temperatures just below the ash fusion point. When oxidation temperature is too low, a fluffy ash relatively rich in residual carbon is obtained; it is difficult to handle in disposal systems. At too high a temperature ash fuses into a liquid impossible to handle on a supporting grate mechanism. Thus the object of careful, artful control is to

ISBN 0-201-08300-0

produce an ash that has just reached the temperature of incipient fusion, making it grainy in character and containing minimum residual carbon.

Since ash character and ash content vary from point to point in a single coal deposit, it is clear that adequate process control requires the constant, careful attention of an experienced operator. Coals of low reactivity (bituminous, for example) generally require higher temperatures in the combustion zone than do coals of high reactivity. A rough idea of the temperature profiles required is given in Figure 4.2, where it is clear that less reactive coals actually reach higher maximum temperatures, thus imposing a requirement for high ash fusion temperature, and consequently greater steam dilution of oxygen to the gasifier. This can be a serious constraint upon candidate feed coals to a dry ash fixed bed gasifier.

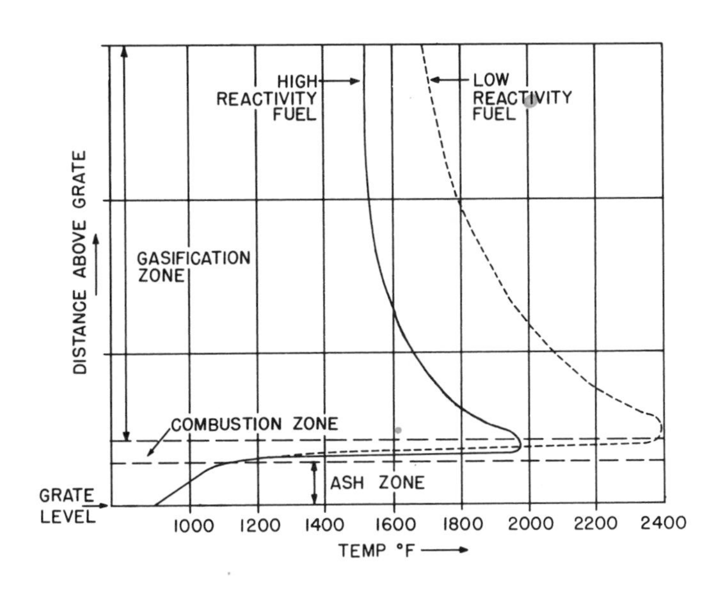

Figure 4.2. Temperature distribution when gasifying fuels of differing reactivity in a fixed bed under identical conditions.

Operation of a dry ash fixed bed gasifier on a highly caking coal was successfully demonstrated in the Westfield, Scotland, trials of four American coals during 1973 and 1974 [78]. Pittsburgh seam coal, the most difficult because of its tendency to swell and agglomerate, was processed successfully without ash dilution at reduced agitator rate, employing the maximum freeboard available between coal bed and feed distributor. While throughput rates were relatively slow because of the fact that the equipment was designed for a more reactive noncaking coal and because of downstream product handling limitations, the trials clearly demon-

ISBN 0-201-08300-0

strated the operability of an agitated fixed bed gasifier on a highly caking coal feed.

Slagging bottom fixed bed gasifiers overcome many of the limitations of the dry ash fixed bed gasifier, primarily by allowing high temperatures to be attained in the combustion zone. The actual temperature is typically about 3500°F (1927°C), where any ash is molten and flows freely. The viscosity of the molten slag is dependent upon temperature and upon composition; it is an important property for proper flow from the fixed bed and for removal from the gasifier.

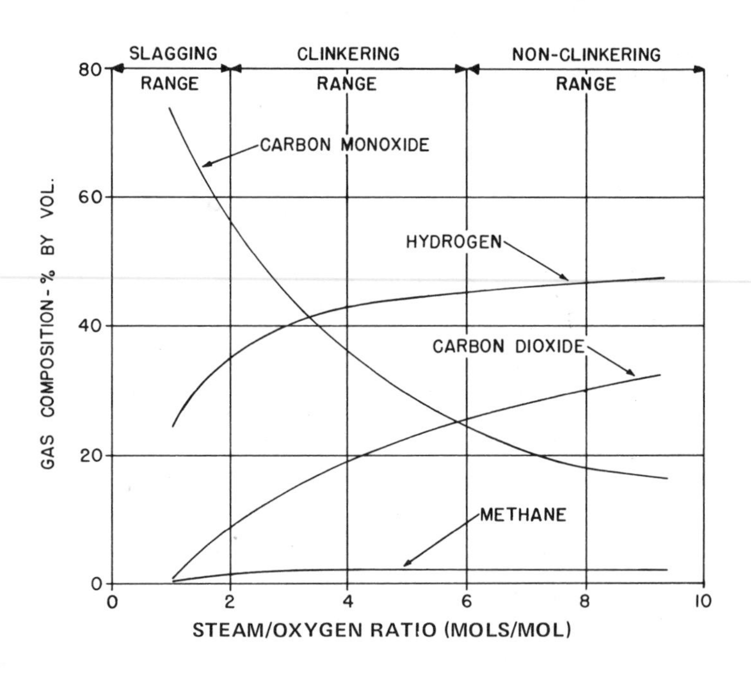

Figure 4.3. Influence of steam/oxygen ratio on gas composition.

The essence of slagging gasification is that the steam supplied per unit volume of oxygen is only that required for gasification. Thus temperatures hundreds of degrees centigrade above ash fusion temperatures are generated, and the ash melts to a liquid slag. The importance of the steam/oxygen ratio is clearly illustrated in Figures 4.3 and 4.4 as it affects product gas composition and undecomposed steam passing through the gasifier. Consequential advantages of slagging gasification include (1) high thermal efficiency; (2) high throughput; (3) fuel choice unconstrained by low ash fusion temperature or poor reactivity; (4) absence of a mechanical grate; (5) product gas richer in CO and leaner in H_2 and CO_2, hence with higher calorific value; (6) less liquor to be treated in the product condensate;

ISBN 0-201-08300-0

and (7) nearly complete carbon utilization, as evinced by little or no carbon residue in the slag. Slagging gasification entails the hazards of high temperature at high pressures, however, and the design of gasifiers must provide for these effects in addition to those of corrosion and erosion at high process severity.

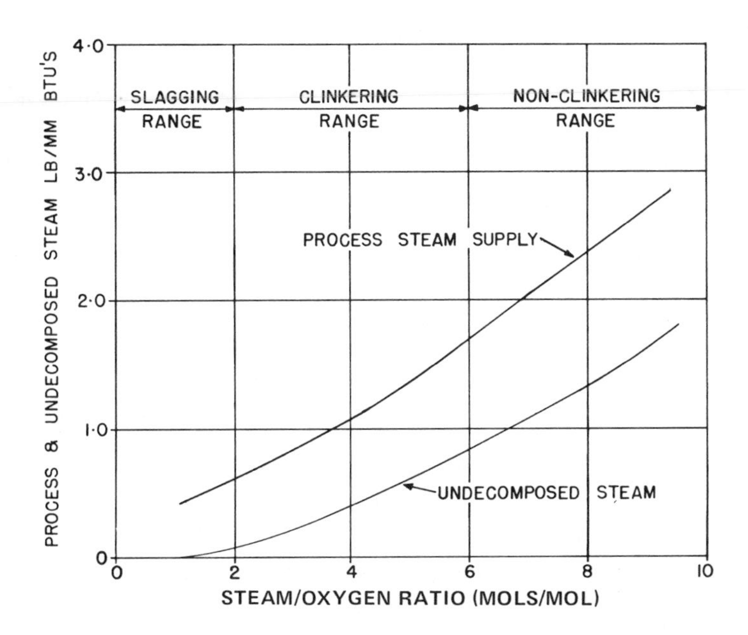

Figure 4.4. Influence of steam/oxygen ratio on process steam consumption and undecomposed steam (per million Btu of product gas).

Experimental work on slagging gasifiers was undertaken by the British Gas Council at its Midlands Research Station at Solihull in 1955. The British effort was motivated by a desire to gasify a wide range of solid fuels with high efficiency and at low cost. Development work ceased about 1964, however, with the advent of oil gasification and the discovery of natural gas in the North Sea. Little further development occurred until the middle 1970s, when Continental Oil Company and 13 others undertook a slagging gasifier development program cooperatively with the British Gas Corporation at the Westfield Development Center near Edinburgh, Scotland. Results of this program are proprietary, making it necessary to rely principally on the reports of British work at Solihull in the period 1955–1964 for technological information.

Lacey [58] provides a good overall description of operation and results from a slagging fixed bed gasifier having a fuel bed 3 ft in diameter by 10 ft deep. A

ISBN 0-201-08300-0

diagram of the gasifier is shown in Figure 4.5; it was later modified in the slagging area, shown in Figure 4.6. The gasifier was a refractory-lined pressure vessel equipped at the top with a water-cooled agitator to break up any agglomerations in the fuel bed. Coal, premixed with a suitable flux, entered the top of the gasifier and flowed through the stirrer by gravity for distribution on top of the fuel bed. Near the bottom of the fuel bed the steam-oxygen mixture was injected through four water-cooled tuyeres at a velocity of about 200 ft/sec. The hot zone was confined to the center of the gasifier and away from refractory walls by projecting the tuyeres 6 in into the fuel bed.

The hearth of the gasifier was contained in a water jacket and supported above a water-filled quench chamber, into which the molten slag was discharged to form a black glassy frit. Water, circulated at high rate through the quench chamber, quenched and broke the slag into small particles.

Product gas was removed from above the fuel bed through a duct fitted with a scraper to keep it free of tar and dust deposits. Hot raw gas was quenched by recirculating liquor from the base of the waste heat reboiler, cooled in the waste heat boiler and an aftercooler, and then passed through the pressure regulating valve to maintain process pressure at about 300 psi. Product gas was finally incinerated, as it could not be used commercially.

Mostly because of the small reactor size, continuous removal of slag was plagued with a high rate of heat loss and consequent problems with high slag viscosity, solidification, and stalactite ("beard") formation. Intermittent tapping of slag at a high rate of discharge for short intervals provided a solution to these problems, but increased residence time encouraged separation of liquid iron by reduction of the iron oxide in the slag. Removal of iron oxide adversely affects slag flow (viscosity), and the production of liquid iron aggravates corrosion of the metal parts of the slag tapping system.

The hearth of Figure 4.5 was replaced by that of Figure 4.6 at an early stage of development because of iron attack. The new hearth refractroy walls sloped toward the axis at 45° to a carbon steel slag tap assembly having a center slag tap tube 1.5 in in diameter by 2 in long. Below were mounted two swinging burners, one of which burned air, oxygen, and town gas vertically upward into the slag tube at a linear velocity high enough to hold back the slag in the hearth. The second burner, fitted with a refractory tunnel, was normally retracted from the slag tube position. It was used only to clear the tap hole in the event of blockage, a rare occurrence. To minimize iron attack, water-cooled coils of copper tubing were installed in the hearth.

Intermittent slag tapping was provided automatically, based on the liquid slag level in the hearth, as an interrupter of a collimated beam of gamma rays from a level detector about 1 ft above the tap hole. The tap burner was swung clear, and a control valve on the vent line from the quench chamber opened to reduce the pressure in the quench chamber. A controlled differential pressure across the hearth was maintained for a preset time to force the slag to flow from the hearth; the control valve was then closed, the quench chamber pressure built up, the slag flow

ISBN 0-201-08300-0

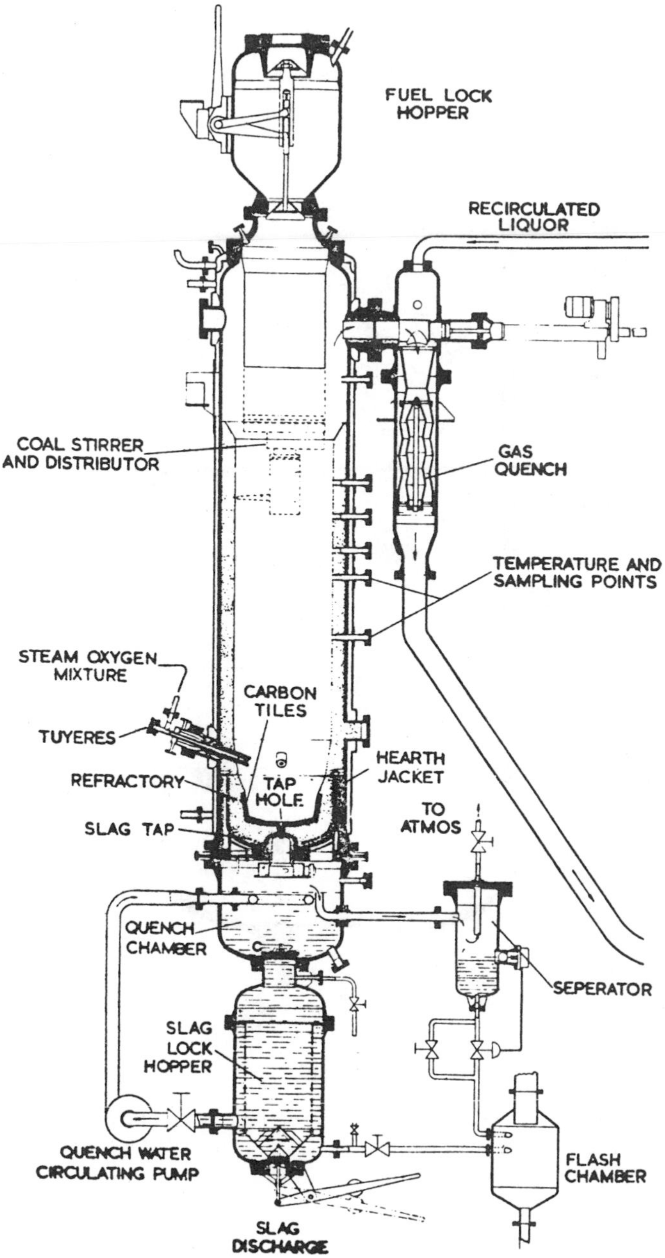

Figure 4.5. Slagging gasifier. Reproduced with permission from *Fuel Gasification*, F. C. Schora, Ed., *Advances in Chemistry Series* No. 69, American Chemical Society, Washington, D.C., 1967; copyright by American Chemical Society.

ISBN 0-201-08300-0

stopped, and the tap burner swung back to its position at the tap hole. Typically, slag was tapped for 20 sec every 4–6 min at rates reaching 10,000 lb/hr under a pressure differential of 1 psi. No carbon could be detected in the slag so produced.

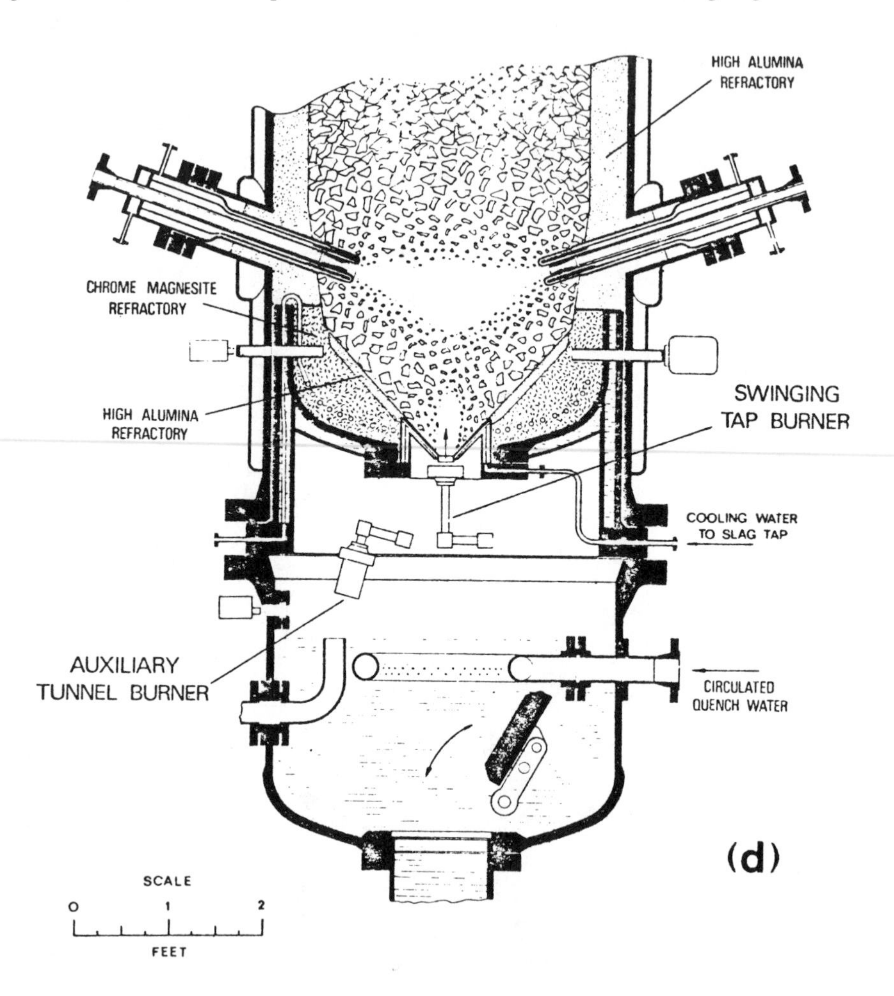

Figure 4.6. Modified hearth and quench chamber. Reproduced with permission from *Fuel Gasification,* F. C. Schora, Ed., *Advances in Chemistry Series* No. 69, American Chemical Society, Washington, D.C., 1967; copyright by American Chemical Society.

ISBN 0-201-08300-0

The coal used in the slagging gasifier tests was similar to that used at the Westfield facility when it produced commercial town gas contemporarily from dry ash Lurgi [8, 78, 82] gasifiers. It is thus possible to make direct comparisons between the performance characteristics of the two different gasifiers, as shown in Tables 4.1 and 4.2. Improvements in the Westfield Lurgi gasifiers produced about 20% more crude gas than the guaranteed maximum of 12 million scf/day, but only under exacting operating conditions bordering on clinker formation. In contrast, the slagging gasifier throughput was increased 50% with no loss in performance (not shown in Tables 4.1 and 4.2). Examination of Table 4.1 shows the Lurgi gasifiers to consume nearly five times as much steam as the slagging gasifier, and the crude gas composition is markedly different, with the slagging gasifier producing much higher concentrations of CO, somewhat lower hydrogen concentration, and only about one tenth the concentration of CO_2. The product gas fuel value is 20% greater for the slagging gasifier product. Although operated at lower pressure, the 3 ft slagging gasifier produced ($CO + H_2$) at over one half the rate for the commercial 10 ft diameter Lurgi gasifier. When measured in terms of solid fuel gasified (dry, ash-free), the slagging gasifier output was at least four times that of the dry ash gasifier. However, the slagging gasifier consumed about 10–12% more oxygen per therm of crude gas because of the smaller proportion of exothermic products (CO_2 and methane) and the loss of high grade heat to the hearth tap and tuyere cooling water. Oxygen consumption on a product volume basis showed negligible difference between the two gasifiers.

Table 4.1

Comparison of the Slagging Gasifier and a Commercial Lurgi Plant

Parameter	Lurgi Gasifier	Slagging Gasifier
Pressure, psia	355	300
Fuel		
Rank	902	902
Size range, in.	1½–3/8	1½–1/4
Ash, including flux, %	14.6	11.4
Moisture, %	15.6	14.7
Steam/oxygen ratio, vol/vol	5.4	1.1
Crude gas, vol %		
CO_2	24.6	2.5
C_nH_m	1.1	0.45
CO	24.6	60.5
H_2	39.8	27.75
C_nH_{2n+2}	8.7	7.6
N_2	1.2	1.0
	100.0	100.0

ISBN 0-201-08300-0

Table 4.1 (Continued)

Parameter	Lurgi Gasifier	Slagging Gasifier
Product, Btu/ft^3	309	371
Steam, lb/therm crude gas	11.1	2.56
Steam, lb/1000 ft^3 ($CO + H_2$)	56.1	10.7
Oxygen, ft^3/therm crude gas	49.5	55.2
Oxygen, ft^3/1000 ft^3 ($CO + H_2$)	238	236
($CO + H_2$) output, scf/hr ft^2	4930	26,700
DAF feed, lb/hr ft^2	210	981
Efficiency (HHV gas/HHV coal)	81	82.5

If a slagging bottom fixed bed gasifier could be successfully scaled up to commercial proportions and retain the properties observed and measured in the 3 ft gasifier at Solihull, it could be characterized as shown in Table 4.2, where it is compared with the fixed bed dry ash Lurgi gasifier.

Table 4.2

Relative Fixed Bed Gasifier Characteristics

Characteristic	Lurgi Dry Ash	Experimental Slagging
Feed size, minimum, in.	¼	¼
Control of bed channeling	Required	Required
Countercurrent efficiency	Yes	Yes
Ash fusion temperature required	High	–
Relative throughput	1.0	4
Relative product gas HHV	1.0	1.2
Relative steam/oxygen ratio	1.0	5.0
Relative oxygen/unit fuel value produced	1.0	1.1
Relative steam/unit fuel value produced	1.0	0.22
Relative gas volume produced	1.0	5.4
Cold gas efficiency, %	81	82.5

ISBN 0-201-08300-0

4.2. Fluidized Bed Gasifiers

These operate with finely divided coal suspended in a vertically rising current of reacting steam and oxygen or steam and air. Gas superficial velocities, from 0.5 to 1.5 ft/sec generally, are chosen to cause the bed of coal particles to expand to roughly twice its settled height, where vertical drag forces are balanced by gravitational force. Although the bed of solids as a whole is stationary, the particles themselves are in an extreme state of agitation. Fluidized beds of solids are characterized by their resemblance to a boiling liquid and by their lack of thermal or concentration gradients caused by extreme turbulence and backmixing within the bed. This technology has been used with impressive success to dry grain and to provide huge catalyst interfacial surface for such chemical operations as catalytic cracking of petroleum fractions.

When applied to coal gasification however, certain new factors should be borne in mind. For example, as coal is gasified the character of the residual particle is changed dimensionally, as well as in terms of particle density, porosity, chemical properties, and probably frictional drag coefficient. As carbon gasification advances, the carbonaceous matrix binding adventitious mineral matter disappears until a catastrophic collapse of one particle into several smaller ones occurs. In general, the mineral fraction of coal is more dense than the hydrocarbon (sp. gr. $\geqslant 2.3$ vs. about 1.3), leading to a tendency for only the finest mineral particles to elutriate from the fluidized bed of char. Thus there is a tendency for the mineral content of the gasifier bed to build up and ultimately to interfere with the gasification rate and with the fluidization characteristics of the bed, or to reach an equilibrium concentration in the bed by virtue of a steady removal of resident bed material. It is a mistake to expect all mineral residue to elutriate from a fluidized bed of char because mineral matter occurs in coal adventitiously and in a wide range of particle sizes and particle densities.

Fluid bed gasifiers have encountered serious difficulties with swelling and caking coals. Fresh coal suddenly injected into an active fluidized gasifier bed becomes heated to reactor temperature almost instantaneously. Pyrolysis reactions and steam-coal reactions are much too slow to maintain that thermal pace; hence the particles melt and, upon touching others of like kind, agglomerate to form larger particles. Very little of this is needed to change adversely the fluidization behavior of the char bed. Defluidization and collapse of the bed follow rapidly, with severe coking of the collapsed material to form hard, difficult-to-remove deposits and an inoperable system.

Much effort has been expended to combat this tendency, largely by preoxidation of the coal to destroy its caking tendency. Gasior and co-workers [27] reported success with 20 ft of free fall through steam containing 5.5 mol % oxygen at coal temperature of 310–430°C and at 250–330 psig. They converted Pittsburgh seam coal (1/4–3/8 in.) from an initial free swelling index (FSI) of 8.0 to final values of

ISBN 0-201-08300-0

1.5–2.0; the product remained uncaked when exposed to hydrogen at 600°C. Pretreatment reduced the volatile content from 35.6% to about 25% and the particle density by about 50%. Kavlick and Lee [54] reported success with fluidized bed pretreatment of coal at 725–750°F, 1 atm pressure, and about 1–2 hr residence time, consuming 1.0–1.5 scf oxygen per lb coal treated. Their work showed a similar decrease of volatile content. Petrographic examination by Mason and Schora [62] showed a skin of high reflectance formed by surface oxidation. The skin thickness was roughly uniform. Particles had been inflated to cenospheres, with the reflective skin found in cracks and in the interior of some vesicles (unconnected cavities). Preoxidation of caking coals does indeed destroy their caking qualities, but at the sacrifice of about one third of the hydrogen-rich volatile matter in the raw coal and a decrease of solids bulk density of nearly 50%. The economic penalty of this has been estimated variously at from 5 to 19¢/million Btu of fuel gas product.

Other approaches have been suggested, and some have been tested with varying degrees of success. Premixed char dilution of fresh feed coal, using char/feed ratios from 2.0 upward, has maintained operability in fluid bed atmospheres of hydrogen at 1000°F and over 500 psi, but it is deemed safer to use larger premix ratios. If premixing is not contemplated, incoming fresh coal must be completely mixed nearly instantaneously into a large volume of resident char. If mixing is rapid and thorough, melting fresh feed particles can impinge upon and agglomerate with only a few refractory char particles; in a short time pyrolysis and steam-hydrocarbon reactions will convert the molten material to additional resident char. Since this approach increases particle size, it is in competition with gasification reactions trying to decrease particle sizes. It is clear that large excesses of resident char over incoming fresh feed are needed to maintain a stable fluidized solids system. This, in turn, implies a need for larger reactor vessels to accommodate the char for feed dilution, but the resulting cost is generally a mild, incremental one unless major changes in vessel dimensions are required.

When subdivided solids are suddenly introduced into a fluidized bed of solids at gasification temperature (say 1700°F), heat is transferred nearly instantly to each new particle, largely by radiation. If the fresh coal particle has caking properties, it quickly melts. At the same time initial pyrolysis reactions begin to evolve volatile matter, first from the particle surface and then from the interior as heat absorption continues. Larger particles can develop a firm skin of pyrolysis residue while decomposition is continued in the interior. Such particles can develop enough internal pressure to explode into many smaller particles, thus producing fines that could be elutriated from the fluidized bed and lost to the process.

Very small particles, upon similar sudden introduction into the gasifier bed, receive heat to their centers so rapidly that the entire particle can be molten during pyrolysis. These particles swell to only a mild degree because volatile matter can escape with relative ease. Particulate feed of an intermediate size offers more resistance to the evolution of volatile matter, but not enough to cause explosive rupturing. These particles swell in size substantially, having large voids when

ISBN 0-201-08300-0

pyrolysis is complete. Since these have some vague resemblance to an inflated rigid-wall balloon with at least one hole to the interior, they are called cenospheres. They have low particle and bulk densities and therefore are fluidizable at lowered gas velocity. To avoid elutriation of these solids low gas velocity is required, and thus throughput rate for given reactor dimensions is reduced. If a reactor is designed to accept cenosphere formation, the low gas velocity will require an increase in reactor diameter at some increase in cost for the larger vessel, larger foundation, and so on.

It is clear that cenosphere formation should be minimized; it is equally clear that explosive shattering to produce feed fines must be avoided. These two requirements applied to a specific coal feed for a specified thermal shock impose bounds on the maximum feed particle size. Feedstock fines, because they elutriate from a fluidized bed, are also undesirable and must be held to a minimum unless there exists a good way to use them elsewhere in the process. Thus the desirable particle size range is dictated by the nature of the process and the nature of the coal feed. A typical feed particle size range for gasification in a fluidized bed gasifier is 100% through 8 mesh, 90% retained on 200 mesh (standard U.S. sieve).

Surprisingly, potassium carbonate has been found capable of preventing the swelling of an Illinois No. 6 coal. By impregnating the coal with 20 wt % K_2CO_3 before gasification at 1300°F (500 psig), Eakman and co-workers [18] determined that slight *shrinkage* in volume occurred in place of the usual melting and swelling as trapped gases tried to escape from the pores.

4.3. Transport Reactors

Transport (also called suspension or entrained flow) reactors for coal gasification carry the finely divided coal (70+%, -200 mesh) in a stream of gas (steam-oxygen, steam-air, or hydrogen) into a reaction zone where gasification occurs more or less completely, generally at high temperature. Partial oxidation processes can feed raw coal to the gasifier for conversion in a single step at slagging temperatures (2700–3500°F) to CO, CO_2, H_2, H_2S, and unconverted steam [8, 102]. Alternatively, gasification can be accomplished in two steps by feeding fresh coal entrained in steam into the 3000°F combustor flue gas for cooling and devolatilization with some steam reforming of the fresh coal and its devolatilization products. The devolatilized coal (char) is separated for use as fuel for the partial oxidation step in steam-oxygen at the combustor [8, 11, 32]. The latter method has three desirable characteristics: (1) the raw gas heating value is enriched with the volatile matter not subjected to high combustor temperature; (2) less oxygen is consumed in the combustion of fixed carbon only for process heat than in the partial combustion of full range coal; (3) raw gas leaves the gasifier at a lower temperature, leading to reduced process heat demand and consequent reduction in oxygen consumption.

Solids can be entrained into a reactor in a liquid medium, followed by simple evaporation of the liquid (recycle oil, for example) or by evaporation and direct reaction with the solids (water). In one form of transport reactor a water slurry of feed coal (or char) is injected into the gasifier with oxygen for partial combus-

ISBN 0-201-08300-0

tion under slagging conditions. This procedure has the advantage of simplicity of process and equipment; it carries the penalty of large oxygen demand, mainly to generate steam from the liquid water slurry medium (very costly steam). In an interesting transport reactor for coal hydrogenation reported by Rosen et al. [79], ground coal (50–100 mesh) at ambient temperature is mixed into hot hydrogen at 2790°F and 4000 psig for brief contact times on the order of 10 msec. Cold hydrogen is used to quench the reaction at predetermined (short) contact time to produce large yields of benzene, toluene, and xylene (1–2.6 bbl/ton maf coal) and large yields of methane-ethane mixtures (3000–12,000 scf/ton maf coal).

Transport reactors, especially when operating at high pressure, have the advantage of high throughput rates. This is somewhat offset by the cocurrent flow of reactants, or the lack of countercurrent contacting. A further disadvantage generally arises from the production of raw gas products at relatively high temperatures, which frequently leads to excess steam production for the recovery of heat, a process penalty difficult to minimize. Quenching the hot raw gas with a water spray or by mixing with cold gas is a highly irreversible (and thus inefficient) thermodynamic process. The partial combustion-steam gasification of coal in transport reactors (or combustors) normally entails removal of ash as molten slag. Cocurrent flow of gas and solids eliminates the need for any oxidative pretreatment of the coal feed.

4.4. Liquid Medium Gasifiers

These fall into two general categories, operating at distinctly different temperature levels. Molten salt reactors, generally employing molten sodium carbonate, operate at about 1800°F; molten iron gasifiers, at about 2700°F. Such temperature levels permit no hydrocarbons to survive long enough to appear in the product gas, especially in the molten iron system.

When molten salt is used as a reaction medium, it is chosen for its catalytic effect in promoting the steam-carbon reaction to produce CO and H_2 [8, 11, 14]. The reactant coal, finely divided, is entrained in air or in steam-oxygen and conveyed into the molten salt bath with entrained makeup salt. Reaction occurs in the salt bath at pressure levels prechosen at 10–80 atm. Since gas and solids flow cocurrently, no oxidative pretreatment is required. The mineral constituents (ash-formers) of the coal feed are only partially dissolved in the liquid salt medium, where their accumulation is limited by steady withdrawal of molten salt and "ash" to separation and salt recovery. This part of the process is a substantial one, representing a major fraction of capital and operating expense. Since molten sodium carbonate is a good solvent for many materials of construction and insulation, the confining vessel wall is cooled to maintain an equilibrium layer of frozen salt as a corrosion protection measure. Reaction rates in molten sodium carbonate are said to be about five times as fast as those attained in competing solid-gas reaction systems.

ISBN 0-201-08300-0

Molten iron has been proposed [8, 60] as a coal gasification medium with certain advantages, the primary one being the removal of sulfur from the raw gas product by the molten iron and an overlying layer of molten slag. Finely divided coal entrained in steam is blown into the molten iron bath, there to be pyrolyzed and reacted with steam. Residual char is dissolved in the hot iron for further reaction with steam, hydrogen, or the oxygen which also is blown into the bath. Enough oxygen is used to maintain the molten iron bath. Sulfur is believed to dissolve first in the molten iron and then to be transferred to the slag, which is removed for sulfur recovery. At the high temperature of the molten iron bath all hydrocarbons and other organic molecules are completely destroyed to yield a raw gas composed of CO, CO_2, H_2, H_2O, and probably some H_2S. The interaction of iron with steam and reducing gases is believed to play an important role in the reaction mechanism.

4.5 *In situ* Gasification

Numerous attempts have been made over the years to gasify coal directly in place to minimize the cost of fuel for electric power generation. Included among the attempts of the past are projects in Russia and in the United States with less than resounding success. Interest has recently been revived in the United States, partly as a result of a report by G. H. Higgins of the Lawrence Livermore Laboratory [41]. While the concept proposed was aimed at high Btu gasification, mere substitution of air for oxygen provides a basis for low Btu gasification of coal in place. Most *in situ* gasification projects are aimed at low Btu gas production for electric power generation [92, 93].

Figure 4.7 shows generally the mode of operation proposed for most *in situ* coal gasification projects. Although details vary from project to project, it is universally necessary to admit mixtures of oxygen and steam or air and steam into a hot zone in the coal seam for gasification to occur, and to withdraw the raw product gas from the coal seam for whatever surface processing is required to make the fuel environmentally and otherwise technically acceptable. Nadkarni et al. [69] provide a detailed discussion of underground coal gasification procedures and problems and some estimates of cost.

Some questions must be resolved before *in situ* gasification processes can be regarded as viable:

1. Do suitable coal deposits exist?
2. Can adequate permeability be created by explosive fracturing, and maintained through the gasification process?
3. It is possible to maintain and control a reaction zone?
4. Can the process be scaled up to commercial size?
5. What are the effects of by-product tar, tar oils, naphtha, and phenols upon steady performance of the production wells?

ISBN 0-201-08300-0

6. What will be the efficiency of use of the coal resource?

7. What is the blasthole fracture efficiency, or the optimal radius of permeability generation, which can be attained by detonation within a blasthole?

8. How does the thickness of the coal seam affect the yield of fractured coal and gas production performance?

9. How can the detrimental effects of fissuring and channeling be controlled or minimized?

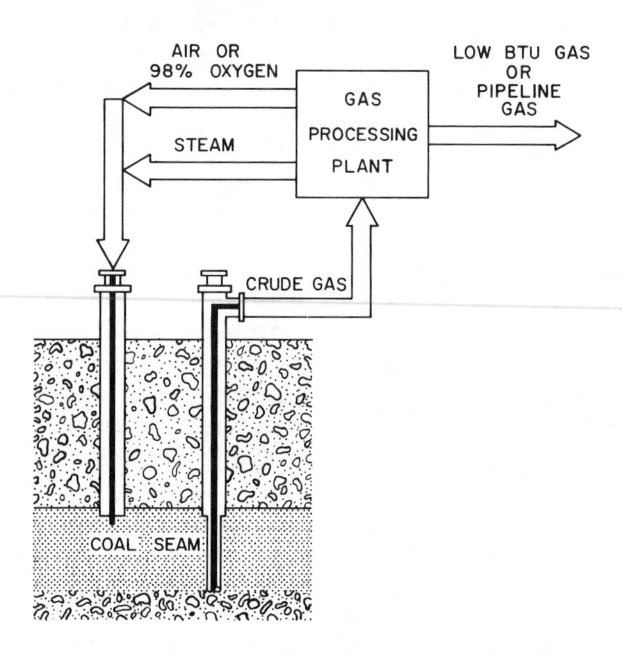

Figure 4.7. *In situ* coal gasification.
Source: [28].

The situation for *in situ* coal gasification can be summarized as being in its early stages of development. Earlier attempts did not produce satisfactory results, generally because of abnormally low heating values and generally poor economics for the gas products obtained. It is likely that the new *in situ* gasification projects will overcome many of the deficiencies of the past; it is improbable, however, that *in situ* coal gasification will arrive on a commercial scale until late in the twentieth century.

4.6. Degasification of Coal Beds

Although the subject of this chapter is the gasification of coal by normal processing technologies, it would be remiss not to mention the potential for de-

ISBN 0-201-08300-0

gasification of coal beds to produce commercial quantities of methane. Intensive research conducted by the U.S. Interior Department's Bureau of Mines on degasification of coal beds for improved mine safety indicates that commercial quantities of pipeline quality natural gas can be produced. Coal bed degasification by draining gas from the coal bed before it is mined can be accomplished by drilling vertical holes into coal beds with subsequent stimulation by hydrofracturing or by drilling horizontal holes into virgin coal bed areas from small, specially designed shafts. Deul et al. [17] estimate that the minable coal beds (less than 3000 ft deep) of the conterminous United States may contain 260 trillion ft^3 of natural gas. A productive gas well was effectively established on the Pittsburgh coal bed, where a small vertical shaft was drilled from the surface to provide access to the coal bed; the shaft bottom was widened to permit drilling eight holes radially into the coal bed. Seven of these holes, ranging in depth from 500 to 850 ft, were connected to a manifold to produce gas at the rate of 700,000–720,000 scf/day. After 326 days of continuous production, the cumulative total volume of gas piped to the surface for disposal was 182 million ft^3.

Typical of production characteristics from bore holes drilled into coal beds is the observed high initial production rate, followed first by a minimum rate and then by a gradual but continuous increase. The ultimate decline depends largely on the permeability of the coal bed, but the initial decline is known to be due to water blockage. As the coal bed around the bore holes is dewatered, the gas flow rate continues to increase.

Gas production from a single 6 in. diameter vertical hole in the Marylee coal bed in Jefferson County, Alabama, rose from its maximum of 5000–83,000 scf/day after stimulation by hydrofracturing. Generally the production of gas from wells in the Pittsburgh coal bed is higher than that from other coal beds.

Measured methane content in seven coals tested ranged from 29 ft^3/ton for the Illinois No. 5 coal bed in Jefferson County, Illinois, to 387 ft^3/ton for the Pocahontas No. 3 coal bed in Buchanan County, Virginia. These coal beds vary in rank and in depth, but most are less than 1000 ft deep, with only one sample from the Pocahontas No. 3 coal bed reaching a depth of 1500 ft. It is assumed that an average of 200 ft^3 of gas/ton of coal is a reasonable estimate for coal below strippable depths. The coal resources of the United States from minable coal beds with less than 3000 ft of cover, excluding strippable coal beds, are estimated to be 1.3×10^{12} tons. This means that the total gas resource in minable coal beds to a depth of 3000 ft is on the order of 260×10^{12} ft^3 of gas. Regardless of how this estimated quantity is revised upward or downward, it is a large resource to consider, especially because no significant exploration is required to find the gas. Coal beds are important marker beds, and many oil and gas drillers know where even unmapped coal beds occur because of their distinctive response to gamma ray and hydrocarbon emission logging. With minor investment for exploration costs, the gas production industry could give major attention to production and stimulation techniques.

ISBN 0-201-08300-0

5. GASIFIER HEAT SUPPLY METHODS

It is fair to say that gasifiers are distinguished from one another principally on the basis of the means used for supplying the endothermic heat of steam gasification.

Wherever gasification of coal is to be accomplished by reaction with water or steam, substantial high level thermal energy demands must be satisfied at the gasifier. Numerous schemes have been proposed and tried in addition to the employment of pure oxygen mixed with steam. When low Btu gas (120–180 Btu/scf) is required, air instead of oxygen is used for a very considerable reduction in capital equipment and operating costs by carrying out partial combustion with steam gasification. If the objective is the production of substitute natural gas (SNG), however, nearly pure methane must be produced. Nitrogen in the product gas, having no fuel value, degrades the product fuel value and combustion characteristics. Therefore its concentration in the product must be minimized. One way to accomplish this, of course, is to separate pure oxygen from air in a tonnage air distillation plant before using the oxygen for coal gasification. The price paid for this convenience is the capital and operating expense of the air distillation plant, which is considerable, and the incremental expense (also considerable) of *removing all process CO_2 in the gas purification part of the system.* (It should be remembered that more than 1 ft^3 of CO_2 is produced for each cubic foot of SNG manufactured.)

Alternatively, the gasification plant could use air instead of oxygen if a suitable nitrogen barrier or nitrogen removal system could be provided. The preheating of steam hot enough to supply the heat demand is impractical in any present day material of construction, thus ruling out highly superheated steam as a candidate method. Heat pipes for this purpose are being investigated, and for fluidized solid systems there is some possibility of success if the difficult materials problems can be resolved (reducing atmosphere at the cold 1800°F end; oxidizing conditions at the hot 2000+^{o}F end; a suitable, stable operating fluid).

A more likely means for fluid bed gasification systems is the continuous circulation of a heat carrying solid from a heating zone (combustor) to the cooling zone (gasifier), as shown in Figure 5.1. The solid can be a chemically inert refractory (alumina, an ash-derived solid, etc.) or a chemically active material such as limestone, or resident char. Each of these has its own peculiar advantages and disadvantages, in view of the requirements, as follows:

1. Inert refractory. Such a solid must be physically strong, be abrasion and attrition resistant, be resistant to oxidizing and to reducing atmospheres at operating temperatures, pass through no significant phase or crystal habit change within operating temperature cycles, possess high specific heat, be capable of fabrication in small spheroidal form to minimize attrition, be inexpensive, and be cleanly separable from coal char via fluidized solids technology. Except for ash-derived solids, an inert refractory represents a substance foreign to the system, with attendant potential for forming unwanted (perhaps low melting) intercompounds

ISBN 0-201-08300-0

with coal ash or with refractory linings of the system. Adequate, reliable separation from char is essential and can be difficult. Because of the large heat carrier solids circulation rate, small amounts of entrained char could exceed the process fuel requirement at the combustor. If excess air is used for combustion, excess fuel could lead to loss of control; excess fuel not burned would elutriate from the circulating refractory and become a loss of carbon from the process.

2. Limestone. The best known example of a chemically active heat carrier solid is limestone, which is calcined to lime at the combustor (endothermic) and carbonated in the gasifier (exothermic). The relatively large heat of calcination-carbonation reduces the circulation rate required below that for an inert solid:

$$CaO + CO_2 \rightarrow CaCO_3; \quad \Delta H = -76{,}200 \text{ Btu/lb-mole}$$

The raw gas product contains less CO_2, thus reducing the magnitude of the CO_2 removal task in the gas purification system [11, 22].

All the requirements listed for the inert refractory apply to limestone as well, although there is less freedom for control of properties. Because $CaCO_3$, CaS, and $CaSO_4$ can form low melting liquids, operation of the combustor must provide for slightly reducing conditions at all times to avoid encrustation and plugging. Flue gas from the combustor must contain a few percent of carbon monoxide. Because of the tendency to form these and $Ca(OH)_2$–CaO encrustations, use of limestone as a circulating heat carrier is restricted to steam partial pressures below 9 atm and to temperatures in the gasifier of 1500°F and lower. These restrictions limit the use of this heat carrier to the gasification of reactive coals like western subbituminous and lignites gasifiable at lower temperatures, to the exclusion of less reactive bituminous coals requiring higher gasifier temperatures.

Apparently because of gradual changes in the crystal structure of $CaCO_3$, the circulating heat carrier loses its activity for calcination-carbonation. Means have been found to reconstitute the limestone for full recovery of its activity.

3. Char. Circulation of resident char as a heat carrier dispels any possibility of controlling or imposing the desired properties for an inert refractory. But char, although not inert, possesses some very convenient properties such as high specific heat, no need for separation from gasifier bed material, and a state of subdivision characterized by large interfacial surface and good thermal conductivity for heat transfer. It has the disadvantage that the whole body of circulating solids cannot be exposed at once to combustion air, but must be split into fuel and circulation solids. Fuel is burned, preferably under slagging conditions with minimal excess air; hot flue gas then transfers its heat by direct contact with the circulating char. Experience indicates that char reactivity toward steam gasification is enhanced if the char is circulated as a heat carrier from combustor to gasifier.

Contact time between flue gas and char must be kept brief, and solids-gas mixing must be rapid and thorough to minimize loss of carbon through CO_2 reaction with circulating char. Generally, CO in the flue gas is a carbon loss and an endothermic heat of gasification loss to the process that cannot be easily recovered.

ISBN 0-201-08300-0

Any hydrogen in the fuel or mositure in combustion air will promote steam gasification of circulating char, with attendant loss of carbon into the flue gas in addition to the endothermic heat of steam gasification lost from the process. Combined steam and CO_2 gasification potentials represent a possibly serious loss of process efficiency, requiring careful design to minimize the effect.

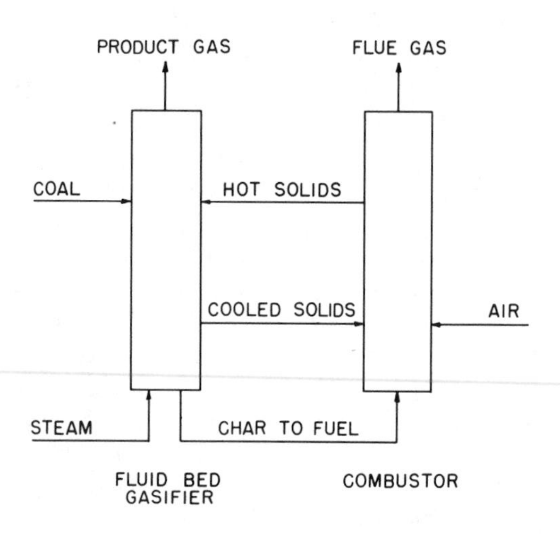

Figure 5.1. Fluidized bed gasification: heat transfer by solids circulation.

The high cost of pure oxygen for coal gasification makes process heat via air for combustion a strongly attractive alternative if nitrogen exclusion from the product can be realized. The circulating heat carrier solid accomplishes nitrogen exclusion reasonably well, but is imperfect because it entrains flue gas into the gasifier. Since nearly pure carbon is burned, the entrained flue gas entering the gasifier contains nearly 79% nitrogen, which becomes an inseparable part of the gas product. The flue gas is entrained to the gasifier in the interstitial spaces and pore structures of the circulating solids. Similarly, product gas (CO and H_2) is entrained into the combustor, constituting a loss of product to the flue gas. These effects can be mitigated by stripping the entrained gas from interstices by displacement with a suitable gas, possibly CO_2. Unfortunately, the latter gas can react with hot char, making it a questionable choice if char is to be used as the heat carrier. Steam stripping of heat carrier char is even less acceptable, since steam is more active than CO_2 for gasification of char. If the heat carrier solids are inert, steam stripping is appropriate, or CO_2, if available, can be used.

Circulating solids for heat transfer with no stripping of entrained gases can have a serious effect on the quality of high Btu gas produced. It is estimated that entrained nitrogen from combustor flue gas (char circulation) could amount to

ISBN 0-201-08300-0

14.6% of the finished gas product, reducing the heating value from 1000 to 854 Btu/scf. Even poor stripping efficiency would provide substantial improvement in product quality. By comparison, circulation of an impervious inert solid of the same heat capacity (but higher density) could reduce the entrained nitrogen in the product methane to 2% with a gross heating value of 980 Btu/scf. This material could easily be stripped of all its nitrogen from the flue gas; the stripping of char would be more difficult.

Fluidized char beds exhibit electrical characteristics suitable for resistance heating, preferably with direct current. Extensive studies were carried out at the Institute of Gas Technology to evaluate this method for producing hot syngas from char and steam for direct hydrogasification of coal in the Hygas process (43, Vol. 2, Part IV). Work was carried out successfully on a 300 kW (6 in.) electrothermal unit and later on a 2.25 MW (30 in.) unit. Desired synthesis gas yields and carbon gasification were consistently obtained at char residence times of 10–20 min and at steam/char weight ratios of 1.0–1.5. The electrothermal gasifier operated at $1900^{o}F$ and 1000 psi. It employed electrodes made of type 316 stainless steel and was designed for operation at 1500 psi, feeding 4000 lb/hr of char of bulk density 25 lb/ft^3. Refractory char, suitable for power generation fuel, was discharged at 2500 lb/hr at bulk density of 20 lb/ft^3. Maximum char residence time was 45 min with steam input of 200–300 lb-mol/hr to provide fluidization velocity of 0.2 ft/sec at operating conditions.

Electrothermal gasifier studies were initiated when electric power generation from the refractory char appeared to be economically feasible. Rapidly rising costs of electric power made this process method less economical than the use of steam-oxygen, and development of the electrothermal gasifier was suspended. Renewal of interest could occur, however, if magnetohydrodynamic (MHD) power generation becomes economically competitive, since electrothermal Hygas char would be a good MHD fuel.

Proposals have been made to build a high temperature gas-cooled nuclear reactor (HTGR) for power generation, using helium for the transfer of high level heat because of its good thermal characteristics and independence from radiation effects. The circulating helium could be tapped off to supply thermal energy for the production of hydrogen from water or for direct heat supply for the gasification of coal. One of the principal advantages this proposal offers is the substantial reduction in the consumption of fossil resources for the production of hydrogen or of methane. The economic prospect for this approach depends upon excess heat availability from a nuclear power source, since the cost of nuclear heat dedicated to coal gasification or hydrogen production would be prohibitive.

6. LOW BTU GASIFICATION

One of the interesting facts about fuel gas combustion is the remarkably constant quantity of flue gases produced for the same quantity of heat released.

ISBN 0-201-08300-0

Thus, for example, stoichiometric combustion produces the following flue gas volume per 1000 Btu of fuel used: methane, 10.5; carbon monoxide, 9.0; hydrogen, 8.9; producer gas, 12.35. Producer gas, with a heating value of only 168 Btu/scf and a nitrogen content of 50%, produces only 17.7% more flue gas products than are obtained by combustion of methane at 1000 Btu/scf. The importance of this conclusion lies in the capability to use different fuel gases in a given piece of heating equipment with little concern about adequacy of heat exchangers, flues, stacks, and the like. Only a change in the burners themselves, and probably in the fuel distribution system, would be required in making a change from one fuel gas to another.

6.1. Historical: Wellman-Galusha and Lurgi Gasifiers

Before the advent of natural gas into the industrial fuel market, fuel gases were made at the industrial site by gasification of coal, and industrial needs could occasionally be met by purchase of manufactured gas from a local utility. Coal-derived fuel gases have been used in chemical plants, glass plants, steel mills (for normalizing, annealing, atmosphere work, etc.), independent research industries, magnesium manufacture, silk mills, bakeries, wire mills, foundries, potteries, aluminum and stainless steel manufacture, pipe annealing, ordnance factoreis, tin-plate mills, lime plants, brick plants (for firing both periodic and tunnel kiln), zinc smelting, iron ore processing, baking carbon electrodes, and fuel for stationary internal combustion engines, as well as for synthetic fertilizer manufacture from synthesis gas. One of the more popular U.S. gas producers was the atmospheric pressure Wellman-Galusha gas producer, manufactured by the McDowell Wellman Engineering Company of Cleveland, Ohio [97].

The Wellman-Galusha gas producer is a self-contained unit and requires no investment for boiler plant or other accessory. Adequate provision for steam for gas making is included in the engineering design of the plant. Ample fuel and ash storage bins are provided as an integral part of the unit. The plant consists of a continuous, automatic gravity feeding system for the fuel, a specially designed revolving grate, and an elevated ashpit. The plant is rugged in construction, and its simple design makes for low maintenance and depreciation costs. High gasification efficiency minimizes fuel consumption, and this effect is enhanced by the use of cheap grades of fuel and the increased throughput capacity, measured in pounds per hour per square foot of grate area. The Wellman-Galusha gas producer is built in two styles, with and without agitation of the fixed bed of coal in the gasifier, as shown in Figures 6.1 and 6.2.

A two-compartment fuel bin forms the top of the unit. The upper section serves as a storage bin, fed by any suitable device for fuel handling. The lower compartment is separated from the upper by disk valves through which fuel is fed as required. Similar valves cover the entrance to each of the heavy steel feed pipes connecting to the lower bin with the fire chamber. Fuel from the lower bin flows continuously through these feed pipes to fill the fire chamber, and revolving grates

ISBN 0-201-08300-0

discharge the ash from below the fire at the same rate at which it is formed. There are no moving parts in this fuel feeding system, thus eliminating the repair costs common to machines where mechanical devices are used on highly abrasive fuels.

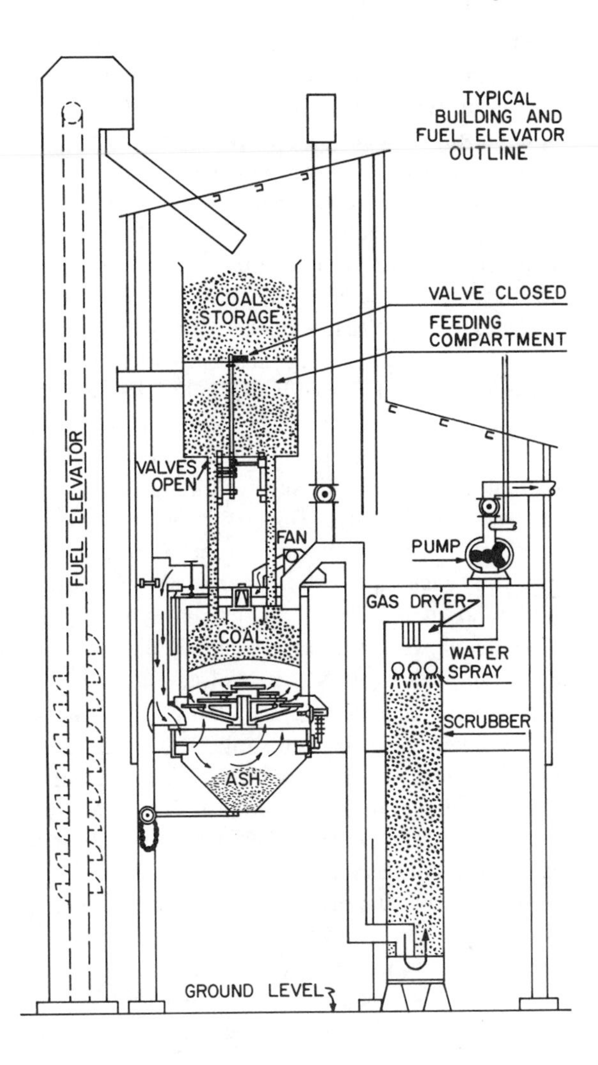

Figure 6.1. Wellman-Galusha fixed bed gasifier. Redrawn from McDowell-Wellman Company brochure no. 576 with permission.

ISBN 0-201-08300-0

The gas making chamber is completely water jacketed. The inner wall is made of 1 in. thick steel plate and requires no brick lining. Waste heat in the water jacket generates the steam (actually water vapor at about 180°F) required for making gas. A direct-driven fan supplies the air required by the gasifier. This air acquires mositure on its way to the fire bed by passing over the steaming water at the top of the jacket. Saturation is automatically controlled by regulating the rate of supply of jacket water to get and hold the desired saturation temperature. A thermostat controlling the water supply valve keeps the ratio of air to moisture at the desired point. Steam can be diverted from the water jacket if desired to avoid the natural formation of hydrogen from the steam, and CO_2 can be substituted for steam, depending upon the type of gas required. Blast mixtures of air plus CO_2 or oxygen mixed with either steam or CO_2 can be used.

The grates are made of circular, heavy steel plate rings, flat and without perforations. They are set one above the other with edges overlapping, so that ash cannot escape unless pushed horizontally through the vertical space between the step plates. The grates are eccentric with the center support, and the entire assembly is rotated very slowly. Because of eccentricity the space between the grate plates and the shell of the gas making chamber varies constantly as the grates revolve. As the space increases it fills with ash, and as the space decreases the ash is forced over the inner edge of the straight platforms and falls into the cone-shaped ash hopper below. In the area of greater eccentricity and least space, ash is crushed between the bosh and the grate plates, and is thoroughly broken up as it is pushed through the space between the grates. This action occurs progressively throughout the entire grate area, delivering a constant stream of loose, readily flowing ash to the hopper below. The cone-shaped ash hopper is elevated sufficiently high above ground level to allow a truck, railroad car, or conveyor to be placed under the discharge gate. The free-flowing ash can be quickly and conveniently discharged for disposal. No interruption of operation is involved.

Figure 6.2 shows the general nature of the Wellman-Galusha agitated fixed bed gas producer. A slowly revolving horizontal arm, which also spirals vertically below the surface of the fuel bed, retards channeling and maintains a uniform fuel bed for the maximum production of high Btu gas. The agitator arm and its vertical drive shaft are made of heavy water-cooled steel tubing with the wearing parts protected by heat- and wear-resistant castings. The arm can be revolved at varying speeds, and its position within the fuel bed may be changed as desired for different fuels and operating rates. The agitated producer can gasify about 25% more anthracite or coke of the same size than is possible with the standard unit. A Wellman-Galusha agitator gas producer has operated on high volatile bituminous coal for several months without flue cleaning. The agitator spiraling within the fixed bed of fuel under gasification retards to a very significant degree the carry-over of coal dust with the outgoing bituminous coal gases. Lowered gas offtake temperature prevents the lampblack fallout common with conventional gasifier top temperatures.

The standard producer (Figure 6.1) can gasify anthracite or coke; the agitator producer (Figure 6.2) can gasify anthracite, coke, or bituminous coal. Each of these producers can also manufacture a high concentration of CO from calcined petroleum coke or from coal by employing a blast of oxygen and CO_2. When a high

ISBN 0-201-08300-0

percentage of hydrogen and CO is required, steam and oxygen in the blast are used. For normal heating requirements, air and self-made steam are used in the blast. About 25% of the steam necessary for synthesis gas can be self-made. These producers operate efficiently on fuels of small size or high ash content, preferring to operate on sizes from 2 in. lumps down to particles as small as 3/16 in. The smaller sized fuels are usually cheaper, resulting in minimum cost of the gas produced.

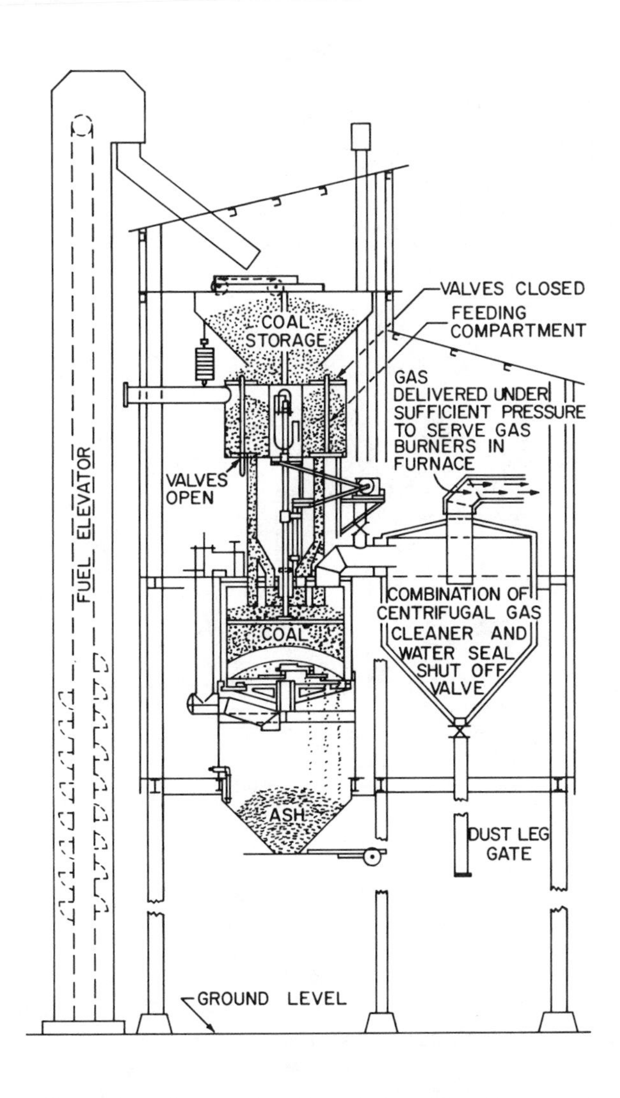

Figure 6.2. Wellman-Galusha agitated fixed bed gasifier. Redrawn from McDowell-Wellman Company brochure no. 576 with permission.

ISBN 0-201-08300-0

Wellman-Galusha gas producers have been built in sizes from 18 in. inside diameter to 10 ft inside diameter. Normally throughput capacities are rated at about 20–30 lb/hr ft^2 of grate area, but a gas producer has operated at rates as high as 99 lb of coal/hr ft^2 of grate area. A producer has also operated at a low throughput rate of 7.5 lb of coal/hr ft^2 of grate area. This is an extremely wide range of process operability, well suited to the fluctuating demands of many industries.

The Wellman-Galusha gas producer normally operates at pressures slightly above atmospheric, but a gas producer for operation at 300 psi was built for the Bureau of Mines Research Center, Morgantown, West Virginia, equipped with an agitator to allow the handling of caking quality bituminous coals. In 1974 McDowell Wellman Company submitted a report to the Office of Coal Research describing the preliminary engineering design of a minimal cost, 25 ft diameter, high pressure gas producer [59]. At its nominal design capacity the producer would gasify 70 tons/hr of highly caking bituminous coal at a pressure of 300 psi. Air blown, it would produce about 170,000 scf/min of low Btu gas, rated at 165 Btu/scf. The proposed gas producer has not been constructed at the time of this writing.

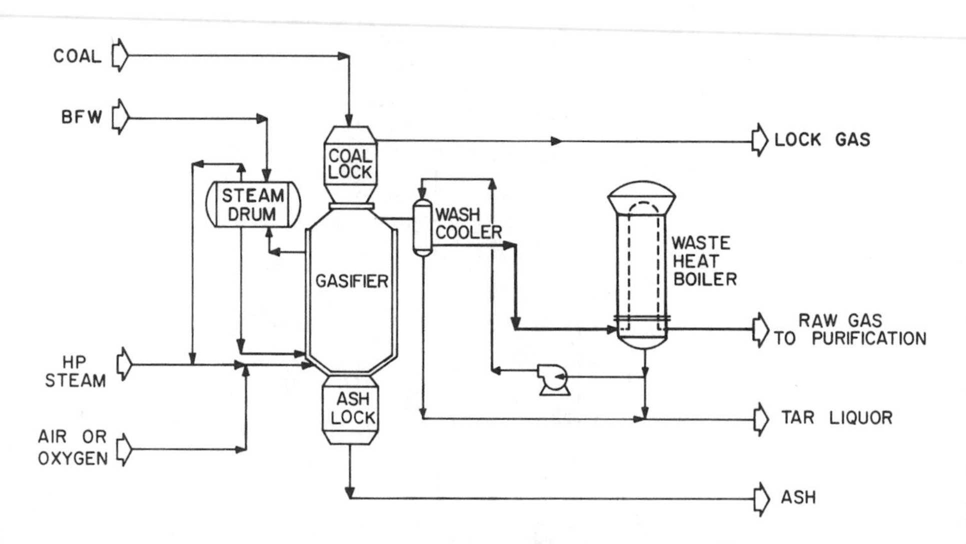

Figure 6.3. Lurgi process.

Lurgi pressure gasification [55, 82] is still in use outside the United States. It is carried out at pressure levels from 300 to 450 psig, and in accordance with the discussion earlier in this chapter, relating to Figure 4.1. Figure 6.3 shows a flow diagram of the process, indicating the principal elements and streams surrounding

ISBN 0-201-08300-0

the gasifier. Coal is introduced into the gasifier through a lockhopper; ash is similarly removed from the bottom of the gasifier. The raw gas is washed and cooled before going for purification. Condensates collected are sent for further processing as tar liquor. Figure 6.4 shows the nature of the gasifier, which is a water-jacketed vessel generating steam under pressure control to the steam drum of Figure 6.3. Upon release from the pressured coal lock above the gasifier, coal is distributed to the top of the fixed bed of resident material by a slowly rotating distributor. Ash residues are discharged from the grate, a procedure which consists of slowly moving circular rings in an orbital pattern to displace ash material consecutively from the upper central portions to the lower outer portions, thus keeping open the spaces through which steam and air are fed. Residues ultimately are discharged at the periphery of the lowest portion of the grate, where clinkers, if any, are broken to passable size.

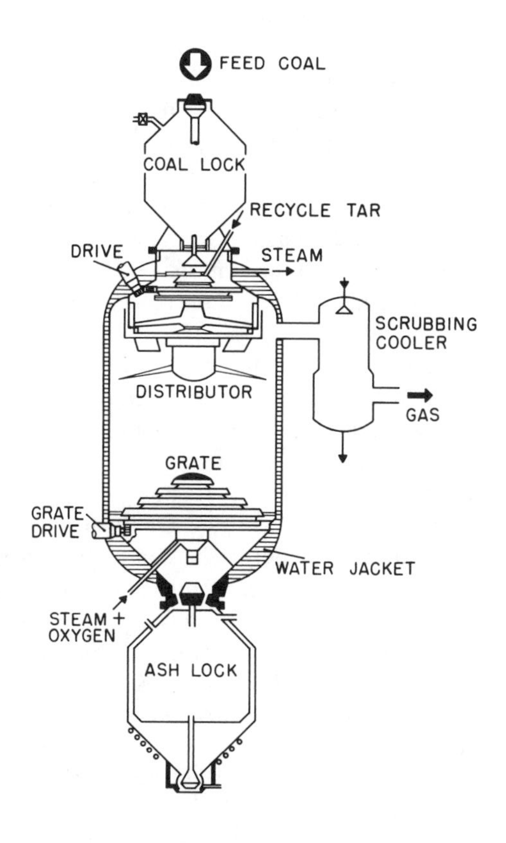

Figure 6.4. Lurgi fixed bed pressure gasifier.

ISBN 0-201-08300-0

Lurgi gasifiers have been built and operated at pressures from atmospheric to 450 psig at numerous locations around the world. In general, each installation has been relatively small in size, with gasifier inside diameters of 12 ft or smaller. Gasifiers of appreciably larger size suffer mechanical limitations believed to confine diameters, in practical terms, to about 15 ft. Caking coals present difficulties in these gasifiers.

6.2. Modern Combined Cycle Power Generation

Low Btu gaseous fuel has its greatest attraction for use in gas turbine power generation. The gas turbine is a relatively inexpensive engine, capable of rapid start-up when needed. It consumes fuel directly without the need for an expensive boiler, feedwater treatment, and the like; but it requires a fuel essentially free of particulates and free of certain corrosives. The gas turbine operates at high air/fuel ratios on the order of 100, depending upon the nature of the fuel employed. Its mode of operation employs a rotary air compressor and the power generating turbine on a common shaft with the electric power generator. Ambient air is compressed and fed to a combustion chamber, where fuel is injected to heat the air to a turbine inlet temperature of about 1300°F. Hot, pressured flue gas then expands through the turbine to produce shaft work of air compression and electric power generation. Exhaust gases, because they are at relatively high temperature (about 850–900°F), are cooled by heat exchange with the compressed air to be used for combustion and are then discarded to the atmosphere. This *open cycle* combustion gas turbine recovers an appreciable fraction of the exhaust gas heat to conserve fuel; it operates in a *regenerative cycle.*

Turbine exhaust gas heat can be more efficiently used in alternative ways, the best known being the *combined cycle* power generation scheme, in which gas turbine exhaust is used to generate steam to a steam turbine. The two turbines drive separate generators. When the fuel for such a system is generated from coal as low Btu gas, the configuration of the system is as shown in Figure 6.5. It should be observed that all air compressed into the system ultimately leaves by expansion through the gas turbine. However, some of the compressed air is diverted to the coal gasifier through a booster compressor. Steam and air are used in the gasifier to make a typical producer gas (about 120 Btu/scf) which is subsequently cooled, cleaned of particulates, and fed to the combustor. Combustor flue gas is expanded through the gas turbine and cooled in a heat recovery steam generator before exhausting to the atmosphere. Steam so generated is expanded through a steam turbine; exhaust steam is condensed and used to cool compressed air to the booster compressor before returning as warm boiler feedwater to the heat recovery steam generator (HRSG in Figure 6.5). The gas turbine operates on *open cycle;* the steam turbine, on *closed cycle.* Frei [24] has computed the thermal efficiency of power generation by combined cycle to be 31.4%, and that for a steam boiler-steam turbine system to be 20.0%.

ISBN 0-201-08300-0

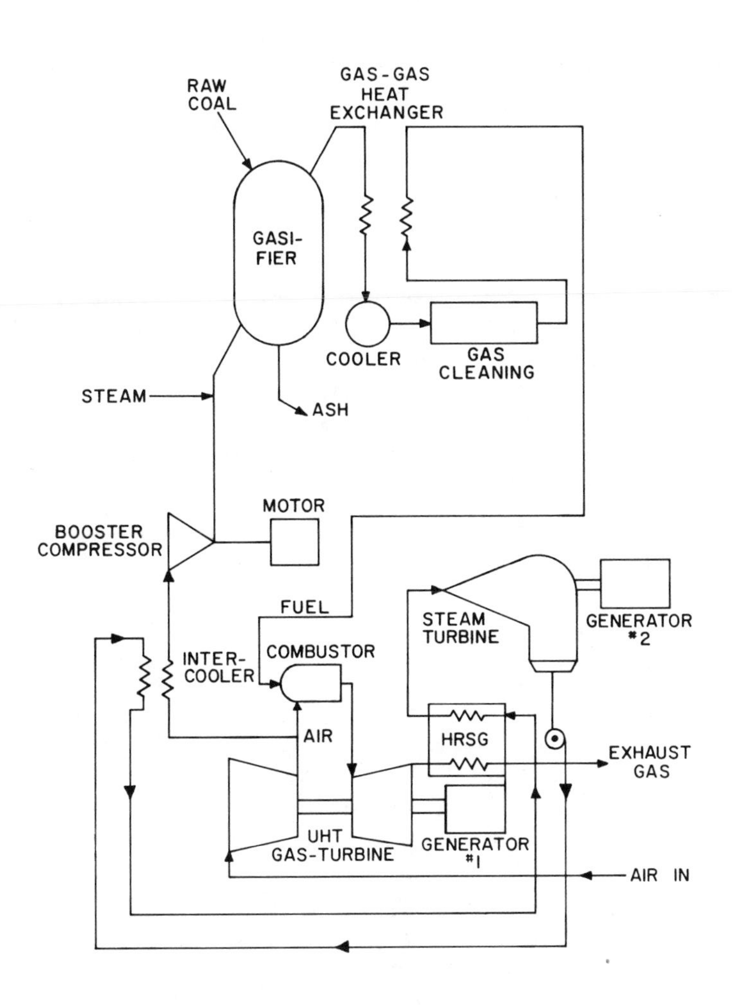

Figure 6.5. Low Btu coal gasification in combined cycle power generation.

A 170 MW demonstration combined cycle plant was built at Lünen, Germany, in 1972, employing Lurgi gasification equipment and technology [55]. It has suffered from various technical problems, among them the following:

1. The dynamics of gasifier start-up prevented simultaneous start-up of all plant units: caused by a drop in gasifier product heating value during start-up; solved by adding oil-fired start-up burners.

2. The gasifier gas outlet flange cracked because of temperature excursion to nearly 400°C above design: caused by inexact coal flow through the gasifier; solution not specified.

ISBN 0-201-08300-0

3. After failure of the gasifier flange, it was necessary to install an oil burner for 75% of full load and use it to fuel the pressurized steam boilers.

4. Gas entrained particulates were not reduced to the 1.5 mg/m^3 level specified, requiring improvements in scrubber design.

5. Gasifier operation was faulty when fed with a slightly caking coal. This might have been responsible for the cracked flange. Noncaking coal feed relieved the problem.

Since the combined cycle power generation concept is essentially independent of the nature of the coal gasifier, U.S. development efforts are aimed largely at (1) removal of sulfur compounds from the fuel gas, and (2) removal of dust particles to meet gas turbine erosion and corrosion standards. To a considerable extent these efforts affect the choice and design of gasifiers to be used.

7. INTERMEDIATE AND HIGH BTU GASIFICATION

7.1. General Aspects

Low Btu gas production is limited to on-site power production applications almost exclusively because of the relatively high cost of energy transportation. High Btu gas, on the other hand, has from 5 to 10 times the heating value, making energy transportation and distribution costs very substantially less. Since it is undesirable to pay the cost of transportation for inert materials or other substances of low heating value, high Btu gasification seeks to minimize the presence of such substances in favor of the production of, as nearly as possible, pure methane. It is thus clear why the presence of nitrogen is so deleterious to the quality of a pipeline gas, and why it is impractical to use a low Btu gasification process in connection with the production of a pipeline quality product. One of the distinguishing features separating low Btu gasification processes from those producing pipeline quality gas is the need either to use pure oxygen to provide process heat by combustion, or to provide heat by a means that excludes nitrogen from the product. This subject has been reviewed thoroughly in earlier pages, and needs no further elaboration here. The comparative properties required for pipeline gas and for power plant fuel gases are summarized in Table 7.1.

Nearly all processes for the manufacture of pipeline quality gas can be fairly represented by the overall process scheme shown in Figure 7.1, in which the nature of each of the process steps is not detailed. Thus the coal preparation step is peculiar in any given process to the nature of the gasification step, which may be carried out in a fixed bed gasifier requiring lump coal with minimum fines, a fluidized bed gasifier requiring crushed coal 100% through 8 mesh with only 10% passing through 200 mesh, or a transport reactor requiring ground coal so fine that 70% will pass through a 200 mesh screen. In any event, the raw gas leaving the gasification step will contain primarily carbon monoxide, carbon dioxide,

ISBN 0-201-08300-0

hydrogen, excess steam, some methane, and an amount of hydrogen sulfide depending upon the nature of the coal being processed. Numerous other substances may also be present in the raw gas, depending strongly upon the nature of the gasifier. Generally, raw gases are cooled somewhat and sent through a shift reaction, in which CO and steam interact to produce additional CO_2 and hydrogen. This reaction normally occurs at about 550–750°F over a solid catalyst of various possible compositions, including cobalt-molybdenum and nickel-molybdenum on a solid support. The converted gas, now richer in CO_2 and hydrogen and leaner in CO (see Table 3.10), is sent to gas purification, where CO_2 and H_2S are removed. Carbon dioxide is emitted to the atmosphere, and H_2S is processed for recovery of elemental sulfur. The stream of gas leaving purification has been adjusted to approximately 3 mols hydrogen/mol CO and is fed to the methanation unit. The methanation reaction is highly exothermic (see Table 3.8), and since it occurs on a thermally sensitive supported nickel catalyst, heat removal from the methanator becomes an important design problem. Normally the heat is removed at a temperature level of about 800–900°F by generation of steam to be used elsewhere in the process. The water generated in the reaction is nearly pure (small amounts of methanol, acetone, etc.) and may be retained for recycle to steam generation for the process. The moist methane product leaving the methanator must subsequently be dried and, if necessary, compressed to pipeline pressure specification.

Table 7.1

Fuel Gas Properties Required: Pipeline Gas versus Combined Cycle Power Fuel[a]

Property	Pipeline Gas	Power Plant Fuel
Fuel value, Btu/scf	$\sim$1000	200 > HHV > 90
Pressure, atm	>60	10–20
Temperature, °F	$\sim$70	70–1800[b]
Composition	>90% CH_4	CO, H_2, N_2, CO_2, H_2O, CH_4
Sulfur, ppm	<1[c]	$\sim$550[d]
Particulates, lb/million Btu	<<0.01	0.1
Cost (relative)	1.0	0.5

[a] Source: Archer et al. [3].

[b] High temperature can offset low fuel value and is desirable.

[c] Limit is dictated by process requirements.

[d] Equivalent = 1.2 lb SO_2/million Btu.

ISBN 0-201-08300-0

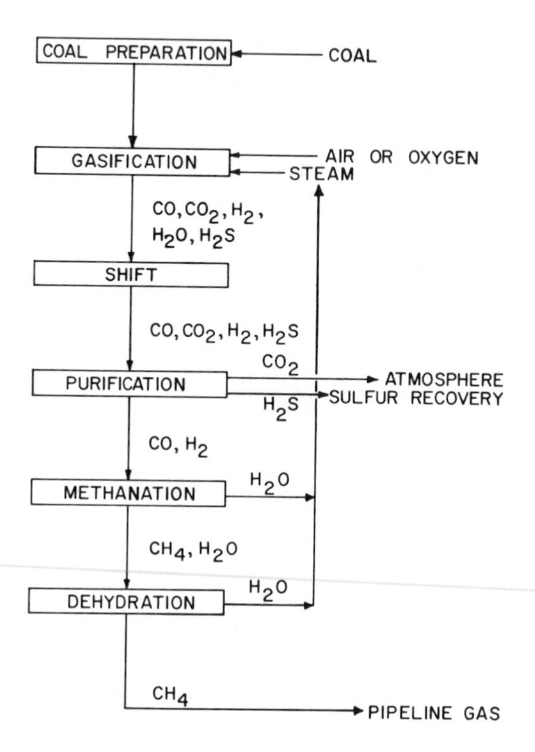

Figure 7.1. General high Btu gasification process scheme.

The majority of high Btu gasification processes depend upon the use of air or oxygen and steam; a few processes employ the separate production of hydrogen followed by hydrogasification. The latter procedure has the characteristic of eliminating the shift and methanation parts of the process scheme, and employs somewhat altered techniques for purification of the product methane. Since the majority of the processes employ shift, purification, and methanation steps which are sensibly independent of the nature of the gasification step, the nature of the gasifier becomes the major factor in distinguishing the various processes for the production of pipeline quality gas.

When steam gasification is used in conjunction with air or oxygen for process heat, an intermediate Btu gas can be produced by eliminating the shift and methanation sections of the plant process scheme of Figure 7.1. Intermediate Btu gas has a nominal heating value of about 300 Btu/scf, derived from the fact that the gas is a mixture of CO in hydrogen with some inert gases. Intermediate Btu gas has the interesting property of producing about 15% less flue gas/million Btu than is generated by the combustion of pure methane. Intermediate Btu gas, although not suited to long distance pipeline transmission because of its low heating value, has

ISBN 0-201-08300-0

substantial promise as an industrial fuel gas (IFG) in situations where distribution distances are limited in an industrial area of large consumptive capacity. It is clear that nearly any of the gasification techniques or processes based on steam or CO_2 gasification of carbon is capable of producing industrial fuel gas by employing the minimum gas purification steps necessary to reduce sulfur content to the levels required for industrial process consumption and to reduce CO_2 content to the optimal level for suitable product gas heating value. The smaller capital and operating costs associated with the production of industrial fuel gas make interesting the proposition for construction of a central industrial fuel gas production facility, with relatively short distribution lines to a heavily concentrated industrial consuming area. Operation of such a facility could provide manufacturers with an uninterruptible supply of fuel gas highly suitable to their requirements, and natural gas consumers would benefit from the reduced demand for natural gas and the automatic extension of limited natural gas reserves.

Since the principles of operation of fixed bed, fluid bed, and transport reactors were discussed earlier, little needs to be said in this regard with respect to the many processes. Most of the discussion to follow deals with process elements downstream of the gasifier, in which the raw product of the gasifier is upgraded to pipeline quality gas (SNG). Table 7.2 summarizes typical raw gas characteristics for 12 processes.

Table 7.2

High Btu Gasification Processes: Typical Gasifier Raw Product Characteristics[a]

	Product, mol %									Higher Heating Value	Gasifier	
Process	CO	CO_2	H_2	H_2O	CH_4	C_2H_6	H_2S	COS	N_2, Etc.	Btu/scf	Pressure, psia	Temp., °F
Lurgi	9.2	14.7	20.1	50.2	4.7	0.5	0.6	–	–	302	450	700+
Koppers-T	50.4	5.6	33.1	9.6	0	–	0.3	–	1.0	298	15	2750
Winkler	25.7	15.8	32.2	23.1	2.4	–	0.25	0.04	0.8	275	15	1500+
Hygas oxygen	18.0	18.5	22.8	24.4	14.1	0.5	0.9	–	0.8	374	1000	600
Hygas steam-iron	7.4	7.1	22.5	32.9	26.2	1.0	1.5	–	1.4	565	1000	600
Hygas elec.	21.3	14.4	24.2	17.1	19.9	0.8	1.3	–	1.0	437	1000	600
CO_2 Acceptor	14.1	5.5	44.6	17.1	17.3	0.37	0.03	–	1.0	440	150	1500
Synthane	10.5	18.2	17.5	37.1	15.4	0.5	0.3	–	0.5	405	1000	1800
Bi-Gas	22.9	7.3	12.7	48.0	8.1		0.7	–	0.3	378	1000	1700
Hydrane	3.9	–	22.9	–	73.2	–	–	–	–	826	1000	1475
Molten salt	26.0	10.3	34.8	22.6	5.8	–	0.2	–	0.3	329	1200	1700
Molten iron	69.7	–	9.6	–	20.0	–	–	–	0.7	457	15	2500

[a] Source: Bodle and Vyas [8].

ISBN 0-201-08300-0

Hot gas leaving the gasifier (600–2800°F) is laden with dusts, gas impurities, and, frequently, condensable tars, phenols, light hydrocarbons, and the like. Each process has its peculiar combination of these under its characteristic temperature and pressure of operation. Reference to Table 7.2 shows the broad range of conditions and raw gas compositions to be considered, which makes a truly detailed discussion of this topic beyond the scope of this book.

Since temperatures are high, it is clearly necessary to cool the raw gas for subsequent processing. This is usually done by a quench water spray into the raw gas, causing cooling by evaporation of water and collection of particulates by impingement on liquid surfaces. A popular device for this purpose is the venturi scrubber, in which water (or other liquid) is sprayed radially into the throat of a venturi through which the gas to be treated is passing axially at high velocity. The effluent mixture is passed through a separator, from which is withdrawn a stream of cooled particulate-free gas and a liquid slurry of particulate matter. If condensable components are present in the raw gas, the liquid slurry stream may contain these in solution or, more probably, as immiscible liquid phase droplets with an affinity for the combustible particulates. The latter condition sometimes causes serious difficulty in separation of two liquid phases when an "oil" phase, floating on an aqueous phase, preferentially wets ungasified hydrocarbonaceous solid matter to form a dense clod which sinks into the aqueous phase and plugs outlet lines.

Table 7.3

Some Coal Gasification Processes

Process	Capability, Btu gas product			Liquid Coproducts	Feed			Reactor Conditions		Gasifier Characterization					Commercial
	H	M	L		Lig/ Subb.	Bit.– No Caking	Bit.– Caking	t, °F max.	Pressure, psia	Heat Source	Circ. Solid	Reactor	Solid Residue	Raw Btu/ scf	
Lurgi standard	√	√	√	√	√	√	No	1800	400	O_2	–	Fix	Ash	302	√
Lurgi slag	√	√	√	√	√	√	No	3500	400	O_2	–	Fix	Slag	371	–
Winkler	√	√	√	–	√	√	√	1500	15	O_2	–	FLB	Ash	275	√
Bi-Gas	√	√	√	–	√	√	√	2700	1200	O_2	–	TR	Slag	378	–
Synthane	√	√	√	–	√	√	√	1800	1000	O_2	–	FLB	Char	405	–
Molten iron	√	√	√	–	√	√	√	2600	20	O_2	–	MLT	Slag	457	–
Hygas steam-oxygen	√	√	√	–	√	√	√	1800	1000	O_2	–	FLB	Ash	374	–
Hygas steam-iron	√	√	–	–	√	√	√	1800	1000	Air	–	FLB	Ash	565	–
Hygas elec.	√	√	–	–	√	√	√	1800	1000	EL	–	FLB	Char	437	–
Hydrane	√	√	–	–	√	√	√	1500	1000+	H_2	–	TR/FLB	Char	826	–
Koppers-Tzk	√	√	–	–	√	√	√	2750	20	O_2	–	TR	Slag	298	√
CO_2 Acceptor	√	√	–	–	√	No	No	1500	150	Air	Lime	FLB	Ash	440	–
Kellogg salt	√	√	–	–	√	√	√	1700	1200	O_2	–	MLT	Ash	329	–
Cogas	√	√	–	√	√	√	√	1600	75	Air	Char	FLB	Slag	300	–
Battelle	√	√	–	–	√	√	√	1800	1000	Air	Ash	FLB	Ash	300	–
Combustion Engineering	–	–	√	–	√	√	√	3500	20	Air	–	TR	Slag	127	–
Westinghouse	–	–	√	–	√	√	√	2000	180	Air	–	FLB	Ash	130	–
Atomics Intl.	–	–	√	–	√	√	√	1800	150	Air	–	MLT	Ash	150	–
U-Gas	–	–	√	–	√	√	√	1900	350	Air	–	FLB	Ash	150	–
Foster-Wheeler	–	–	√	–	√	√	√	2800	500	Air	–	TR	Slag	146	–

ISBN 0-201-08300-0

7.2. Carbon Monoxide Shift

Cooled, clean, humid gases can now be reheated and "shifted," if necessary, to adjust the H_2/CO ratio to 3+ for methanation. Some processes, such as the Hygas steam-iron and CO_2 Acceptor processes (see Table 7.2), provide a suitable ratio without a separate shift conversion. Adjustment of the ratio by shift conversion entails passage of part of the gas stream over a catalyst at about 550–750°F to effect the mildly exothermic reaction (Table 3.10)

$$CO + H_2O \rightleftharpoons H_2 + CO_2$$

The shift converter product is blended with the unshifted portion of the total gas product to obtain the desired 3 : 1 ratio of H_2/CO. Depending upon actual choice of catalyst, pressures for shift conversion may lie in the range from 200 to 1000 psi; lower pressures can be employed at the cost of reduced catalyst activity. For the Girdler catalyst G-93 it is said that activity increases with increasing pressure to a maximum at 300–400 psig. Phenol acts as a catalyst poison, but the effect is small.

Much debate has occurred with respect to shift conversion before or after acid gas removal. Gas shift before purification is used in some Lurgi gas processing systems to convert organic sulfur compounds in the gas to H_2S, presumably making it easier to remove sulfur from the product gases before methanation. When gasifier raw gas is sensibly free of higher hydrocarbons, it is questionable that shift before purification can be justified. One study of these alternative schemes indicates that shift after gas purification may cost about 8¢/million Btu less, with about 1¢ saved in the shift section and about 7¢ saved in the gas purification section. However, shift after gas purification implies the presence of large amounts of CO_2 which must be removed, generally after methanation to take advantage of the gas volume reduction. The commercial concepts for the Hygas steam-oxygen, Bi-Gas, and Synthane processes employ shift conversion before acid gas removal; no shift is required for the CO_2 Acceptor or the Hygas steam-iron process.

7.3. Gas Purification

Gasifiers operating on pure oxygen produce large amounts of CO_2 that must be removed in the gas purification section of the plant. This is so because oxygen is used to supply the endothermic heat of gasification and all process heat losses, a carbon consumption in excess of that for the overall ideal process requirement:

$$2C + 2H_2O \rightarrow CO_2 + CH_4$$

An oxygen-based gasification process thus produces somewhat in excess of 1 ft^3 CO_2 for each cubic foot of methane, and the gas purification system must remove essentially all the CO_2 from the total synthesis gas stream before methanation.

When air is used instead of oxygen to provide process heat requirements, a barrier is employed to prevent dilution of the product gas with nitrogen. Thus in

ISBN 0-201-08300-0

this case all carbon burned for process heat materializes as CO_2 in the flue gas, and is discarded directly to the atmosphere without adding to the gas purification task. Hence the gas purification systems for these processes have a lighter duty to perform, and this is reflected in lowered capital cost.

Many (if not most) processes employ gas purification after shift conversion, as suggested in the general high Btu gasification scheme of Figure 7.1. Gas purification means primarily the removal of H_2S and CO_2, but encompasses also the removal of certain minor constituents, some of which may be toxic: COS, NH_3, HCN, phenols, and some hydrocarbons. It is not desirable that the purification process remove hydrocarbons from the gas being treated, in part because there is a loss of product fuel value, but more importantly because these absorbed hydrocarbons tend to be emitted to the atmosphere with discarded CO_2 and become environmentally objectionable. Most of the minor components of the gas purification can (and should) be removed from the product stream to avoid poisoning the methanation catalyst and to produce a nontoxic substitute for natural gas. For processes making a raw gasifier product containing appreciable amounts of hydrocarbons heavier than methane, a "hydrocarbon wash," or absorption into a hydrocarbon solvent, is advised before acid gas removal is attempted. Sulfur is the most serious poison for the nickel catalysts used in the methanators. Its concentration must be reduced to levels below 1 ppm (part per million) in the gas purification step.

Many processes are available for removal of H_2S and CO_2. Only three will be presented here to characterize the process types.

A. BENFIELD

The proprietary Hot Carbonate process (Benfield Corporation, Berwyn, Pennsylvania) is a chemical absorption system based upon the reversible exothermic reaction

$$K_2CO_3 + CO_2 + H_2O \rightleftharpoons 2KHCO_3; \quad \Delta H = -13{,}500 \text{ Btu}$$

Although referred to as a chemical absorption system, it is restricted to relatively high pressure for absorption because CO_2 and particulary H_2S must be physically absorbed into the solution before it can react chemically. Undisclosed promoters are used to hasten absorption and reaction rates so that H_2S is much more rapidly absorbed than is CO_2. Benfield Corporation offers a version of its process which claims selective absorption of H_2S in preference to CO_2.

The Benfield Hot Carbonate process has been used extensively for bulk CO_2 removal in the manufacture of hydrogen, ammonia, and methanol and in natural gas treating plants. Undisclosed promoters are employed to enhance the CO_2 absorption rates and to control corrosion. In a typical application, feed gas (400 psig, 44.3% CO_2) enters the bottom of an absorption column while lean hot carbonate solution enters at the top. Treated gas leaves the absorber containing 1%

ISBN 0-201-08300-0

CO_2 on a dry gas basis. The rich absorber liquid leaves the bottom of the absorber through a hydraulic expansion turbine, recovering a part of the energy required for solution circulation. The rich solution enters the carbonate regenerator, where part of the CO_2 is removed by flashing and the rest by stripping with steam. Overhead vapor from the regenerator, saturated with steam, is cooled and dehumidified in an overhead condenser. Carbon dioxide is vented to the atmosphere, and condensate is pumped back to the regenerator to maintain the carbonate solution-water balance.

The Benfield process absorbs both CO_2 and H_2S, with little potential for controlling selectivity between them. Thus a stream containing a high ratio of CO_2 to H_2S is likely to produce a dilute H_2S stream to sulfur recovery, where at least 20% H_2S in the feed is required. Under such conditions the Benfield process is not likely to be suitable. It may also be unsuitable if the stream to be treated contains organic sulfur compounds such as carbonyl sulfide or light mercaptans. The process has proved highly successful, however, for removal of CO_2 in the absence of sulfur compounds as, for example, in the synthesis of ammonia from natural gas.

B. SELEXOL

Selexol is a physical gas absorption process employing a proprietary solvent patented by Allied Chemical Corporation. As a physical absorbent Selexol has several favorable features making it desirable for purification of gasifier raw product gas with or without light condensable hydrocarbons:

1. High solubility for RSH, H_2S, COS, and C_3+ hydrocarbons.
2. High selectivity for RSH, H_2S, and COS versus CO_2.
3. Low solubility for H_2.
4. Low vapor pressure and hence low solvent losses.
5. Low corrosiveness; carbon steel equipment is used.
6. Chemical and thermal stability; no solvent degradation loss.
7. Convenient temperature of operation: absorption 30–70°F; regeneration 30–70°F with air or inert gas stripping, or 200–300°F with steam stripping.
8. Low permissible operating pressures; units have operated successfully as low as 150 psig.

The Selexol process may be used for the following:

1. Simultaneous removal of H_2S and CO_2.
2. Partial removal of CO_2 with substantial removal of H_2S and organic sulfur components.
3. Selective removal of H_2S.
4. Removal of organic sulfur, ammonia, hydrogen cyanide, higher hydrocarbons, water, and other impurities from a coal gasifier.

ISBN 0-201-08300-0

In a typical application cool lean Selexol solvent enters the top of an absorber at 500 psia; the sour gas (10% CO_2, 1.5% H_2S, 100°F) to be treated enters at the bottom. Treated gas containing almost no H_2S and about 8–9% CO_2 leaves at the top. The rich solvent leaves the absorber bottom and enters a reboiled stripper for removal of H_2S. An overhead condenser provides reflux to the top tray of the reboiled stripper. The H_2S stripped thus is separated from condensed water and sent to sulfur recovery at about 20–40 mol % H_2S.

When the same case is considered with small amounts of COS added to the gas to be treated, substantial increases in solvent flow rates are required to remove the toxic COS to acceptably low levels. Increased solvent circulation leads to greater absorption of other gas constituents along with H_2S and COS, causing a need for additional equipment to reduce light fuel gas losses to the regenerator overhead. Additional equipment is also needed to enrich the sulfur in the absorbed acid gases, which contain more diluting CO_2 as a result of absorption of less soluble COS. The effect of COS is greater than just the addition of costly equipment and higher operating expense in gas purification: the concentration of H_2S in the gas to sulfur recovery is reduced substantially, causing increased costs for sulfur recovery. Selexol clearly suffers a disadvantage when carbonyl sulfide is present in the gas to be treated.

With Selexol, as with nearly all physical absorbents, the results depend upon the principal variables, temperature and pressure. The absorptive capacity of the solvent increases with decreasing temperature and with increasing pressure. But solvent viscosity increases as temperature is lowered, adversely affecting the kinetics of absoprtion. Increased pressure improves absorption kinetics as well as absorptive capacity, resulting in reduced solvent circulation rates for a given duty; but offsetting this gain is the increased cost of feed gas compression.

C. RECTISOL

Gas purification can be accomplished with the cold methanol physical absorption process, known commercially as the Rectisol process. Under its trade name it is a proprietary process developed by Lurgi and Linde, and is licensed in the United States by American Lurgi and by Lotepro. Rectisol plants built by Lurgi treat over 1 billion scf/day of raw gas.

The refrigerated solvent enters the absorber at −40°F, and rich solution leaves the absorber at about +15°F. Treated gas leaves the absorber at about 500 psia, containing about 1 ppm of total H_2S + COS + organic sulfur compounds. The rich solution is dropped from 500 psia through a hydraulic turbine for power recovery; part of the dissolved gases is removed by flashing, and the rest by stripping with methanol vapors in a reboiled stripper. Gas to sulfur recovery contains about 20 mol % H_2S and some methanol, but the process can be designed to produce up to

ISBN 0-201-08300-0

40 mol % H_2S in the gas to sulfur recovery. Treated gas contains minor concentrations of methanol which, because of the large volume of gas treated, constitute a significant loss of solvent.

Treated gas leaving the purification system contains sulfur compounds at about the 1 ppm level when purification is operating correctly. Because sulfur is a serious catalyst poison in the methanators, precaution must be taken against the accidental entry of sulfur into the methanation section. This is usually done by means of two chambers having fixed beds of zinc oxide, one in service while the other is down for recharging. The zinc oxide lowers the sulfur content of the syngas well below 1 ppm, and is expendable at incipient breakthrough of sulfur.

7.4. Sulfur Recovery

The acid gases removed as overhead product from the gas purification system contain H_2S and COS diluted with CO_2 and varying amounts (at low levels) of H_2, CO, and CH_4. The sulfur contained in these gases is recovered by the Claus process, which is based on two consecutive exothermic reactions for conversion of H_2S at 400+°F:

$$(1)\ H_2S + 3/2O_2 \rightarrow H_2O + SO_2$$

$$(2)\ 2H_2S + SO_2 \rightarrow 2H_2O + 3S$$

where the second reaction is catalyzed with bauxite. Carbonyl sulfide undergoes similar reactions. The heat of reaction is absorbed by generating steam. Tail gases leaving the Claus unit must be treated for removal of sulfur to environmentally acceptable levels.

7.5. Methanation

Methanation is the final synthesis step used to upgrade the purified syngas to pipeline quality gas. Considerable research and development has been carried out by coal gasification process developers, catalyst manufacturers, and various research agencies to develop reliable and economic methanation schemes for commercial plant design.

In earlier pages of this chapter the calculation of methanation equilibria and the thermodynamic potential for carbon deposition were discussed. However, in the methanator several different reactions can occur between hydrogen and carbon oxides, some producing small amounts of oxygen-bearing organic substances such as alcohols, aldehydes, and ketones. The more important reactions are given below, accompanied by the heat of reaction (expressed in Btu/lb-mole and in kcal/g-mole):

ISBN 0-201-08300-0

					Btu/lb-mole	kcal/g-mole
(1)	$CO(g) + 3H_2(g)$	$\rightarrow$	$H_2O(g) + CH_4(g)$		− 88,700	−49.28
(2)	$CO_2(g) + 4H_2(g)$	$\rightarrow$	$2H_2O(g) + CH_4(g)$		− 71,000	−39.44
(3)	$CO(g) + H_2O(g)$	$\rightarrow$	$CO_2(g) + H_2(g)$		− 17,700	− 9.83
(4)	$2CO(g) + 2H_2(g)$	$\rightarrow$	$CO_2(g) + CH_4(g)$		−106,400	−59.11
(5)	$2CO(g)$	$\rightarrow$	$C(s) + CO_2(g)$		− 74,200	−41.22

Methanation reactions are highly exothermic, requiring efficient heat removal as a major consideration in the development of a new methanation scheme. Heat is removed either from the catalyst or from the methanator effluent, or from both. Reactions 1 and 2 are the preferred methanation reactions, and it is desirable to inhibit reactions 3–5. Reactions 3 and 4 will produce undersirable CO_2, which will reduce the heating value of the pipeline gas product or will require further CO_2 removal in a final gas purification step. Reaction 5 is the Boudouard reaction; this produces carbon, which in turn deactivates the catalyst and tends to cause reactor plugging. By using feed gas with hydrogen/CO ratios above 3.0, reactions 1 and 2 take precedence over the other reactions under normal operating conditions.

Carbon deposition in accordance with reaction 5 purports to lay down a deposit of graphite; it is known, however, that carbon deposits actually found may have properties substantially different from those of graphite, as mentioned earlier in this chapter. Such deposits are referred to as nonideal carbon and are sometimes called Dent carbon after F. J. Dent, who discovered this form of carbon in his researches [15, 16]. Dent describes experiments at 1 and 20 atm on the decomposition of CO and CH_4, where the gas after carbon deposition assumed a final composition that was higher in CH_4 and CO than was predicted by graphite thermodynamics. This led to the concept of a nonideal carbon and to further investigation on the form and composition of the solid phase. Pursley et al. [77] later confirmed qualitatively the formation of Dent carbon. Rostrup-Nielson [81] reported experimental data on CO and CH_4 decomposition over nickel and cobalt catalysts at atmospheric pressure. The equilibrium was approached from both directions: precipitation of carbon from a saturated gas, and vaporization of the carbon into an unsaturated gas. His data indicate that the free energies of the carbon deposits can be correlated to the crystallite size of the catalyst. He used nickel and cobalt catalysts, and obtained results similar to those of Dent. Both carbon deposit forms were temperature dependent.

White et al. [101] provide a good discussion of carbon formation zones in methanators in terms of the C-H-O ternary diagram at pressures to 400 psia. Nonideal carbon formation boundaries are shown as isotherms from 600 to 2700°F. Whalen [100] has made a detailed study of the formation of graphite and of nonideal carbon in methanators under all conditions pertinent to methanation processes. He concludes that for practical purposes the carbon formation boundary is far more sensitive to the atomic fraction of carbon than it is to temperature,

ISBN 0-201-08300-0

pressure, or the atomic fractions of oxygen and hydrogen. For methanator design Whalen suggests choosing an atomic fraction of carbon as a fraction of the equilibrium atomic fraction of carbon at the formation boundary, at the same temperature, pressure, and hydrogen/oxygen ratio.

Figure 7.2 shows the graphite and nonideal carbon deposition boundaries as reported by Whalen at 600°F and for pressures from atmospheric to 1000 psia. Carbon deposition is controlled primarily by the decomposition of CO and CH_4.

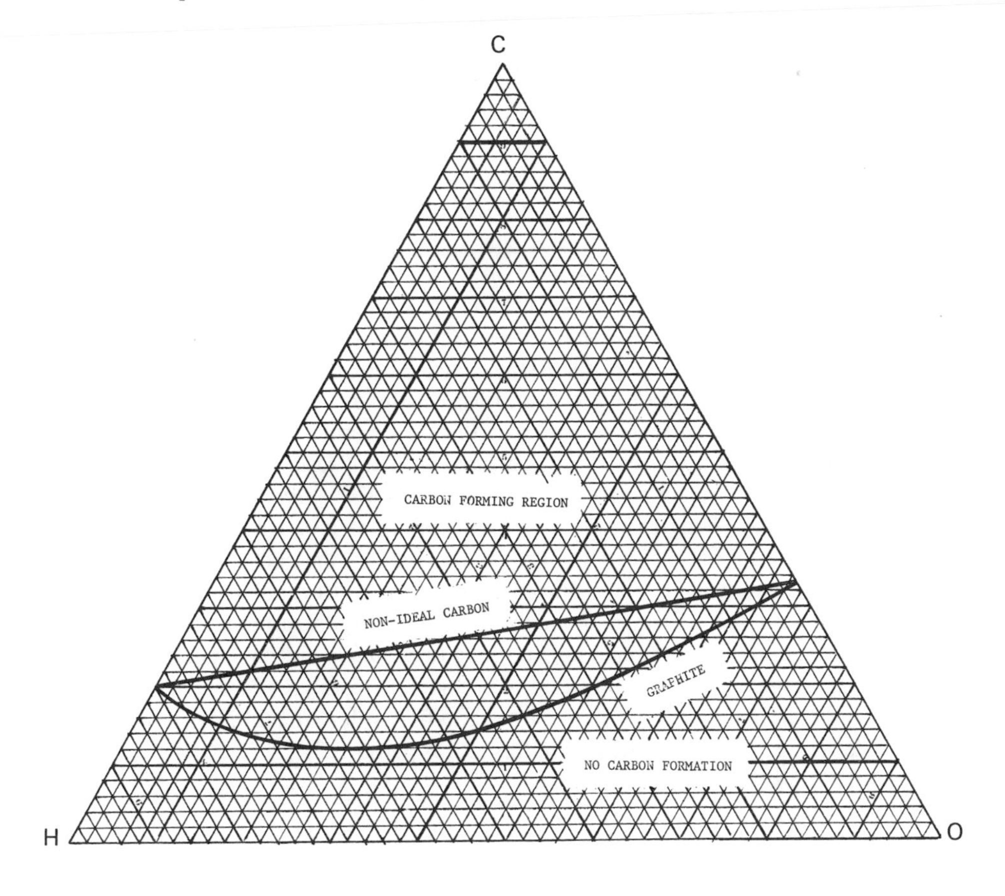

Figure 7.2. Atomic carbon-hydrogen-oxygen fraction: comparison of nonideal and graphite boundaries at 600°F, 14.7–1000 psia.

ISBN 0-201-08300-0

Carbon monoxide decomposition is favored by low temperature, and CH_4 decomposition by high temperature. Increasing pressure suppresses decomposition of CH_4 and promotes decomposition of CO. The low temperature graphite isotherm of Figure 7.2 exhibits a minimum between the terminal points for pure CH_4 and pure

CO_2, indicating that mixtures of these two pure components are less stable than are the pure components themselves. The effect of water or hydrogen addition is to raise stability, but CO_2 addition lowers stability. But the nonideal carbon formation isotherm is sensibly a straight line on Figure 7.2, indicating that all mixtures of the two gases CH_4 and CO_2 remain stable below 600°F.

Methanation has long been practiced as a means of eliminating carbon oxides from hydrogen streams to be used for conventional ammonia and hydrogen gas plants. Such processes characteristically convert only minor amounts of carbon oxides in the presence of overwhelming quantities of hydrogen, while employing a mild-activity nickel catalyst. The heat capacity of the excess hydrogen is adequate to maintain minor temperature increases caused by the exothermic conversion of CO to methane. Operating conditions for methanation units in coal gasification plants are different from those experienced in conventional hydrogen manufacturing gas plants, primarily in the higher CO/H_2 ratio in the methanator feed. Because of the low activity of most of the conventional commercial catalysts, a large catalyst volume is required in a coal gasification methanation reactor, entailing increased equipment and operating costs for a coal gasification plant. More suitable methanation catalysts for coal gasification are being sought. Nickel supported on alumina, silica, kieselguhr, and the like has been studied extensively for methanation catalytic activity. Hausberger et al. [36], in the development of methanation catalysts for SNG processes, have shown that (1) methanation activity increases with increasing nickel content, but decreases with increasing particle size; (2) increased steam/dry gas ratio in the feed increases CO shift conversion without affecting the rate of methanation; (3) trace impurities in the process gas, such as H_2S and HC1, poison the catalyst (sulfur remains on the catalyst; chlorine does not); (4) hydrocarbons in low concentration do not affect methanation activity significantly but are reformed to methane; and (5) hydrocarbons at higher concentration levels inhibit methanation and tend to promote carbon deposition.

Consistently with over 30 years of commercial catalyst practice, Hausberger and co-workers adopted a very simple psuedo-first-order equation for preliminary converter design (great changes of conditions not permitted):

$$Kw = (SVW) \log_{10} \left(\frac{X_{in} - X_{eq}}{X_{out} - X_{eq}} \right) = (SVW) \log_{10} A$$

where Kw = rate constant at specific pressure and temperature

SVW = total wet gas space velocity, scf/hr ft³ bulk catalyst

X = constituent, lb-mol/hr

- X_{in} = entering section of catalyst, for which space velocity is measured
- X_{out} = leaving the section of catalyst for which space velocity is measured
- X_{eq} = passing through the same section under equilibrium conditions

ISBN 0-201-08300-0

For reactions with high heats it is necessary to divide the catalyst bed into a number of sections so that each section is essentially isothermal. This equation is the specific solution of

$$\frac{-dX}{dV} = (6.07)(10^{-3})\ (Kw)\left(\frac{X}{\text{gas flow}} - \frac{X_{eq}}{\text{gas flow}_{eq}}\right)$$

The effect of temperature on *Kw* was introduced through an activation energy term, and showed the normal pattern of very high activation energy below the threshold temperature and lower activation energy at higher temperature levels when methanating CO. Hausberger et al. thus found and reported the following:

Average t, °C	Kw	Activation Energy, cal/g-mol
160	170	–
183	710	24,000
198	6,020	61,000
213	11,850	21,000
276	24,300	6,100

At low pressure and temperature activation energies are generally 15,000–30,000 cal/g-mol; at higher pressure and temperature, using commercial catalysts of 1/8–1/4 in. particle size, the values fall to the range 0–10,000 cal/g-mol.

Bridger and Woodward [9] discuss the formulation and operation of nickel methanation catalysts. They point out that high activity is associated with high nickel surface area, and stability is obtained by maximizing strength and resistance to sintering. They report as follows on the poisoning effects of the gas purification systems:

Process	Chemicals	Effect
Benfield	K_2CO_3(aq)	Blocks pores with K_2CO_3
Vetrocoke	K_2CO_3–As_2O_3(aq)	Blocks pores with K_2CO_3; As_2O_3 also poisons
Benfield DEA	K_2CO_3(aq) + 3% diethanolamine	Blocks pores with K_2CO_3; DEA is harmless
Sulfinol	Sulfolane, water, di-2-propanolamine	Sulfolane decomposes to give sulfur poisoning
MEA, DEA	Mono- or diethanolamine(aq)	No poisoning effect
Rectisol	Methanol	No poisoning effect

ISBN 0-201-08300-0

Bridger and Woodward also report a graphical technique for measurement of the effective position of complete conversion in the fixed catalyst bed. The method depends on very accurate temperature measurement and is useful for predicting the future useful life of the catalyst bed.

Haynes et al. [38] have reported their pilot plant studies of Raney nickel catalysis using coated parallel plates and space velocity to 3000 hr^{-1}. Carbon monoxide concentrations in the product ranged from below 0.1 to 4% and catalyst service reached 32,000 scf methane/lb Raney nickel during 2307 hr of operation. Carbon and iron deposition and nickel carbide formation were suspected causes of catalyst deactivation.

Bair et al. [5] reported on 561 total hr of methanation in the Hygas pilot plant. No significant deterioration was observed in the service behavior of the supported nickel catalyst operating at 3200–4400 scf/hr ft^3 catalyst upon an essentially sulfur-free feed gas. Hydrogen/carbon monoxide ratios ranged from 2.9 to 16.8.

Eisenlohr and Moeller [19] examined methanation experience in two semi-commercial pilot plants, one a Lurgi coal-to-SNG plant (Sasolburg, South Africa) and the other a naphtha-to-SNG plant (Vienna, Austria). In addition to long term test runs of 4000 and 5000 hr that established a catalyst service life in excess of 16,000 hr in a commercial plant, numerous tests demonstrated the effects of operating temperature and of steam in reactor feed gas on catalyst activity and catalyst deactivation. Effects of feed gas composition, CO_2 content, trace components, and catalyst poisons are reported. All tests were performed with a special methanation catalyst developed by BASF, Ludwigshafen, for the process. There was no sign of carbon deposition, as pressure drop remained constant and the discharged used catalyst carbon content was lower than the amount of carbon added to the catalyst as a pelletizing aid. The authors report that variation of the H_2/CO ratio will not affect the operability of an SNG plant using a recycle system for methanation, as demonstrated in the SASOL plant.

Blum et al. [7] reported on their liquid phase methanation process development, employing an inert fluidizing liquid for attainment of nearly isothermal methanation. Upflowing liquid fluidizes the solid-particulate-supported nickel catalyst, and feed gas flows upward cocurrently. The exothermic reaction heat is taken up by the liquid mainly as sensible heat and partially by vaporization. Product gases are cooled to condense water and recover vaporized liquid. A separate heat exchanger is used for cooling the liquid medium in a loop external to the reactor. Catalyst particle size was reduced, in a series of steps, from 1/8–3/16 to 1/32 in. in an attempt to improve catalyst utilization. The authors hope to attain a vapor hourly space velocity (VHSV) of 4000 hr^{-1} at 1000 psig and 343°C with a feed gas containing 20% CO, 60% H_2, and 20% CH_4. They predict for this that CO conversion will be 95–98%. Data from the process development unit confirm the first-order reaction rate model proposed as a result of earlier bench-scale work.

A bewildering array of methanation process schemes have been proposed and studied, primarily to deal with the large exothermic heat of reaction and the

ISBN 0-201-08300-0

attendant problems of temperature control. Good temperature control is necessary because most catalysts tend to deactivate more rapidly as temperature levels are increased. It is beyond the scope of this book to deal in detail with such a large variety of detailed process schemes. The reader is referred to Sirohi's report [88] describing 10 distinct methanation processes, differentiated from one another primarily by the configuration for temperature control.

7.6. Dehydration

After cooling the methanator product stream and separating out liquid water, the gas remains saturated with water vapor. Dehydration of the methane product is required to prevent solid hydrates of methane from forming in the distribution system. Condensation of water must also be avoided, since solids or liquids in a gas distribution system can cause major problems in control instrument response, compressor performance, and simple blockage of lines. Various hydrates can form, depending upon the availability of the candidate components:

Methane	$CH_4 \cdot 7H_2O$
Ethane	$C_2H_6 \cdot 8H_2O$
Propane	$C_3H_8 \cdot 18H_2O$
Carbon dioxide	$CO_2 \cdot 7H_2O$

Butanes and hydrogen sulfide also form some hydrates. All of the above are solids and have crystalline structures.

The primary conditions for hydrate formation are as follows: (1) the gas at or below its water dew point with "free" water present; (2) low temperature; (3) high pressure. Secondary conditions promoting hydrate formation are (a) high velocities; (b) pulsating pressure; (c) any type of agitation; and (d) introduction of a small crystal (seed) of the hydrate. The hydrates are all phase change products (like vapor-liquid) for which calculable equilibrium relationships exist. Thus, given a gas composition, the pressure necessary to cause hydrate formation at any temperature can be calculated; or, given a pressure and a lowest temperature of exposure (as in a transportation or distribution pipeline), the maximum water content of the gas can be calculated that will maintain a water dew point below the lowest temperature the system will encounter. Thus are the specifications set for dehydration of the product gas from the methanation section of a plant producing SNG.

Available techniques for dehydration include (1) absorption using liquid desiccants; (2) absorption using solid desiccants; (3) inhibition by injection of hydrate point depressants; (4) dehydration by expansion refrigeration. Space limitations do not permit detailed treatment of any of these methods. Economic considerations generally favor the first method, in which triethylene glycol (TEG), diethylene glycol (DEG), or ethylene glycol (EG) is the working fluid.

ISBN 0-201-08300-0

The essential components of a glycol dehydrator are an inlet gas scrubber, a glycol-gas contactor, a glycol regenerator, glycol heat exchangers, a filter, and a glycol pump. Process flow is simple. Cool, regenerated glycol is pumped to the top tray of the contactor (absorber) to contact gas countercurrently at system pressure. Water-rich glycol is removed from the bottom of the absorber through a glycol-glycol heat exchanger to the regenerator. The regenerator, a reboiled stripper, removes water overhead by operation at 1 atm pressure, and takes hot, water-lean glycol as a bottom product to the glycol-glycol heat exchanger. The cooled, regenerated glycol is pumped to the top tray of the absorber to complete the cycle.

Glycol circulation rates vary from about 2 to 5 gal glycol/lb water to be removed, depending to some extent on the number of trays in the absorber. Absorber tray efficiencies generally are about 25%. Glycols can absorb about 1 ft^3 methane/gal at 1000 psig absorber pressure, requiring venting of the regenerator overhead to a safe area to avoid fire and personnel hazards. Regenerator temperatures range from 375 to 400°F. Glycol losses range from 0.05 gal/million ft^3 of gas treated (for low temperatures) to as much as 0.30 gal/million scf for lower pressure, higher temperature gases. Water in the product gas is typically limited to 7 lb/million scf for high pressure, low temperature gas.

8. MATERIALS OF CONSTRUCTION

Coal gasification conditions are severe: high temperature, high pressure, corrosive gases, corrosive liquids, and erosive solids are present in one or another part of the plant, and frequently occur in combinations that are devastating to common materials of construction. Not only are the raw gasifier gaseous components themselves severely corrosive, but also they are frequently accompanied by high velocity particulates whose erosive behavior can inhibit or prevent passivation of metal surfaces by corrosion products. The synergism of gas, liquid, and solid phases with respect to corrosive and erosive attack on metals, ceramics, and refractories constitutes a major challenge to commercial coal gasification at tolerable cost.

Testing programs have been undertaken at numerous laboratories with generally discouraging results [1, 66, 84]. A detailed summary of the extensive programs is beyond the scope of this chapter, but a few general statements can summarize some of what is known:

1. Gasifier atmospheres are severely corrosive to all but the high nickel-high chromium alloys.

2. Pilot plant combustor-regenerator atmospheres are highly corrosive to all but the aluminized high chromium-nickel alloys.

ISBN 0-201-08300-0

3. Abrasion resistance in virgin refractory materials increases with decreasing porosity.

4. Changes in abrasion resistance due to exposure to coal gasification conditions increase with increasing initial porosity of refractories.

5. Aqueous-gaseous interface corrosion is severe and is promoted linearly by the concentrations of CO and CO_2, but the effect of phenol is inhibitory. Only welded U-bends of 410 SS showed significant stress corrosion cracking; copper alloys showed very high corrosion rates; stainless steels, hastelloys, high nickel alloys, and titanium were the most corrosion resistant alloys. Temperatures have the greatest effect on corrosion rates; protective corrosion products are suggested as the concentrations of CO_2, HCN, and chlorine increase; pH, pressure, and the concentrations of hydrogen and light aromatics appear to increase corrosion rates; ammonia at low concentrations appears to have little effect. Corrosion rates appear higher in the liquid than in the gaseous phase of some compositions; all alloys pitted to some extent. Carbon monoxide has a strong positive corrosion effect on carbon steel; hydrogen sulfide is somewhat inhibitory to CO-promoted corrosion.

6. Particulate erosion-corrosion of stainless steels is strongly aided by the formation of low melting eutectics of metal-metal sulfide (Fe, Co, Ni) to allow continuous exposure of new material to further erosion-corrosion; but *aluminized* stainless 310 and IN 800 (47 Fe, 31 Ni, 21 Cr) were well protected from this slagging corrosion.

A very large body of information has been generated, but it is clear that a great deal more work is needed to develop the data required for the design of safe, long lasting plant vessels, piping, reactors, and supporting structures.

9. ENVIRONMENTAL CONSIDERATIONS

To produce 250 million scf/day of pipeline quality gas, about 15,000 tons/day of coal must be processed. The magnitude of this operation gives rise to special environmental difficulties, for in addition to basic coal conversion, additional facilities are needed for a self-sufficient plant. These must be included and considered in the environmental picture. For example, utilities such as steam, water, and electric power must be supplied as part of the overall project. Auxiliary facilities include a utility boiler, cooling tower, wastewater treating, makeup water treating, a sulfur plant, and frequently an oxygen plant. In considering the potential environmental impact of such a plant, the following must be considered: gas emission, liquid effluents, solids disposal, water consumption, trace elements, thermal effluent, odor and taste, noise, visual impact, and land use. Specific areas of concern are (1) solids disposal: refuse from coal cleaning plus ash, and sludge from water treating; (2) emissions to air: those from furnaces, coal dryers, and cooling towers, sulfur recovery plant tail gas, and carbon dioxide from acid gas removal;

ISBN 0-201-08300-0

(3) water effluent: gas liquor (phenols, sulfur, and nitrogen; total dissolved solids), and rain runoff from coal storage; (4) trace metals: volatiles from the gas clean-up system, from leaching of refuse and ash, and from oil products.

Some perspective on the magnitude of the environmental problems can be gained by consideration of the following [51]. The cleaning of 15,000 tons/day of coal produces refuse amounting to approximately 3000 acre-ft/year, enough to cover an area of 3000 acres 1 ft deep for each year the plant operates. From gasifier product gas quenching operations and the like, a total gas liquor stream amounting to about 16,000 tons/day is generated. This includes unreacted water or steam from the gasification and shift reactions, together with ammonia, phenols, sulfur compounds, cyanides, and thiocyanates which are side products formed in the gasification reactions. In addition it is known that a large number of elements are volatile under gasification conditions, and many of these will appear in the gas liquor.

The carbon dioxide vent stream is very large, amounting to about 14,000 tons/day. It is clearly important that the hydrocarbon and sulfur contents of this huge effluent stream be held very low to avoid contamination of a community. If only 0.1% of hydrogen sulfide remained in the CO_2 effluent stream, about 14 tons/day of sulfur would be emitted to the surrounding community. While this amount may appear not to be very large when distributed over a large area, it is true that carbonyl sulfide is formed in considerable proportion to H_2S in the gasifier, and is more difficult to remove in gas treating operations. Carbonyl sulfide is extremely toxic and would be a most undesirable constituent for release to the ambient air of any community.

The sulfur recovery plant in this typical case contributes somewhat more than 3000 tons/day of tail gas, while the flue gas from the utility boiler amounts to about 32,800 tons/day, more than 10 times as much. By far the largest effluent to the air, however, is from the cooling tower. At first thought this would appear to pose little if any difficulty; however, drift losses or mists of water droplets carried out with the air can settle on equipment or roads and cause deposits or icing in the winter time. Moisture rising from a water cooling tower can cause a plume which can present a fog problem if it affects public highways. While it may be expected that cooling water should be clean, experience shows that it can be contaminated with acidic components, oily substances, and the like. Finally, effluent water from a wastewater treating operation is likely to amount to about 6000 tons/day and obviously needs to be cleaned thoroughly if it is to be discharged to a river. Even if cleaned to the point of no biological oxygen demand, the effluent water may require treatment to reduce dissolved solids to an acceptable level.

Jahnig and Bertrand have summarized the environmental aspects of coal gasification [51]. Tables 9.1–9.3 are taken from their paper and are presented here to help round out the picture of the nature and magnitude of the environmental problems associated with a full-scale commercial coal gasification operation.

ISBN 0-201-08300-0

Massey has reported on the status of effluent characterization and regulation in coal gasification [65]. He points out that measured effluent *production* data on all phases of process effluent production (gas, liquid, and solids) are extremely limited, that the effluent production data which are available were collected primarily on small-scale equipment and, although valuable, must be interpreted with great care, and that projection of potential process *emissions* is complicated both by weaknesses in the effluent production data base and by lack of actual operating experience with existing treatment technologies on coal gasification effluent streams.

Table 9.1

Gas Emissions[a]

Stream	Million scf/day	Relative Volume
Synthetic natural gas	250	1.0
Coal dryer vent gas	100	0.4
CO_2 from acid gas removal	275	1.1
N_2 from oxygen plant	425	1.7
Claus plant tail gas	85	0.3
Flue gas from utility furnace	770	3.1
Air from cooling tower	55,000	220.0
H_2O evaporated in cooling tower	1,270	5.1

[a] Source: Jahnig and Bertrand [51].

Massey and co-workers were active in measuring effluents from the Hygas and the CO_2 Acceptor pilot plants [63, 64]. Measurements were made on the Synthane 4 in. reactor, where it was demonstrated that production of phenols could be drastically reduced by converting the feed point from free fall above the fluidized bed in the gasifier to deep injection below the surface of the fluidized bed [70]. Observations made at the CO_2 Acceptor pilot plant confirm that deep feeding below the surface of a fluidized bed results in negligible amounts of condensable hydrocarbons in the raw gas produced. These observations strongly suggest that deep feeding of fresh feed material permits adequate vapor residence time to

ISBN 0-201-08300-0

accomplish steam reforming of condensable substances, thereby minimizing the production of objectionable constituents that must be removed by gas and liquid treating operations.

Table 9.2

Some Contaminants in Sour Water, ppm[a]

Item	Gas Liquor	To Biox[d]	From Biox[d]
Phenol	2,000–4,000	100–500	0.1–0.3
Fatty acids	–	500–1500	9
Ammonia	8,000–11,000	200–1000	5–10
Thiocyanates	5–10	1	0.1
Fluoride	–	56	6
BOD[b]	–	2500	75
COD[c]	10,000–20,000	1100	82

[a] Source: Jahnig and Bertrand [51].
[b] BOD = biological oxygen demand.
[c] COD = chemical oxygen demand.
[d] Biox is biochemical oxidation.

ISBN 0-201-08300-0

Table 9.3

Estimated Volatility of Trace Elements[a]

Element	Typical Coal, ppm	% Volatile[b]	Amount Volatile, lb/day[c]
Cl	1500	90+	32,400
Hg	0.3	90+	6
Se	1.7	74	30
As	9.6	65	150
Pb	5.9	63	89
Cd	0.8	62	12
Sb	0.2	33	2
V	33	30	238
Ni	12	24	69
Be	0.9	18	4
Zn	44	e.g., 10	106
B	165	e.g., 10	396
F	85	e.g., 10	204
Cr	15	Nil	Nil

[a] Source: Jahnig and Bertrand [51].

[b] Volatility based mainly on gasification experiments [83], but chlorine taken from combustion tests, and zinc, boron, and fluorine taken at 10% for illustration in absence of data.

[c] Estimated amount volatile for 12,000 tons/day of coal to gasification.

ISBN 0-201-08300-0

Table 9.4

Summary of Observed Process Liquid Effluent Characteristics[a]

Component	Measured Effluent Production, lb/ton coal, maf		
	Synthane Process Development Unit[b]	Hygas Pilot Plant[c]	By-Product Coke Manufacture
Tar	74.1 ± 27	Negligible	∿93
Light oils	NA	NA	∿33
Phenols	11.9 ± 1.3	11.4 ± 2.4	0.9–1.0
Ammonia	19.5 ± 3.0	13.1 ± 0.3	∿8.5
Total organic carbon	22.0 ± 3.3	39.1 ± 15.4	1.6–2.0
Chemical oxygen demand	77.7 ± 14.4	NA	4.0–5.5
Aqueous cyanide	Negligible	Negligible	0.02–0.05
Aqueous thiocyanate	0.05 ± 0.08	2.5 ± 0.2	0.3–0.4
Component	**Production, tons/day[d]**		
Tar	556	–	698
Light oils	–	–	248
Phenols	89	86	7
Ammonia	146	98	64
Total organic carbon	165	293	12
Chemical oxygen demand	583	–	36
Aqueous cyanide	–	–	0.3
Aqueous thiocyanate	0.5	19	2.6
Sulfur	∿450	∿450	∿225

[a] Source: Massey [65].

[b] Five 25 lb/hr tests on North Dakota lignite.

[c] One 2 ton/hr test on Montana lignite.

[d] Basis: plant feed 15,000 tons/day, 3% sulfur.

ISBN 0-201-08300-0

Measured effluent production data are meager indeed. Table 9.4, taken from Massey's report to the ERDA General Technology Advisory Committee [65], reports essentially the only measured information available at this writing. The upper half of Table 9.4 presents his data measured in pounds per ton of maf coal, while the lower half of the table gives the same figures converted to tons per day for a plant feeding on 15,000 tons/day of coal bearing 3% sulfur. Data reported for the Synthane process are the result of five tests at the rate of 25 lb/hr while gasifying North Dakota lignite in the Synthane 4 in. reactor. Data for the Hygas pilot plant were obtained from a test on Montana lignite feeding at the rate of 2 tons/hr. By-product coke manufacturing data are provided in the last column for comparison, since this represents a current commercial operation. Although one must be cautious in using these data because of the variation in reliability caused by the scale of operation, the indications appear to be clear. It is evident from Table 9.4 that considerable process improvement will be required to minimize, if not eliminate, liquid effluent problems. There remain a great need for many more data of higher quality, and a great need for more attention to the environmental aspects of coal gasification.

10. PROSPECTS FOR COAL GASIFICATION

Production of high Btu gas from coal is indeed a complex and expensive prospect. Obstacles to commercial coal gasification range from technical to economic to political in nature. Ample provision must be scheduled for procurement of permits from governmental bodies such as the Federal Power Commission and multilevels of local, state, and federal regulatory, environmental, and other affected agencies. Further delays can arise from actions of various citizens' groups concerned with environment and local ways of life, as well as from landowners in the affected region.

If all political resistance could be resolved, commercial coal gasification would still face the inherent deterrent of *cost,* which follows directly from the common nature of nearly all known processes. Readers should note (Figure 7.1) that SNG processes carry out a highly endothermic reaction (gasification) at high temperature, purify the product gas after cooling, and then carry out a highly exothermic reaction (methanation) at intermediate temperature levels. Because real-world economics requires it, methanation heat must be recovered and in effect pumped uphill for use in the process. This is analogous to a refrigerator operating at enormous scale and at very high temperature.

Since all processes employ this concept in one way or another, all must suffer its economic consequences. These are embodied in the large amount of

ISBN 0-201-08300-0

expensive capital equipment required and the associated operating expense implied. Analysis of coal gasification capital cost distribution shows less than 10% of total capital cost in any one part of the process except gas purification (16%). General facilities and steam and utility systems taken together account for about 30% of total capital cost. It is evident that even total elimination of the cost of any process element is incapable of making large reductions in product cost.

Projections of U.S. requirements for natural gas produced by various agencies, including the Institute of Gas Technology, indicate a deficiency of supply growing approximately linearly from 1975 to 2000 if no SNG plants are built. The deficit would grow to about 20 trillion scf per year by 2000, creating a demand for SNG of, say, 1 trillion scf of new capacity each year beginning in 1980. One plant producing 250 million scf/day and on-stream 90% of the time would produce 0.082 trillion scf/year. It thus would take 12 new plants per year to satisfy the demand.

Could it be done? Probably not, starting before about 1986, even with a proven, commercially accepted process because lead time for engineering design, procurement, construction, mine development, and water supply development dictates an interval of 8–10 years from commitment to production. Other factors can further limit the attainable rate of construction of U.S. coal gasification facilities. Some examples are the national ability to finance at this rate, the national steel making capacity in the face of competing demands, the national metal fabrication capacity, and the national availability of managerial, operator, and maintenance skills of all kinds at all levels. There are many more possible bottlenecks, making it appear improbable that the United States will build 12 SNG coal gasification plants per year. An analysis of the nation's capacity to meet its natural gas and other energy requirements to the year 2000 would be very interesting indeed.

While these speculations cast suspicion on the suitability of currently known processes for the future needs of the country, they also suggest the strong incentive that exists for innovative thought and the probable economic value of a successful search for better alternatives. Among the possibilities deserving attention is the prospect of deashing the coal before conversion or combustion, a physical rather than chemical process to extract pure hydrocarbonaceous material free of mineral matter. Where production of methane from coal and water is the necessary objective, a catalytic process is indicated to take maximum advantage of the thermochemical near-neutrality of the reaction $2C + 2H_2O \rightarrow CO_2 + CH_4$ at or near ambient temperature. Where on-site combustion near a coal supply is required, means should be sought to burn the coal with minimum prior treatment. No prior treatment would be needed if an effective and inexpensive flue gas clean-up process were available. Expensive chemical conversion of coal to a fuel for combustion should be avoided in favor of inexpensive direct combustion processes wherever possible.

It is said that the United States possesses more energy reserves in the form of coal than the Arab world possesses in the form of petroleum. We need efficient, innovative ways to take advantage of this natural wealth.

ISBN 0-201-08300-0

ACKNOWLEDGMENTS

The author acknowledges with gratitude the support of Consolidated Natural Gas Company in the preparation of this manuscript. Special thanks are due to Mrs. Marilyn Grabell, who skillfully and patiently performed the extensive typing duties, and to Messrs. J. C. Janik, Wayne Clark, and Daniel Vatavuk of The East Ohio Gas Company, who prepared all the graphics for this chapter. Additional thanks are due to Miss Deborah Gaetano and Mrs. Julie Rikon for their last-minute assistance with manuscript preparation.

PROBLEMS

1. For a fluidized bed gasifier, circulation of char is used to transfer heat from combustor flue gas (see Figure 5.1). Char has a true solid density of 90 lb/ft^3 and a bulk density of 23.6 lb/ft^3. The specific heat of char = 0.3; the temperature rise in circulating char is 150°F. The system operates at 50 psig, and the char is heated to 1800°F before leaving the combustor. The heat of combustion of char is ΔH_c = −94.3 kcal/g-mol; the heat of steam gasification of char is ΔH_R = +32.4 kcal/g-mol. Assuming that there are no side reactions, and assuming 100% removal of carbon dioxide and water from the product methane, (a) calculate the gross heating value and show the nitrogen content of the product methane; (b) repeat for circulation of impervious particles of quartz: specific heat = 0.3, true density = 164.7 lb/ft^3, bulk density = 102.1 lb/ft^3.

2. Coal is fed to a gasification process at the rate of 15,000 tons/stream day through lockhoppers having an L/D ratio of 2.0 and operating on a 1 hr cycle. Coal has a bulk density of 50 lb/ft^3 and a true density of 81.1 lb/ft^3. The lock-hoppers operate at 100°F and have 25% excess space over coal volume handled. Assume that the gas compressed to the lockhoppers has a ratio of specific heats K = 1.395, and that the gas behaves ideally, entering the compressor at 100°F and 1.0 atm. Compute the power consumption for gas compression if the process must operate at (a) 100 psia; (b) 500 psia. Propose a way to reduce power consumption, if possible.

3. Calculate the equilibrium composition for the CO shift conversion at 700°K for the following ratios of steam to CO in the feed: (a) 0.75; (b) 1.00; (c) 1.50; (d) 2.0. Plot percent conversion of CO as a function of feed consumption. What is the effect of pressure on equilibrium conversion?

4. Using the data of Table 3.10, plot heat of reaction as a function of temperature. Evaluate and plot the coefficient $d(\Delta H)/dT$ at 298.16, 400, 500, 600, 800, 1000, 1500°K. How does this temperature sensitivity coefficient differ from heat capacity?

5. A gas purification unit sends to a Claus sulfur recovery unit 1000 lb-mol/hr of waste gas containing 30 mol % H_2S and 70 mol % CO_2, at 500 psia (34.0136 atm). The Claus unit uses exactly enough dry air for combustion to obey

ISBN 0-201-08300-0

the overall reaction, which goes to completion: $H_2S + 1/2\ O_2 \rightarrow H_2O + S$. The vapor pressure of sulfur (*Chemical Rubber Handbook* No. 47, p. D-113) is given:

t, °C	Pressure, mm Hg
183.8	1
243.8	10
288.3	40
327.2	100
399.6	400
444.6	760

The Claus unit recovers molten sulfur as a condensate and sends the residual gas at 400°F to a tail gas incinerator and stack gas clean-up unit. Assuming ideal gas behavior, calculate (a) the sulfur recovered, tons/day; (b) total gas to the incinerator, millions scf/day; (c) sulfur emission to the atmosphere, lb SO_2/day.

6. A high pressure fluid bed gasifier vessel is to be fabricated of 0.5% molybdenum steel clad internally with 18-8 stainless steel. The vessel will be 200 ft long by 19 ft in outside diameter and has walls 10 in. thick. Estimate the weight (tons) of the vessel exclusive of internals, assuming it to be a cylinder closed with flat heads (sp. gr. of steel = 7.80).

If fabricated alloy steel of this type costs $10/lb (field fabrication required), report the installed cost of this vessel.

7. Bulk removal of CO_2 from CH_4 is to be accomplished in a gas absorption operation at 300 psia. Fresh absorber liquid is fed at the top of the absorber: 70°F, 70 lb-mol/hr, mole weight 142, specific heat 0.4 Btu/lb °F. The absorber removes no methane, but removes 95% of the CO_2 entering. Treated gas leaves the top of the absorber at 70°F. Calculate the composition and the temperature of the rich liquid leaving the absorber if the enthalpy change of CO_2 is −131 Btu/lb absorbed.

8. Calculate the equilibrium temperature and composition for *adiabatic* methanation of the following at 34 atm and a feed temperature of 327°C: (a) CO = 1 mol %, H_2 = 99 mol %; (b) CO = 20%, H_2 = 80%; (c) CO = 20%, H_2 = 58%, H_2O = 22%.

Does carbon deposition occur?

9. Construct a table similar to Table 3.7 for (a) hydrogenation of hexane to form methane; (b) hydrogenation of benzene to form methane. Data can be found in Rossini et al. [80]. Use the table as a basis for your written discussion of the relative susceptibilities of hexane and benzene to hydrogenation. Discuss their differences and their similarities.

10. Develop a general equation for combustion of hydrocarbons in air, relating flue gas volume (scf) per thousand Btu (HHV) to fuel composition.

11. Explain in detail why real steam-oxygen gasification of carbon uses more than the ideal quantity of oxygen per unit of carbon gasified. Could real gasifiers be made to use less than the ideal quantity of oxygen?

ISBN 0-201-08300-0

12. Pure methanol is fed to the top of a gas absorber at $-40^{\circ}C$ for removal of acid gas components from 500,000 scf/hr of synthesis gas at 300 psia. Assuming 95% removal of CO_2 and 100% removal of H_2S from the feed gas, compute the loss in pounds per day of methanol from the absorber by evaporation into the treated gas.

Feed gas:

H_2	29%
CO	50
CO_2	20
H_2S	1
	100%

Methanol vapor pressure:

1 mm Hg at $-44.0^{\circ}C$
5 mm Hg at $-25.3^{\circ}C$

Assume that there is no absorption of H_2 or CO, and that treated gas leaves the absorber at $-40^{\circ}C$.

REFERENCES

1. "Abstracts on Materials Program," from Conference on Materials for Coal Conversion and Utilization: Energy Research and Development Administration, Electric Power Institute and The American Gas Association, Sept. 30–Oct. 1, 1976.
2. Anastasia, L. J. and Bair, W. G., Chapter 10 in *Clean Fuels from Coal, Symposium II Papers,* Institute of Gas Technology, Chicago, Ill., October 1975, p. 177.
3. Archer, D. H. et al., Chapter 32 in *Clean Fuels from Coal, Symposium II Papers,* Institute of Gas Technology, Chicago, Ill., October 1975, p. 577.
4. Attari, A., "The Fate of Trace Constituents of Coal during Gasification," Rep. EPA 650/2-73-004, August 1973.
5. Bair, W. G. et al., Chapter 7 in *Methanation of Synthesis Gas,* L. Seglin, Ed., *Advances in Chemistry Series* 146, American Chemical Society, Washington, D.C., 1975.
6. Banchik, I. N., Chapter 21 in *Clean Fuels from Coal, Symposium II Papers,* Institute of Gas Technology, Chicago, Ill., October 1975, p. 359.
7. Blum, D., Sherwin, M. B., and Frank, M. E., Chapter 9 in *Methanation of Synthesis Gas,* L. Seglin, Ed., *Advances in Chemistry Series* 146, American Chemical Society, Washington, D.C., 1975.
8. Bodle, W. W. and Vyas, K. C., Chapter 2 in *Clean Fuels from Coal, Symposium II Papers,* Institute of Gas Technology, Chicago, Ill., October 1975.
9. Bridger, G. W. and Woodward, C., Chapter 4 in *Methanation of Synthesis Gas,* L. Seglin, Ed., *Advances in Chemistry Series* 146, American Chemical Society, Washington, D.C., 1975.

ISBN 0-201-08300-0

10. Chao, R. E., "Thermochemical Water Decomposition Processes," *Ind. Eng. Chem. Prod. Res. Dev.*, **13**, 94–101, (June 1974).
11. "Coal Gasification," Quarterly Report April–June, 1975, ERDA 76-30-2, Office of Fossil Energy, Energy Research and Development Administration, Washington, D.C. 20545.
12. Coates, R. L., Chen, C. L., and Pope, B. J., Chapter 7 in *Coal Gasification,* L. G. Massey, Ed., *Advances in Chemistry Series* 131, American Chemical Society, Washington, D.C., 1974.
13. "COGAS Process," a descriptive brochure available from COGAS Development Co., P.O. Box 8, Princeton, N.J. 08540.
14. Cover, A. E., Schreiner, W. C., and Skaperdas, G. T., "Kellogg's Coal Gasification Process," *Chem. Eng. Prog.,* **69**, 31–36 (March 1973).
15. Dent, F. J. et al., "An Investigation into the Catalytic Synthesis of Methane by Town Gas Manufacture, " 49th Report of the Joint Research Committee of the Gas Research Board and the University of Leeds, GRB20, 1945.
16. Dent, F. J. et al., "An Investigation into the Catalytic Synthesis of Methane by Town Gas Manufacture, " Communication GRB51, The Gas Research Board, Beckenham, England, 1949.
17. Deul, M., Fields, H. H., and Elder, C. H., in Chapter 33 *Clean Fuels from Coal, Symposium II Papers,* Institute of Gas Technology, Chicago, Ill., October 1975, p. 615.
18. Eakman, J. M., Kalina, T., and Nahas, N. C., "Catalytic Coal Gasification for SNG Production," Exxon Research and Engineering Co., presented to Southwest Catalysis Society, Houston, Tex., Mar. 29, 1976.
19. Eisenlohr, K. H., and Moeller, F. W., Chapter 6, *Methanation of Synthesis Gas,* L. Seglin, Ed., *Advances in Chemistry Series* 146, American Chemical Society, Washington, D.C., 1975.
20. *Energy Res. Digest,* **II**, No. 18 (Aug. 30, 1976).
21. Feldman, H. F. et al., Chapter 8 in *Coal Gasification,* L. G. Massey, Ed., *Advances in Chemistry Series* 131, American Chemical Society, Washington, D.C., 1974.
22. Fink, C., Curran, G., and Sudbury, J., Chapter 15 in *Clean Fuels from Coal, Symposium II Papers,* Institute of Gas Technology, Chicago, Ill., October 1975, p. 243.
23. Foster, J. F. and Lund, R. J., *Economics of Fuel Gas from Coal,* McGraw-Hill, New York, 1950.
24. Frei, D., *Gas Turbine Int.,* **17**, 38 (May/June 1976).
25. Funk, J. E. et al., "Evaluation of Multi-Step Thermochemical Processes for the Production of Hydrogen from Water," Paper S11-1 in *Proceedings of the Hydrogen Economy Miami Energy Conference,* University of Miami, Coral Gables, Fla., Mar. 18–20, 1974.
26. Gardner, N., Samuels, E., and Wilks, K., Chapter 13 in *Coal Gasification,* L. G. Massey, Ed., *Advances in Chemicstry Series* 131, American Chemical Society, Washington, D.C., 1974.
27. Gasior, S. J., Forney, A. J., and Field, J. H., Chapter 1 in *Fuel Gasification,* F. C. Schora, Ed., *Advances in Chemistry Series* 69, American Chemical Society, Washington, D.C., 1967.
28. Goddin, C. S., Hall, D. E., and Mason, R. Z., "Economics of *in Situ* Coal Recovery," *Hydrocarbon Processing,* **55** (109) (July 1976).

ISBN 0-201-08300-0

29. Goldberger, W. M., "Collection of Fly Ash in a Self-Agglomerating Fluidized Bed Coal Burner," paper presented at Fuels Division, ASME Annual Meeting, Pittsburgh, Pa., Nov. 12–17, 1967.
30. Goldberger, W. M., "The Union Carbide Coal Gasification Process: Status of the Development Program," *Fourth Synthetic Pipeline Gas Symposium,* October 1972, American Gas Association Catalogue No. L11173.
31. Goring, G. E. et al., *Ind. Eng. Chem.,* **45**, 2586 (1953).
32. Grace, R. J., Chapter 13 in *Clean Fuels from Coal, Symposium II Papers,* Institute of Gas Technology, Chicago, Ill., October 1975, p. 207.
33. Gray, J. A. and Yavorsky, P. M., Chapter 9 in *Clean Fuels from Coal, Symposium II Papers,* Institute of Gas Technology, Chicago, Ill., October 1975, p. 159.
34. Gruber, G., Chapter 2 in *Methanation of Synthesis Gas,* L. Seglin, Ed., *Advances in Chemistry Series* 146, American Chemical Society, Washington, D.C., 1975.
35. *Handbook for Authors of Papers in the Journals of the American Chemical Society,* 1st ed., American Chemical Society, Washington, D.C., 1967.
35(a). *Handbook for Authors of Papers in American Chemical Society Publications,* American Chemical Society, Washington, D.C., 1978.
36. Hausberger, A. L. et al., Chapter 3 in *Methanation of Synthesis Gas,* L. Seglin, Ed., *Advances in Chemistry Series* 146, American Chemical Society, Washington, D.C., 1975.
37. Haynes, W. P., Gasior, S. J., and Forney, A. J., Chapter 11 in *Coal Gasification,* L. G. Massey, Ed., *Advances in Chemistry Series* 131, American Chemical Society, Washington, D.C., 1974.
38. Haynes, W. P. et al., Chapter 5 in *Methanation of Synthesis Gas,* L. Seglin, Ed., *Advances in Chemistry Series* 146, American Chemical Society, Washington, D.C., 1975.
39. Haynes, W. P. and Forney, A. J., Chapter 8 in *Clean Fuels from Coal, Symposium II Papers,* Institute of Gas Technology, Chicago, Ill., October 1975, p. 149.
40. Hickman, R. G. et al., "Thermochemical Hydrogen Production Research at Lawrence Livermore Laboratory," Paper S11-23 in *Proceedings of the Hydrogen Economy Miami Energy Conference,* University of Miami, Coral Gables, Fla., Mar. 18–20, 1974.
41. Higgins, G. H., "A New Concept for *in Situ* Coal Gasification," *Lawrence Livermore Lab. Rep.* UCRL-51217, Rev. 1, Sept. 17, 1972.
42. Hougen, O. A., Watson, K. M., and Ragatz, R. A., *Chemical Process Principles.* Part II: *Thermodynamics,* 2nd ed., John Wiley & Sons, New York, 1964.
43. "Hygas: 1964 to 1972 Pipeline Gas from Coal-Hydrogenation," *R & D Rep.* 22, Vols. 1, 2, 3, 4; Final Report by IGT Process Research Division for Energy Research and Development Administration, Washington, D.C. 20545, July 1975.
44. *International Gas Technology Highlights,* Vol. VI, No. 14, July 5, 1976, Institute of Gas Technology, Chicago.
45. *International Gas Technology Highlights,* Vol. VI, No. 19, Sept. 13, 1976, Institute of Gas Technology, Chicago.
46. *International Gas Technology Highlights,* Vol. VII, No. 3, Jan. 31, 1977, Institute of Gas Technology, Chicago.

ISBN 0-201-08300-0

47. *International Gas Technology Highlights,* Vol. VII, No. 6, Mar. 14, 1977, Institute of Gas Technology, Chicago.
48. *International Gas Technology Highlights,* Vol. VII, No. 13, June 20, 1977, Institute of Gas Technology, Chicago.
49. *International Gas Technology Highlights,* Vol. VII, No. 16, Aug. 1, 1977, Institute of Gas Technology, Chicago.
50. *International Gas Technology Highlights,* Vol. VII, No. 19, Sept. 12, 1977, Institute of Gas Technology, Chicago.
51. Jahnig, C. E. and Bertrand, R. R., "Environmental Aspects of Coal Gasification," *Chem. Eng. Prog.,* **72** (51) (August 1976).
52. Johnson, J. L., Chapter 10 in *Coal Gasification,* L. G. Massey, Ed., *Advances in Chemistry Series* 131, American Chemical Society, Washington, D.C., 1974.
53. Kalina, T. and Marshall, H., "Hydrogasification Process," South African Patent 7,302,075, Mar. 26, 1973.
54. Kavlick, V. J. and Lee, B. S., Chapter 2 in *Fuel Gasification,* F. C. Schora, Ed., *Advances in Chemistry Series* 69, American Chemical Society, Washington, D.C., 1967.
55. Kreib, K. H., Chapter 5 in *Clean Fuels from Coal, Symposium II Papers,* Institute of Gas Technology, Chicago, Ill., October 1975, p. 101.
56. Krukonis, V. J. et al., Chapter 3 in *Coal Gasification,* L. G. Massey, Ed., *Advances in Chemistry Series* 131, American Chemical Society, Washington, D.C., 1974.
57. Kydd, P. H., Chapter 7 in *Clean Fuels from Coal, Symposium II Papers,* Institute of Gas Technology, Chicago, Ill., October 1975, p. 131.
58. Lacey, J. A., Chapter 4 in *Fuel Gasification,* F. C. Schora, Ed., *Advances in Chemistry Series* 69, American Chemical Society, Washington, D.C., 1967.
59. "Large Diameter 300 PSI Gasifier, Preliminary Engineering Report," *Natl. Tech. Inf. Serv.* PB-238360, U.S. Department of Commerce, December 1974.
60. LaRosa, P. and McGarvey, R. J., Chapter 14 in *Clean Fuels from Coal, Symposium II Papers,* Institute of Gas Technology, Chicago, Ill., October 1975, p. 227.
61. Martin, J. R., Chapter 12 in *Clean Fuels from Coal, Symposium II Papers,* Institute of Gas Technology, Chicago, Ill., October 1975, p. 869.
62. Mason, D. M. and Schora, F. C., Chapter 3 in *Fuel Gasification,* F. C. Schora, Ed., *Advances in Chemistry Series* 69, American Chemical Society, Washington, D. C., 1967.
63. Massey, M. J., Dunlap, R. W., McMichael, F. C., and Nakles, D. V., "Environmental Analysis: Characterization of Effluents from the Hygas Pilot Plant," a report to the Institute of Gas Technology, Apr. 12, 1976 (available from Carnegie-Mellon University, Pittsburgh, Pa.).
64. Massey, M. J., Dunlap, R. W., McMichael, F. C., and Nakles, D. V., "Environmental Analysis: Characterization of Effluents from the CO_2 Acceptor Pilot Plant," a report to Conoco Coal Development Co., Aug. 13, 1976 (available from Carnegie-Mellon University, Pittsburgh, Pa.).
65. Massey, M. J., "Environmental Considerations in the Construction and Operation of Commercial Coal Gasification Facilities," paper presented before the ERDA General Technical Advisory Committee, Oak Ridge, Tenn., Aug. 19, 1976.

ISBN 0-201-08300-0

66. "Materials and Components in Fossil Energy Applications," *ERDA Newsl.* 4, Apr. 1, 1976.
67. Mentser, M. et al., Chapter 1 in *Coal Gasification,* L. G. Massey, Ed., *Advances in Chemistry Series* 131, American Chemical Society, Washington, D.C., 1974.
68. Mosely, F. and Patterson, D. J., *Inst. Fuel,* **38**, 378 (1965).
69. Nadkarni, R. M., Bliss, C., and Watson, W. I., Chapter 33 in *Clean Fuels from Coal, Symposium II Papers,* Institute of Gas Technology, Chicago, Ill., October 1975, p. 625.
70. Nakles, D. V., Massey, M. J., Forney, A. J., and Haynes, W. P., "Influence of Synthane Gasifier Conditions on Effluent and Product Gas Production," *NTIS Rep.* PERC/RI-75/6, December 1975.
71. Nelson, E. T., Warrall, J., and Walker, P. L., Jr., Chapter 38 in *Coal Science, Advances in Chemistry Series* 55, American Chemical Society, Washington, D.C., 1966.
72. Norman, O. E., *The Romance of The Gas Industry,* A. C. McClurg, Chicago, 1922.
73. Palmer, P. M., "High Pressure Gasifier and Sulfur Removal from Coal," presented at The Second Symposium on Coal Gasification, Liquefaction, and Utilization: Best Prospects for Commercialization, University of Pittsburgh, School of Engineering, August 1975.
74. Pangborn, J. B. and Schorer, J. C., "Analysis of Thermochemical Water Spitting Cycles," Paper SII-36 in *Proceedings of the Hydrogen Economy Miami Energy Conference,* University of Miami, Coral Gables, Fla., Mar. 18–20, 1974.
75. Patel, J. G. and Loeding, J. W., Chapter 11 in *Clean Fuels from Coal, Symposium II Papers,* Institute of Gas Technology, Chicago, Ill., October 1975, p. 193.
76. Perry, H. and DeCarlo, J. A., "The Search for Low Sulfur Coal," *ASME Publ.* 66-PWR-3, United Engineering Center, New York, 1966.
77. Pursley, J. A. et al., "Reaction Kinetics," *Chem. Eng. Prog. Symp. Ser.* (1958), **48** (4), 51–58.
78. "Report on the Trials of American Coals in a Lurgi Gasifier at Westfield, Scotland" for the American Gas Association and Office of Coal Research, Department of the Interior, U.S. Government; Woodall-Duckham, Engineering and Construction, June 1974.
79. Rosen, B. H., Pelofsky, A. H., and Greene, M., U.S. Patent 3,960,700, June 1, 1976.
80. Rossini, F. D. et al., *Selected Values of Physical and Thermodynamic Properties of Hydrocarbons and Related Compounds,* API Project 44, Carnegie Press, Pittsburgh, Pa., 1953.
81. Rostrup-Nielson, J. R., "Equilibria in Decomposition Reactions of Carbon Monoxide and Methane over Nickel Catalysts," *J. Catal,* **27**, 343-56 (1972).
82. Rudolph, P. F. H. and Bierback, H. H., Chapter 4 in *Clean Fuels from Coal, Symposium II Papers,* Institute of Gas Technology, Chicago, Ill., October 1975, p. 85.
83. Russel, J. L. and Porter, J. T., "A Search for Thermochemical Water Spitting Cycles," Paper SII-49 in *Proceedings of the Hydrogen Economy Miami Energy Conference,* University of Miami, Coral Gables, Fla., Mar. 18–20, 1974.

ISBN 0-201-08300-0

84. Schaefer, A. D. et al., "A Program to Discover Materials Suitable for Service under Hostile Conditions Obtaining in Equipment for the Gasification of Coal and Other Solid Fuels," *NTIS Rep.* FE-1784-24, Metal Properties Council, Inc., to Energy Research and Development Administration, Feb. 4, 1977.

85. Segeler, C. G. et al., *Gas Engineers Handbook,* Industrial Press, New York, 1965.

86. Seglin, L. et al., Chapter 1 in *Methanation of Synthesis Gas,* L. Seglin, Ed., *Advances in Chemistry Series* 146, American Chemical Society, Washington, D.C., 1975.

87. Sheer, C. and Korman, S., Chapter 4 in *Coal Gasification,* L. G. Massey, Ed., *Advances in Chemistry Series* 131, American Chemical Society, Washington, D.C., 1974.

88. Sirohi, V. P., "Methanation Processes," Preliminary Report to ERDA/AGA, C. F. Braun and Co., Project 4568-NW, Mar. 11, 1975.

89. Skamser, R., "Coal Gasification Commercial Concepts Gas Cost Guidelines," C. F. Braun and Co. for Energy Research and Development Administration, Contract E949-18-1235 (available from National Technical Information Service).

90. Stephens, F. M., Jr. and Goldberger, W. M., "Combustion of Carbonaceous Solids," U.S. Patent 3,171,369, Mar. 2, 1965.

91. Stull, D. R. and Prophet, H., *JANAF Thermochemical Tables,* NSRDS-NBS 37, 2nd ed., U.S. Government Printing Office, Washington, D.C., 1971 (Catalogue No. C13.48:37).

92. *Synthetic Fuels Summary* **I** (29-32) December 1975, Cameron Engineers, Inc., Denver, Colo.

93. *Synthetic Fuels Summary* **II** (4-8 et seq.) December, 1974, Cameron Engineers, Inc., Denver, Colo.

94. Tarman, P. B., in *Proceedings of the Fifth Synthetic Pipeline Gas Symposium,* Oct. 29–31, 1973, American Gas Association Catalogue No. L51173, pp. 19–30.

95. Virk, P. S., Chambers, L. E., and Woebcke, H. N., Chapter 14 in *Coal Gasification,* L. G. Massey, Ed., *Advances in Chemistry Series* 131, American Chemical Society, Washington, D.C., 1974.

96. Walker, F. E. and Hartner, F. E., "Forms of Sulfur in U.S. Coals," *U.S. Bur. Mines Inf. Circ.* 8301, U.S. Government Printing Office, Washington, D.C., 1966.

97. "Wellman-Galusha Gas Producers," Brochure Form 576, McDowell Wellman Engineering Co., Cleveland, Ohio.

98. Wen, C. Y. et al., Chapter 16 in *Fuel Gasification,* F. C. Schora, Ed., *Advances in Chemistry Series* 69, American Chemical Society, Washington, D.C., 1967.

99. Wen, C. Y. et al., Chapter 2 in *Coal Gasification,* L. G. Massey, Ed., *Advances in Chemistry Series* 131, American Chemical Society, Washington, D.C., 1974.

100. Whalen, M. J., "Study Report: Carbon Formation in Methanators," C. F. Braun and Co., Rep. FE-2240-10 to Energy Research and Development Administration and the American Gas Association, July 1976.

ISBN 0-201-08300-0

101. White, G. A., Roszkowski, T. R., and Stanbridge, D. W., Chapter 8 in *Methanation of Synthesis Gas,* L. Seglin, Ed., *Advances in Chemistry Series* 146, American Chemical Society, Washington, D.C., 1975.

102. Whiteacre, R. C., Farnsworth, J. F., and Mitsak, D. M., Chapter 6 in *Clean Fuels from Coal, Symposium II Papers,* Institute of Gas Technology, Chicago, Ill., October 1975.

103. Willson, W. G. et al., Chapter 12 in *Coal Gasification,* L. G. Massey, Ed., *Advances in Chemistry Series* 131, American Chemical Society, Washington, D.C., 1974.

104. Woebcke, H. N., Chapter 29 in *Clean Fuels from Coal, Symposium II Papers,* Institute of Gas Technology, Chicago, Ill., October 1975, p. 511.

105. Zahradnik, R. L. and Grace, R. J., Chapter 9 in *Coal Gasification,* L. G. Massey, Ed., *Advances in Chemistry Series* 131, American Chemical Society, Washington, D.C., 1974.

106. Zielke, C. W. and Gorin, E., *Ind. Eng. Chem.,* **47**, 820 (1955).

107. Zielke, C. W. and Gorin, E., *Ind. Eng. Chem.,* **49**, 396 (1057).

ISBN 0-201-08300-0

5. Coal Liquefaction

E. Stanley Lee

I. INTRODUCTION

1.1. Coal Liquefaction

Among the various approaches for converting coal into an improved, non-polluting energy source, liquefaction has advantages in terms of both thermal efficiency and economics. Both of these advantages are derived from the fact that fewer chemical changes are required to convert solid coal into liquids than into gases, and the process conditions are much milder. Furthermore, liquid fuels have far higher energy density than gaseous fuels and therefore are cheaper to store and transport. Another favorable factor is that the processing water requirements are much lower than those for gasification.

Various schemes have been introduced to convert solid coals into liquids. These schemes can be roughly divided into four broad categories, namely, non-catalytic liquid phase dissolution or solvent extraction, direct catalytic hydrogenation or hydroliquefaction, Fischer-Tropsch type catalytic synthesis, and pyrolysis.

Pyrolysis is covered in Chapter 2. This chapter is principally concerned with the other three categories. Most of the commercial developments and researches during recent years have been concentrated on solvent extraction and catalytic hydrogenation. Thus most of our discussion will also be concerned with these two types of processes.

The development of catalyst is intimately connected with coal liquefaction. For example, Fischer-Tropsch synthesis contributed enormously to the understanding of the behavior of catalysts and their manufacture. In fact, the catalysts used for gasoline upgrading or processing have their origin in coal conversion. On the other hand, most of the novel second-generation liquefaction processes are based on some new catalyst concepts.

C. Y. Wen and E. Stanley Lee, Editors, Coal Conversion Technology, ISBN 0-201-08300-0

ISBN 0-201-08300-0

Since liquefaction processes were commercialized during World War II, it is desirable to give a brief discussion of these earlier processes. Most of the rest of this section, therefore, will be devoted to these earlier historical developments.

1.2. Historical Development

The history of the development of coal hydrogenation is closely related to needs or projected needs. Because of the shortage of petroleum reserves from the late 1920s until the end of World War II, extensive commercial developments for coal liquefaction were carried out in Germany. In 1943 12 plants were hydrogenating coal and tar and supplied 98% of aviation gasoline, or 47% of the total hydrocarbon products consumed in Germany. However, for economic reasons most of these plants were shut down after the war.

During the years 1940–1944 the reduced rate of discovery of petroleum reserves and the fantastic increase in consumption in the United States led to serious concern about the future supply of oil and resulted in the passage of the Synthetic Liquid Fuels Act in 1944. From 1946 to 1953 a large amount of development and research effort was carried out by the U.S. Bureau of Mines and several private companies. The process was based primarily on the German designs.

Starting in 1950, the Middle East discoveries of petroleum reserves began to grow at a fantastic rate. According to costs obtained from U.S. Bureau of Mines data, gasoline from coal cost two to three times more than gasoline from petroleum crude [8]. With the change in administration in 1953 in the U.S. Government, the commercial development of coal liquefaction was discontinued.

In view of the current gas shortage, it is interesting to note that natural gas was an undesirable product and was flared or used for the production of carbon black until the long gas pipelines were built.

In recent years, with the shrinking reserves in petroleum crude in the United States, it has gradually become clear that our vast coal reserves must be utilized in ways that will cause the least damage to our environment. The amount of coal liquefaction, gasification, and other coal utilization research has been accelerating since the late 1960s.

A. SOLVENT EXTRACTION

In earlier work, solvent extraction was used primarily as a tool for studying the composition of coal. De Marsilly [1] extracted coals with various solvents as early as 1860. However, most of the important coal extraction researches for analyzing the chemical constitution of coal were not performed until after 1913. Several comprehensive reviews [2–5] on coal extraction or dissolution can be found in the literature.

A tremendous amount of work has been carried out on coal extraction. Almost all organic solvents have been tried. However, commercial significance was not realized until the investigations of Bergius in 1913 and of Pott and Broche in

ISBN 0-201-08300-0

1933. Pott and Broche's classical work was the forerunner of the noncatalytic liquid phase dissolution process. Bergius' work with later developments helped to build the German coal hydrogenation processes during World War II.

B. BERGIUS PROCESS

Historically, Berthelot [6] probably was the first investigator to convert coal into liquids by hydrogenation with hydriodic acid at a temperature of 270°C. However, Bergius was the first researcher to show the commercial value of coal hydrogenation. In 1913 Bergius and Billwiller [7] obtained liquid products from coal under high pressure and temperature in the presence of hydrogen. During World War II oil made from coal in Germany reached an output of 5 million metric tons per year.

Coal hydrogenation developments until a decade ago are summarized in the literature [8–11]. The products obtained from Bergius' original process were of poor quality with high sulfur, oxygen, and nitrogen contents. The process involves the mixing of approximately equal amounts of powdered coal and heavy recycle oil with a hydrogen pressure of 200 atm and with approximately 5% of "Luxmasse," which is essentially a mixture of iron and aluminum oxides with a small amount of titanium oxides. It is interesting to note that Luxmasse was employed merely as a reagent for converting the sulfur in coal to iron sulfide. The catalytic effect of Luxmasse in the hydrogenation of coal was not realized until later.

The commercialization of Bergius' process was mainly due to the work of I. G. Farben. The first commercial coal hydrogenation plant was constructed at Leuna, Germany, and started to operate in April 1927 to hydrogenate brown coal in a two-step process. In this arrangement coal is mixed with catalyst and hydrogenated in liquid phase to produce a gasoline fraction, a middle oil, boiling at 180–325°C, and a heavy oil, boiling above 325°C. The heavy oil slurry is usually centrifuged to separate the solids. In the second step or phase, also known as the vapor phase, the middle oil is further hydrogenated to gasoline in vapor phase over a fixed bed of catalyst. The heavy oil is recycled and used as pasting oil for the coal hydrogenation step. If fuel oil is the desired product, the middle oil is recycled in place of a large part of the heavy oil and the excess heavy oil is taken from the system and processed further with some middle oil. Because hydrogenation is less severe for producing fuel oil than for producing gasoline, lower hydrogen consumption and higher throughput can be achieved when fuel oil is produced. This two-step arrangement was originally used to avoid the fouling of the catalyst due to the decomposition or coking of high molecular weight material.

The development of coal hydrogenation catalysts represents another major effort. I. G. Farben succeeded in finding an effective catalyst which greatly increased the hydrogenation velocity. During the first year of operation of the Leuna plant, brown coal was hydrogenated without a catalyst. Only 60% of the coal was

ISBN 0-201-08300-0

hydrogenated, and the asphaltene content was very high [8]. To increase efficiency, molybdenum oxide catalyst was used. However, molybdenum compounds were expensive and not available in Germany in sufficient quantities. Iron salts were found to be as effective as molybdenum compounds. In 1934 iron catalyst replaced molybdenum, and the former became the standard catalyst for liquid phase hydrogenation in Germany.

According to the excellent summary of Wu and Storch [8], both liquid phase and vapor phase hydrogenations were operated at a pressure of 300 atm or lower during the earlier years of commercial operation. However, for plants constructed after 1936, 700 atm was adopted first for vapor phase and later for liquid phase hydrogenation. Both brown coal and bituminous coal were hydrogenated. However, a high pressure of 700 atm was found necessary for the hydrogenation of German bituminous coal.

Outside Germany the only commercial plant in operation before the end of World War II was the Imperial Chemical Industries plant at Billingham, Great Britain, for hydrogenating bituminous coal. This plant has the capacity to produce 150,000 tons/year of gasoline at a pressure of 250 atm. Greatly improved catalysts were obtained in Great Britain. The iron catalyst 321 was used to replace I. G. Farben's tungsten catalyst in vapor phase hydrogenation.

After World War II the U.S. Bureau of Mines built a demonstration plant in Louisiana, and Union Carbide built a semicommercial plant at Institute, West Virginia. The latter plant was mainly used to produce chemicals and thus had a milder hydrogenation condition.

The Bureau's demonstration plant was patterned after the German operations. The design was developed with the aim of applying modern U.S. engineering practices.

C. POTT-BROCHE PROCESS

Pott and Broche began their research on coal hydrogenation in 1927 [8]. Their U.S. patent was granted in 1938 [12].

For this process the coal was hydrogenated under very mild conditions in a mixture of tetrahydronaphthalene and cresols with or without the addition of naphthalene. In commercial plant operations, middle oil was used as the donor solvent. The only plant constructed, at Welheim, Germany, had an annual capacity of 26,000 metric tons of the hydrogenation product. However, this process never went to commercial operation because of filtration problems in the solid separation stage.

Only coal with more than 25% volatile matter was considered suitable for this process. The grounded coal was mixed at 60–100°C with twice its weight of middle oil. The mixture entered the preheater at 100°C and went to the extraction vessel at about 430°C. The pressure used was approximately 100–150 atm.

ISBN 0-201-08300-0

1.3 Clean Burning Fuel as the New Objective

As can be seen from the very brief summary given above, the hydrogenation of coal to liquid products has been investigated by many researchers over the past 50 years. However, until very recently the objective was to produce coal liquids which could be processed into high octane gasoline. The degree of hydrogenation needed to achieve this objective is very costly and wasteful.

In recent years the principal objection to coal utilization has been that it is very dirty and environmentally unacceptable. Furthermore, solid fuel is less convenient to handle than the liquid form. Thus an alternative and more limited objective has emerged, namely, to add just enough hydrogen to the coal to produce a pumpable oil of low sulfur and mineral contents. This oil can be used for industrial and commercial heating and for electrical power generation.

The broader objective calls for new directions of investigation and research. For example, certain coals may be more suitable for certain conversion processes. Furthermore, certain components or macerals of the coal may be more suitable for conversion into certain forms of fuels. Thus more attention should be paid to coal characterization and to the removal of pollutants and ashes in coal on the basis of this characterization.

1.4. Commercial Development

Various commercial processes are currently under development by both the federal government and the private sector. Because of the state of flux, no effort will be made to summarize these commercial processes.

Because coal is a solid and contains large amounts of mineral matter, and also because severe processing conditions, such as high pressure and temperature, are involved, many mechanical and operative problems exist in the development of large scale commercial processes. Two of the most common problems are equipment erosion and abrasion due to the presence of strong chemicals formed during processing, as well as to the presence of mineral matters in coal at high speeds, and equipment plugging due to coking, polymerization, or simply, solidification.

Another difficulty in the development of commercial processes is the fact that almost none of the intermediate products can be stored, and thus the process cannot allow a shutdown at certain stages of the processing train.

ISBN 0-201-08300-0

2. COAL LIQUEFACTION

2.1. Chemical Composition

As Chapter 1 on coal characterization points out, the two principal problems with coal utilization are that it is dirty and environmentally unacceptable, and that it is a solid which is difficult to handle and which is limited in application. To illustrate this, Table 2.1 [73] lists typical compositions of several coals, asphaltene, toluene, petroleum crude, gasoline, and methane. The chief distinguishing features of coal are its low hydrogen and high sulfur contents. The contents of nitrogen and oxygen are also fairly high. For lignite the oxygen content is often 20% or higher. Since coal is principally composed of aromatic ring clusters, toluene serves to contrast the change in hydrogen contents between solid coal and liquid hydrocarbons. Methane has been added to the table to emphasize the enormous amount of hydrogen needed for converting solid coal into high Btu gas.

Table 2.1

Typical Compositions of Coals and Liquid Hydrocarbons[a] (Mills [73])

Element	Anthracite	mv Bit.	hvB Bit.	Lignite	Asphaltene[b]	Toluene	Petroleum crude	Gasoline	Methane
C	93.7	88.4	80.3	72.7	87	91.3	83-87	86	75
H	2.4	5.0	5.5	4.2	6.5	8.7	11-14	14	25
O	2.4	4.1	11.1	21.3	3.5				
N	0.9	1.7	1.9	1.2	2.2		0.2		
S	0.6	0.8	1.2	0.6	0.37		1.0		
H/C atom ratio	0.31	0.67	0.82	0.69	0.9	1.14	1.76	1.94	4

[a] Coal analysis on moisture- and ash-free basis; ash content of coal 3–15%.

[b] From Ref. 81. The asphaltene was obtained from a Synthoil process with hvA bituminous coal at 450°C and 2000 psi hydrogen pressure.

There are two ways to increase the hydrogen/carbon ratio in coal: the addition of hydrogen or the rejection of carbon. For example, for bituminous and lignite coals we must adjust the chemical formula from approximately $CH_{0.6 \sim 1.0}$ for solid coal to approximately $CH_{1.2 \sim 2.0}$ for liquid products. This chemical adjustment can be accomplished in several different ways. Figure 2.1 illustrates four of the most investigated and most advanced approaches.

ISBN 0-201-08300-0

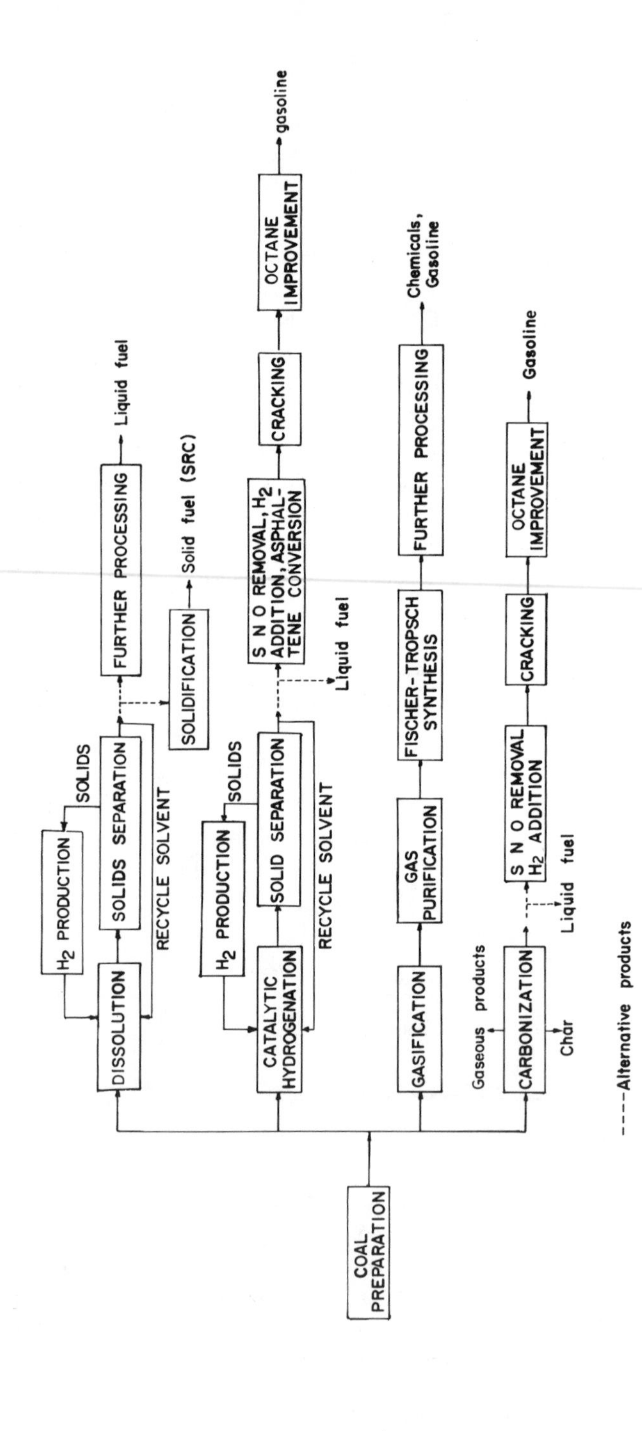

Figure 2.1. Alternative routes for coal liquefaction.

ISBN 0-201-08300-0

2.2. Direct Liquefaction

The first two approaches (see Figure 2.1), namely, hydrogen-donor solvent dissolution and catalytic hydrogenation, represent the hydrogen addition approach. Most of our discussion in this chapter will be concerned with these two approaches, which will be referred to as direct hydrogenation or simply as hydrogenation. The hydrogen is generally added to the coal through a donor solvent, and the reaction can be represented by

$$CH_{0.8} + 0.4H_2 \rightarrow CH_{1.6} \tag{2.1}$$

or, more realistically,

$$CH_{0.8}S_{0.2}O_{0.1}N_{0.01} + H_2 \rightarrow (RCH_x) + CO_2, H_2S, NH_3, H_2O \tag{2.2}$$

where the reaction may be carried out with or without catalysts. The first terms on the left-hand sides of both equations represent the coal used, and the first terms on the right-hand sides represent the liquid produced. Obviously, the chemical formulas are approximate. This is especially true for the liquid, which depends on the degree of hydrogenation; this, in turn, depends on the operating conditions, the nature of the donor solvent, the presence or absence of hydrogenation catalyst, and the presence or absence of hydrogen pressure.

A. SOLVENT EXTRACTION

When catalyst is not used, the process is generally known as solvent extraction or liquid phase dissolution and is represented by the top approach in Figure 2.1. This process follows basically the process of Pott and Broche, and the degree of hydrogenation is relatively low. The purpose of solvent extraction is to produce, with minimum treatment, a relatively clean burning fuel from coal. The fuel can be either in solid form, known as solvent-refined coal (SRC), or in liquid form. These fuels are primarily designed for industrial use or for electric power generation.

Hydrogenation not only increases the hydrogen content and removes the ash in coal, but also reduces the undesirable heteroatoms in coal, such as sulfur, nitrogen, and oxygen, by combining them with hydrogen. The degree of removal of these undesirable chemical elements depends on the degree of hydrogenation. In general, in solvent-refined coal all the inorganic sulfur and part of the organic sulfur are removed, and the sulfur content is reduced to below 1%.

Because of the low degree of hydrogenation in solvent dissolution, a relatively large amount of solid is still present in the effluent of the dissolver or extractor. Since one of the purposes of solvent extraction is to remove the undesirable ashes

ISBN 0-201-08300-0

or minerals, it is important that a fairly complete separation between the solution and the undissolved solids be achieved. Since the sizes of the solids in residues are very small (in the micron range) and also since the solution is fairly viscose, solid separation is often a troublesome step for these processes. In fact, Pott and Broche's process never went to commercial operation in Germany because of difficulties in the solid separation step.

Two important factors in solvent extraction are the nature of the donor solvent and the presence of hydrogen pressure. To increase the hydrogen donor capability, the solvent is frequently hydrogenated before use. In commercial practice the solvent is obtained by recycling part of the oil product stream.

B. CATALYTIC HYDROGENATION

When catalyst is added to the coal-solvent slurry, the process is known as direct catalytic hydrogenation, hydroliquefaction, or liquid phase hydrogenation. During World War II the process was well developed and supplied nearly all of the aviation gasoline in Germany. However, the process developed during the war is not economically competitive. With the development of better catalysts and materials of construction, much less severe operating conditions can now be used.

The catalytic hydrogenation process is represented by the second process in Figure 2.1. One of the best catalysts is cobalt molybdate. Other catalysts, such as tungsten and molybdenum sulfide and an impure iron oxide ("Bayermasse"), were also used but required higher pressures. The operating conditions are approximately $450^{\circ}C$ and 2000–4000 psia. In recent years processes using entrained flow, fluidized bed, or fixed bed catalysts have been proposed.

The degree of hydrogenation is much higher than that obtained with solvent extraction, and thus the problem of solid separation is much less severe. Furthermore, most of the sulfur in coal is converted to H_2S, the oxygen to H_2O or CO_2, and the nitrogen to NH_3. These compounds leave with the gas stream, resulting in a much cleaner product than solvent-refined coal. The coal is converted to liquids ranging from heavy to light oils and gases; however, the main product is liquid fuel. Alternatively, these liquid products can be further processed and upgraded to gasoline.

2.3. Hydrodesulfurization of Coal and Hydrogen Requirement

Most of the earlier research investigations on catalytic hydrogenation of coal were concerned with the degree of dissolution of coal. Very little attention was paid to the question of hydrodesulfurization. In recent years, because of environmental concerns, the emphasis has been on the removal of sulfur and other pollutants in coal and thus has resulted in the development of solvent extraction processes and the investigation of the effectiveness of catalysts for desulfurization as well as for hydrogenation.

ISBN 0-201-08300-0

One of the most expensive factors in direct coal liquefaction is the need for large quantities of costly hydrogen. In general, 5000–10,000 ft^3 of hydrogen is needed to produce one barrel of oil from coal. The production of hydrogen is very expensive; hence considerable improvement of process economics can be obtained if the requirement for hydrogen can be reduced.

There are many processes for producing hydrogen. In coal conversion the needed hydrogen is generally produced by gasifying the solid residue with steam and oxygen, followed by shift conversion. Detailed descriptions of these processes are given in Chapter 4 on coal gasification.

2.4. Pyrolysis

The second way to increase the hydrogen/carbon ratio is by the rejection of carbon atoms. Pyrolysis is one method whereby a large number of carbon atoms are rejected as solids, with the liquid and gaseous products containing a much higher hydrogen/carbon ratio. This process is represented by the bottom approach in Figure 2.1. One characteristic of this approach is that it produces significant quantities of by-product gas and char which must be disposed of economically. The liquid product, which is known as coal tar, can be further processed by hydrogenation, desulfurization, and quality improvement to produce gasoline or, with much less processing, liquid fuel.

Pyrolysis has been discussed in detail in Chapter 2. For our purpose it is sufficient to say that the quantities of gas, liquid, and char produced depend on the type of coal, the rate of heating, the nature of the gas atmosphere surrounding the coal, and the ultimate temperature achieved. Yields of liquid are maximized by minimizing the time during which the product is exposed to elevated temperatures, thus avoiding further decomposition of the gas.

2.5. Degree of Hydrogenation, Dissolution, or Liquefaction

The degree of hydrogenation or liquefaction will be used to indicate the effectiveness of the process. Unfortunately, there exists no uniform definition for hydrogenation or dissolution.

Since solvent extraction was used principally as an analytical tool in the earlier days of investigation, the amount of coal extract dissolved in a given organic solvent is frequently employed as an indication of the degree of hydrogenation. The most frequently used solvents are benzene and hexane. Frequently, in laboratory practice, the reaction mixture from the reactor or dissolver, after venting the gaseous products, is first subjected to extraction by benzene. Any material insoluble in benzene is considered as unhydrogenated residue. The "conversion" is usually calculated from the difference in weight between the original coal and the insoluble residue, making any appropriate corrections such as for mineral matters and catalyst. After the removal of benzene from the extract, hexane is used to separate the extract into two fractions: a benzene-soluble and hexane-insoluble fraction, which is known as "asphaltene," and a hexane-soluble fraction, which is generally

ISBN 0-201-08300-0

called "oil." Other solvents occasionally used for this purpose are pyridine and cresol.

Although "oil" has been defined as the hexane-soluble fraction, some authors in the literature also call the total benzene-soluble liquid "oil." Therefore the reader must watch out for the definition. In this work "oil" will designate the hexane-soluble fraction, and the total benzene-soluble liquid will be referred to as "liquid" or "total liquid."

In the last few years it has been found that for solvent extraction or dissolution the benzene-insoluble material is not composed totally of the unreacted coal. Another fraction, which is pyridine soluble and benzene insoluble, is obtained. This fraction, which has been called preasphaltenes by Sternberg and co-workers [25, 26] and asphaltols by Farcasiu et al. [66], can be clearly distinguished from asphaltenes and oils by its high viscosity.

Another way to express the degree of hydrogenation is by the use of filtration. The effluent from the autoclave reactor or dissolver, after venting the gaseous products, is essentially a slurry product which can be filtered to separate the solids from the solution. The filtrate can be further processed into different fractions by organic solvent extraction, distillation, chromatographic techniques, centrifugation, or a combination of these methods. The "conversion" or, more appropriately, the "solvation" can be calculated from the difference in weight between the original coal and the dried filter cake.

Only a general definition of solubility or degree of hydrogenation has been given above. The exact definition varies with different investigators. The reasons for this complication are the use of solvent which frequently presents in the extract product or in the filter cake, the moisture and ash content of the coal, and the temperature and pressure used. Thus, when comparing results, the exact definition used must be borne in mind.

Another complicating factor is that, in order to obtain the exact degree of hydrogenation, the filtrate or benzene extract must be further analyzed for hydrogen content. For example, low hydrogenation frequently results in a high viscose product which causes significant problems in solid separation. Preasphaltenes and asphaltene are fairly high viscose products and are undesirable.

2.6. Indirect Liquefaction

The production of hydrocarbons from carbon monoxide and hydrogen in the presence of Fischer-Tropsch catalyst is generally known as indirect hydrogenation. The process, shown as the third approach in Figure 2.1, was originally developed by Fischer and co-workers. Full scale commercial production was carried out in Germany during World War II. A commercial plant was also built in the 1950s in South Africa by Sasol, and this is the only commercial liquefaction plant in the world today.

One advantage of this approach is the complete separation from the original coal minerals and the complete destruction of the original coal structure. Thus a

ISBN 0-201-08300-0

clean burning fuel with negligible amounts of sulfur and ash can be obtained. However, the complete destruction of the original coal structure involves a large amount of processing and many chemical changes which are very expensive in terms of thermal efficiency.

In actual practice a medium Btu synthesis gas containing a hydrogen-carbon monoxide mixture is used as the raw material. To produce this medium Btu gas, the process is generally combined with the gasification of coal, followed by gas purification. In the Sasol plant the Lurgi gasifier is used. The coal gasification and purification steps are discussed in detail in Chapter 4 on gasification.

The two most important parameters in the design of this type of reactors are the catalyst used and the methods employed to remove the large amounts of heat generated during the reaction. Catalyst poisoning is a problem, and thus the feed to the reactor must be purified with considerable care.

An advantage of the process is its flexibility. By adjusting the composition of the catalyst, the hydrogen/carbon monoxide ratio, and the operating conditions, fairly high selectivity can be obtained for a wide variety of products such as gas, LPG, gasoline, kerosene, diesel fuel, and fuel oil. Furthermore, by the use of different catalysts, other products such as methanol and acetone can also be produced. The types of catalysts typically used are Fe, Co, Ni, Ru, ZnO_2, and ThO_2. The behaviors of these catalysts depend on the presence of chemical and structural promoters, on the procedure of catalyst manufacture, on catalyst surface conditions, and other factors.

The product selectivity spectrum in the Fischer-Tropsch reaction has been studied by many investigators. Dry [115] has found that by simultaneous manipulation of the catalyst basicity and gas composition the product selectivity can be varied over a wide range, from 1 to over 70% methane, or from 0 to over 50% hard wax. The maximum gasoline fraction obtained was 40%. The selectivity primarily depends on two factors: catalyst composition and gas composition. The important gas composition is the H_2/CO ratio in low temperature fixed bed reactors and the partial pressure of CO_2 in high temperature entrained bed reactors. Both of these types of reactors are used in the Sasol plant.

A. HYDROCARBON SYNTHESIS

Early developments of the Fischer-Tropsch process are summarized in the literature [13-16]. Fischer-Tropsch used alkalized iron turnings as catalysts at high temperature and pressure in its initial work. It was found later that at atmospheric pressure and 200°C cobalt and, to a lesser extent, iron were suitable catalysts for the synthesis of higher hydrocarbons. Nickel promotes the formation of methane and, when diluted with kieselguhr, can also be used for synthesis of higher hydrocarbons.

In terms of hydrogenation power, cobalt, nickel, and ruthenium are most powerful and orient the hydrogenation of carbon monoxide almost entirely toward hydrocarbons. Iron is less active and also produces some alcohols in the product.

ISBN 0-201-08300-0

Oxide catalysts such as zinc oxide have a low activity and favor the formation of alcohols.

In addition to catalyst activity, the mechanical strength of the catalyst and its shape, which influences the pressure drop through the catalyst bed, are other important factors. Furthermore, because of the large amounts of heat generated during the reaction, heat removal must also be considered in selecting a catalyst. For example, if external cooling is used to remove the heat, a fluidized bed is more efficient than a fixed bed. However, to use a fluidized bed, the catalyst particles must be strong mechanically. Thus the structure support of the catalyst, the procedure of manufacturing the catalyst, and the catalyst surface conditions are other important factors.

The U.S. Bureau of Mines has studied these problems for many years [17, 110–114]. To increase the rate of heat removal, external cooling was abandoned and internal cooling was proposed. Oil circulation through a fixed bed of granular fused iron oxide catalyst was first used. However, because of the gradual cementation of catalyst particles during reactor operation, an expanded bed was next proposed, but this causes attrition of the catalyst. To avoid all these problems, the Bureau of Mines finally developed lathe turnings catalyst in a fixed bed. Because this catalyst has a large void volume, cementation of the catalyst is not a problem.

The oil circulation system was also abandoned, and hot gas recycle was adopted. This was done because the oil is fairly volatile and unstable at temperatures above 300°C and thus the reaction temperature is limited. The hourly space velocity is fairly low with the oil circulation system. The hot gas recycle process uses recycled gas to remove the reaction heat as the sensible heat of the gas. Combined with the lathe turnings catalysts, this process yielded satisfactory operation.

Another catalyst, the parallel plate type, which gives a much lower pressure drop than the lathe turnings catalysts, was finally adopted [17]. Catalysts are first flame sprayed onto plates. These plates are then assembled into parallel plate modules and stacked in the reactor. The catalyst used is Alan Wood magnetite impregnated with K_2CO_3. Another advantage of the flame-sprayed catalyst on the parallel plates with extended heat exchange surface is the increased thermal efficiency.

In the Fischer-Tropsch synthesis section of the Sasol plant [18–20], H_2 + CO mixtures react over iron catalysts to produce mixtures of hydrogen carbons and oxygenates. Two types of reactors are used. One is a fixed bed reactor containing a pelletized precipitated iron catalyst and operating near 220°C (the Arge system). The second, the "Synthol" system, is an entrained circulating bed reactor operating at 330°C and using a promoted iron catalyst powder. Some typical operating conditions and product distributions are listed in Table 2.2 [72]. As can be seen, there exist significant differences in product distribution for the two types of reactors. This confirms earlier discussions that product distribution or product selectivity depends heavily on the type and structure of the catalysts and on the operating conditions.

ISBN 0-201-08300-0

Table 2.2

Product Distribution at Sasol from Fixed and Entrained Beds on Iron Catalysts[a]

	Fixed Bed		Entrained Bed	
Temperature, °C:	220–240		320–340	
Pressure, bars:	26		22	
H_2/CO in feed gas:	1.7 : 1		3 : 1	
Products	Wt % Total	Wt % Olefins	Wt % Total	Wt % Olefins
C_1	7.8	–	13.1	–
C_2	3.2	23	10.2	43
C_3	6.1	64	16.2	79
C_4	4.9	51	13.2	76
C_5-C_{11}	24.8	50	33.4	70[b]
C_{12}-C_{20}	14.7	40	5.1	60[b]
C_{21}+	36.2	~15	–	–
Alcohols, ketones	2.3	–	7.8	–
Acids	–	–	1.0	–

[a] Source: Ref. 72 by Wender.

[b] These fractions also contain appreciable amounts of aromatic compounds.

B. LIQUEFACTION THROUGH METHANOL

Recent research on a shape-selective zeolite catalyst suggests another route for making gasoline from coal. Researchers at Mobil Oil [117] discovered that methanol can be converted into gasoline in one step by the use of a new zeolite catalyst. Furthermore, the hydrocarbons thus obtained are predominantly in the gasoline boiling range of C_4–C_{10} hydrocarbons with an unleaded Research Octane Number of 90–100.

Thus the production of gasoline from coal can be carried out in three principal steps: gasification of coal, synthesis of methanol, and conversion of methanol to gasoline. Since the technology needed for large scale production of methanol from methane by catalytic conversion has been commercially demon-

ISBN 0-201-08300-0

strated, and since the gasification step is already undergoing pilot plant development, this approach appears to be a promising one.

Methanol, in addition to gasoline, has also been suggested as an automobile fuel [116]. The Mobil process makes this approach even more attractive. To use methanol as an auto fuel, extensive investment is required and modifications are needed on the internal combustion engine and on other facilities. This is due to the particular properties of methanol, such as its affinity to water, its corrosiveness, its toxicity, and its boiling point. Furthermore, since a gallon of gasoline has twice the energy content of methanol, the volume handled would be doubled if methanol were used. It has been estimated that, using the zeolite catalyst, the cost of producing 1 gallon of gasoline from coal-derived methanol is about 5¢. This appears to be a fairly small price to pay, compared to the investments required if methanol is used directly as an auto fuel.

The Mobil process to convert methanol to gasoline can be represented by

$$xCH_3OH \rightarrow (CH_2)_x + xH_2O \qquad (2.3)$$

A conversion of 99% in a single pass was obtained. The product distribution based on the CH_2 feed to the reactor in weight percent is as follows: 3% methane and ethane and 25% hydrocarbon gases, including methane and ethane; 75% of the products are in the C_5^+ gasoline fraction.

3. INFLUENCE OF COAL PROPERTIES ON LIQUEFACTION

3.1 Introduction

In the study of coal liquefaction, one of the most unpredictable problems is the influence of the characteristics of coal on its liquefaction behavior. Because of this unpredictability the different coals used in the comparison of any results must be borne in mind.

The properties of coal greatly influence liquefaction. It is known that some types of coal cannot be hydrogenated under liquefaction conditions whereas other types can be hydrogenated easily. Since coal will be used to produce various forms of clean fuels by means of different processes, it will be desirable to use only coals with optimum properties for hydrogenation. In this way the yield of oil can be increased, the yield of inert residue can be reduced, operating difficulties and

ISBN 0-201-08300-0

maintenance costs can be minimized, and the capital investment cost per unit product can be lowered.

3.2. Rank and Composition of Coal

As was discussed in Chapter 1 on coal characterization, the chemical and physical behavior and characteristics of coal depend on its rank and its petrographic composition or type. These two properties, in turn, depend on two basic factors operating during the formation of coal. The first basic factor, which is known as the biochemical or peat stage, is the set of conditions existing during the accumulation of peat and before its covering by geological sedimentation. This basic factor determines the type of vegetation and the nature and extent of the degradation which results in varying assemblages of vegetation components, or macerals, and inorganic minerals. The petrographic type of composition is determined during this stage.

The second factor is operative during the metamorphic stage, where the covering of the peat deposit by a succession of sediments causes the development of high temperature and pressure. These temperatures, pressures, and burying conditions determine the extent of changes and the rank in a coal seam.

Since different petrographic types are usually present in the same seam of coal, which is typically composed of the same rank, the liquefaction behavior can be quite different for the same rank of coal at different geological locations and hence different petrographic compositions.

The liquefaction behavior is further complicated by the presence of different amounts of minerals in the coal. These minerals, which can be either derived from the original plant material or accumulated from other sources, not only result in different approaches being used for purification and processing of the coal, but also exert profound catalytic effects during hydrogenation, thus making the interpretation of data extremely difficult. These catalytic effects will be discussed in greater detail in subsequent sections.

Coal liquefaction behavior depends not only on the rank of coal, but also on the type or petrographic composition. Fisher and associates [21] found that fusain was unsuitable for hydrogenation, the presence of opaque attritus also reduced the yield, and anthraxylon was beneficial for conversion. Table 3.1 gives their results based on benzene-insoluble residue. The petrographic constituent and the researchers' ultimate analyses are also given. The inert character of fusain is clearly indicated. However, because of the limitations of the petrographic constituent classification method used, these workers' results were very scattered and inaccurate.

ISBN 0-201-08300-0

Table 3.1
Effect of Petrographic Constituents on Hydrogenation, % maf basis (Fisher et al. [21])

Constituent	Starting Coal C	H	O	N	S	Volatile Matter	Benzene Insoluble
Anthraxylon	89.9	4.8	2.9	1.8	0.6	20.4	4.6
	90.5	4.7	2.7	1.4	0.7	17.2	10.1
Splint coal	88.0	4.8	5.1	1.5	0.6	28.3	34.8
(high in	85.4	4.7	8.1	1.4	0.4	31.9	40.2
opaque attritus)	86.9	4.4	7.0	1.3	0.4	24.1	21.6
Fusain	89.5	3.1	6.4	0.7	0.3	7.3	75.3
	88.1	3.5	5.8	0.7	1.9	18.9	77.3

3.3. Macerals

Instead of macroscopic differentiation, microscopic or macerals should be used. As was discussed in Chapter 1, coal macerals can be separated into three groups based on the Stopes-Heerten system of nomenclature: vitrinite, exinite, and inertinite. Both vitrinite and exinite are susceptible to liquefaction, whereas inertinite is very resistant. Both micrinite and fusinite belong to the inertinite group. The macroscopic rock type anthraxylon has vitrinite as its main constituent. The main constituent of opaque attritus is micrinite, and fusain has fusinite as its main constituent.

Given and co-workers [22–24] carried out a fairly systematic investigation of the influence of coal macerals on liquefaction. On the basis of the fact that coal macerals differ in their grindability, these authors separated different size fractions of a crushed coal with different petrographic compositions. These different maceral fractions were further separated into different coal components by the float-and-sink separation technique and by using mixtures of halogenated hydrocarbons as solvent. The different fractions of maceral compositions from a subbituminous coal from the Blue Seam in McKinley Mine, Gallup, New Mexico, are shown in Table 3.2. This coal has a carbon content on dmmf basis of 78.3% and 45% volatile matter. The percentage of solvation, which is essentially an indicator of yield, is also shown in Table 3.2 and is plotted in Figure 3.1. Since the amounts of exinite and micrinite are small for this particular coal, solvation or yield of hydrogenation is essentially inversely proportional to the amount of fusinite in the sample. Figure 3.1 indicates a fairly good correlation between the yield and the reactive macerals,

ISBN 0-201-08300-0

which are vitrinite and exinite. Thus vitrinite is very beneficial for hydrogenation, and fusinite contributes little toward liquefaction, thereby confirming the general belief that fusinite is essentially inert for liquefaction.

Table 3.2

Effect of Coal Type on Liquefaction[a]

Fraction No.	Coal Maceral,[b] %, mmf				Ash, % dry basis	Yield, % dmmf			
	V	E	F	M		Solvation[c]	B.I.	Asphaltene	Viscosity[d]
1	96	1	2	1	2.5	93.6	56.0	14.7	32
2	91	3	4	2	2.1	95.1	53.2	21.4	25
3	91	2	7	0	3.0	82.6	50.8	12.3	37
4	89	3	7	1	3.2	91.5	60.6	11.3	27
5	89	2	8	1	3.5	72.7	48.2	21.5	46
6	85	1	12	2	3.6	87.9	49.2	17.0	23
7	83	2	14	1	4.4	83.4	40.6	20.7	17
8	81	2	16	1	4.1	73.3	43.3	18.7	49
9	70	3	23	4	7.8	63.7	49.8	18.9	38
10	61	5	33	1	24.0	42.6	44.1	15.3	13
11	63	2	31	4	23.3	64.1	32.6	18.1	7

[a] Hydrogenation condition: 385°C and 238 atm in the presence of a Gulf proprietary catalyst, using partly hydrogenated anthracene oil as a vehicle.

[b] V: Vitrinite; E: exinite, also includes sporinite and resinite; F: fusinite, also includes semifusinite; M: micrinite; B.I.: benzene insoluble.

[c] Solvation is defined as percent of coal which is not in the solid filter cake.

[d] Filtrate viscosity at 100°C, mm^2/sec.

Reproduced from Given et al., *Fuel*, **54**, 1975, by permission of the publishers [24].

The hydrogen-rich sporinite, resinite, cutinite, and alginite belong to the exinite maceral group, the members of which are very desirable components for liquefaction. Note that resinite, sporinite, cutinite, and alginite are coalified remains of resins, spores, cuticles, and algae bodies, respectively. The high activity of exinite for liquefaction is indicated in Figure 3.2, where the results of Given et al. are reproduced. Note that in Figure 3.2 conversion is defined as the difference in weight between the original coal on a mineral-matter-free basis and the benzene-insoluble residue. Since in some samples shown in this figure the vitrinite is fairly low but the exinite is relatively high and the liquefaction yield remains high, the only explanation is that exinite is a desirable component for hydrogenation. The inertness toward hydrogen for fusinite and micrinite is also clearly indicated in Figures 3.1 and 3.2. However, in a more detailed examination of the solid residue from liquefaction, Mitchell et al. [68] found that some parts of micrinite may be reactive for liquefaction.

ISBN 0-201-08300-0

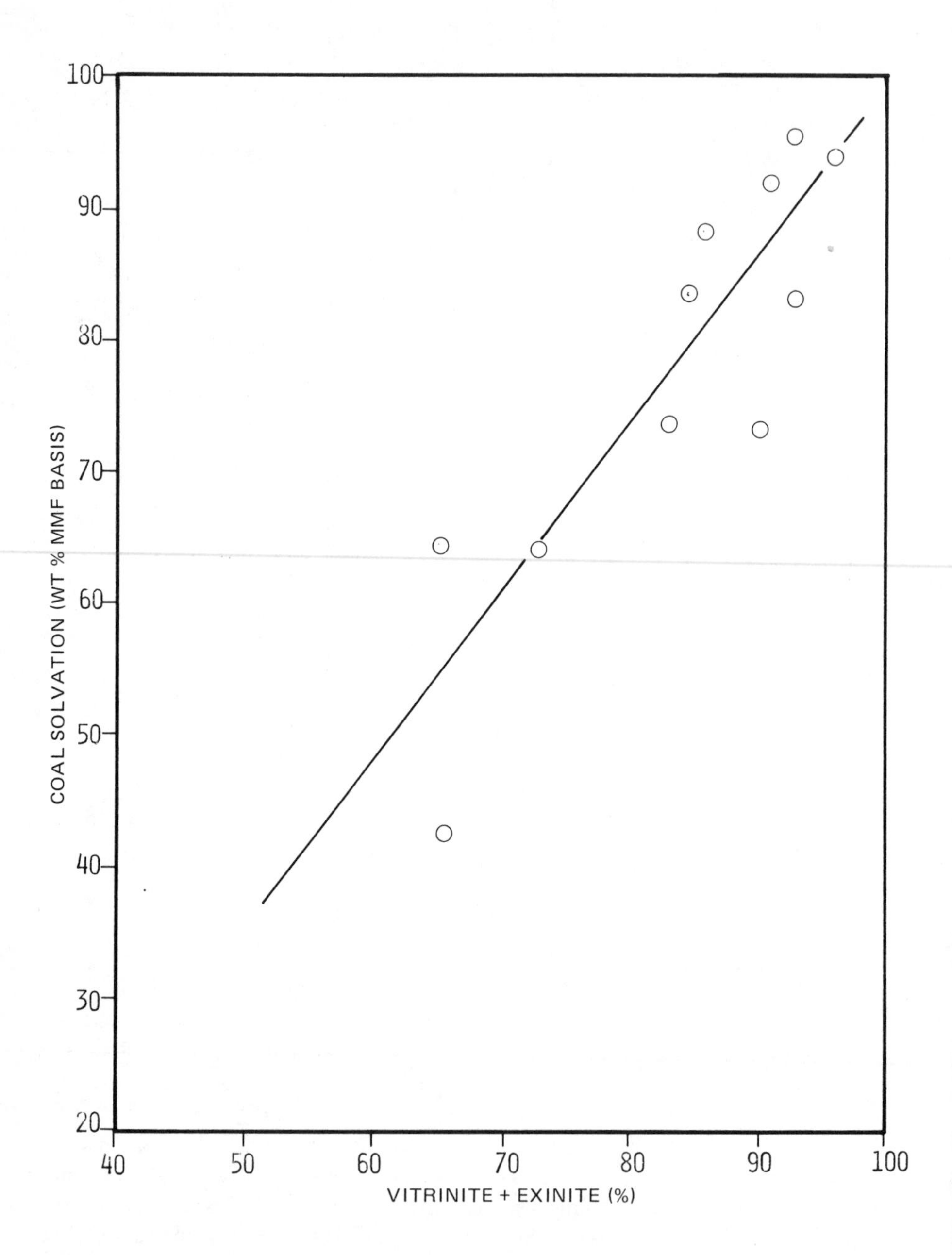

Figure 3.1. Effect of reactive materials on liquefaction. From Given et al., *Fuel,* **54**, 1975. (Reproduced by permission of publishers [24].)

ISBN 0-201-08300-0

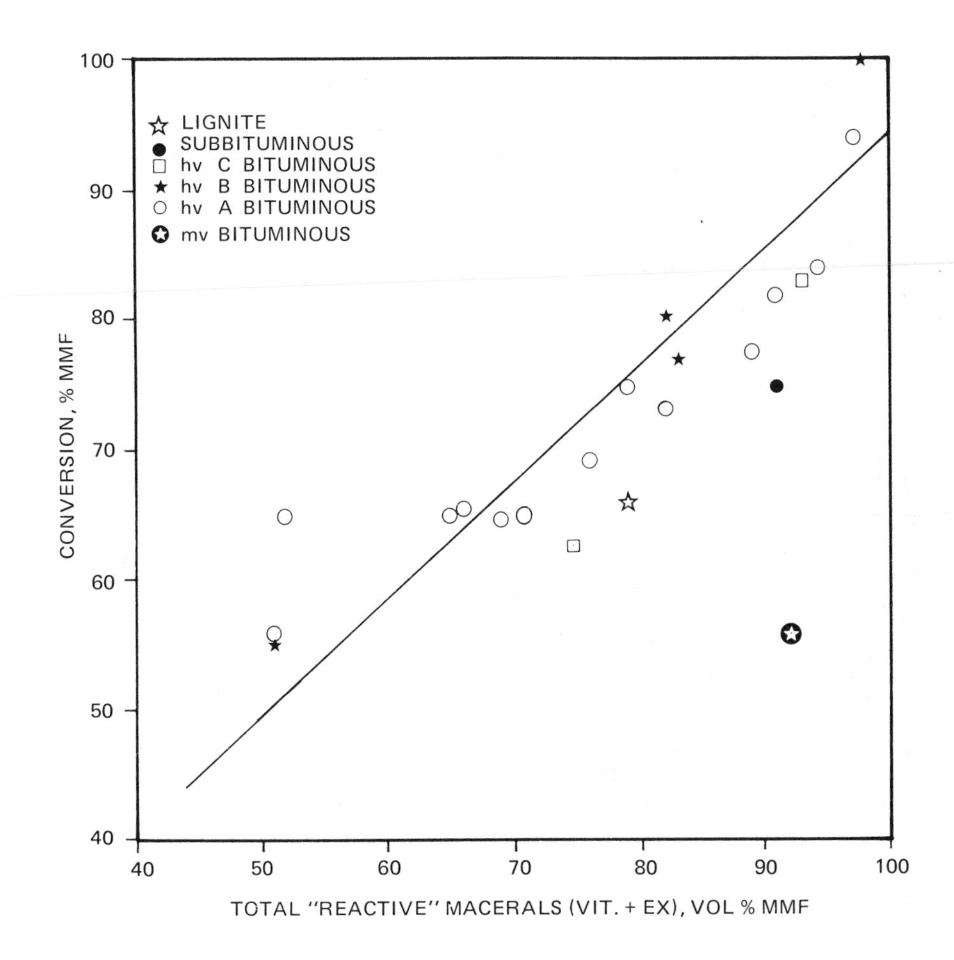

Figure 3.2. Conversion versus reactive macerals. (Davis et al. [22])

In predicting the coking properties of coal, it was found useful to recognize more than one vitrinite: vitrinite and pseudovitrinite. It has been found that pseudovitrinite is not active in the coking process, while vitrinite is the most important ingredient for good coking properties. According to the work of Given et al., pseudovitrinite is just as beneficial for liquefaction as is vitrinite. In fact, there is some indication that pseudovitrinite may be more active than vitrinite.

Another petrographic constituent of coal is semifusinite, which is intermediate in appearance and properties between vitrinite and fusinite. Semifusinite

ISBN 0-201-08300-0

belongs to the inertinite group. Recently, Mitchell et al. [68] petrographically examined the solid residue derived from coal liquefaction and concluded that fractional parts of semifusinite may react with hydrogen and contribute to liquid products.

3.4. Lignite

Since the vegetable materials altered far less because of coalification in lignite than in higher ranks of coals, there exists a much greater degree of petrographic heterogeneity in lignite. Because of this low coalification, the nature and properties of the macerals in lignite are fundamentally different from those of the macerals of higher rank coals.

Given and co-workers [24] obtained different fractions of coal macerals for lignite coal from Savage, Montana, based on size differences and the float-and-sink method. These fractions were then hydrogenated with an anthracene oil and catalyst under hydrogen pressure in an autoclave. Although, based on the results for higher rank coals, inertinite macerals should not be as active for hydrogenation as are other macerals, this experiment did not yield conclusive evidence concerning the influences of different macerals on hydrogenation. It appears that in the float-and-sink method the mineral materials have a tendency to concentrate in the heaviest (sink) fractions. For example, in Table 3.2 samples 10 and 11 are heaviest (sink) fractions and have the highest percentages of ash. Thus catalytic effects or other unknown factors may have influenced the results.

3.5. Coal Rank

Earlier investigators of the influence of the rank of coal on hydrogenation were handicapped by lack of knowledge of the petrographic composition or type of coal; thus their results were partly obscured by the differences in petrographic compositions.

Fisher et al. [21] made a fairly systematic investigation of the influence of coal rank on conversion with consideration of the different types. Some of the results were as follows: (a) coals with more than 89% carbon content are unsuitable for hydrogenation and give a fairly low liquid yield, (b) high volatile bituminous coals appear to be the most desirable rank for liquefaction, and (c) low rank samples such as lignite and sub-bituminous coal also give lower liquid yields than high volatile bituminous coals. Furthermore, these low rank coals were more sensitive to experimental conditions; higher pressures and temperatures gave higher yields.

ISBN 0-201-08300-0

The results reported by Fisher et al. [21] on the continuous hydrogenation of coals in the U.S. Bureau of Mines experimental plant are reproduced in Figure 3.3. The liquid obtained in the experimental plant is removed continuously and contains about 20% gasoline and 70% oil, distilled between 200 and 330°C. This figure shows that the yield of oil increases rapidly with increasing carbon content of coal to about 81%. Although the number of data is limited and some of the samples, such as the one at 87% carbon, contain a fairly high amount of fusain, Figure 3.3 indicates the general trend for the influence of coal rank on hydrogenation.

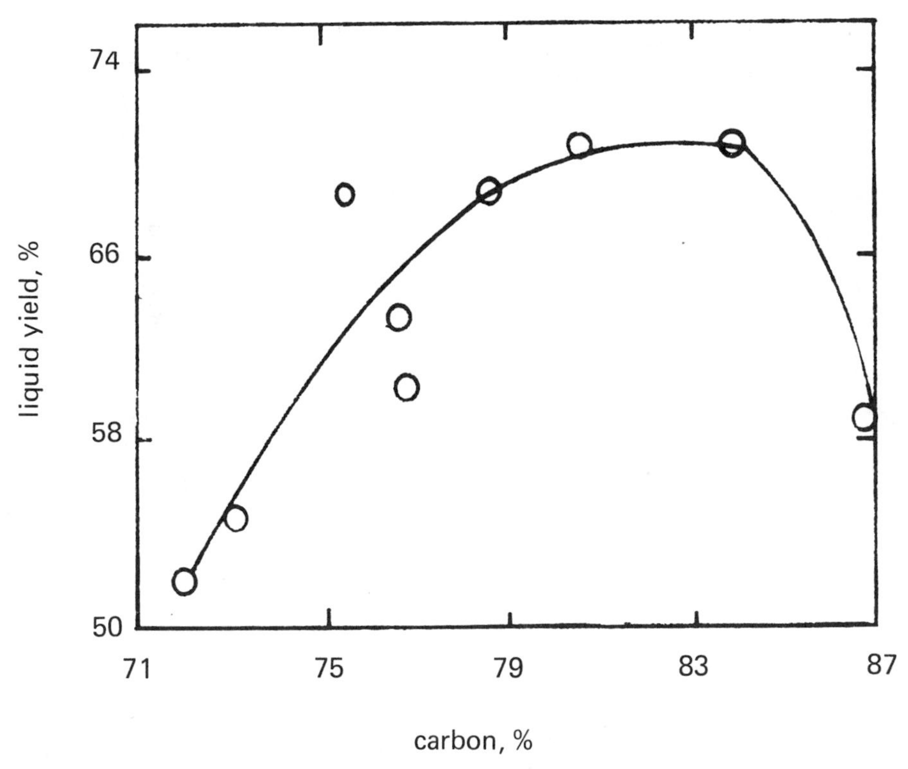

Figure 3.3. Effect of carbon content of maf coal. (Fisher et al. [21])

ISBN 0-201-08300-0

The results of Given et al. [24] are shown in Figures 3.2, 3.4, and 3.5. In Figures 3.2 and 3.4 conversion is plotted against the total reactive macerals, with ranks indicated by various symbols. These figures confirm earlier results in that high volatile bituminous coal gives the highest yield.

As was discussed in Chapter 1, the reflectance of vitrinite in oil can be used as a fairly good index of coal rank. For high volatile bituminous coals the reflectance lies approximately within the limits of 0.49–1.0%. Another important index is the product liquid's viscosity, which gives a clear indication of the extent of hydrogenation. Furthermore, high viscosity liquid products cause difficulties in solid separation. The filtrate viscosity is plotted against reflectance in Figure 3.5, which clearly indicates the general trend of increasing filtrate viscosity with increasing rank for the high volatile bituminous and lower rank coals.

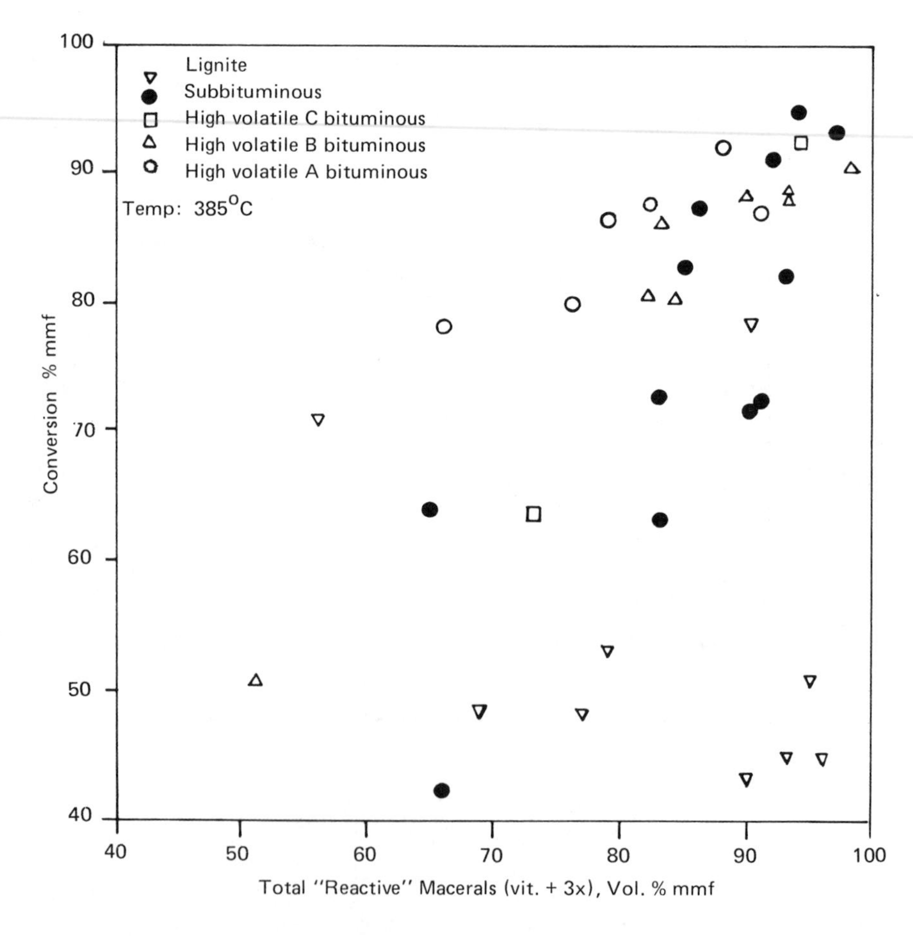

Figure 3.4. Effect of reactive macerals and rank on conversion. (Davis et al. [22])

ISBN 0-201-08300-0

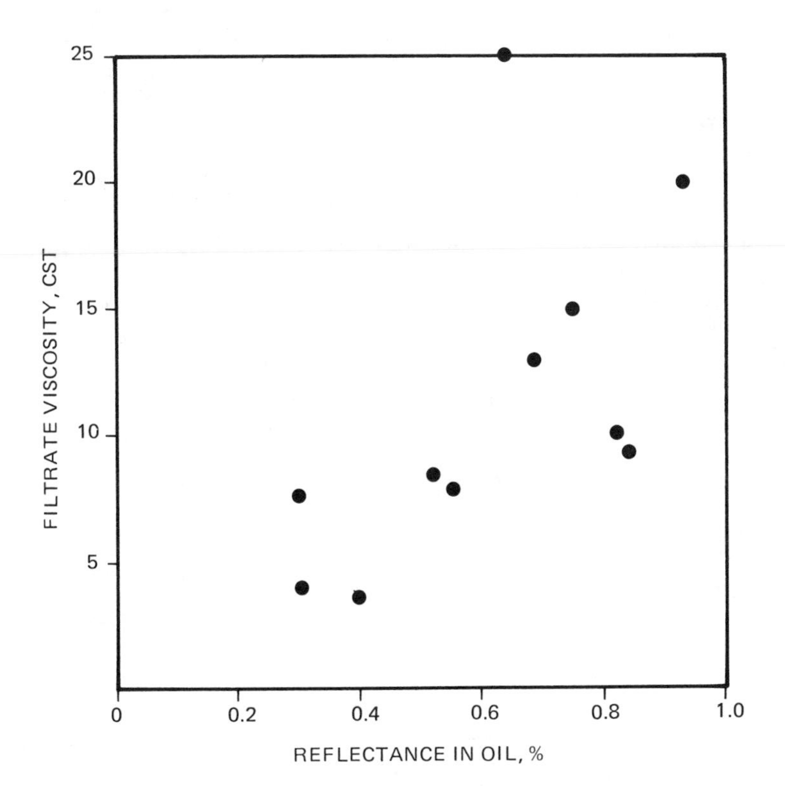

Figure 3.5. Effect of rank on filtrate viscosity. (Davis et al. [22])

3.6. Rate of Conversion

The rank of coal influences both the final yield and the rate of conversion. Most of the preceding discussion has been concerned with the final yield. The rate of conversion increases in general with decreasing rank. The results of Neavel [30] are reproduced in Figure 3.6, where the conversion is defined on the basis of benzene insolubles. Hydrogenated creosote oil was used as a donor vehicle, and the reaction was carried out at 400°C. At the beginning of each run atmospheric pressure was used. The reaction rate increases rapidly from low volatile bituminous coal to lignite.

Anthracite is an expensive coal and is in great demand for combustion and for metallurgical purposes. It has a large, condensed polynuclear structure, usually contains little ash, and has low oxygen and sulfur contents. Furthermore, because of its structure it cannot be processed easily and is almost impossible to convert into liquid. Thus anthracite is not used in liquefaction.

ISBN 0-201-08300-0

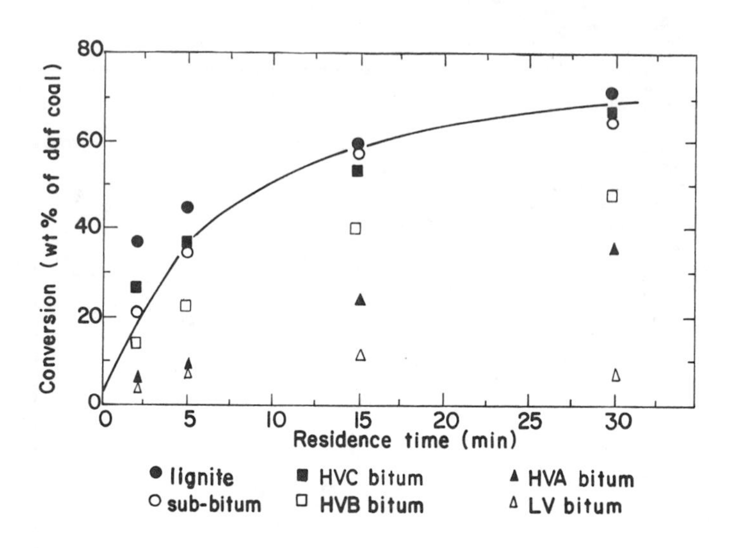

Figure 3.6. Effect of rank on reaction rate and conversion. (Reproduced from Neavel, *Fuel,* **55**, 1976, by permission of publishers [30].)

On the other side of the spectrum, lignite contains a large amount of water and oxygen and requires a great deal of expensive hydrogen during liquefaction. Because its characteristics are quite different from those of bituminous coal, different processing and handling is frequently needed. For example, fresh lignite can be liquefied easily. However, after lignite is exposed to air, it is much more difficult to liquefy. A synthesis gas, instead of hydrogen, is frequently used to obtain solvent refined lignite.

3.7. Coal "Structure"

Any chemical processing of coal represents an attack upon the molecular "structure" of the coal. In liquefaction, where the original molecular structure is not completely destroyed, a knowledge of the structure is especially useful to elucidate the physical and chemical processes involved in liquefaction.

Coal consists of two-dimensional aromatic lamellae. The individual lamellae consist of mono-, di-, tri-, and perhaps tetracyclic aromatic monomers with mean molecular weights of 230–350, connected by methylene linkages into generally linear polymers with some cross linking through oxygen to carbon and sulfur to oxygen bonds [32, 71]. The average molecular weight of the coal polymer is probably 2000–3000. There are seldom more than three aromatic rings per cluster.

Coal liquefaction is believed to be directly related to the breaking of the methylene linkages and to the cleavage of C-O, C-N, and, to a lesser degree, C-C bonds which originally cross-linked the large polynuclear coal structure.

ISBN 0-201-08300-0

Given developed a structure for vitrain with a carbon content of 82% [71]. But for our purposes Wender's representation [72], reproduced in Figure 3.7 for a low volatile bituminous coal, a hvA bituminous coal, a sub-bituminous coal, and a lignite, is more convenient and meaningful. The corresponding typical compositions of each coal, both on a dry and on a maf basis, are given in Table 3.3. From an examination of Figure 3.7, it is not surprising that, although coal is a very complex material, it often behaves as one would expect an aromatic or hydroaromatic organic compound to behave.

lv bituminous

hva bituminous

subbituminous

lignite

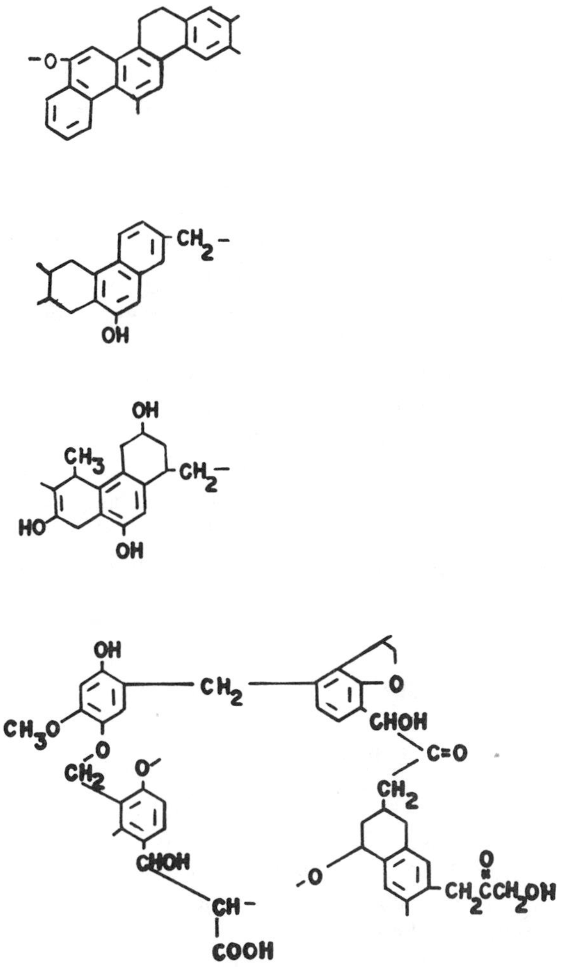

Figure 3.7. Typical structures of some coals. (Wender [72])

ISBN 0-201-08300-0

Table 3.3

Compositions of Coals Represented by Figure 3.7 (Wender [72])

	Pocahontas No. 3 Bed, W. Va. (lvB)		Pittsburgh Bed (hvAb)		Mammoth Bed, Wyo. (Subbituminous A)		Beulah-zap, N.D. (Lignite)	
Constituent	dry	maf	dry	maf	dry	maf	dry	maf
C	83.8	90.7	77.1	84.2	72.9	76.7	64.5	72.6
H	4.2	4.6	5.1	5.6	5.3	5.6	4.3	4.9
O	2.6	2.8	6.4	6.9	14.8	15.5	18.0	20.2
N	1.2	1.3	1.5	1.6	1.2	1.3	1.0	1.1
S	0.6	0.6	1.5	1.7	0.9	0.9	1.1	1.2
Ash	7.6	–	8.4	–	4.9	–	11.1	–
Volatile matter	17.3	18.7	36.5	39.9	41.5	43.6	40.8	45.9
Btu/lb		15,660		15,040		13,490		12,150

3.8. Other Factors

Coal is a very complex solid. In addition to chemical structure, rank, and type, many other factors can influence the hydrogenation results. Furthermore, even for the above three, our understanding of the factors is limited and is certainly not adequate to predict the exact behavior of a given coal during liquefaction. Thus it is always advantageous to use the same coal in order to facilitate comparison and elucidation of the experimental results. If different coals are used, the unknown factors in these coals must always be kept in mind.

Some of the other factors that are known to be important are the catalytic and other influences of the minerals or ashes in coal, the pretreatment involved before using the coal, the conditions during storage and transportation such as possible oxidation due to exposure to air, the location and geological history of the coal, and the nature of the porosity in the coal.

4. SOLVENT EXTRACTION OR DISSOLUTION

4.1. Introduction

As discussed earlier, hydrogenation of coal in the presence of solvent, but not catalyst, is generally known as extraction or dissolution. The purpose of dissolution is to provide a minimum degree of processing to produce a clean burning fuel. In fact, a large amount of the basic original structure of coal is still intact in the solvent refined coal (SRC).

There are two significant factors in coal dissolution. First, the coal must be heated to a temperature above its plastic or softening temperature to provide thermal decomposition or cleavage. The plastic temperature depends on the coal

ISBN 0-201-08300-0

used and is generally between 325 and 375°C for bituminous coals. Second, readily available atomic or free radical hydrogen must be present to prevent polymerization of the thermally decomposed product.

4.2. Laboratory Multiphase Reactor Systems

One important factor complicating the processing of coal is its solid state. Processing this solid material under high pressure and high temperature can cause a lot of difficulty, both in the interpretation of results and in the mechanical handling and construction.

When coal is slurried with solvent under hydrogen pressure, we are dealing with a three-phase system. The types of reactors which can effectively handle this complicated three-phase system under severe operating conditions are limited. The most frequently used batch reactor for this high pressure and temperature system is a bomb autoclave. The autoclave is first loaded with the solid reactant, slurried with the solvent. It is then filled with hydrogen and heated to the reaction temperature. After being held for a specified time at this temperature, the contents are cooled and the reaction products are analyzed.

There are many disadvantages in using this type of reactor to obtain the desired data. For example, the work is usually done in a massive rocking autoclave which, after loading, requires a long time to heat to the reactor temperature, making it impossible to obtain accurate reaction rate data.

Recently, various investigators have devised better autoclaves to reduce the thermal inertia of the reactor. For example, Curran et al. [29] have used a high pressure batch autoclave which can be rapidly heated by immersing a low thermal inertia reactor in a heated sand bath, and can be cooled by submersing in water. The volume of the reactor is fairly small, and it can be heated to 400°C in 2.5 min and cooled in 0.5 min. More recently, the research group at Mobil Oil has developed a 300 ml batch autoclave system which can be used to study reactions with a contact time as short as 15 sec [46].

4.3. Some Experiments

To start our discussion, let us consider three series of experiments in coal dissolution carried out by Guin et al. [28]. In these experiments coal was dissolved in different solvents under different gaseous atmospheres with different temperatures. The solution, after a specified time of reaction, was examined photomicroscopically. In other words, photomicroscopy was used to monitor the changes that occurred in the sizes of coal particles during different stages of dissolution. The coal particles were magnified 50 times.

In the first series of experiments, a mixture consisting of six parts creosote oil and one part crushed and dried bituminous coal, a Kentucky No. 9 coal, was heated under 1000 psig hydrogen atmosphere. Upon reaching the desired temperature, the mixture was immediately quenched to room temperature and the resulting mixture

ISBN 0-201-08300-0

was examined photomicroscopically. The various temperatures used were 330, 340, 350, 360, and 400°C. The results of these experiments are as follows:

At 330°C: The coal particle in the mixture is essentially the same as the unreacted original coal. No disintegration is visible.

At 340°C: Some disintegration has taken place, although some original coal particles can still be seen.

At 350°C: Considerable disintegration has taken place and no original coal particles remain intact.

At 360°C: Disintegration appears complete.

At 400°C: *Allow 2 hr of reaction before quenching.* No additional change in physical appearance as compared with the results at 360°C.

In the second series of experiments, four different solvents – creosote oil, tetralin, decalin, and paraffin oil – were used. These solvents and coal mixtures, in a 6:1 ratio, were heated under 1000 psig nitrogen atmosphere to 400°C. The mixtures were allowed to stay at this temperature for 2 hr before quenching to room temperature. The results are as follows:

Creosote oil and tetralin: The disintegration of coal particles appears complete.

Decalin: The disintegration of coal particles is not complete. Only partial disintegration is achieved.

Paraffin oil: The coal particle in the mixture is essentially the same as the unreacted original coal. No disintegration is visible.

The major components of the creosote oil used are given in Table 4.1.

In the third series of experiments, two paraffin oil and coal mixtures, both with a 6:1 ratio of oil to coal, were heated to 400°C and allowed to stay at this temperature for 2 hr before quenching. One of the mixtures was heated under 1000 psig nitrogen pressure, and the other under 1000 psig hydrogen pressure. The results are as follows:

ISBN 0-201-08300-0

Hydrogen atmosphere: Considerable disintegration has taken place, and no original coal particles remain intact. However, the disintegration does not yet appear to be complete.

Nitrogen atmosphere: The coal particle in the mixture is essentially the same as the unreacted original coal. No disintegration is visible.

From the above experiments it can be seen that there are three important factors in coal dissolution: temperature, property of the solvent, and presence or absence of hydrogen atmosphere. A fourth factor, pressure, is also implicitly indicated in the last series of experiments where relatively high hydrogen pressure also promotes coal dissolution.

Table 4.1

Major Components in a Typical Creosote Oil (Guin et al. [28])

Component	Amount, %
Naphthalene	10
2-Methylnaphthalene	8
1-Methylnaphthalene	3
1,2-Dimethylnapthalene	9
Acenaphthalene	5
Fluorene	5
Anthracene and phenanthrene	17
Carbazole, fluoranthene, pyrene	5
Other (about 30 identified)	38
Total	100

ISBN 0-201-08300-0

4.4. Hydrogen-Donor Solvent

It is apparent from the above experiments that, in order for coal to disintegrate in the solvent, the solvent must possess a certain property, known as the hydrogen-donor property. In other words, the solvent can donate its hydrogen to the coal and hence make the coal particle disintegrate and partially dissolve in the solvent. Consider tetralin (1,2,3,4-tetrahydronaphthalene), which has the capability of donating four hydrogens as follows:

H H liquefaction solvent H hydrogenation H + 4H (4.1)

Tetralin Naphthalene Donor hydrogen

For hvA bituminous coal we may represent the liquefaction reaction schematically as follows:

$-CH_2^{(+)}$ + H $\longrightarrow$ $-CH_3$ (4.2)

where tetralin is the donor molecule in the solvent and naphthalene is the spent molecule. Since naphthalene cannot donate any more hydrogen, it is considered a nondonor solvent. Paraffin oil is a nondonor solvent. Decalin has a donor activity between that of tetralin and that of paraffin oil.

A wide variety of solvents function as hydrogen donors as long as they contain mobile hydrogen bonds. In general, a solvent with a high boiling point, with a hydroaromatic structure (such as tetralin with easily available hydrogen atoms), and with polarity (such as amine or phenolic hydroxyl) is a better hydrogen donor.

4.5. Solvent Classification

It is important to distinguish the various solvents used in coal processing. For example, benzene and pyridine serve for the definition of solvation or conversion. The hydrogen-donor solvents such as tetralin are used for hydrogenation. Note that dissolution using benzene is carried out at a much lower temperature than is hydrogenation using a hydrogen-donor solvent.

ISBN 0-201-08300-0

Oele et al. [31, 32] classified the solvents into four groups with respect to their effects on coal. The two groups which are of direct interest to our discussion are the specific solvents and the reactive or chemical disintegration solvents. The specific solvents dissolve appreciable amounts of coal at low temperatures, preferably below 200°C. These solvents have electron donor capabilities. Apparently, the process is a physical solution, and the nature of the solute is identical with that of the original material. Benzene and pyridine belong to this group.

The reactive solvent dissolves coal by reacting chemically with it. The temperature used is normally high, and the chemical composition of the extract is usually markedly different from that of the original material. In other words, this group of solvents has hydrogen-donor capabilities. Tetralin is a typical reactive solvent.

Because of the enormous commercial interest in reactive or hydrogen-donor solvents in coal processing, many investigators have studied solvents with hydrogen-donor properties. Pott and Broche [27] extracted coal with a tetralin-phenol-naphthene ternary mixture. Orchin and Storch [33] tested various solvents at 400°C and divided them into three classes: the very good, the moderately effective, and the less effective solvents. These experiments were carried out with a hydrogen atmosphere. The initial (cold) pressure was atmospheric. The reaction pressure, due to solvent vapors and permanent gases formed at 400°C, was usually 250 psi. The percentages converted or benzene soluble, using a 0.5 hr reaction time, are listed in the middle column of Table 4.2. The coal used had a petrographic composition of 63% anthraxylon 22% translucent attritus, 12% opaque attritus, and 3% fusain. Recall that opaque attritus and fusain are essentially inert toward hydrogenation. It is interesting to note that in the last two rows the benzene-soluble percentage is increased to 94.4% when anthraxylon is used and decreased to only 4.2% when fusain is used with *o*-cyclohexylphenol.

The three classes of solvents can be correlated with the chemical structure of the solvent. The least effective solvent is a high boiling aromatic compound or a hydroaromatic compound which dehydrogenates only slowly at 400°C and low hydrogen pressure. Some of the least effective solvents are dicyclohexyl, naphthalene, *o*-phenylphenol, diphenyl, and cresol. If the solvent functions as a good hydrogen donor and dehydrogenates readily under the hydrogenation conditions, its solvation effectiveness is enhanced. Tetralin is a good example of a moderately effective solvent which can be easily dehydrogenated. A solvent that has, in addition to a hydroaromatic ring, an aromatic hydroxyl group is an extremely effective vehicle for liquefying coal. The solvents listed in the first two rows of Table 4.2 are very good solvents which hydrogenated coal into over 80% benzene soluble. It should be added that among the less effective group the hydroaromatic compounds are generally superior to the pure aromatic compounds.

ISBN 0-201-08300-0

Table 4.2

Effectiveness of Some Solvents for Hydrogenation
(Orchin and Storch [33])

Solvent	Benzene Soluble (%, maf coal basis)	
	Without Catalyst[a]	With Catalyst[b]
o-Cyclohexylphenol	81.6	90.7
1,2,3,4-Tetrahydro-5-hydroxynaphathalene	85.3	–
Tetralin	49.4	82.8
Cresol	32.1	91.7
Dicyclohexyl	27.2	80.8
Naphthalene	22.2	80.4
o-Phenylphenol	19.6	91.0
Diphenyl	19.4	78.1
o-Cyclohexylphenol[c]	94.4	–
o-Cyclohexylphenol[d]	4.2	–

[a] With 1 atm initial (cold) hydrogen pressure without catalyst. The reaction time is 0.5 hr at 400°C with a 4 : 1 solvent/coal ratio.

[b] With 1000 psi initial (cold) hydrogen pressure and 1% SnS + 0.55% NH_4Cl catalyst. The reaction time is 1 hr at 400°C with a 1 : 1 solvent/coal ratio.

[c] Hydrogenate anthraxylon component of coal only.

[d] Hydrogenate fusain component of coal only.

4.6. Industrial Hydrogen-Donor Solvents

In industrial applications the hydrogen-donor solvent is self-generated downstream and usually is a mixture of solvents from a distillation cut. For example, during World War II the solvent used in Germany was the heavy oil boiling above 350°C. During the initial start-up the solvent was from an outside source but was still a mixture of organic solvents. Some of the easily available industrial mixed solvents from coal are creosote oil, light creosote, and anthracene oil. Anthracene and creosote oils are usually obtained from the manufacture of coal tar and are generally known as tar oil.

One particular property of the hydrogen-donor solvent is that a small quantity of it can increase the hydrogen-donor capability of a nondonor mixture tremendously. This is illustrated in Figure 4.1 [33], where the experiments were carried out with a 4 : 1 ratio of solvent to coal at 400°C. Notice the tremendous

ISBN 0-201-08300-0

increase in the benzene-soluble yield when a small amount of tetralin was added to cresol. Another noticeable fact is that a mixture with an equal amount of tetralin and cresol has a much higher hydrogen-donor capability than the pure solvents. These results show clearly that a mixed solvent as used in industry is probably more effective than the individual pure solvents and also explain the effectiveness of creosote oil in the experiments discussed at the beginning of this chapter.

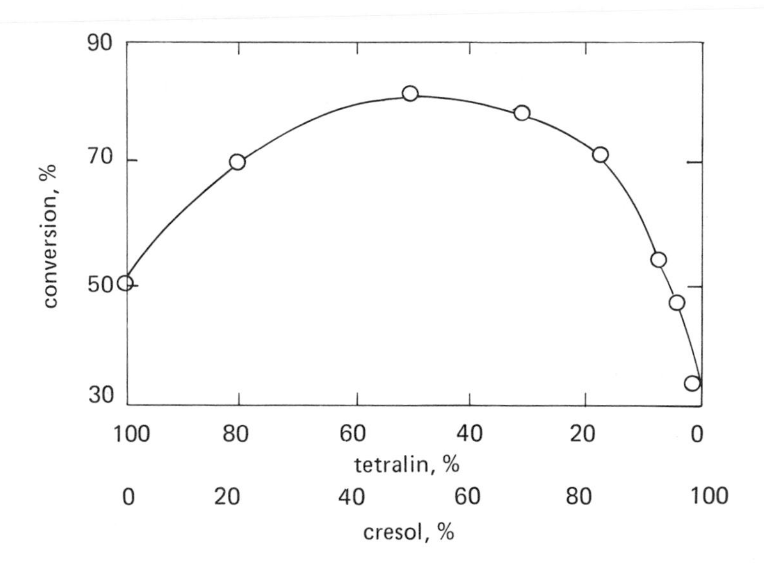

Figure 4.1. Effect of solvent composition on liquefaction. (Orchin and Storch [33])

4.7. Influence of Solvent Properties

Since the solvent plays an integral part in coal dissolution, an understanding of the basic physical and chemical properties of the solvent is essential. Unfortunately, however, most of the knowledge about solvents and polymers is semi-empirical, and the understanding about their behavior is very limited.

For coal dissolution two of the known parameters of a good solvent are that it is thermally stable at the reaction conditions and that it possesses hydrogen-atom donor capacity. However, these are not the fundamental properties in solvent dissolution.

Various investigators have attempted to correlate the effectiveness of solvents with their basic chemical and physical properties. Van Krevelen [5] suggested that, based on available evidence in the extraction of coal studies, an acid-base reaction exists between solvent and coal and the solvent appears to act as a Lewis base which can react with the acid sites in coal. Thus Lewis' basicity of the solvent is an important parameter of the solvent for coal extraction.

ISBN 0-201-08300-0

Dryden [3] studied the physical properties of the solvent and suggested correlating the effectiveness of the solvent by the use of the square of the solubility parameter, which is a measure of the cohesive forces in a solution that has no excess entropy of mixing [64].

Blanks and Prausnitz [63] separated the solubility parameter into two parts for polar compounds: a polar part and a nonpolar part. According to this theory, maximum solubility should occur when the nonpolar parts of the solvent and the solute solubility parameters are equal, provided that the solvents used have similar polarity and hydrogen-bonding tendencies. Because coal is a very complicated material and also because the original coal does not directly dissolve into the solvent, the optimal nonpolar solubility parameter (NSP) for coal dissolution cannot be estimated.

To test this theory, Silver and co-workers [60] recorrelated the data of Kiebler [62] based on the NSP of the solvents. The correlation was reasonably good,and a NSP value of 9.5 $(cal/cm^3)^{1/2}$ for the solvent was found to be the best for the solution of coal. The very effective hydrogen-donor solvent tetralin has a NSP value of 9.4 $(cal/cm^3)^{1/2}$ at 25^oC.

4.8. Solvent Rehydrogenation

One important feature of the hydrogen donor solvent is the fact that spent solvent which contains no more hydrogen donor can be rehydrogenated under hydrogen pressure. Thus a spent and recycled solvent from downstream in an industrial process can be rehydrogenated to nearly its original strength before reuse.

One important point to emphasize is that there is a possibility of overhydrogenating a donor solvent. For example, partially hydrogenated phenanthrene such as 9,10-dihydrophenanthrene is a very good donor solvent and is slightly superior to tetralin in hydrogen donor activity, but perhydrophenanthrene is much less active.

4.9. The Dissolution Mechanism

Now let us look again at the experiments discussed at the beginning of this chapter. No matter what the donor capability is for a given solvent, the temperature must be high enough for this donating activity to be effective. For the particular coal used in the experiments, the minimum temperature for disintegration is 340^oC. Another important fact is that the disintegration, once the required temperature has been reached, is very fast. This is shown clearly from the experiments. The rapidity of disintegration of the coal particle suggests a heat transfer rate controlling mechanism [28], or a purely thermal process. Various other investigators have reached the same conclusion. Curran et al. [29] attempted unsuccessfully to accelerate this hydrogen transfer reaction with various catalysts and hence concluded that the reaction is a thermal process.

ISBN 0-201-08300-0

Neavel [30] measured the rate of conversion of a high volatile C bituminous coal in tetralin solvent at 400°C. The results are reproduced in Figure 4.2. About 10% of the coal is converted to gas and benzene-soluble materials within 20 sec, and approximately 90% of the ash-free coal becomes soluble in pyridine within 2 min. Since 3% of the ash-free coal is the unreactive maceral fusinite, 93% of the potentially soluble coal material is rendered pyridine soluble within only 2 min. All of this reactive coal material becomes pyridine soluble in 10 min. Neavel also observed swelling of the coal particles, accompanied by rapid conversion of coal, during the first few minutes of coal dissolution. This swelling may be attributed to pyrolysis-generated gases.

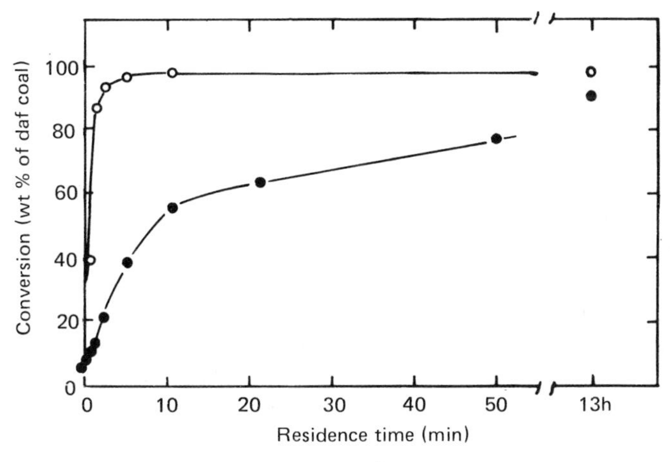

Figure 4.2. Conversion based on benzene and pyridine solubles. (Reproduced from Neavel, *Fuel,* **55**, 1976, by permission of original copyright owners [30].)

In an interesting experiment using gel permeation chromatography, Farcasiu et al. [66] found that early in the conversion process of coal dissolution the product in solution has up to 40% high molecular weight material (molecular weight higher than 2000) or strongly associated molecules. The relative concentration of this material then rapidly declines, producing material in the 300–900 molecular weight range.

In addition to the thermal nature of the process, Curran et al. [29] suggested that the coal extract forms a true solution and not colloidal mixtures. The extracts were separated into three fractions: cyclohexane-soluble oil, benzene-soluble and cyclohexane-insoluble asphaltenes, and a benzene-insoluble and cresol-soluble fraction, with average molecular weights of approximately 400, 750, and 1500,

ISBN 0-201-08300-0

respectively. These molecular weights are too low for colloidal particles.

Wiser et al. [39] have suggested that the dispersion of coal in tetralin appears to be largely a colloidal rather than molecular solution because of the fact that the average molecular weight in the solution is on the order of several thousands. However, as the reaction proceeds, more lower boiling material is produced because of bond rupture and stabilization.

Another generally agreed upon fact is that in the dissolution of coal the rate of extraction is independent of particle size within the particle size range generally used.

4.10. Free Radical Mechanism

Curran et al. [29] suggested that the transfer of hydrogen from a slurry vehicle (such as tetralin) to coal can be described as a free radical reaction in which the coal molecules are thermally cleaved into free radicals, which seek stabilization among the competing possibilities. Obviously, the final stabilization route depends on the energy requirements; the route requiring the least energy constitutes the successful one. Thus, if a donor solvent is present, the available hydrogen atom stabilizes this free radical by hydrogen transfer.

Although free radical formation is a thermal process, the second series of experiments discussed earlier shows that the application of heat alone at this relatively low temperature range is not sufficient to disintegrate the coal particle. For bituminous coals even 400°C is generally too low for any extensive disintegration or pyrolysis. Thus hydrogen-donor solvent or hydrogen transfer promotes the thermal cleavage. This is seen by the fact that, although all the original coal particles disintegrate at 400°C with a hydrogen-donor solvent (tetralin), the same coal at the same temperature with a nondonor solvent (paraffin oil) does not show any disintegration, while with decalin, a not very active hydrogen-donor solvent, even after two hours only partial disintegration is observed.

Thus it is clear that, if a sufficiently active hydrogen-donor solvent is used in sufficient quantity, the hydrogen transfer mechanism is essentially the thermal decomposition of the coal into free radicals.

4.11. Donor and Nondonor Solvents

Another interesting fact was discovered by Neavel [30] by extracting a high volatile C bituminous coal in nondonor vehicles at 400°C. The conversions, defined as benzene solubles and gases on a dry-ash-free coal basis, are plotted in Figure 4.3. The two nondonor vehicles used were dodecane and naphthalene. The results obtained with the donor solvent tetralin are also shown in Figure 4.3.

With the nondonor solvents about 25% of the coal was converted to benzene solubles and gases within 5 min. However, continued reaction in nondonor solvents decreases the yield of benzene-soluble material, indicating a probable mechanism

ISBN 0-201-08300-0

of recombination into high molecular weight substances. It is interesting to note that vacuum pyrolysis of the dry coal at 400°C for 2 hr results in a total weight loss of 27%. Thus approximately 25–30% of the coal can be converted into low molecular weight products without any addition of hydrogen.

From the above observation Neavel [30] concluded that in the extraction of coal "pyrolytic rupture of chemical bonds in the coal results in the formation of highly reactive free radicals. In nondonor vehicles, these free radicals react (combine) with surrounding molecules to form high molecular wieght substances which, though they may remain dispersed in the slurry vehicle, are no longer soluble in subsequently employed extracting solvents such as pyridine or benzene. Alternatively, when a hydrogen-donating vehicle is employed, hydrogen is abstracted from the vehicle and stabilizes pyrolysis-formed free radicals."

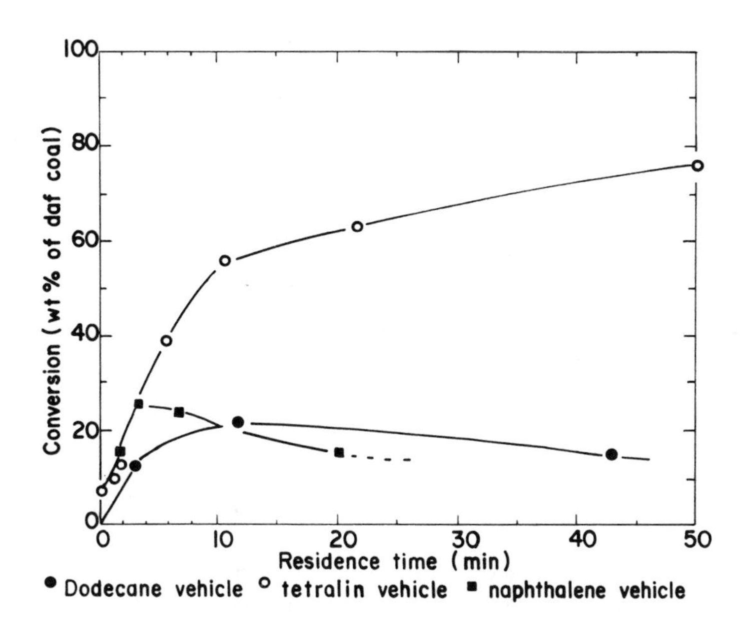

Figure 4.3. Effect of donor and nondonor solvents on conversion. (Reproduced from Neavel, *Fuel,* **55**, 1976, by permission of publishers [30].)

The same conclusion has been reached by various other investigators. For example, Kang et al. [35], in studying the role of solvent in the SRC, concluded that coke formation results when thermal cracking gets ahead of hydrogen-donor-promoted hydrogenation – in other words, when the hydrogen-donor solvent is not active enough.

ISBN 0-201-08300-0

These free radical reactions can be expressed as

$$\text{Coal} \xrightarrow{\text{heat}} 2\text{R} \tag{4.3}$$

$$\underset{\text{(donor H)}}{\text{H}} \; + \; \underset{\text{(free radical)}}{\text{R}} \longrightarrow \underset{\text{(smaller molecular weight products)}}{\text{RH}} \tag{4.4}$$

When there is a deficiency of donor hydrogen, the coal fragments recombine to form coke:

$$\underset{\text{(free radical)}}{n(\text{R})} \longrightarrow \underset{\text{(coke)}}{(\text{R})_n} \tag{4.5}$$

4.12. Pyrolysis and Dissolution

Wiser and co-workers [38, 39] have compared the mechanisms of pyrolysis and dissolution and have concluded that both processes are initiated by thermal cracking at temperatures above 350°C, irrespective of whether other reactants are present in the system. This thermal rupture of covalent bond results in two free radicals. These free radicals can be stabilized via three different routes, depending on the availability of other species and on the free energy of stabilization: (a) addition of atoms or other radical groups to the free radicals, (b) rearrangement of atoms within the free radical, and (c) polymerization of the free radical.

A. ADDITION OF ATOMS OR OTHER FREE RADICALS

This is the case in the dissolution of coal with a hydrogen-donor solvent such as tetralin. Furthermore, it appears that coal and tetralin promote each other in this dissolution reaction.

Another interesting fact was discovered by Heredy and Fugassi [40], who studied the dissolution of coal in phenanthrene. Phenanthrene does not possess the hydrogen-donor property, and yet the yield in this dissolution is very high (80–90% conversion). By replacing the hydrogen in phenanthrene by tritium, these investigators discovered that after the dissolution process the tritium was transferred to the free radicals. The phenanthrene remained the same except that the tritium was replaced by hydrogen. It appears that the free radical formed in coal abstracts a hydrogen from phenanthrene. In turn, the phenanthrene free radical thus formed

ISBN 0-201-08300-0

abstracts a hydrogen atom from the coal, thus producing more new radicals. The net results of this series of reaction can probably be represented as follows [40]:

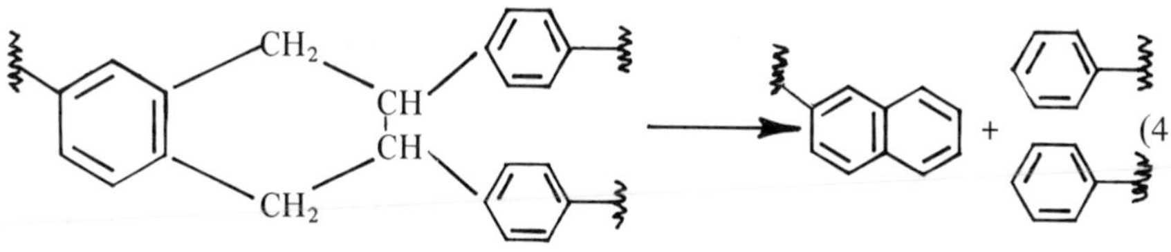

(4.6)

This process of disproportionation results in more aromatic structures of the product by cleavage of the hydroaromatic bonds. The higher aromaticity in the products, as compared to the original coal by infrared analysis, appears to confirm Equation 4.6. Thus phenanthrene probably plays the role of a free radical carrier or hydrogen transfer agent instead of a hydrogen donor. The process involves the thermal breaking of C-C, C-O, and C-H bonds and partial dehydrogenation of the hydroaromatic structure of coal.

B. REARRANGEMENT OF ATOMS

With no hydrogen-donor solvent, rearrangement of the free radical is possible if the latter contains hydroaromatic units or other unstable structures.

C. POLYMERIZATION

If the free radical contains only the stable aromatic units and no other means of stabilization is available, the free radical will polymerize into large, insoluble molecules.

4.13. Influence of Hydrogen

From the third series of experiments discussed earlier in this section, we can conclude that the presence of a hydrogen atmosphere is important for coal dissolution. Coal dissolves, even though fairly slowly, in a very poor hydrogen donor such as paraffin oil if molecular hydrogen is present under pressure in the atmosphere. The exact mechanism of this dissolution is not clear. The hydrogen molecule can react directly with coal, or the hydrogen molecule can transfer the hydrogen atom to coal through the nondonor solvent.

As has been shown by Weller and co-workers [34], the presence of hydrogen pressure is also important, even for hydrogen-donor solvents. These investigators

ISBN 0-201-08300-0

found that, when coal is extracted with tetralin in a 4 : 1 solvent/coal ratio at 455°C for 1 hr, the results are greatly influenced by the presence or absence of hydrogen. With a 1000 psia hydrogen pressure at the beginning of the run, the benzene-insoluble percentage is reduced to 13.8%, whereas with a 1000 psia nitrogen pressure the benzene insoluble is only reduced to 25.3%. (See Table 4.3.)

Table 4.3

Sulfur Contents in Products[a] (Yen et al. [34])

T, °C	Gas[b]	Catalyst,[c] %	Sulfur, wt%		Sulfur,[d] g, in B.I.	Product Distribution, wt % total product			
			Oil	Asphaltene		Liquid	Asphaltene	Oil	B.I.[e]
405	H_2	1	0.41	1.21	1.45	67.2	41.2	18.8	32.8
455	N_2	0	0.34	1.10	1.34	74.7	41.3	24.2	25.3
455	H_2	0	0.25	0.47	0.80	86.2	49.9	25.4	13.8
455	H_2	1	0.21	0.47	0.52	86.2	38.0	36.9	13.8
455	H_2	10	0.093	0.25	0.58	89.4	27.2	50.7	10.6

[a] Tetralin/coal ratio = 3.2 for the run at 405°C and = 4.0 for others.

[b] Initial gas pressure 1000 psia.

[c] 42–60 Mesh Harshaw 0402T catalyst.

[d] Amount of coal used in each run is 75 g, which contains 3.48 g sulfur (1.08 g organic and 2.40 g inorganic).

[e] Benzene insoluble on catalyst-free basis.

4.14. Hydrogen Consumption

Since coal dissolution is essentially a thermal cleavage followed by hydrogen stabilization, each gram mole of hydrogen transferred should be associated with the stabilization of two free radicals. Therefore there exists a direct relationship between hydrogen transferred and amount of coal dissolved. The results of Curran et al. [29] with various donor solvents are plotted in Figure 4.4, where conversion is based on material insoluble in xylenol solvent at 220°F, except for the last two points, with hydrogen transferred over 1.2%. The conversion for these two points is based on material insoluble in cresol. In all runs ordinary atmospheric pressure and regular air above the charge were used before the autoclave was sealed.

From this figure it can be seen that the conversion versus hydrogen transferred is approximately independent of the solvent and temperature used within the range investigated. However, the hydrogen consumption rate increases rapidly with the increase in coal conversion. Only 0.2% or less of hydrogen is needed to obtain the first 50% of coal conversion. To reach 92% conversion, 1.4% hydrogen consumption is required. Thereafter any more increase in hydrogen consumption does not increase the conversion.

ISBN 0-201-08300-0

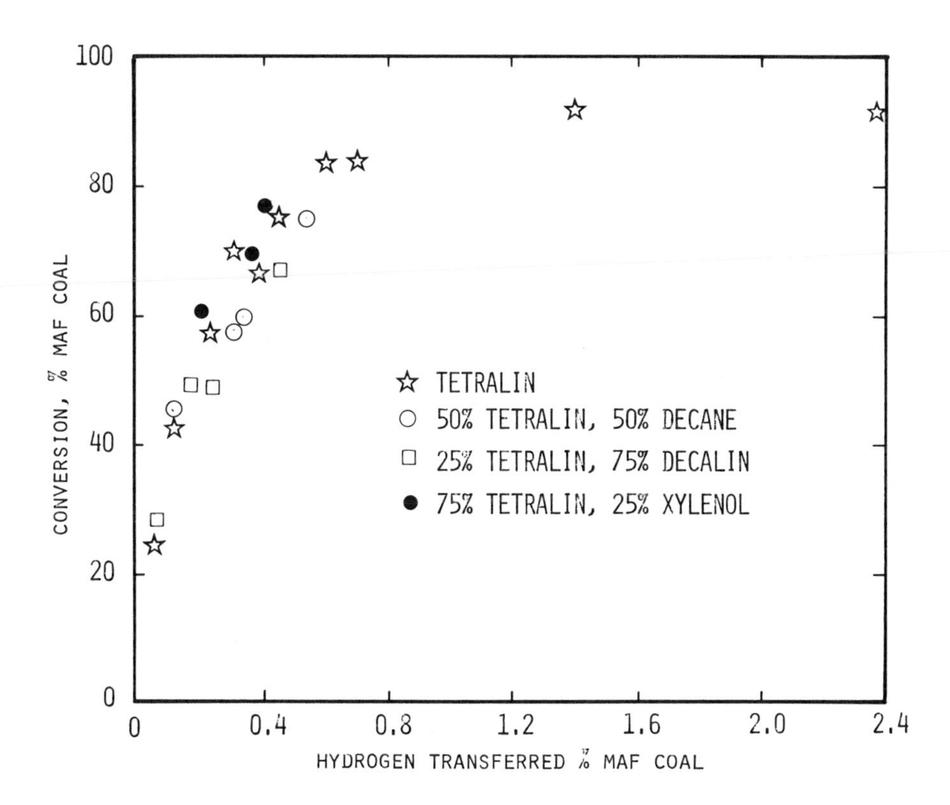

Figure 4.4. Coal conversion versus hydrogen transferred at 355°C to 440°C. (Curran et al. [29])

Curran et al. [29] also estimated the practical maximum quantity of hydrogen that can be transferred in a dissolution process without catalyst. Although more hydrogen can be transferred under more severe reaction conditions, little can be gained in the way of obtaining more liquid products. The maximum limit for coal dissolution is 2.6% hydrogen transferred, based on maf coal.

Recently, Neavel [30] investigated the hydrogen transfer process in dissolution. After reacting a hvC bituminous coal at 400°C with tetralin for 13 hrs, the hydrogen transferred was 2.7% based on maf coal. This result agrees quite well with those given above.

There appears to be an exponential relation between hydrogen transferred and coal conversion. Neavel [30] studied this problem in more detail and concluded that during the later stages of the reaction the hydrogen consumed is predominantly used to produce more benzene soluble. Thus each successive increment of benzene soluble produced consumes more donor hydrogen than the previous incre-

ISBN 0-201-08300-0

ment. The 10% incremental conversion from 40 to 50% consumes 0.23 part by weight of hydrogen, whereas the incremental conversion from 80 to 90% requires about 0.8 part by weight of hydrogen.

Whitehurst and Mitchell [46] also reached essentially the same conclusion. They observed that coal dissolution is very fast in a hydrogen-donor solvent and that the pressure of hydrogen gas in the early stages of dissolution is not essential. The initial products contain significant amounts of high molecular weight (over 2000) material which is rapidly converted into low molecular weight (300–900) products. This is shown in Table 4.4. The experiment was carried out at 800°F. At the early stages up to 40% of the product is high molecular weight material, which then declines rapidly to produce the low molecular weight fraction. The highest yield of solvent-refined coal is obtained early in the reaction. Improvement in the quality of the yield is accompanied by a decrease in yield and a large increase in hydrogen consumption.

Table 4.4

Increase in Low Molecular Weight Material during Dissolution, wt % coal (Whitehurst and Mitchell [46])

Time, min	Low Molecular Weight	High Molecular Weight	Total
1.20	16.86	10.97	27.80
3.60	15.41	6.13	21.53
6.00	44.32	8.48	52.80
19.50	44.99	6.70	51.69
38.00	65.77	6.87	72.64
74.00	66.45	4.79	71.24
137.50	70.54	4.03	74.57

4.15. Coal Dissolution with Carbon Monoxide and Steam

Since hydrogen is very expensive, it would be ideal if some forms of cheaper gas could be used. This is especially true for low rank coals such as lignite, where, because of the high oxygen content of the coal, hydrogen consumption is high.

The use of carbon monoxide and water or synthesis gas and water to liquefy low rank coals has been studied by various investigators. The idea was originally reported by Fischer in 1921. Recently, Appell and co-workers [83–86] did considerable investigation and found that, when carbon monoxide and water are used, conversion is easiest for low rank coals. The highest conversion was obtained with

ISBN 0-201-08300-0

lignite. Furthermore, lignite dissolves very quickly in the presence of carbon monoxide and water. With solvent the dissolution is essentially complete within 10 min at 380°C. At this temperature and time the conversion of lignite with hydrogen is only approximately half of that obtained with carbon monoxide and water. However, the conversion is slow with bituminous coal and, with carbon monixide and water, is seldom over 75%. Subbituminous coal is intermediate in reactivity.

The effectiveness of carbon monoxide and water in liquefying a subbituminous coal is shown in Table 4.5 [86]. Notice that, within the temperature range shown, higher conversion was always obtained when carbon monoxide and water were used; conversion was much lower with hydrogen. No solvent was used for the data listed in Table 4.5, where the initial pressure indicates the initial cold pressure, and the maximum pressure indicates the maximum pressure reached during dissolution. Conversion in this table was defined on the basis of benzene-soluble and gaseous products.

Table 4.5

Comparison of Carbon Monoxide and Hydrogen for Liquefying Subbituminous Coal (Appell et al. [86])

Gas	Water, ml	Pressure, psig		Time, min	Temperature, °C	Conversion, %
		Initial	Maxium			
CO	15	1500	4200	15	375	43
H_2	None	2500	5400	15	375	27
CO	30	1000	3970	30	400	60
H_2	None	2000	4670	30	400	40
CO	30	1000	4200	30	425	70
H_2	None	2000	4650	30	425	63

4.16. Mechanism with Carbon Monoxide and Steam

Formerly the effectiveness of carbon monoxide and water on low rank coals was thought to be due to the presence of nascent hydrogen, which is activated hydrogen formed by the water-gas shift reaction during dissolution. However, it is now believed that the reaction is much more complex and a number of factors may be involved. For example, the reactivity of lignite decreases considerably on aging. Mass spectrometric analysis of the benzene-soluble products from fresh lignite and an aged one revealed that a major difference in composition is an increase in methylated aromatic compounds in the product from the more reactive lignite [85]. Thus the introduction of alkyl groups may be one reason for the effectiveness

ISBN 0-201-08300-0

of carbon monoxide and water on low rank coals. Another reason could be the unique ability of carbon monoxide to cleave certain types of bonds or inhibit condensation reactions leading to benzene-insoluble materials. For example, it has been found that carbon monoxide is highly active and selective for reducing carbonyl groups, whereas hydrogen causes more cracking. Low rank coals contain more carbonyl groups than do high rank coals [86]. Furthermore, low rank coals contain the alkaline materials that are converted to formates, which are the probably active reducing agents for this reaction.

Experimental results with solvents show that both carbon monoxide and water must be present to obtain high conversion. Furthermore, an increase in carbon monoxide pressure has a greater effect upon conversion than does increasing steam pressure. Conversion also increases when the carbon monoxide-steam pressure is raised. The presence of a good solvent is also important for high conversion.

4.17. Synthesis Gas

Instead of carbon monoxide, synthesis gas has been used as the reducing agent. The process with synthesis gas and steam to liquefy lignite is known as the costeam process. The use of synthesis gas instead of carbon monoxide has at least two advantages: (a) synthesis gas is much cheaper and can be easily obtained by coal gasification, and (b) the hydrocracking activity of hydrogen can reduce the molecular weight of the product and hence its viscosity. It has been found that the product obtained with carbon monoxide and water from the liquefaction of lignite is too viscous to be easily filtered. Thus a limited cracking is desirable to obtain a liquid of sufficiently low viscosity. Furthermore, the higher reactivity of carbon monoxide is counterbalanced by the low cost of synthesis gas.

The liquefaction of low rank coals with synthesis gas in an autoclave and with anthracene oil as a solvent has been studied by Appell et al. [86]. Their results are reproduced in Figure 4.5 and Table 4.6. Conversion increases with increasing carbon monoxide in the synthesis gas, as shown in Figure 4.5. However, at high pressure this increase is not too significant; hence hydrogen can be added at high pressure without decreasing conversion, where conversion is defined as benzene solubles and gases on a maf coal basis. The addition of tetralin also increases

ISBN 0-201-08300-0

conversion. It should be mentioned that high conversion was obtained only with high agitation, thus increasing the gas-liquid contact. The ratios of maf lignite to anthracene oil solvent and to water used in Figure 4.5 were 3:6:1, and the temperature was 420°C. The synthesis gas used was composed of carbon monoxide and hydrogen.

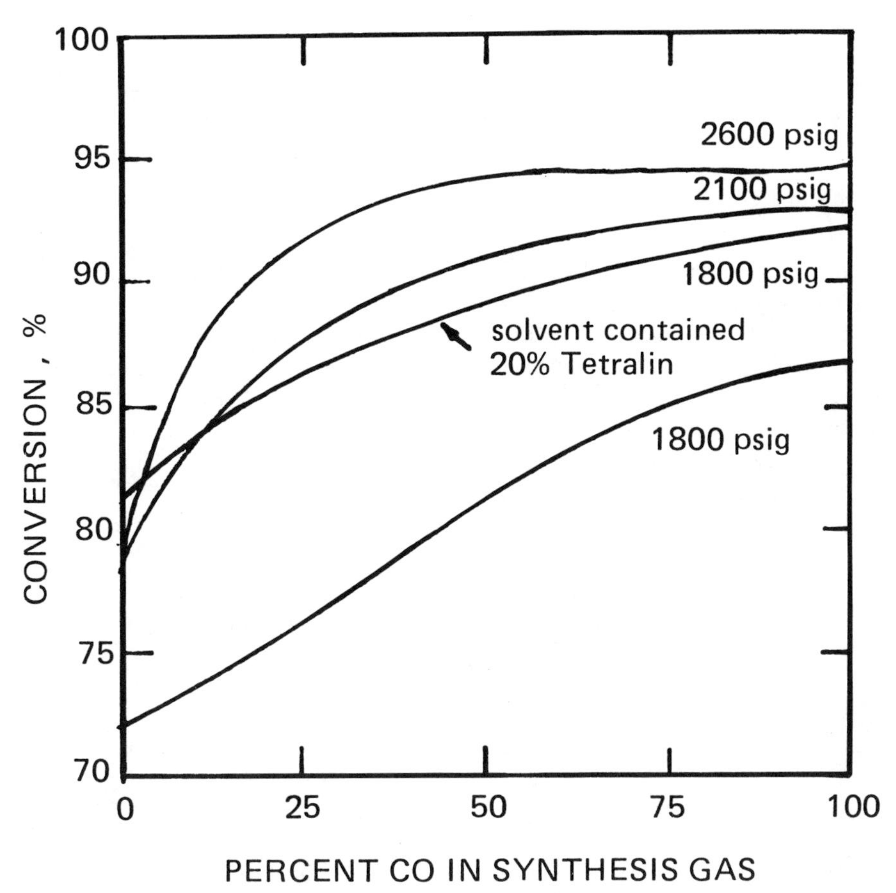

Figure 4.5. Effect of pressure and carbon monoxide on lignite conversion. (Appell et al. [86])

ISBN 0-201-08300-0

Table 4.6

Effect of Temperature and Residence Time on Conversion with Synthesis Gas (Appell et al. [86])

Temperature, °C	Time, min	Conversion, %	Liquid Yield, %	Kinematic Viscosity, cSt at 60°C	Gas Used, ft^3	
					CO	H_2
450	15	96	62	39	0.250	-0.075[a]
460	15	94	43	36	.229	-.014[a]
475	15	94	42	28	.259	.019
450	3	94	56	52	.221	-.018[a]
460	3	92	53	48	.217	-.010[a]
475	3	94	50	44	.222	.031

[a]Minus figures indicate hydrogen formation.

The influence of temperature and reaction time on conversion, liquid yield, and viscosity is shown in Table 4.6, for experiments carried out at 4000 psig pressure. Product viscosity decreases as temperature increases. However, the liquid yield, which is the benzene-soluble oil product on a maf coal basis, decreases with increasing temperature. The amount of carbon monoxide used is nearly constant within the temperature and reaction time considered. The hydrogen produced via the water-gas shift reaction is greater than the hydrogen consumed in the dissolution for all cases except at 475°C.

Table 4.7

Feed Composition to Continuous Reactor, wt % (Del Bel et al. [87])

Gas Composition	CO	$0.7H_2 : 1.0CO$	$3H_2 : 1.0CO$
Maf coal	19.1	20.0	20.7
Ash	2.0	2.1	2.1
H_2O[a]	8.9	9.4	9.7
Anthracene oil	58.9	61.9	63.9
Total gas	11.1	6.6	3.6

[a]Include the water in coal.

ISBN 0-201-08300-0

Del Bel et al. [87] investigated the liquefaction of lignite with anthracene oil as the pasting vehicle in a continuous flow stirred tank reactor. Three different gaseous compositions were used. The feed compositions in weight percentages used for these three different conditions are given in Table 4.7. The liquid yield as a function of temperature is shown in Figure 4.6 for a 1 hr residence time and 4000 psig pressure. When carbon monoxide and water are used, there is an optimum temperature at 400°C. However, when syngas is used, the liquid yield increases with increasing temperature. The liquid yield continues to increase with increasing pressure for all three gas compositions used.

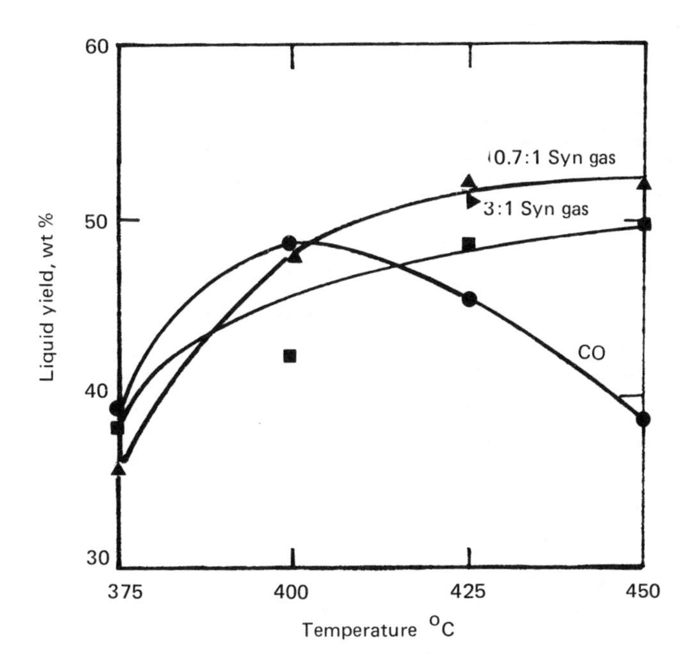

Figure 4.6. Liquid yield versus temperature at 4000 psig and 1 hr residence time. (Del Bel et al. [87])

5. KINETICS OF SOLVENT EXTRACTION

5.1. Kinetics of Hydrogen Dissolution

To understand the coal dissolution mechanism, the mechanism of hydrogen during dissolution must be studied.

Curran et al. [29] studied the mechanism of transfer of hydrogen to coal during dissolution. They found that a simple first-order expression cannot represent the kinetics. For the case of high solvent/coal ratio, a series of simultaneous first-

ISBN 0-201-08300-0

order kinetics was proposed. In correlating the data, only two simultaneous reactions were used. The equation for the two simultaneous reactions can be represented by

$$H_T - H = \alpha H_T e^{-k_1 t} + (1 - \alpha) H_T e^{-k_2 t} \tag{5.1}$$

where H_T = maximum amount of hydrogen that can be transferred
= 2.6% maf coal

k_i = reaction rate constants

H = hydrogen transferred, % maf coal

The above model is based on two types of bonds in the coal, one relatively strong, and the other relatively weak. During the cleavage of the weak bonds, the hydrogen required is relatively low and thus the transfer rate is low. This cleavage of weak bonds occurs during the early coal conversion period, and the faster hydrogen transfer rate associated with the cleavage of strong bonds in coal probably takes over in the coal conversion range of 50–90%. Notice that the fast hydrogen transfer rate corresponds to the slow coal conversion rate and vice versa. Equation 5.1 represents the experimental data quite well.

When the solvent/coal ratio is low, these investigators assumed that the coal or extract also has built-in hydrogen-donor structures. Thus the free radicals can obtain hydrogen from either the solvent or the coal:

$$\begin{aligned}
&M \xrightarrow{k_0} 2R \\
&R + D \xrightarrow{k_1} RH + D^{\cdot} \\
&R + M \xrightarrow{k_2} RH + M^{\cdot} \\
&R + D^{\cdot} \xrightarrow{k_3} RH + (D\text{-}H_2) \\
&R + M^{\cdot} \xrightarrow{k_4} RH + (M\text{-}H_2)
\end{aligned} \tag{5.2}$$

where D and M represent tetralin and coal (extract), respectively; R is the free radical; $D^{\cdot}$ and $M^{\cdot}$ correspond to the radicals produced by abstraction of a hydrogen atom from tetralin and from coal (extract), respectively; and $(D\text{-}H_2)$ and $(M\text{-}H_2)$ are, respectively, dihydronaphthalene and the corresponding dehydrogenated structure derived from coal or the extract. Good fit was obtained by these investigators for the experimental data by using mechanism 5.2.

Guin et al. [45] studied both the solubility of hydrogen in creosote oil and the kinetics of hydrogen dissolution in a bituminous coal by using creosote oil as

ISBN 0-201-08300-0

solvent. These investigators found that the solubility of hydrogen in the coal liquefaction solution is appreciable and may be represented by Henry's law.

Since three phases – solid coal, donor solvent, and hydrogen gas – are present in the coal dissolution system, several mass transfer effects may be important. However, Guin et al. [45] discovered that a first-order homogeneous chemical reaction can represent the data adequately and that none of the possible diffusional effects is controlling.

The possible diffusional resistances are the transfer of hydrogen across the gas-liquid interphase, and the transfer of reactants and solvents into the surface and the pores of coal. Since coal readily disintegrates in the presence of a hydrogen-donor solvent and the particle size of coal appears to have little effect on the rate of dissolution, only mass transfer across the gas-liquid interphase may be important.

Assuming a first-order reaction for the dissolved hydrogen in the liquid phase, these investigators obtained the following hydrogen dissolution and reaction equation:

$$\frac{H_L}{H_{L_0}} = \frac{H_G}{H_{G_0}} = \frac{H_T}{H_{T_0}} = e^{-\alpha k t} \qquad (5.3)$$

where α is a parameter and is related to the solubility of hydrogen, H represents the mass of hydrogen in grams, k is the first-order homogeneous rate constant in the liquid phase, and t is time in minutes. The subscripts T, L, and G represent total, liquid phase, and gas phase, respectively. The subscript 0 indicates the amount at time, $t = 0$.

Equation 5.3 fits the data quite well. An Arrhenius plot gives an activation energy of 21 kcal/mol with a frequency factor of 1.06×10^5 per minute.

5.2. Conversion of Tetralin

Various solvents have been used in coal dissolution. However, except for tetralin, little attention has been given to the conversion of the solvents during dissolution. Even for tetralin only a beginning has been made in the possible reactions of the solvent.

Neavel [30] investigated the conversion of tetralin. When tetralin was heated alone and with charcoal for 0.5 hr at 400°C, no dehydrogenation or cracking was detectable. Thus the presence of coal promotes the dehydrogenation of tetralin under these reaction conditions. The same conclusion has also been reached by various other investigators.

The results of Neavel [30] on the conversion or disproportionation of tetralin when used as a vehicle in the dissolution of hvC bituminous coal are reproduced in Figure 5.1. The reaction was carried out at 400°C and the tetralin/coal ratio was 2:1. The concentrations of the various components are expressed on a total tetralin and coal weight basis. The major reaction products are naphthalene, dihydro-

ISBN 0-201-08300-0

naphthalene, and decalin (decahydronaphalene). At 90% conversion about 35% of the reactor contents is tetralin, 30% naphthalene.

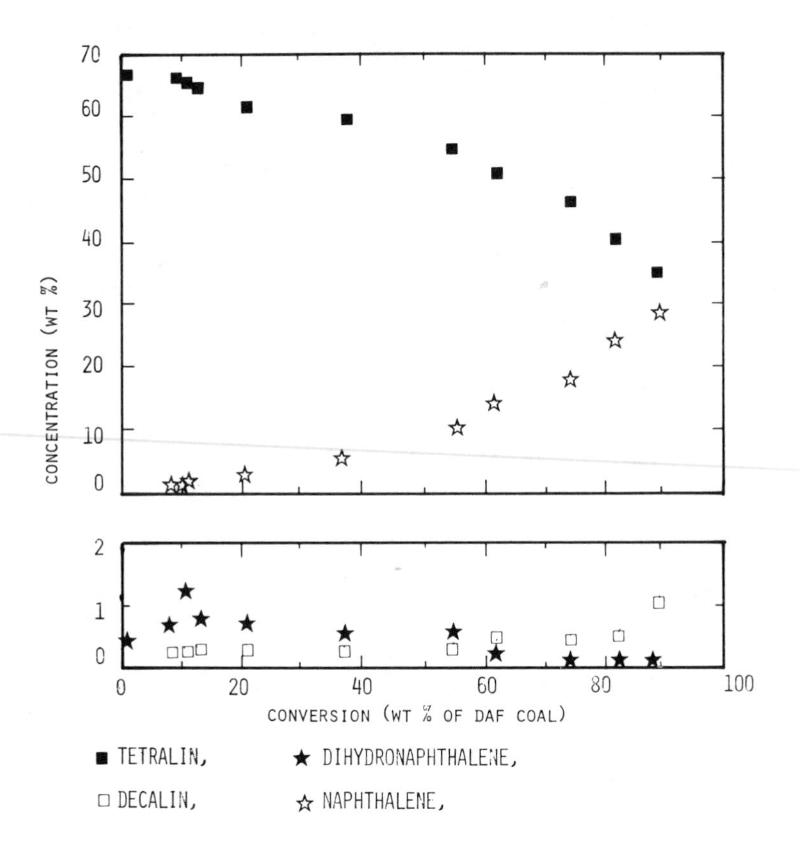

Figure 5.1. Tetralin derived components in reaction products during dissolution. (Reproduced from Neavel, *Fuel,* **55**, 1976, by permission of publishers [30].)

Yen et al. [34] calculated the equilibrium composition for the following disproportionation reaction of tetralin to naphthalene and decalin:

$$5C_{10}H_{12} = 3C_{10}H_8 + 2C_{10}H_{18} \tag{5.4}$$

These investigators compared the equilibrium composition with the actual composition during dissolution of an hvA bituminous coal under 1000 psia initial (cold) pressure. They discovered that at 405°C the disproportionation reaction occurs slowly in dissolution and is far from reaching equilibrium after 1 hr reaction. In

ISBN 0-201-08300-0

the absence of any catalyst, and at 455°C, reaction 5.4 occurs fairly fast and appears to be independent of the presence or absence of a hydrogen or gas atmosphere.

Potgieter [59] has investigated the kinetics of the conversion of tetralin during the hydrogenation of coal in the presence of catalyst and has suggested that the conversion of tetralin, mainly to naphthalene, may be either a reversible or a nonreversible reaction, depending on the catalyst employed.

5.3. Pollutant Removal

Environmentally, one of the main reasons to avoid large scale utilization of coal is the presence of the undesirable pollutants sulfur and nitrogen, and the large amount of ash in coal. Thus the degree of removal of these pollutants during dissolution becomes an important factor to consider.

The ash in coal can be removed easily as solids after the coal is dissolved in solvent. This is true also for most of the pyritic sulfur. However, organic sulfur and nitrogen, which are usually present as heteroatoms, can be removed only by dissolution and by converting them into H_2S and NH_3.

From the pollution standpoint most of the sulfur pollutants come from fuel sources during combustion. However, NO_x pollutants arise from two different sources during combustion: molecular nitrogen in the combustion air is oxidized to form thermal NO_x, and nitrogen, which is present in the fuel, is converted to fuel NO_x. Thus the most effective way to eliminate NO_x emissions is to modify the combustion process.

Economically, one of the most expensive aspects of coal dissolution is the high cost of hydrogen. The economics of the process can be improved tremendously if hydrogen consumption can be reduced.

On the basis of the above discussions, the minimum hydrogen requirement, assuming that all the hydrogen gas used is for the removal of these pollutants, can be estimated. To convert 2.5% sulfur in coal completely into H_2S requires a hydrogen consumption equivalent to approximately 0.15% by weight of hydrogen gas based on total weight of coal, and to convert 0.5% nitrogen in coal completely into NH_3 requires approximately 0.10% of hydrogen gas. However, in actual practice the hydrogen consumption rate is much higher.

Wolk et al. [36] collected hydrogen consumption data for the liquefaction of various coals and concluded that to reduce the sulfur levels in coal to 0.5–1.0% a hydrogen consumption rate equivalent to 1.5–6.0% of the coal is required for bituminous coal. Because of the higher amount of oxygen in subbituminous coal the hydrogen consumption is moderately higher. The oxygen concentration in bituminous coal is usually in the range of 7–12%, and in subbituminous coal is 15–20%. Most of the hydrogen is used to produce gas and to reduce the large molecules to liquid hydrocarbons. Thus, if producing a clean burning SRC (solvent-refined coal) is the sole purpose, improvement can certainly be made, at least

ISBN 0-201-08300-0

theoretically, in reducing the hydrogen consumption rate. One possible approach is the development of a more selective catalyst.

5.4. Hydrodesulfurization

When a hydrogen donor reacts with coal, a fairly large amount of sulfur, oxygen, and nitrogen in the coal is eliminated as hydrogen sulfide, water, and ammonia. This elimination could be a direct result of the cleavage of the hetero-bonds present in coal.

The variables which may influence this reaction appear to be the nature and the amount of the hydrogen-donor solvent, the presence and absence of hydrogen pressure, the level of this pressure, the reaction temperature, the reaction time, and the nature (rank and type) of the coal.

Yen et al. [34] have shown that, when coal is treated with tetralin in the absence of catalyst for 1 hr at 455°C, the presence of hydrogen atmosphere is important to increase the conversion and also to increase the amount of sulfur removed. Their results were reproduced in Table 4.3. When hydrogen pressure (third row of the table) is used instead of nitrogen (second row), the sulfur removal rate is doubled in the product oil and asphaltene. The coal used by these investigators was a high sulfur, high ash, hvA bituminous coal with a total sulfur content of 4.64% (0.61% sulfate, 2.59% pyrite, and 1.44% organic) and an ash content of 14%.

Whitehurst and Mitchell [46] observed that about 40% each of sulfur and oxygen is removed readily and rapidly in a kinetically parallel fashion with little or no consumption of hydrogen. Thereafter, considerably more hydrogen is needed to remove these elements.

Gary et al. [37] have investigated the influence of the various variables on desulfurization in the dissolution process by the use of experimental design. Using anthracene oil as the donor solvent and assuming that the presence of hydrogen pressure is essential, they concluded that the most important variables are the temperature, the partial pressure of hydrogen, and the solvent/coal ratio for a given coal. The reaction time appears not to be significant as long as it exceeds 15 min.

Koltz et al. [44] studied the kinetics of desulfurization during the hydrogenation of a bituminous coal with anthracene oil as solvent under hydrogen pressure. The temperature range used was 360-420°C. However, because of the complex reactions involved, efforts at correlating the data into an *n*th-order reaction were unsuccessful, and the reaction appears to change order as conversion and temperature change. Following the approach of Hill and co-workers [42], Koltz et al. used a first-order reaction with variable reaction rate constants:

$$\frac{dx}{dt} = k(1 - x) \tag{5.5}$$

ISBN 0-201-08300-0

It was found subsequently that

$$k = k_0(1 - ax)$$

Thus the reaction can be represented by a pseudo-second-order equation:

$$\frac{dx}{dt} = k_0(1 - ax)(1 - x) \tag{5.6}$$

Equation 5.6 represents the data quite well at low conversions, for which an activation energy value of 33 kcal/mol was obtained.

Koltz et al. concluded that the desulfurization of coal is affected to a large extent by the nature of the coal. Even coal taken from the same sample will show large variations in the percent of desulfurization if it is not carefully mixed.

5.5. Kinetics of Dissolution

Various investigators have studied the kinetics of hydrogenation without catalyst. However, because of the complexity of the process, only a start has been made, and much more work needs to be done to gain a basic understanding of the reaction mechanisms.

In establishing the kinetics of coal dissolution, it must be remembered that the system is very complex and many reaction steps, both in parallel and in series, may occur. It would be impossible, both experimentally and theoretically, to establish each reaction mechanism among all the reactions. Thus the reaction expression is usually a semiempirical one. For consecutive reactions this expression represents only the slowest step.

Various investigators have studied the kinetics of dissolution of coal. The work of Oele et al. [31] was widely used over a decade ago. These investigators assumed that the reaction proceeds with a zero-order forward reaction and a first-order backward reaction. The reaction rate equation can be represented by

$$\frac{dx}{dt} = k_f - k_b x \tag{5.7}$$

where x is the fraction of coal converted, t is the time of reaction, and k_f and k_b are the forward and backward reaction rate constants, respectively.

Hill et al. [41] proposed that the dissolving of the organic material from coal could be considered as first order with respect to both the coal and the solvent. However, the hydrogen transfer reaction from tetralin to coal was regarded as second order. The kinetic expression used was

$$\frac{dx}{dt} = k_1(a - x) + k_2(a - x)(b - x) \tag{5.8}$$

ISBN 0-201-08300-0

where a is the initial concentration of coal, and b is that of tetralin. These investigators discovered that the constant k_1 for the "chemical" dissolution is an average value and that the process involves a series of reactions with increasing activation energies. It was proposed that at the beginning of the reaction the materials which enter the liquid phase are those that were trapped in the coal pores and thus require the lowest activation energy to dissolve. The rest of the coal dissolves with the breaking of chemical bonds and consequently requires more activation energy; as the extraction continues, the activation energy of extraction increases. Thus a series of reactions can be proposed:

$$\begin{aligned}
&\text{Coal} + \text{solvent} \xrightarrow{k_1} R_1 + L_1 + G_1\\
&R_1 + \text{solvent} \xrightarrow{k_2} R_2 + L_2 + G_2\\
&R_2 + \text{solvent} \xrightarrow{k_3} R_3 + L_3 + G_3\\
&\quad\vdots\\
&R_n + \text{solvent} \xrightarrow{k_n} R_{n+1} + L_{n+1} + G_{n+1}
\end{aligned} \tag{5.9}$$

where R_i is the solid coal residue, L_i is the extract in solution, and G_i is the gaseous products, with

$$k_1 > k_2 > k_3 \ldots > k_n \tag{5.10}$$

At the beginning of the dissolution process, the first reaction with k_1 is the main route. When reaction 1 is well advanced, reaction 2 becomes the main route with the extraction of R_1, and so on. Based on this concept, a first-order rate expression with variable rate constant k_v is proposed:

$$\frac{dx}{dt} = k_v(1 - x) \tag{5.11}$$

It was found that k_v changes linearly with the conversion, x, thus:

$$k_v = C_1 - C_2 x = k_0(1 - ax) \tag{5.12}$$

Equation 5.11 now becomes

$$\frac{dx}{dt} = k_0(1 - ax)(1 - x) \tag{5.13}$$

ISBN 0-201-08300-0

where a was found to be the reciprocal of the maximum possible conversion, x_m. Rearranging the equation with x_m, we have

$$\frac{dx}{dt} = k(x_m - x)(1 - x) \tag{5.14}$$

A pseudo-second-order rate expression is obtained, where x_m is the possible maximum value of x in weight fraction. This expression fits the kinetic data quite well for most of the conversions except fairly high ones. At high conversions a first-order rate expression fits much better than the second-order equation [42]. The enthalpy of activation for the second-order region was obtained as 28.8 kcal/mol, and for the first-order region as 15.6 kcal/mol.

Wen and Han [43] obtained a much better semiempirical correlation by using the experimental data obtained by the University of Utah and the Pittsburgh and Midway Coal Mining Company. The rate of dissolution was correlated as a function of undissolved solid organics and the coal/solvent ratio. The Arrhenius temperature dependence and an exponential hydrogen partial pressure dependence are also included in the expression.

The rate expression is represented by

$$\frac{dx}{dt} = r = kC_{So}(1 - x)\frac{C}{S} \tag{5.15}$$

where
- x = conversion, defined as solid organics dissolved, divided by solid organics in the untreated coal
- r = rate of dissolution, g/hr cm^3 of reactor volume
- k = rate constant, g/hr cm^3 of reactor volume
- C_{So} = weight fraction of organics in untreated coal
- C/S = solvent/coal ratio

Integrating Equation 5.15, we have

$$\ell n(1 - x) = -k\theta \tag{5.16}$$

where

$$\theta = C_{So}\frac{t}{C_{ao}}\frac{C}{S} \text{ for batch reactor} \tag{5.17}$$

and

$$\theta = C_{So}\frac{V}{F}\frac{C}{S} \text{ for plug flow reactor} \tag{5.18}$$

ISBN 0-201-08300-0

with t = reaction time, hr

c_{ao} = ash-free coal, g/cm^3 of reactor volume

V = volume of plug flow reactor, cm^3

F = mass flow rate of solid organics fed into the reactor, g/hr

A batch reactor was used at the University of Utah with an experimental range of 400–500°C temperature, 0–2000 psig hydrogen pressure, and 2–3 min residence time. The hydrogen-donor vehicle was a coal-derived oil.

The Pittsburgh and Midway Coal Mining Company's experiments were carried out in a tubular flow reactor with and without recycle. Assuming plug flow, the results without recycle can be represented by Equations 5.16 and 5.18.

The results with recycle can be expressed as

$$\ell n \frac{1 - x_2}{1 - x_1} = -k\theta \tag{5.19}$$

where θ is represented by Equation 5.18, and x_1 and x_2 are the conversions at the entrance and the exit of the reactor, respectively.

Assuming an exponential dependence of the reaction on hydrogen partial pressure, the rate constant can be expressed as

$$k = k_0 e^{-E/RT} e^{\alpha p_{H_2}} \tag{5.20}$$

where p_{H_2} is the hydrogen partial pressure (in psia), and the constant α is found to be

$$\alpha = 0.000684 \tag{5.21}$$

By using Equations 5.16, 5.17, 5.20, and 5.21, the conversion, x, can be calculated. The calculated values are compared with the experimental values in Figure 5.2. As can be seen, the predicted values agree very well with the experimental ones. The calculated values agree equally well with the plug flow reactor case with or without recycle.

The activation energy is 11 kcal/mol for the batch reactor and 4.5 kcal/mol for the plug flow reactor.

ISBN 0-201-08300-0

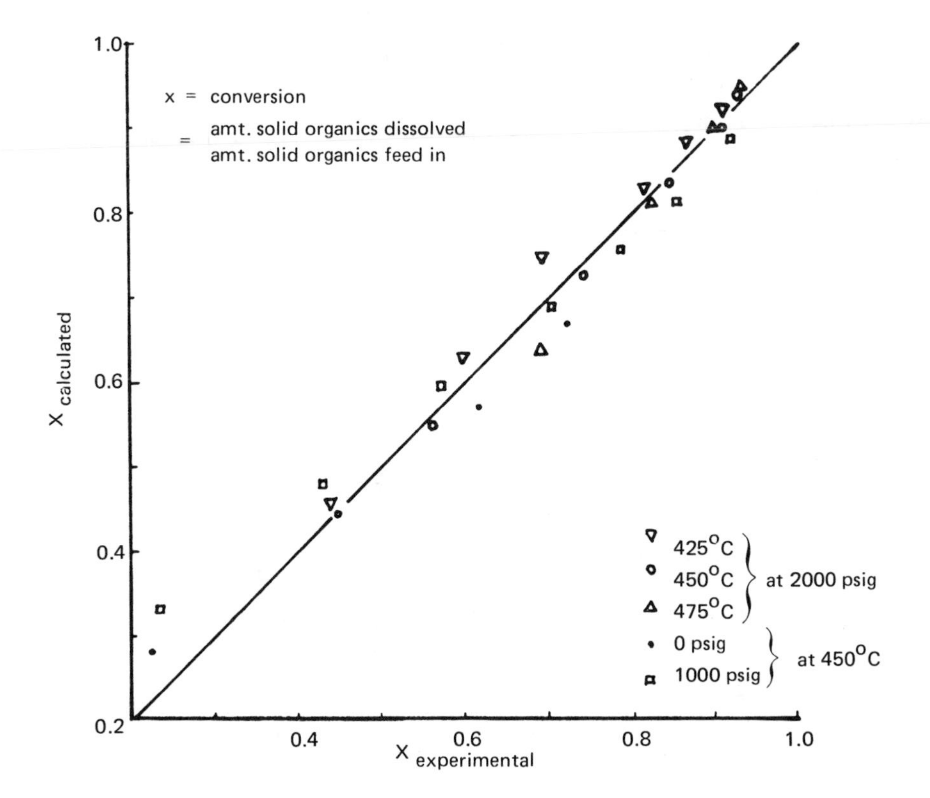

Figure 5.2. Comparison of calculated and experimental conversion. (Wen and Han [43])

5.6. Kinetics of Dissolution Based on Asphaltene

All of the representations given above are essentially empirical. It would be interesting to establish the rate equations based on the mechanism of reaction. However, since the basic mechanism is very complicated and is not completely understood, it is impossible to represent the kinetics theoretically. Weller and co-workers [56, 65, 118, 119] proposed an approach based on the extraction

ISBN 0-201-08300-0

fraction asphaltene. A typical composition of asphaltene was given in Table 2.1. The molecular weight of asphaltene varies greatly, depending on the coal and the experimental conditions. Generally, this molecular weight falls between 300 and 600, but much higher values have been observed.

The results of Weller and co-workers, as summarized by Farcasiu et al. [66], are reproduced in Figure 5.3, which shows approximately that asphaltene does form first, before oil, and that a maximum exists on the asphaltene curve as conversion increases.

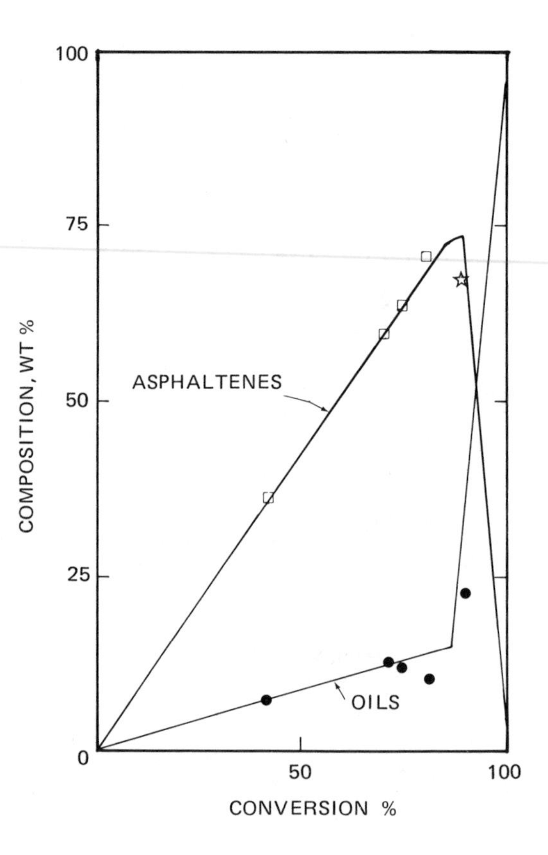

Figure 5.3. Asphaltenes as intermediate products without hydrogen-donor solvent. (Farcasiu et al. [66])

Weller and co-workers concluded that asphaltenes are intermediate products in the conversion of coal to fuel oil:

$$\text{Coal} \xrightarrow{k_1} \text{asphaltenes} \xrightarrow{k_2} \text{oil} \qquad (5.22)$$

ISBN 0-201-08300-0

Clearly, this is an oversimplification of the process. However, experimental evidence does suggest that appreciable amounts of coal convert to oil according to the above schedule and that free radicals probably should be added between each of the above conversions.

Liebenberg and Potgieter [57] studied the kinetics of coal dissolution and found that the mechanism proposed by Equation 5.22 cannot fit the experimental data. These investigators proposed the following, more complicated representation:

$$\text{Coal} \xrightarrow{k_1} \text{asphaltenes} \xrightarrow{k_2} \text{oil} \tag{5.23}$$

$$\text{Coal} \xrightarrow{k_3} \text{asphaltenes} \tag{5.24}$$

$$\text{Coal} \xrightarrow{k_4} \text{oil} \tag{5.25}$$

Yoshida et al. [58] have proposed this reaction mechanism:

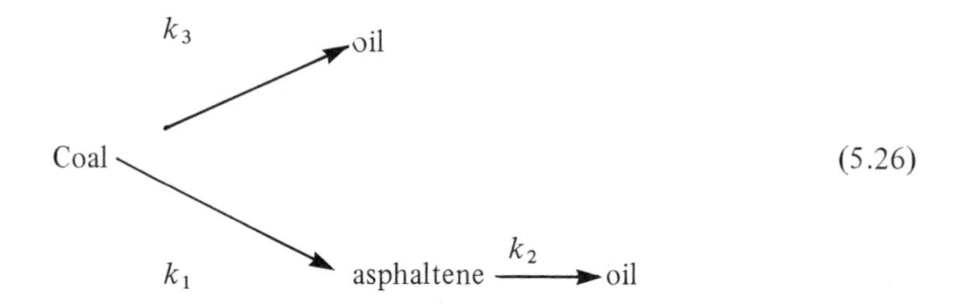

and a reasonably good fit was obtained for the Taiheiyo and Oyubari coals at 400°C by assuming first-order reactions.

Various investigators have discovered that the reaction of coal to asphaltene is very rapid, whereas the reaction of asphaltene to oil is extremely slow. This suggests that molecular weight reduction for the conversion of coal to asphaltene is associated with breaking weak bonds such as some of the heteroatom bonds, while the formation of oil is associated with breaking strong bonds, probably c-c bonds.

A more detailed representation of expression 5.22 may be that shown in Figure 5.4 [56, 65], where the transformation of coal to asphaltene is principally a thermal reaction. The highly reactive fragments (free radicals) can either polymerize or, if a reactive hydrogen is available, stabilize by the addition of hydrogen atoms or radicals. This representation, although oversimplified, furnishes some very useful concepts in the interpretation of kinetic data.

Recently, Sternberg and co-workers [25] have proposed another improvement on the mechanism represented by Figure 5.4. They discovered that in the initial stages of coal hydrogenation a substance called preasphaltene (insoluble in benzene and soluble in pyridine) is formed instead of asphaltenes. This preasphaltene is different both in physical properties and in viscosity from asphaltenes.

ISBN 0-201-08300-0

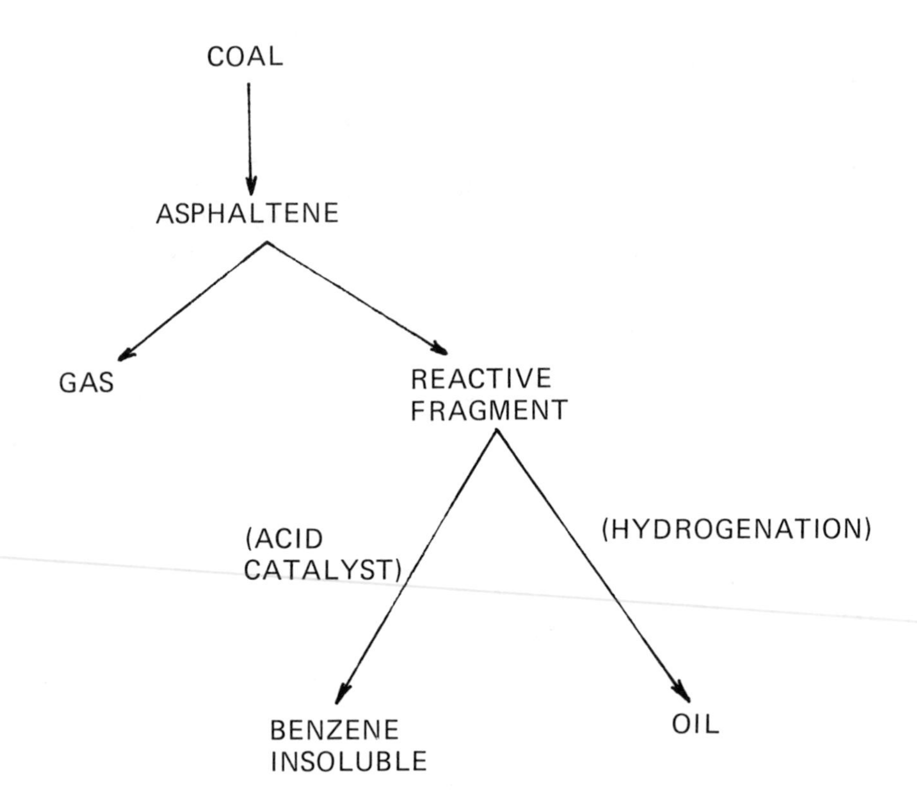

Figure 5.4. Probable mechanism of coal conversion.

Farcasiu et al. [66] have investigated the nature of the benzene-insoluble material in solvent-refined coal. They called the pyridine-soluble and benzene-insoluble materials "asphaltols" because of their high content of -OH groups per molecule. These investigators discovered that with the presence of hydrogen-donor solvent the asphaltenes in Equation 5.22 should be replaced by asphaltols and that the approximate molecular weight of asphaltols is over 600 and ranges up to several thousands. The results of Farcasiu et al. are reproduced in Figure 5.5. As can be seen, asphaltols form a maximum as the conversion increases.

Now Figure 5.4 can be further improved by adding preasphaltene and by adding free radicals just after preasphaltene and asphaltene.

ISBN 0-201-08300-0

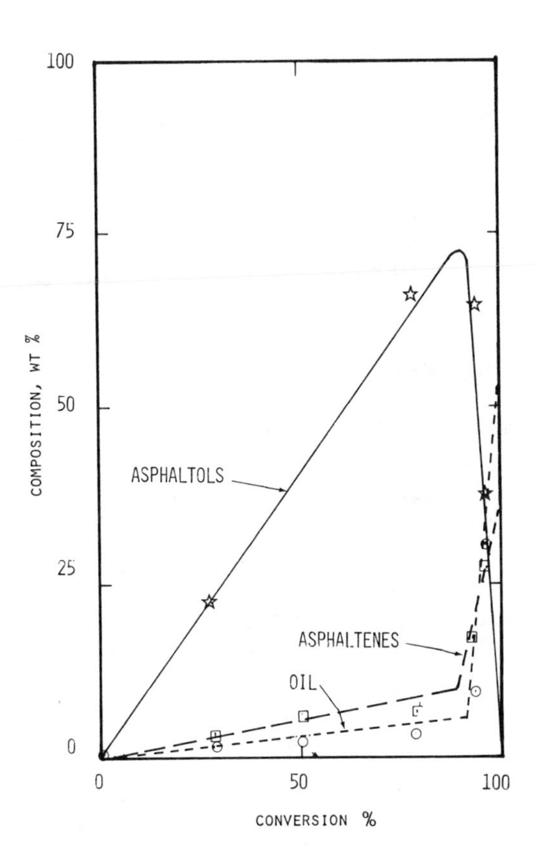

Figure 5.5. Asphaltols as intermediate products with hydrogen-donor solvent. (Farcasiu et al. [66])

6. CATALYTIC HYDROGENATION

6.1. Introduction

The addition of catalyst to coal hydrogenation processes has several advantages: (1) less expensive hydrogen is required than for noncatalytic or dissolution systems for producing fuel with sulfur contents below 0.6%, (2) lower temperature is required for the same throughput, (3) because of the efficiency of

ISBN 0-201-08300-0

the catalyst, lesser amounts of gases and heavy asphaltenes are produced, along with a greater amount of the desirable middle range oil, and (4) the product is much more amenable to different types of solid or mineral matter separation schemes.

Of course, catalytic hydrogenation also has many disadvantages. The most important ones are the complexity of the reaction system and the catalyst cost. However, as research continues, the catalyst systems offer the best hope for process improvement. This is especially true if some very effective catalyst with high selectivity and a high degree of resistance to poisoning can be developed.

6.2. Multiphase Catalytic Reactor System

Various reactor systems – for example, integral reactor, differential reactor, and pulsed microreactor – have been used over the years to study catalytic processes. According to the flow pattern, these reactors can also be classified as batch reactors, plug flow reactors, continuous stirred tank reactors, and recycle reactors. However, almost all the theoretical work on catalytic reactors was based on a system which consisted of a solid catalyst with the reactant, generally gas, flowing through the reactor. The operation of a four-phase coal liquefaction system – gas, liquid, solid coal, and catalyst phases – is certainly much more complicated and much less understandable.

In determining the kinetics of the reaction and in testing catalyst life and deactivation, careful choice of reactor systems and test conditions is essential. This is particularly true with complex processes such as those encountered in coal liquefaction.

In the study of coal hydrogenation kinetics in the laboratory, the reactors commonly encountered are the batch autoclaves, which were discussed in connection with coal dissolution. In using the autoclaves, special care must be taken to minimize the heating and cooling time.

For the study of catalyst life and activity maintenance, a flow system is essential to obtain meaningful data. A flow system is also used in large scale commercial operations for economic and operational reasons.

In the laboratory the autoclave can be converted into a continuous flow stirred tank reactor, with the reactants flowing into the reactor and a product stream with spent catalyst leaving the reactor. However, because the fluid consists of a slurry system, various operating problems and plugging of the line may occur.

To study the kinetics and catalyst life in such a reactor, the reactor can be preheated with just catalyst, solvent, and gaseous reactant. Then the solid reactant, as a slurry, is introduced, and initial reaction rates can be obtained on fresh catalysis. By varying the feed rate and assuming complete mixing, catalyst activity data can also be obtained. Furthermore, different conversion levels and the effect of these levels on catalyst selectivity can be studied.

ISBN 0-201-08300-0

In industrial catalytic liquefaction various types of flow reactors have been used. Some of these are fixed bed catalytic reactors, continuous flow stirred tank reactors, and the ebullated bed reactor.

In the ebullated bed reactor the coal-oil slurry is charged continuously with hydrogen to a reactor containing a bed of ebullated catalyst, wherein the coal is catalytically hydrogenated and converted to liquid and gaseous products. In this reactor the upward passage of the solid, liquid, and gaseous materials maintains the catalyst in a fluidized state. The relative size of the catalyst (1/32–1/16 in.) and coal is such that only the unconverted coal, the ash, and the liquid and gaseous products leave the reactor. This reactor has several advantages: (a) the catalyst can be added or withdrawn continuously to maintain activity; (b) the system has a high efficiency for heat transfer and for providing better contact between catalysts and reactants; and, most importantly, (c) the reactor has the ability to operate a catalyst system continuously with a feed consisting of solids, liquids, and gases.

6.3. Catalysts

Various hydrogenation and hydrodesulfurization (HDS) catalysts have been developed. These include the catalyst developed during World War II in Germany and the much more efficient hydrogenation and HDS catalyst developed in the petroleum industry. It should be noted that the structures of the sulfur-bearing molecules in coal liquids are fairly similar to those present in petroleum residues. In Germany the catalysts used were principally iron ones. Some of the other metals known to be active in coal hydrogenation are Mo, Sn, Ni, and Co.

Kawa et al. [47] did a fairly extensive evaluation of the effectivness of various pelleted catalysts for the hydrogenation and HDS of high volatile bituminous coal. They found that a commercial silica-promoted Co-Mo on alumina catalyst (Harshaw 0402T) is the most effective one, both for HDS and for hydrogenation. This catalyst has a composition of 3% CoO, 15% MoO_3, 5% SiO_2, and 77% Al_2O_3 by weight. These authors also found that presulfidation of the catalyst is beneficial for batch reactors. For continuous operations, however, presulfidation of the catalyst has been reported to be undesirable [48].

Since coal is generally accepted to be a highly condensed polynuclear aromatic substance with heteroatoms such as sulfur, nitrogen, and oxygen, the principal objective of coal liquefaction is to selectively crack this structure with a minimum consumption of hydrogen and a maximum removal of sulfur. Thus, whether the objective is to produce boiler fuel or high octane gasoline, catalyst selectivity is probably more important than catalyst activity for controlling hydrogen utilization and product distribution. It was found that the commerical Co-Mo catalyst on silica-alumina mentioned above has a much better capacity for the selective HDS of coal. Kawa et al. also found, in regard to single-component

ISBN 0-201-08300-0

catalysts, that molybdate catalysts were the best for sulfur removal, and tin catalysts were the best for the conversion of coal to oil. Maximum desulfurization required high surface area, but high surface area was not essential for high conversion of coal to oil.

One of the processes of liquefaction still in the early stages of development is the short-residence-time reactor (residence time is of the order of 5 sec) or free fall reactor. A variety of catalysts have been examined by the use of this reactor. It was found that stannous chloride is probably a superior catalyst, confirming early findings with long residence time.

Since both the coal and the hydrogen donor solvent are involved in the liquefaction reaction, the influences of the various factors must be known before the effectiveness of any given solvent can be predicted. For example, Potgieter [59] has investigated the kinetics of the conversion of tetralin during coal hydrogenation and discovered that stannous chloride has an efficacy for hydrogenating tetralin to decahydronaphthalenes, which are inferior to tetralin in hydrogen-donor capabilities. Thus Potgeiter concluded that, although stannous chloride is considered to be one of the best coal hydrogenation catalysts, it appears to be inferior in its influence on hydrogen donor capabilities.

Catalyst activity is important for minimizing the hydrogen consumption rate and, at the same time, increasing the removal rate of sulfur. Wolk et al. [36] did an interesting study on sulfur and nitrogen removal rates. These investigators compared various hydrogenation and dissolution processes with various bituminous coals and concluded that current coal liquefaction processes require substantial amounts of hydrogen. Furthermore, there is little difference in the amount of hydrogen consumed to reach a moderate sulfur level of about 0.9% in the fuél oil products from catalytic processes and in the fuel oil products from large scale dissolution processes.

6.4. Factors Influencing Catalyst Activity

In the study of catalyst activity and selectivity in the liquefaction of coal, three facts must be kept in mind. First, catalysts only hasten and selectively promote a reaction which proceeds in an appreciable amount without catalyst. Thus the influence of catalyst on liquefaction is not as dramatic as in many reactions that do not occur at all without catalyst.

A second important fact to consider, which is common to all hetrogeneous catalytic reactions, is the contact or distribution between the catalysts and the reactants. This distribution depends on the method of applying the catalyst. Impregnation of the catalyst on the coal reactant gives the best contact, and mechanical mixing without solvent gives the worst contact. Ball milling of catalyst and coal gives a contact between the two just mentioned. However, the degree of difference between these two methods of application depends greatly on the catalyst used. Table 6.1 illustrates the results of liquefying an hvC bitaminous coal at 450°C and 1000 psi hydrogen initial (cold) pressure for 1 hr without any solvent

ISBN 0-201-08300-0

or vehicle [8]. Conversion is defined on the basis of benzene-soluble and gaseous products. As can be seen, the influence of impregnation is much less for stannous chloride than for the other catalysts; apparently, stannous chloride has a better contact when mechanically mixed. Hydrogen-donor solvent aids the distribution and contact between catalysts and reactants.

A third factor, generally known as the "wall effect" or "memory effect," is caused by the left-over residue of catalyst on the wall of the reactor or autoclave. This effect can be avoided by the addition of a glass liner inside the autoclave so that it can be thoroughly cleaned.

Table 6.1

Effect of Catalyst Contact (Wu and Storch [8])

Catalyst	Maf Coal, %			
	Conversion	Asphaltene	Oil	Gases
$SnCl_2$				
Powdered	82.3	26.5	29.2	14.5
Impregnated	88.3	19.9	41.4	15.5
$NiCl_2$				
Powdered	44.2	6.8	13.2	13.4
Impregnated	88.3	15.5	45.3	18.0
$FeSO_4$				
Powdered	38.9	6.9	8.1	13.1
Impregnated	84.9	38.9	21.7	15.0
$(NH_4)_2MoO_4$				
Powdered	33.7	1.0	13.8	8.1
Impregnated	92.7	27.2	41.1	13.6
No catalyst	33.4	2.8	10.4	9.0

6.5. Catalytic Effects of Coal Minerals

The catalytic effects of the minerals in coal on liquefaction are well known. Wright and Severson [49] have discovered that, when anthracene oil is used as solvent in the solvent-refined coal process, the rehydrogenation of this solvent appears to depend on the catalytic activity of the coal minerals. In the liquefaction of lignite by carbon monoxide and water, the hydrogen production from carbon

ISBN 0-201-08300-0

monoxide and water appears to depend on the catalytic effect of the alkali content in the lignite. Note that it is the hydrogen produced from carbon monoxide and water which liquefies the lignite, not the carbon monoxide.

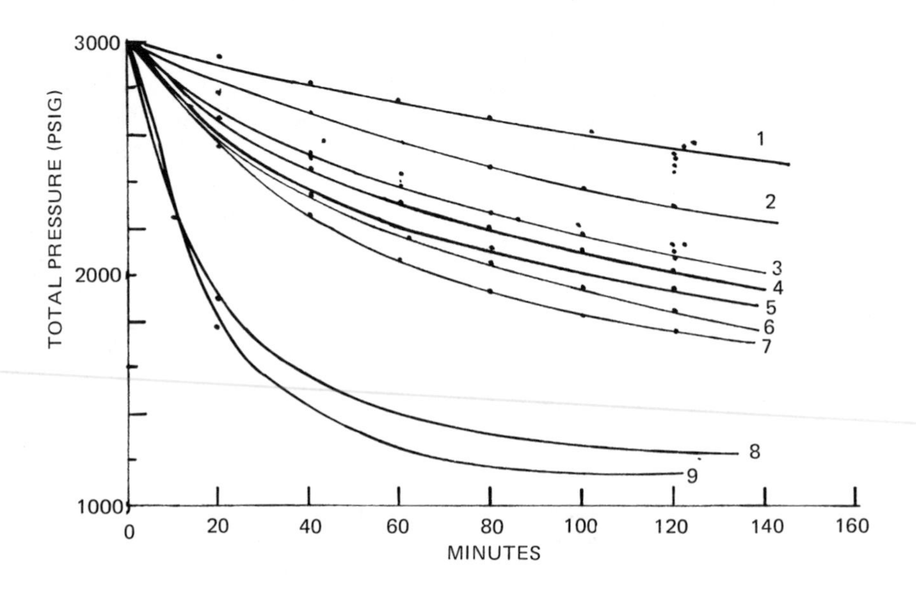

Figure 6.1. Catalytic effect of coal minerals. (Tarrer et al. [50])

1. no catalyst, ankerite, calcite, quartz, dolomite, kaolin
2. muscovite
3. pyrite (80+150 mesh), reduced pyrite, iron (325 mesh)
4. coal ash
5. siderite
6. SRC residue (Wilsonville)
7. pyrite (325 mesh)
8. Co-Mo-Al (80+150 mesh)
9. Co-Mo-Al (325 mesh)

Tarrer et al. [50] have suggested that coal minerals serve to catalyze the hydrogenation of the solvent, which has become hydrogen deficient because of its donor activities. Thus the catalytic effect of the minerals is indirect, by hydro-

ISBN 0-201-08300-0

genating the donor solvent and thus promoting the hydrogen transfer rate from solvent to coal. The catalytic effects of various minerals on the hydrogenation of creosote oil are shown in Figure 6.1 [50]. The reactions were carried out at 425°C temperature and 3000 psig hydrogen pressure with a charge of 100 g creosote oil and 15 g minerals or catalyst. The catalytic activity is indicated by the amount of pressure drop during the reaction, and the top line represents a blank run without catalyst or minerals. As expected, the Mo-Co-Al commercial catalyst is the most active one.

Mukherjee and Chowdhury [75] investigated the influence of coal minerals in an Indian coal and discovered that both iron and titanium in the minerals have catalytic effects on hydrogenation. The relationship between conversion (based on benzene-insoluble material) and total mineral matter is shown in Figure 6.2 [75]. The same plots for iron and titanium are shown in Figure 6.3. Notice the similarity in the shapes of the curves. These investigators also concluded that the total iron present in the coal, not the pyritic iron alone, acts as a catalytic constituent.

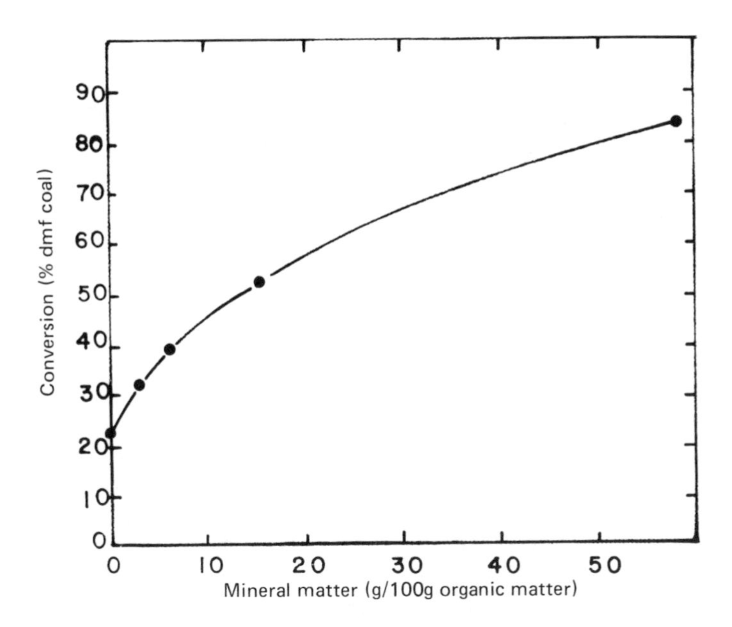

Figure 6.2. Conversion versus mineral matter. (Reproduced from Mukherjee and Chowdhury, *Fuel,* **55**, 1976, by permission of publishers [75].)

ISBN 0-201-08300-0

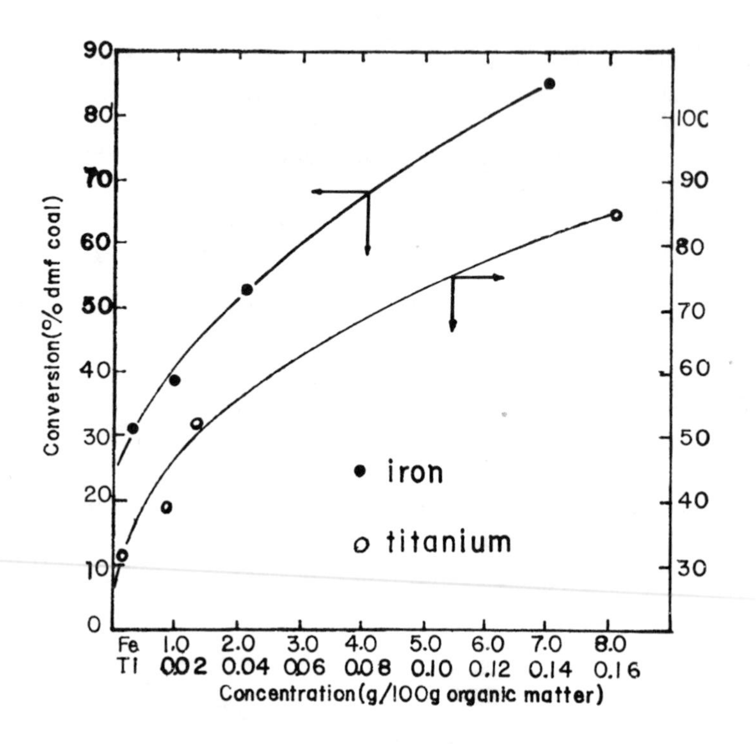

Figure 6.3. Effect of iron and titanium contents on conversion. (Reproduced from Mukherjee and Chowdhury, *Fuel,* **55**, 1976, by permission of publishers [75].)

6.6. Catalyst Deactivation and Poisoning

One of the most critical problems in catalytic coal liquefaction is the fact that coal is extremely dirty and contains a large amount of minerals. Although these minerals have certain advantages due to their catalytic effects under certain conditions, they can deactivate the catalyst rapidly under severe liquefaction conditions by depositing metal on the catalyst and by causing more severe attrition. Other important factors causing catalyst deactivation are carbon deposition and sintering.

Typical catalyst deactivation curves are shown in Figure 6.4 [51], where the hydrogen consumption rate and the sulfur content in the fuel oil are used as approximate indications of catalyst activity. These curves, which were obtained from the H-coal process development unit with Co-Mo catalyst, are typical in that they have a steep initial deactivation rate, followed by a more gradual decline.

Kang and Johanson [51] found that up to 10–35% of carbon, based on the total weight of catalyst, was deposited on the catalyst. In general, low pressure

ISBN 0-201-08300-0

operations yield more carbon deposit. These authors further concluded that in the H-coal process with a Co-Mo catalyst the initial catalyst deactivation is caused by carbon deposition for bituminous coal. For subbituminous coal both carbon deposition and sintering contribute to the initial deactivation. Metal depositions such as titanium cause the gradual decline in catalyst activity. Other major metal deposits found by these investigators are Fe, Ca, Na, and Al.

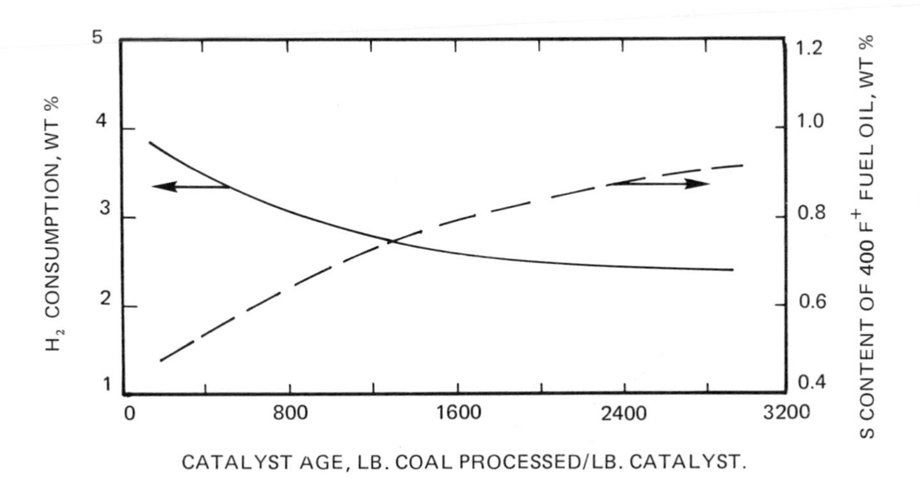

Figure 6.4. Decline in catalyst activity. (Kang and Johanson [51])

For some catalysts, such as gamma-alumina-supported catalyst, catalyst sintering in coal liquefaction is principally caused by steam. In subbituminous coal, because of its high moisture content, sintering is much more pronounced. Williams et al. [52] have studied the sintering effect on Ni-alumina catalyst, a nickel catalyst supported by gamma alumina. Their results on the effect of temperature and steam on sintering, as indicated by the surface area, are reproduced in Figure 6.5.

Kovach and Bennett [53] studied the various poisons to catalyst in coal by using Nalcomo 471 catalyst, which is a Co-Mo-alumina catalyst with 12.5% MoO_3, 3.5% CoO, 0.3% SiO_3, and small amounts of Na_2O and Fe. These investigators divided the poisons into two different types: permanent and temporary poisons. The temporary poison is the carbonaceous deposit, which can be regenerated to the original activity level of the catalyst. The permanent poisons are the coal minerals or coal ash. These investigators impregnated Nalcomo 471 catalyst with the various compounds present in the coal ash through the use of water-soluble salts, and tested the resulting catalysts for activity (see Figure 6.6).

ISBN 0-201-08300-0

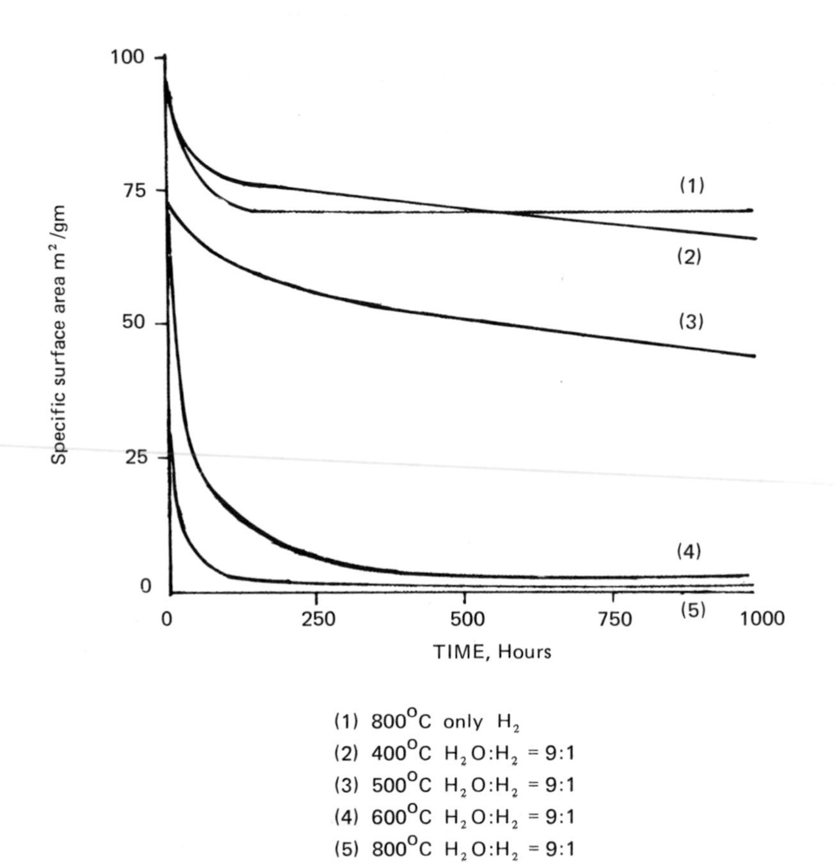

Figure 6.5. Effect of temperature and steam on sintering of gamma alumina catalyst. (Williams et al. [52])

As can be seen from Figure 6.6, the alkali salts (Na, Ca, Mg) gave the highest degree of deactivation, whereas the acidic components (B, Ti, Si) gave little or no deactivation. Kovach and Bennett also investigated the influence on deactivation if the amount of a single ash compound is increased in the coal-solvent slurry. This was accomplished by suspending the fine powder of a compound in tetralin. The

ISBN 0-201-08300-0

catalyst activities after processing 60 lb coal/lb catalyst are shown in Table 6.2. The amounts of metal oxide deposited on catalyst are also listed. For comparison, the last column, which was obtained from Figure 6.6, in Table 6.2 shows the catalyst activities if the amounts of metal oxide have been impregnated on the catalyst.

Stanulonis et al. [54] studied the spent Co-Mo catalyst (Harshaw 0402T) from the Synthoil pilot plant by using a scanning electron microscope and an electron microprobe. These investigators found accumulated inorganic and organic contaminants both on the exterior surface and in the pore mouths. Upstream of the fixed bed reactor the inorganic deposit is principally ferrous sulfide; downstream of the bed the deposit is mainly aluminum and silicon. Ferrous sulfide penetrated into the pores to a depth of about 100 μm, and titanium penetrated about 150 μm. All of these deposits caused a 40% loss of pore volume.

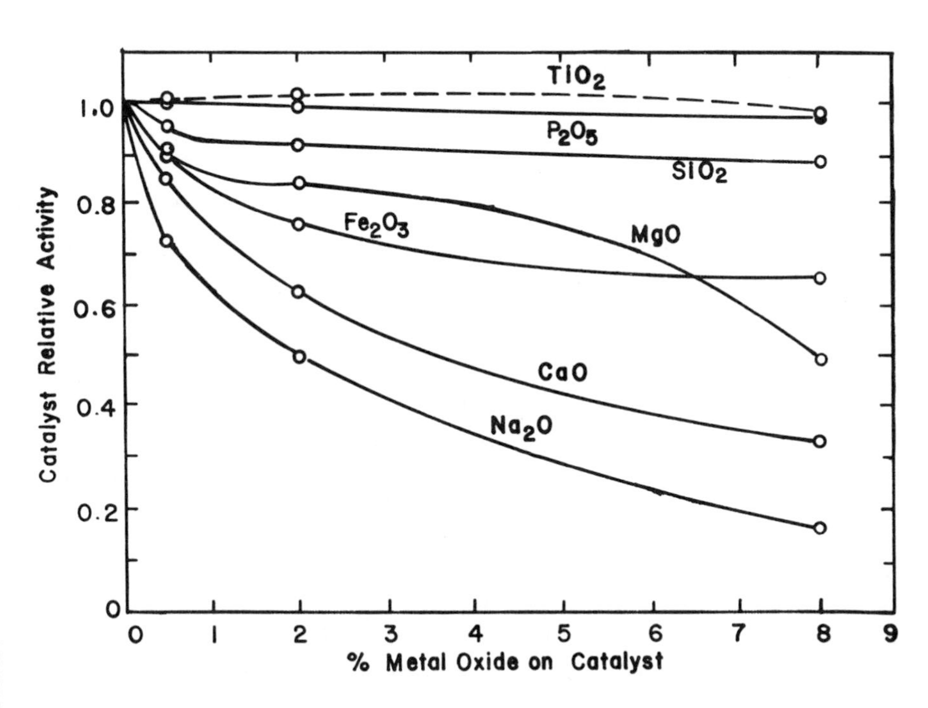

Figure 6.6. Effect of coal minerals on catalyst activity. (Kovach and Bennett [53])

ISBN 0-201-08300-0

Table 6.2

Effect of Mineral Matter on Conversion (Kovach and Bennett [53])

Ash Constituent	Ash/Run, g[a]	Catalyst[b] Activity	Metal Oxide on Catalyst, %	Catalytic Activity with Impregnation
$NaHCO_3$	1.0	0.53	2.2	0.51
$CaCO_3$	1.8	0.69	0.44	0.79
$MgCO_3$	1.3	0.60	3.0	0.72
Fe_2O_3	8.0	0.34	10.0	0.38
$TiCL_2(CP)_2$	2.0	0.82[c]	2.2	0.98
NH_4PO_4	0.5	0.80[c]		
Na_2SiO_3	1.0	0.75	1.1 Total	0.9 – SiO_2
			0.45 Na	0.80 – Na
$SiO_2 - Al_2O_3$	50	0.69	7.3	0.90

[a] Ash per run; each run uses 300 g of coal.

[b] Coal processed: 60 lb coal/lb catalyst.

[c] Catalyst regenerated in air to remove carbon or ammonia.

6.7. Process Variables

The process variables in catalytic liquefaction are the nature and amount of solvent, the activity of the catalyst and its concentration, the hydrogen pressure, the reaction temperature, and the residence time.

Another variable which is not within our control, but is very important, is the characteristics and behavior of coal. It is this variable which causes the most difficulty in the interpretation and correlation of data. Thus it is important to use the same coal in experiments in order to obtain some meaningful results.

It is interesting to compare the liquefaction product without solvent. Hawk and Hiteshue [11] did this by hydrogenating a high volatile A bituminous coal for 1 hr at 450°C and 1000 psig cold (initial) pressure. The results are shown in Table 6.3. The conversion listed in this table is based on benzene-soluble materials and gases. Notice the large increase in conversion, but the relatively small increase in asphaltenes and gases, as catalyst was added to the system.

ISBN 0-201-08300-0

Most of the rest of this section is devoted to the influences of these process variables.

Table 6.3

Liquefaction without Solvent (Hawk and Hiteshue [11]

Reactor Condition	Weight Percent of Coal		
	Conversion	Asphaltene	Gases
Nitrogen atmosphere	11	1	3
Hydrogen atmosphere	54	21	14
Hydrogen and catalyst	83	27	15

6.8. Catalyst Concentration

The influence of catalyst concentration on conversion without solvent has been studied by various investigators. The results of Hawk and Hiteshue [11] are reproduced in Figures 6.7 and 6.8. In Figure 6.7 the catalyst used is a tin catalyst with a high volatile A bituminous coal; results for a high volatile C bituminous coal are shown in Figure 6.8. The temperature used in these figures is 450°C, and the pressure is 1000 psig hydrogen pressure at the beginning of each run. The residence time is 1 hr.

It is interesting to note the remarkable effectiveness of the small amounts of catalyst on conversion. A 0.1% concentration of molybdenum or tin catalyst causes more than 50% conversion of the high volatile C bituminous coal. However, high concentration of catalyst does help to reduce the amount of asphaltene and sulfur (see Table 4.3) in the products.

Yen et al. [34] investigated the influence of the amount of Co-Mo catalyst on liquefaction of a high volatile A bituminous coal in tetralin solvent (see Table 4.3). The conversion is fairly high and is essentially complete even without any catalyst. However, the amount of sulfur removed is much higher and the amount of asphaltene produced is much lower with a 10% catalyst. Thus catalyst is important for the conversion of asphaltene to oil and for increasing the extent of desulfurization.

ISBN 0-201-08300-0

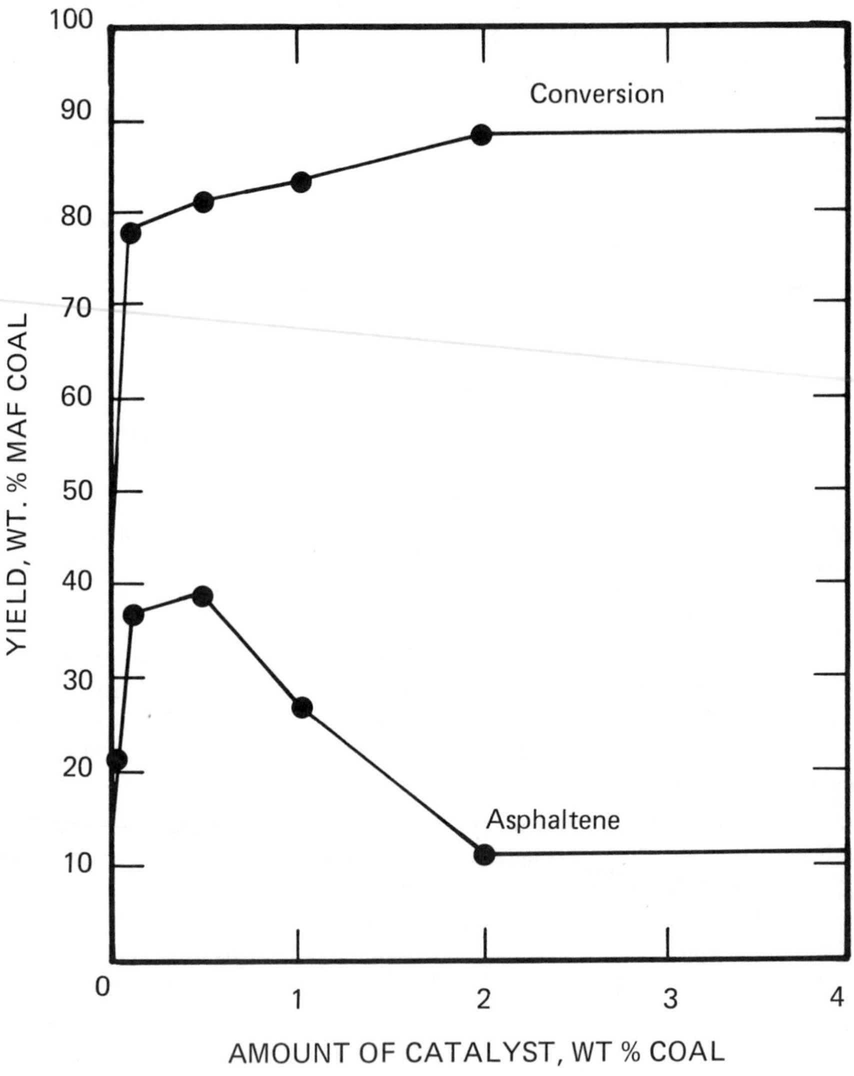

Figure 6.7. Effect of catalyst concentration on hydrogenation without solvent, hvA bituminous coal. (Hawk and Hiteshue [11])

ISBN 0-201-08300-0

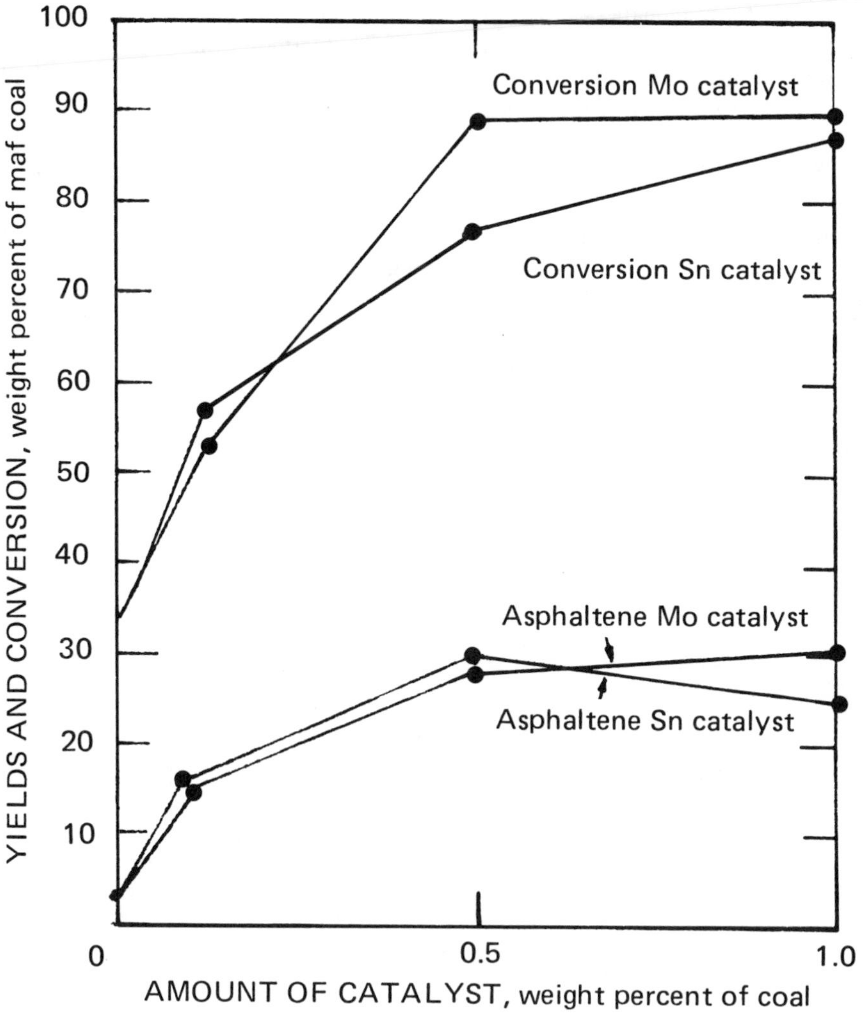

Figure 6.8. Effect of catalyst concentration on hydrogenation without solvent, hvC bituminous coal. (Hawk and Hiteshue [11])

ISBN 0-201-08300-0

6.9. Effects of Temperature and Residence Time

It can be seen from earlier discussions that coal can be hydrogenated to a very appreciable extent without any catalyst if the temperature is high enough, but not too high. It would be interesting to study the liquefaction of coal from a purely thermodynamic standpoint.

Thermodynamically, reaction of coal with hydrogen at low temperature is favored to form aliphatic and alicyclic liquid products; at higher temperatures the production of gases and aromatics is favored. Furthermore, the desired products, liquids, become increasingly less likely as temperature increases, and the undesired products, coke and gases, are increasingly favored. However, for the reaction rate to be appreciable, the temperature must be reasonably high.

The influence of temperature on conversion for a high volatile A bituminous coal is shown in Figure 6.9 [11]. This experiment was carried out without any solvent and with 1% tin catalyst. The reaction time was 1 hr, with an initial (cold) hydrogen pressure of 1000 psig. The change in conversion, based on benzene-soluble material, from 450 to 500°C is negligible, but the undesirable increase in gas produced, and hence the corresponding decrease in desirable liquid product, is certainly noticeable.

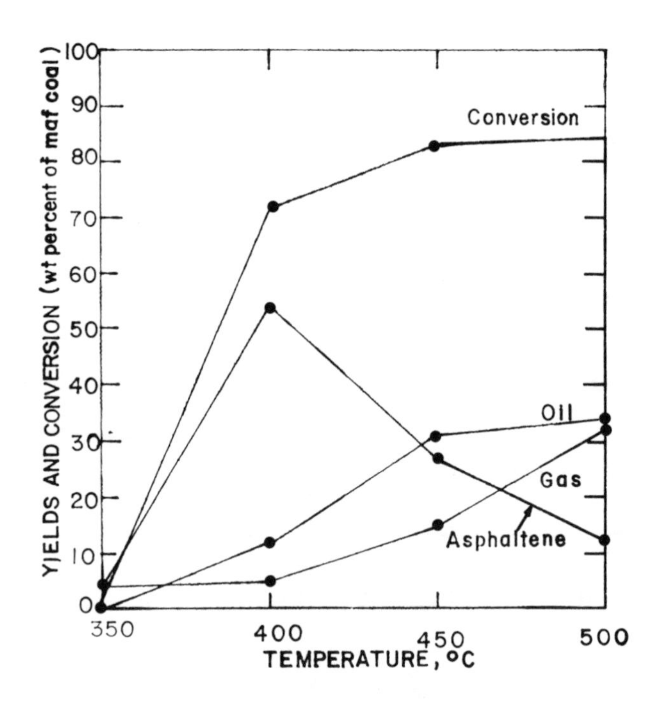

Figure 6.9. Effect of temperature on conversion without solvent. (Hawk and Hiteshue [11])

ISBN 0-201-08300-0

Yen et al. [34] (see Table 4.3) found that, although a fairly high conversion is obtained at 405°C, both the amount of oil produced and the amount of sulfur removed are very small at this relatively low temperature.

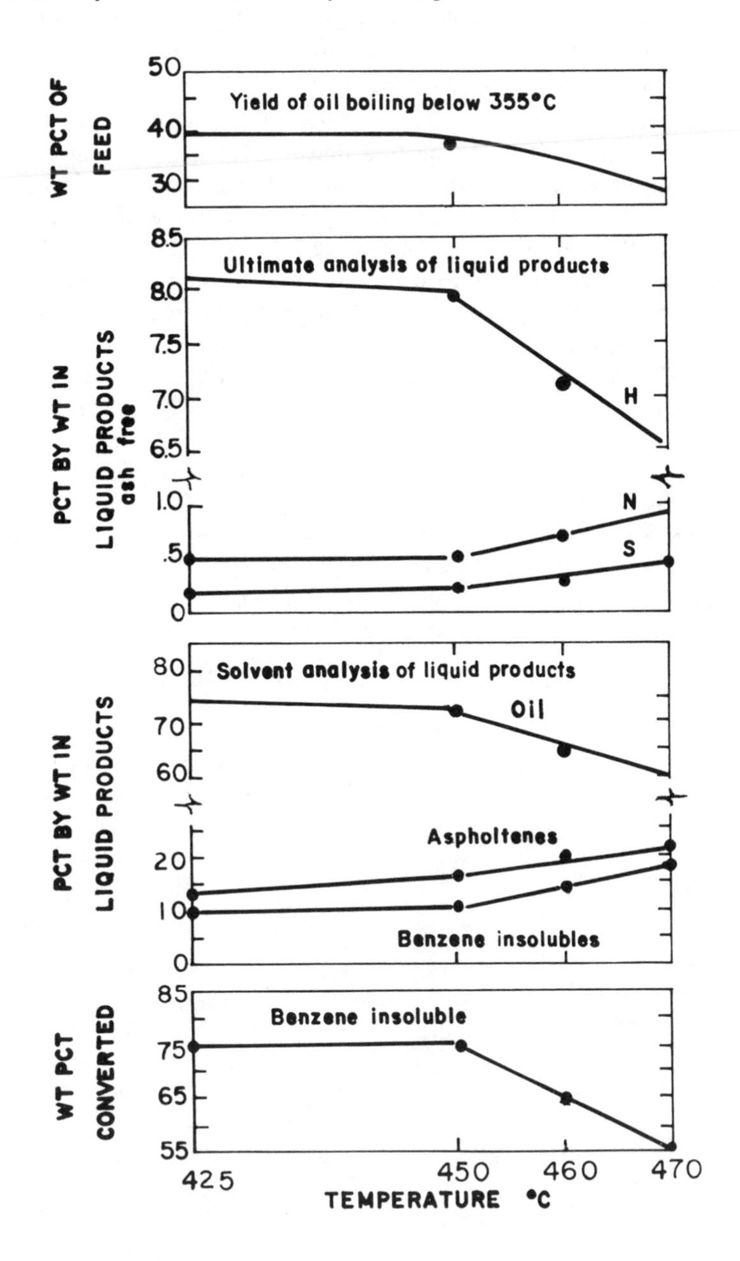

Figure 6.10. Effect of temperature on conversion. (Akhtar et al. [55])

ISBN 0-201-08300-0

Akhtar et al. [55] have liquefied a hvA bituminous coal in a fixed bed of silica-promoted cobalt molybdate catalyst with high temperature tar as solvent. The results with 30% coal in the slurry and with 20 lb/hr ft^3 of empty reactor coal paste feed rate are shown in Figure 6.10 [55]. Notice the decrease in conversion and in oil produced, and the increase in sulfur and nitrogen content in the liquid products, as the temperature increases from 450 to 470°C.

Coal hydrogenation has a fairly low rate of reaction. In general, 1 hr reaction time in batch reactors is fairly standard practice. Although long reaction time is necessary to reduce both the asphaltene and sulfur contents, long contact works against the efficient conversion of coal to oil. In other words, the longer the reaction time, the greater will be the amounts of products thermally decomposed into gases and chars. The residence time has exactly the same effect as temperature. In fact, time and temperature can be partly interchanged, under certain restrictions, as variables for controlling conversion and product distribution.

Fu and Batchelder [67] studied the liquefaction of coal in a batch autoclave by syngas and by hydrogen; some of their results are reproduced in Table 6.4. As temperature or reaction time increases, the asphaltene content decreases and hydrogen consumption increases. The only benefit in increasing reaction time and temperature is the improvement in sulfur content in the liquid and the decrease in viscosity. However, the costs of achieving these improvements are high in hydrogen consumption and in the decomposition of products, resulting in less liquid yield. It should be noted that the difference in conversion between the lowest and the highest is only 6%.

Table 6.4

Effects of Various Processing Variables on Liquefaction[a] (Fu and Batchelder [67])

	Gas Used												
Variable	H_2					Syngas							
Temperature, °C	450	450	450	425	400	450	450	450	425	400	400	425	450
Reaction time, min	15	30	60	30	60	15	30	60	30	60	60	30	15
Pressure, psi	2800	2700	2800	2900	2900	3000	3000	3000	3100	3000	3100	3100	3100
H_2/CO in syngas	–	–	–	–	–	2	2	2	2	2	1	1	1
Conversion, %	94	95	95	95	89	95	94	95	94	89	89	95	93
Liquid yield, %	71	59	48	71	71	68	65	42	70	71	73	65	64
Asphaltene, %	30.3	14.8	14.7	43.7	54.4	42.2	38.2	32.2	53.7	48.9	57.2	51.4	53.2
Sulfur in liquid, %	0.39	0.31	0.26	0.41	0.40	0.40	0.43	0.32	0.51	0.53	0.54	0.47	0.47
Viscosity, cSt[b]	15	10	7	23	31	20	16	9	28	38	45	25	21
Gas consumed[c]	10.7	13.7	16.1	10.3	9.1	8.6	11.2	16.2	9.2	7.5	7.8	9.9	10.3
CO/H_2[d]	–	–	–	–	–	2.1	1.4	0.8	2.2	3.2	15	23	6.1
H_2/CO in off gas	–	–	–	–	–	4.1	4.3	3.2	4.4	3.4	1.8	2.1	2.1

[a] Solvent/coal ratio = 2.3:1; data are in weight percent maf coal; catalyst used: 2 parts cobalt molybdate on alumina-silica catalyst per 100 parts coal plus solvent for H_2, and 3 parts Co-Mo-K_2CO_3 catalyst per 100 parts coal plus solvent for syngas; vehicle used: SRC liquid boiling between 270 and 630°C.

[b] Kinematic viscosity, cSt at 60°C.

[c] In scf H_2 or syngas consumed/lb maf coal.

[d] CO/H_2 consumption ratio.

ISBN 0-201-08300-0

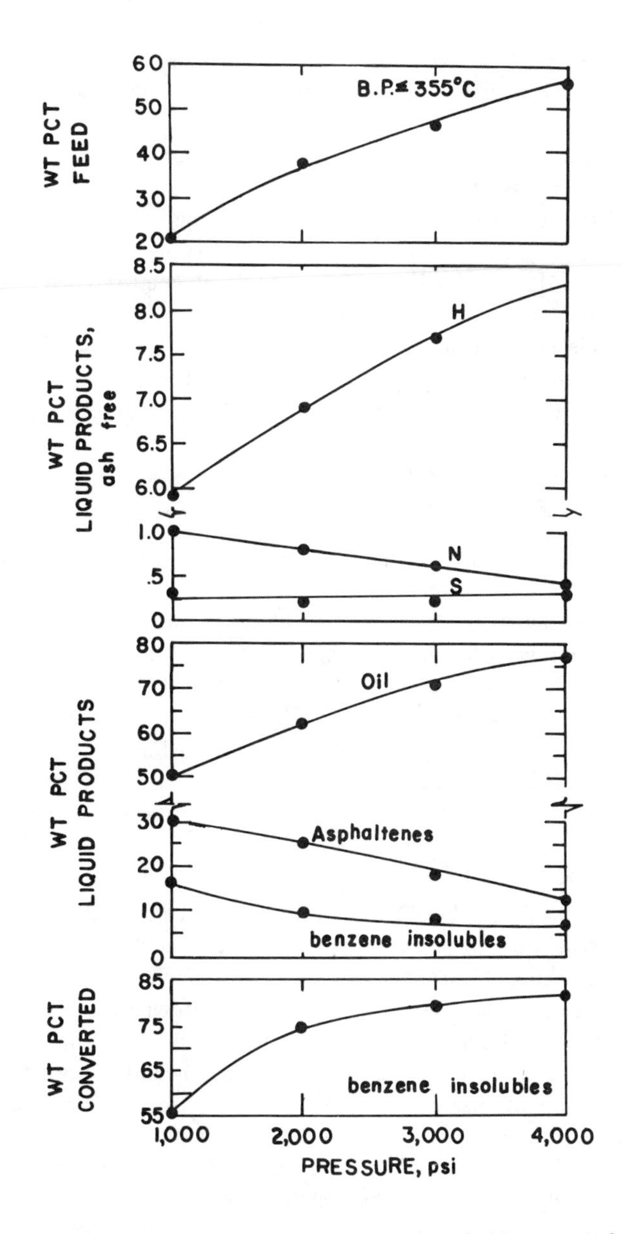

Figure 6.11. Effect of pressure on conversion. (Akhtar et al. [55])

ISBN 0-201-08300-0

6.10. Effects of Hydrogen Pressure

High hydrogen pressure increases the rates of reaction between coal and hydrogen. For hydrogen to be effective, it must first dissolve in the liquid phase.

According to Henry's law, high hydrogen partial pressure favors this dissolution. As far as is now known, pressure can be increased indefinitely with increasing benefit to liquefaction reactions.

High hydrogen pressure also retards the mechanism of coke and gas forming processes, which are essentially dehydrogenation reactions. Furthermore, increasing the hydrogen pressure also reduces the residence time required and allows higher reaction temperature without increasing thermal decomposition.

Akhtar et al. [55], using the same coal and the same fixed bed reactor, found that as the pressure increased from 2000 to 4000 psi the conversion based on benzene-insoluble material increased only from 74 to 81%. However, the amount of oil produced and the increase in hydrogen content in the liquid products were substantial (see Figure 6.11). The temperature used for this experiment was 425°C.

6.11. Effects of the Properties of the Solvent

The effects of the properties of the solvent on coal dissolution were discussed in Section 5. The influence of these properties when catalysts are used will be discussed here.

Pastor and co-workers [61] studied the interaction between the amount of catalyst and the nonpolar solubility parameter (NSP) of the solvent. Three solvents with different NSP values were used: benzene [NSP = 9.2 $(cal/cm^3)^{1/2}$ at 25°C], tetralin [NSP = 9.4 $(cal/cm^3)^{1/2}$ at 25°C], and coal tar fraction [NSP = 9.8 $(cal/cm^3)^{1/2}$ at 25°C]. Their results are reproduced in Figure 6.12. A subbituminous coal was used. To increase contact, the ferrous ion catalyst in the form of ferrous sulfate was impregnated on the ground coal. The slurry mixture, with a 4:1 solvent/coal ratio, was reacted for 3 hr at 440°C. The conversion in Figure 6.12 is based on benzene-insoluble ash-free coal.

As can be seen from Figure 6.12, the presence and the amounts of catalyst are very important when weak solvents such as benzene and coal tar are used. However, the influence of catalyst when tetralin is used is minimal. Furthermore, even with a large amount of catalyst, a weak solvent cannot reach the maximum conversion level of a strong one. A maximum conversion at a NSP value of 9.5 and the shape of the curve in Figure 6.12 were suggested by other results and by the parabolic equation expressing the relation between solubility (conversion) and NSP for solvents of similar polarity and hydrogen bonding tendencies.

According to the free radical mechanism, the desirable role of both the solvent and the catalyst would be to stabilize the free radicals and then to make these stabilized free radicals go into solution. Tetralin accomplishes both of these processes. However, when a nondonor solvent is used, a stabilizing catalyst appears to be necessary to maximize the conversion.

ISBN 0-201-08300-0

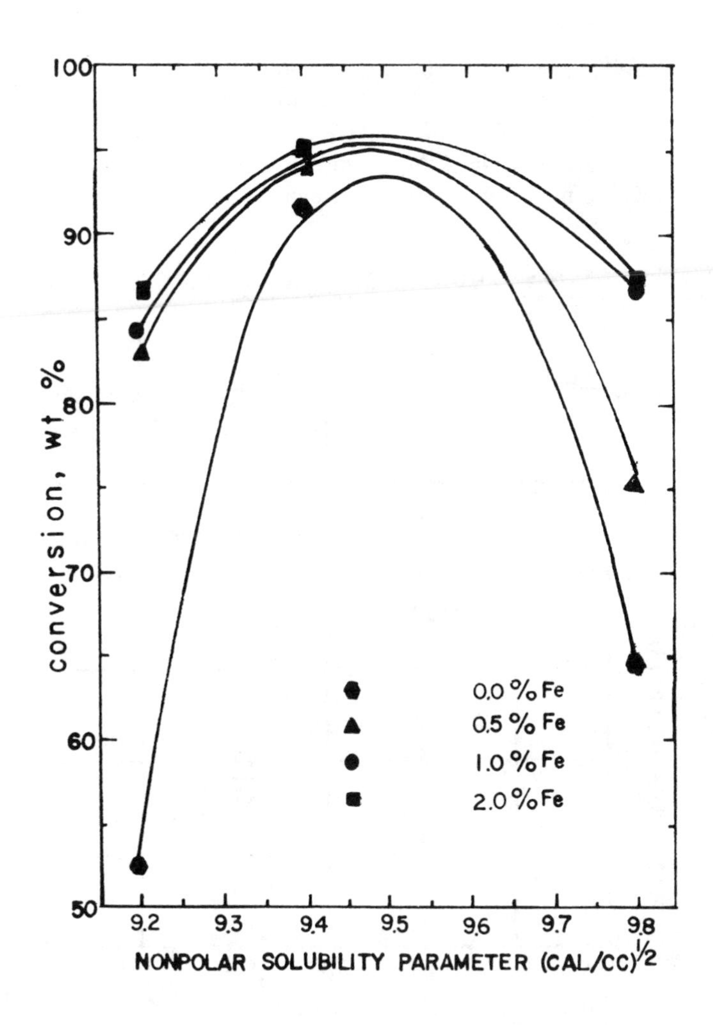

Figure 6.12. Effect of catalyst and solvent on conversion. (Pastor et al. [61])

6.12. Hydrodesulfurization

One of the most important aims in using catalyst is to reduce the sulfur content in the products. The effectiveness of some catalysts, such as the Co-Mo catalyst, in removing sulfur has been discussed in various places. For example, Akhtar et al. [70] studied the sulfur removal rate in a continuous flow fixed bed

ISBN 0-201-08300-0

reactor at 450°C and 2000-4000 psi pressure (the Synthoil process). In the presence of Co-Mo supported by SiO_2-Al_2O_3 catalyst, all of the pyritic sulfur and 60-80% of the organic sulfur were eliminated. However, if the catalyst packing was replaced with inert glass pellets, the pyritic sulfur was again completely eliminated, but only 10-20% of the organic sulfur was eliminated.

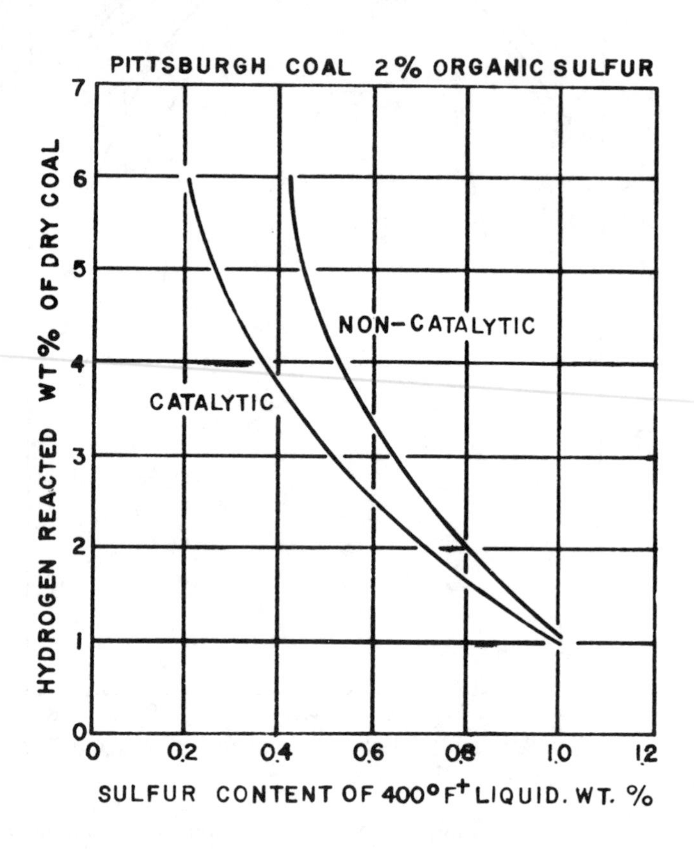

Figure 6.13. Effect of catalyst on sulfur removal. (Stotler [74])

Akhtar et al. [70] have investigated the sulfur compounds present in liquefied coal without catalyst. They identified 14 organic sulfur compounds, all but one being thiophene derivatives. It appears that other sulfur compounds are unstable under the hydrogenation conditions. Among the 14 sulfur compounds, dibenzothiophene is the most difficult to decompose, followed by benzothiophene and naphthobenzothiophene, when Co-Mo catalyst is used with 450°C and 4000 psi pressure. However, it should be emphasized that the sulfur compounds present in the products depend heavily on the coal used. For example, among the 14 sulfur compounds identified, only 4 are present simultaneously in both coals used.

ISBN 0-201-08300-0

Thus extraction or dissolution is fine for coals with low organic sulfur content, but for coals high in organic sulfur a catalyst is almost essential for economic and environmental reasons. This is shown clearly in Figure 6.13 [74], where the catalytic curve was obtained in a continuous flow ebullated bed of catalyst (H-coal process) and the noncatalytic curve was obtained by solvent dissolution. The coal used had a 2% organic sulfur content.

6.13. Reaction Mechanism

Because of the complexity of the catalytic liquefaction system, very little is known about the mechanism of reaction.

In the absence of catalyst, molecular hydrogen does not react appreciably with coal at temperatures below 500°C even with high hydrogen pressure [39]. The yield under these conditions is essentially due to the pyrolysis of coal, not to the presence of hydrogen. It should be mentioned that coal minerals do promote the reaction between coal and hydrogen, and thus a higher yield than that from pure pyrolysis is frequently obtained.

If only dry coal and solid catalyst are mixed in the presence of hydrogen, the contact is limited. However, if a hydrogen-donor solvent is used, the coal can be brought into contact with the catalyst under hydrogen pressure. Another way to produce intimate contact is impregnation of the catalyst on the coal (see Table 6.1).

The mechanism of dissolution was discussed in Section 5. In the presence of catalyst, it appears that additional active hydrogen is made available to stabilize the free radical fragments. In addition, the hydrogen in the presence of catalyst appears to assist the processes of bond rupture [39] of the coal. Thus the average molecular weight in the benzene-soluble fraction is much lower than in the case of dissolution.

For reaction to occur, at least one reactant must have chemical interaction with the catalyst. It appears that the hydrogen molecule decomposes on the surface of the catalyst to supply hydrogen free radicals.

Ruberto et al. [69] extracted subbituminous coal by using hydrophenanthrene and anthracene oil solvents and concluded that the role of the catalyst is primarily to hydrogenate the solvent (not the coal) to maintain the proper level of hydroaromatics and to upgrade the dissolved coal. These investigators also found that the presence of catalyst is very beneficial in the sense that the coal asphaltenes are converted to resins and oils. However, this conversion is subsequent to the coal dissolution into the solvent.

Like any heterogeneous kinetic reaction, the reaction of coal with hydrogen requires many consecutive steps such as surface diffusion, diffusion into the catalyst pores, and surface reaction. However, since this reaction is so complicated, very little can be said about which step is controlling. In fact, the exact chemical species formed for these diffusions and reactions are unknown. Kawa et al. [47] found that the desulfurization rate can be increased by pulverizing the cobalt molybdate catalyst used. This result indicates that the sulfur removal reactions are

ISBN 0-201-08300-0

intraparticle diffusion controlled. Weller and co-workers [34] have reached essentially the same conclusion, stating that with Co-Mo-Al_2O_3 catalyst the rates of both liquefaction and conversion of asphaltene to oil are strongly pore diffusion limited.

In discussing the effects of the catalyst on coal liquefaction, a distinction should be made between conversion, defined as benzene-solube and gaseous products, and degree of hydrogenation, as indicated by the amount of low molecule weight oil products. With a weak and nondonor solvent such as benzene or naphthalene, the presence of an effective catalyst increases the conversion very effectively. However, with a hydrogen-donor solvent such as tetralin, an effective catalyst increases the conversion, but not as much as with a weak solvent. With a strong hydrogen-donor solvent such as *o*-cyclohexylphenol, the effect of catalyst on conversion is minimal. This conclusion is shown clearly in Table 4.2, where the results in the last column were obtained with catalyst and with 1000 psi initial (cold) hydrogen pressure. The reaction pressure due to the production of gases and the increase in temperature is approximately 2500 psi.

For the degree of hydrogenation or the amount of oil fraction produced and for desulfurization, the presence of catalyst is almost always effective. This is due to the fact that the reaction from asphaltenes to oil is very slow and difficult, whereas the reaction from coal to asphaltenes is very rapid. This is illustrated clearly in Table 6.5, which is reproduced from Fu and Batchelder [67]. The coal conversion to asphaltene is very rapid and the asphaltene formed reached a maximum rapidly, but the asphaltene to oil conversion is very slow. A Kentucky high volatile bituminous coal was used.

Table 6.5

Formulation of Asphaltene during Liquefaction[a] (Fu and Batchelder [67])

Variables	Syngas ($2H_2:1CO$)			H_2		
Reaction time, min	5	15	60	5	15	60
Asphaltene formed, %	42.6	38.0	25.5	47.1	26.7	6.7
Asphaltene analysis, %						
C	84.1	85.8	87.0	85.4	87.1	86.9
H	6.3	6.3	5.0	6.5	6.3	5.7
N	1.6	2.1	2.0	1.5	2.0	2.1
S	1.3	0.15	0.4	0.64	0.23	0.36
O	7.0	5.7[b]	4.0	6.0	4.3[b]	5.1
Molecular weight of asphaltene	387	430	407	457	573	597

[a] Coal/alkylnaphthalene ratio = 1:2·3; 2900 psi; 450°C; data are in wt % maf coal.

[b] By difference.

ISBN 0-201-08300-0

6.14. Catalytic Liquefaction with Synthesis Gas

To save the cost of separate hydrogen processing units, catalytic hydrogenation and desulfurization of bituminous coal with synthesis gas and water, using cobalt molybdate-sodium carbonate as catalyst, have been investigated by Fu and co-workers [88, 67] in an autoclave. Cobalt molybdate catalyst promotes hydrogenation and desulfurization, while sodium carbonate promotes the water-gas shift reaction and the reduction of coal by carbon monoxide. The hydrogen consumed is partly furnished from the water-gas shift reaction. It should be remembered that the process of liquefying lignite with carbon monoxide and water without catalyst does not work well with bituminous coal.

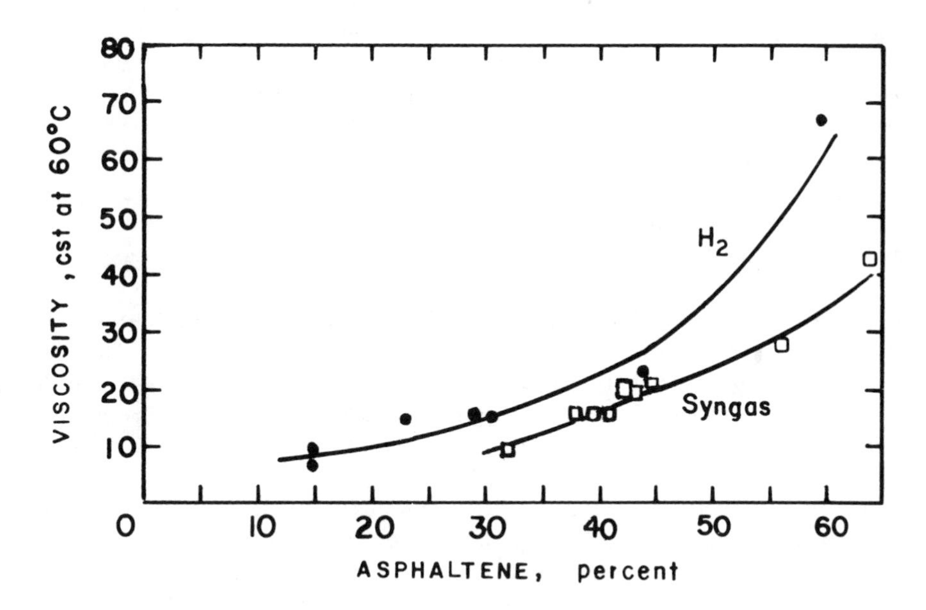

Figure 6.14. Viscosity versus asphaltene content. (Fu and Batchelder [67])

Part of the results of Fu and associates are reproduced in Table 6.4. As can be seen, with increasing time or temperature the syngas or H_2 consumed increases. The asphaltene and sulfur contents decrease with increasing H_2 or syngas consumption. Whether syngas with water or hydrogen is used, the conversion and liquid yield are comparable. The asphaltene content is higher when syngas is used under comparable operating conditions. In Table 6.5 the asphaltenes formed are listed,

ISBN 0-201-08300-0

together with their ultimate analyses and molecular weights. Notice the difference in molecular weights of asphaltenes. The asphaltenes formed had much higher molecular weights when hydrogen gas was used. This may explain the higher viscosity obtained with hydrogen gas (see Figure 6.14).

7. PROPERTIES AND PURIFICATIONS OF LIQUEFACTION PRODUCTS

7.1. Introduction

Liquefaction products are composed of gaseous, liquid, and solid products. The gaseous materials consist of simple mixtures of a few hydrocarbons and common inorganic gases. Most of the heteroatom pollutants such as sulfur, oxygen, and nitrogen are also in the gaseous products, and thus further purification and recovery of valuable products are frequently carried out. Since these purification approaches basically follow the purification of gasification products, no further discussion will be given here.

The liquid products are the desired ones and are discussed in detail in this section. Compared to petroleum crudes, coal liquids generally have a much lower hydrogen/carbon ratio and much higher concentrations of nitrogen and oxygen heteroatoms. The aromatic contents of coal liquids are generally very high.

Depending on the process and the degree of liquefaction, the solid products may or may not contain large amounts of unreacted carbon or char. If the solid residue does contain a large amount of unreacted carbon, this residue, which contains almost all the ashes in coal, is generally called char and is frequently used to produce the hydrogen needed in liquefaction. On the other hand, if the coal dissolution or liquefaction is fairly complete, the residual solid products contain mainly the coal minerals in the original coal. Obviously, the separation or definition between solid products and liquid products depends on the separation technique used. For example, the amount of solid products will be different depending on whether benzene or pyridine serves as solvent, or depending on whether filtration or centrifugation is used for separation.

7.2. Solid Coal Minerals

A large amount of coal minerals becomes available from liquefaction. The disposal or utilization of these coal minerals constitutes an important problem.

The composition of coal mineral depends on the original coal used and on the liquefaction process. Generally, coal mineral contains most of the inorganic sulfur and mineral matter in the original coal. Typical compositions from the Pittsburgh and Midway Coal Mining Company's SRC process are shown in Table 7.1 [89]; these were obtained by first drying the samples overnight at 100°C and then ashing in a furnace at 900°C until no more weight was lost. The ignition losses listed in the table are mainly due to the loss of the carbonaceous material and to the combustion of iron sulfides which changed into iron oxides. The most

ISBN 0-201-08300-0

important components in the ashes are silica, alumina, and iron oxides. Proposals have been made to use coal minerals for the production of iron, sulfur, and mineral wool.

Table 7.1

Typical Compositions of Coal Minerals (Office of Coal Research R & D Rep. 53 Interim Rep. 11 [89])

Component	Kentucky Sample 57215, %	West Virginia Sample 57244, %	North Dakota Lignite Sample 57250, %	Washington Sample 57263, %	Wyoming Sample 57264, %
Ignition loss	55.23	34.64	65.16	26.9	31.0
SiO_2	18.42	28.95	12.50	37.9	9.56
Al_2O_3	10.07	14.19	5.49	4.84	1.79
Fe_2O_3	10.55	16.45	3.13	23.62	5.08
TiO_2	0.20	0.57	0.11	0.87	0.10
MgO	1.66	1.46	2.36	1.73	0.41
CaO	1.09	1.85	4.45	1.56	0.61
K_2O	1.13	0.70	0.25	0.42	0.08
Na_2O	0.78	0.52	1.18	0.48	0.10
B_2O_3	0.21	0.19	0.09	0.05	0.06
CuO	0.62	0.038	0.018	0.04	0.007
V_2O_5	0.066	0.057	0.016	0.054	0.004
ZrO_2	0.068	0.081	0.033	0.03	0.02
BaO	0.074	0.067	0.31	0.21	0.06
MnO	0.035	0.032	0.065	0.01	0.004
PbO	0.022	0.018	0.008	0.026	0.008
SrO	0.027	0.047	0.10	0.36	0.08
NiO	0.027	0.020	0.005	0.009	0.002
Cr_2O_3	0.024	0.025	0.006	0.014	0.003
CoO	0.015	0.013	0.006	0.003	0.002
MoO_3	0.014	0.005	0.004	0. . .	0.001
Ga_2O_3	0.012	0.011	0.006	0.008	0.001
SnO	0.004	0.005	0.0015	0. . .	0.0006
Ag_2O	0.0004	0.0003	0.0005	0.0001	0.0001
GeO_2	–	–	–	–	0.005

7.3. Liquid Slurry Products and Solids Separation

One of the most important bottlenecks in the commercial development of liquefaction is that of deashing and solids separation. This problem is complicated by the high viscosity of the liquid slurry products from liquefaction and by the submicron sizes of the solid particles in the slurry.

Two classes of techniques have been proposed for deashing. One is based on nearly complete separation of the solid particles in the slurry, and the other is based on partial separation. The technique for the nearly complete separation is filtration. There are many partial separation techniques, including the use of hydroclones, centrifugation, flotation and gravity settling, evaporation-distillation, carbonization, coking and solvent precipitation, and solvent washing.

Since one of the most undesirable components in coal is the mineral matter or ash, with its high pyritic sulfur content, complete separation of the solid

ISBN 0-201-08300-0

minerals in coal is most desirable for achieving low sulfur and ash contents. However, because the complete separation process is a mechanical filtration process, many operating difficulties exist.

7.4. Properties of Liquid Slurry Effluents

The most important properties influencing solid separation are the small size of the solid particle, the high viscosity of the slurry, the physicochemical affinity between solid particle and liquid, and the low density difference between particle and liquid.

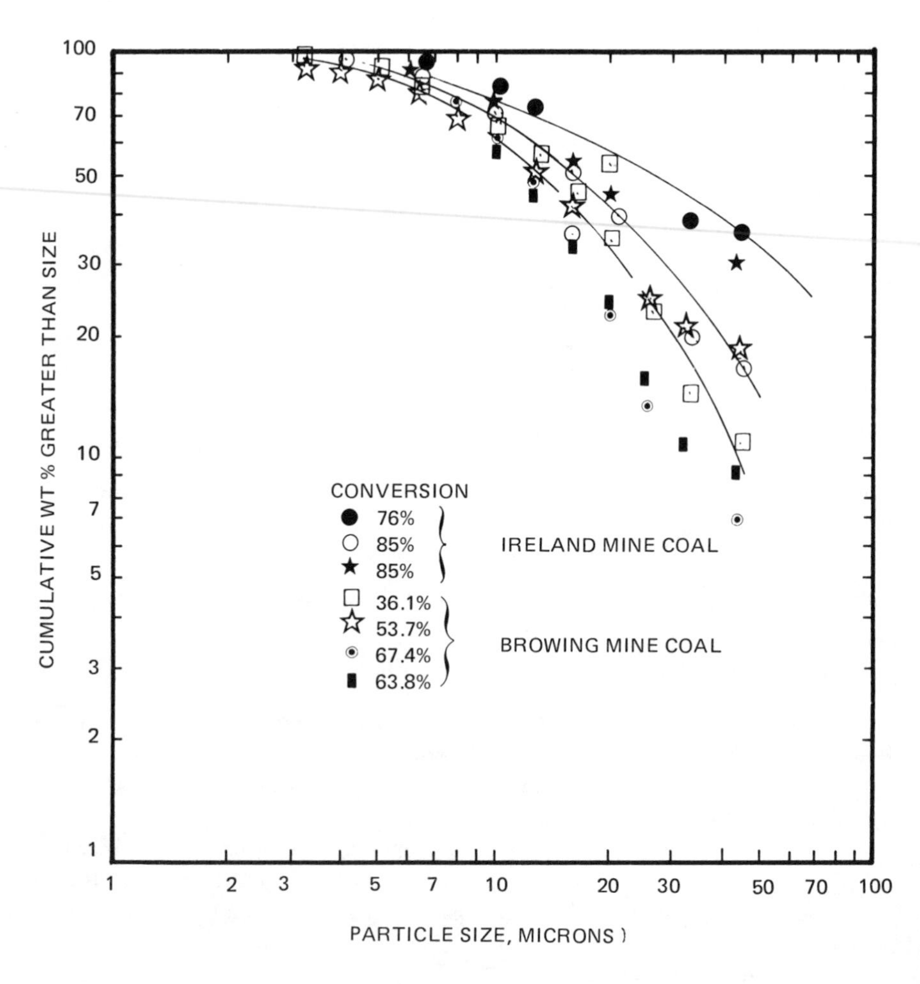

Figure 7.1. Particle size distribution with hydrogen-donor solvent. (Gorin et al. [90])

ISBN 0-201-08300-0

The solid particles are estimated to have mean diameters ranging from 5 to below 0.5 μ, with wide particle size distribution. The particle size depends on the degree of hydrogenation and the severity of the processing conditions. For example, under liquefaction conditions with only hydrogen-donor solvent, the median particle sizes are 28 and 16 μ, with 76 and 85% conversion, respectively, by using Ireland Mine coal, a highly caking bituminous coal [90]. However, this median particle size is much smaller under hydrogen-donor solvent liquefaction conditions with hydrogen pressure, and under these same conditions with catalyst and hydrogen pressure. Some of the typical particle size distributions are shown in Figures 7.1 and 7.2 [90]. These solid particles are obtained based on their insolubility in boiling cresol. Notice the decrease in particle sizes as the conversion increases, as the hydrogen pressure increases, or when catalyst is used.

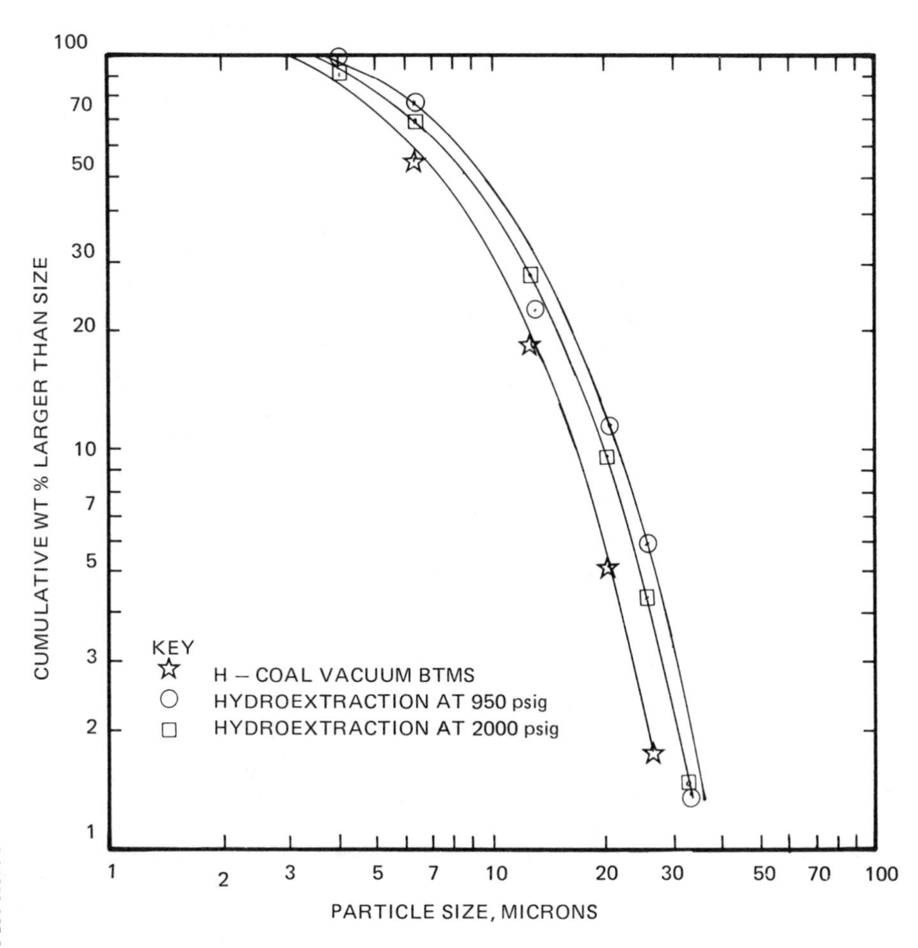

Figure 7.2. Particle size distribution, catalytic hydrogenation. (Gorin et al. [90])

ISBN 0-201-08300-0

Another factor which greatly influences solid separation efficiency is the viscosity of the slurry, which has been discussed in detail in preceding sections. The viscosity decreases as the degree of liquefaction increases. An additional factor which influences viscosity is the presence of asphaltenes and preasphaltenes (see Figure 6.14). To reduce the viscosity of the slurry, solid separation at fairly high temperatures is generally used.

7.5. Agglomeration, Coagulation, and Solid Separation

To decrease the difficulties in solid separation, various methods of agglomeration or coagulation of the solid particles have been suggested. Heat treatment, the use of solvents, and the inclusion of small amounts of additives or promoters are generally suggested.

Gorin and co-workers [90] have studied the deashing of liquefaction products by the use of deasphalting solvents, which were developed for the removal of asphaltenes from petroleum residual oils. However, deasphalting, as practiced in the petroleum industry, is the complete removal of asphaltenes, whereas in the deashing of coal liquids the purpose is to precipitate a minimal amount of benzene insolubles, together with the removal of coal ashes. These investigators found that gravity settling rates are enhanced and product clarity is improved if a minimal amount of precipitation of dissolved asphaltene is obtained by the use of a precipitating solvent such as *n*-decane. It appears that the small solid particles are agglomerated by these precipitating asphaltenes. Both solvent-refined coal slurry and catalytic hydrogenation slurry were investigated. Because of the smaller particle size in the catalytic hydrogenation slurry, the addition of precipitating solvent was necessary to achieve adequate deashing by the use of gravity settling.

Because of the blocking of the capillaries of the filter cakes by the fine solids contained in the liquid slurry, the filtration rate is quite low. Newman et al. [91] found that this filtration rate can be increased considerably by the combined use of antisolvent and thermal treatment. The antisolvent used was a light oil derived from the process. Generally, 50–60% of this antisolvent was used together with a 10 min thermal treatment at 105–115°C. It appears that the small solid particles agglomerated by thermal treatment in the presence of the antisolvent.

Katz and Rodgers [92] did an interesting study on the influence of heat, solvent dilution, and the addition of chemical promoters on the rate of sedimentation of the suspended solids in the solvent-refined coal slurry. The sedimentation test was carried out in a tall metal tube which was divided into 10 equal sections. The bottommost section (section 10) collects most of the solids, while the topmost section (section 1) has the least suspended solid particles as sedimentation progresses. The influences of solvent dilution, with toluene or recycle solvent as

ISBN 0-201-08300-0

the diluent, are shown in Table 7.2 for 1 hr of sedimentation time at 310°C. With 20% dilution the sulfur level after 1 hr of sedimentation is reduced to 0.5% in section 1 of the tube.

Table 7.2

Effect of Dilution on Settling of Suspended Solids
(Katz and Rodgers [92])

Dilution, %	Fraction No.	Ash, wt %	Sulfur, wt %
5	1	1.67	0.63
	4	1.52	0.70
	7	1.44	0.67
	10	3.32	0.70
10	1	1.12	0.59
	4	0.72	0.60
	7	1.22	0.67
	10	3.49	0.75
20	1	0.33	0.50
	4	0.24	0.54
	7	0.03	0.54

The influence of heat was also studied by these investigators. They found that the best heat treatment was obtained at 300–350°C. At higher temperatures, sedimentation settling was poor because of decomposition. Great improvement was obtained by treating the slurry at 310°C for 3 hr or at 350°C for 1 hr.

To find an effective chemical promoter which promotes the settling rate, Katz and Rodgers tested 34 organic and 25 inorganic materials. Only 4000 ppm of the promoters was used. These investigators found that the most effective promoters among those tested are a Tretolite 771-119 organic sample and three inorganic promoters: aluminum sulfate, ammonium sulfate, and ammonium hydrogen sulfate. The comparative settling rates of six runs with and without 4000 ppm of Tretolite are shown in Figure 7.3. As can be seen, when additive Tretolite was used, the sedimentation rates are much faster and the top few sections are relatively free of suspended solids.

ISBN 0-201-08300-0

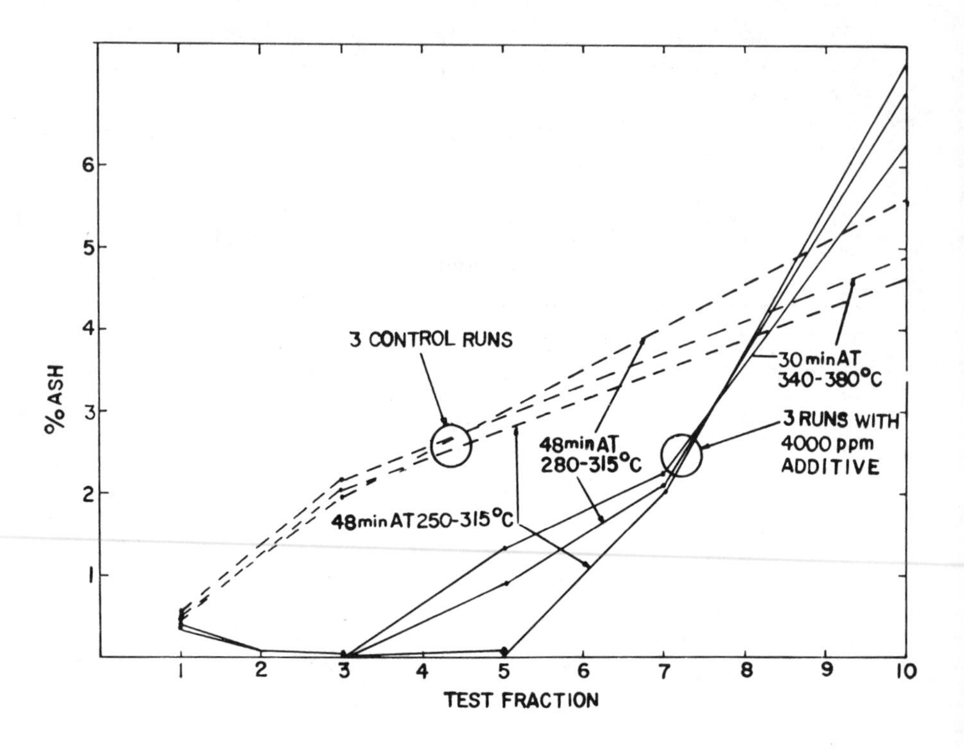

Figure 7.3. Comparative settling tests. (Katz and Rodgers [92])

7.6. Fractionation and Characterization of Coal Liquids

As was discussed earlier, in the laboratory the coal liquefaction products are routinely separated into gaseous products, oils (hexane soluble), asphaltenes (benzene soluble and hexane insoluble), and residues which contain the coal minerals and, depending on the degree of hydrogenation, unhydrogenated organics.

The liquid products consist of complex mixtures of many different organic materials such as aliphatic, olefinic, aromatic, heterocyclic, and polar compounds. To investigate the nature and characteristics of the liquid products, further separation of the sample into a number of well-defined fractions is necessary. It is true that these materials can be analyzed without further separation by the use of such techniques as mass spectrometry, and limited data can be furnished on physical characteristics such as percent oil, asphaltenes, gravity, and viscosity. However, these analyses cannot provide detailed compositional characterizations of the liquids for upgrading the liquids and for fundamental studies.

Depending on the purpose and information desired, coal liquids can be separated into various fractions. Usually both physical and chemical separation

ISBN 0-201-08300-0

techniques, such as distillation, extraction, and gas, liquid, and gel permeation chromatography are used.

Ruberto et al. [76] and Dooley and Thompson [77] have separated and characterized various coal liquids by using procedures developed for petroleum crudes and residues. The approximate procedure is distillation followed by extraction, the use of various chromatographic techniques, and molecular sieves. Ruberto et al. obtained the following fractions: materials boiling below 470°F (243°C), benzene-insoluble residue, asphaltenes, resins, and oils. The oils fraction is further separated into aromatics and saturates. The aromatics can be separated into monoaromatics, di- and triaromatics, and polyaromatics. Typical coal liquid compositions for a Wyoming subbituminous coal and a Pittsburgh seam bituminous coal are listed in Table 7.3 [76]. These liquids were obtained by catalytic hydrogenation followed by filtration to separate the undissolved coal and mineral matter. Any liquefaction products boiling below 130°F are also excluded from the liquids. Notice the very low values for saturates and the very high amounts of aromatics.

Callen et al. [78] used gradient elution chromatography (GEC) to separate the coal liquids into 13 fractions. The GEC was developed to handle highly refractive materials such as vacuum residua from petroleum refining and coal liquids. This separation allows further analysis and characterization of the products.

Table 7.3

Typical Compositions of Liquids from Catalytic Liquefaction (Ruberto et al. [76])

Liquid	Subbituminous	Bituminous
Distillate (b.p. $<$ 470°F)	28.3	25.2
Residues (b.p. $>$ 470°F)	71.7	74.8
Saturates	3.6	1.8
Aromatics	58.3	36.2
Monoaromatics	9.0	5.9
Di- and triaromatics	42.0	26.7
Polyaromatics	7.3	3.6
Resins	8.3	11.1
Asphaltenes	1.5	20.3
Benzene insolubles	$<$0.1	5.5

ISBN 0-201-08300-0

For the purpose of making fundamental studies on the structure of coal and coal liquids, Farcasiu [79] separated the solvent-refined coal liquid into different fractions by the use of a liquid chromatographic column with silica gel as the stationary phase and with a sequence of specific solvents having gradually varying properties. By properly choosing the solvent properties, combined with manipulation of the properties of silica gel, the liquid is separated into fractions which differ mainly in their chemical properties. By using this fractionation technique, combined with various analytical procedures, the kinetics and mechanisms of coal dissolution can be studied in much more detail.

7.7. Properties of Coal Liquids

Since coal liquids will be used principally as substitutes in the various end uses of petroleum, study of their properties based on a comparison between coal and petroleum crude is most useful. On the average, coal liquids are more analogous to petroleum residual fuels than to other petroleum products. Thus the analytical techniques developed for petroleum residua can be modified and used to characterize coal liquids.

The carbon/hydrogen ratio and the concentrations of heroatoms such as nitrogen and oxygen are generally much higher in coal liquids than in petroleum crudes and fuels. The sulfur contents and physical properties of coal liquids usually vary considerably, depending on the type of coal and the processing conditions. However, the molecular weights of coal liquids lie within a surprisingly narrow range, generally between 300 and 1000 [79]. The aromatic contents of coal liquids are very high.

Callen et al. [78] investigated the composition and characteristics of coal liquids by various analytical techniques originally developed for petroleum crude. Three types of analyses were performed: physical property determination, trace metal analysis, and chemical composition and elemental analysis. Their elemental analyses of coal liquids, together with some petroleum fuels, are listed in Table 7.4. The Solvent-refined coal (SRC) process is an extraction or dissolution process without catalyst, and both the synthoil and H-coal processes are catalytic hydrogenation processes.

In comparison with petroleum fuel and petroleum residua, which are listed in the last two columns of the table, coal liquids have higher carbon and much lower hydrogen contents, which suggest both a higher degree of aromaticity and a more highly condensed ring structure for coal liquids. The high aromatic content of coal liquids has been discussed in various places in this chapter. For example, in Table 7.3 the aromatics are very high. This is expected since coal is primarily composed of mono-, di-, and triaromatic rings. Furthermore, the increase in aromaticity during hydrogenation has been detected and discussed earlier. This high aromatic content has two consequences. First, it gives a high octane number and thus forms a good blend stock for gasoline. Second, because of pollution restructions, further processing to decrease the aromatic content may be necessary.

ISBN 0-201-08300-0

Another characteristic of the coal liquids in Table 7.4 is the high content of heteroatoms such as nitrogen and oxygen. However, the sulfur content is lower for these particular coal liquids. The high contents of nitrogen in these liquids are significant because of pollution considerations.

Table 7.4

Typical Elemental Compositions of Coal Liquids and Petroleum Fuels (Callen et al. [78])

Element, wt %	SRC	H-Coal	Synthoil	Fuel Oil[a]	Residuum[b]
C	87.93	89.00	87.62	86.40	83.88
H	5.72	7.94	7.97	11.20	9.97
O	3.50	2.12	2.08	0.30	0.48
N	1.71	0.77	0.97	0.41	0.40
S	0.57	0.42	0.43	1.96	4.19

[a] El Palito No. 6 fuel oil.

[b] $1000^{\circ}F$ + west Texas sour residuum.

Haebig et al. [80] made some preliminary combustion tests of coal liquids and concluded that the higher NO_x emitted from coal liquids is mostly due to fuel NO_x, or comes from the nitrogen in the fuel, not from the nitrogen in combustion air. Thus, to meet pollution standards the nitrogen in coal liquids must be removed by further processing and upgrading.

Using gradient elution chromatography, Callen et al. [78] compared coal liquids with three petroleum fuels. The results for the three coal liquids are shown in Figure 7.4 and those for the petroleum liquids in Figure 7.5. The 13 fractions are obtained by the use of different solvents. In general, fraction 1 is the saturates and melts slightly above room temperature. Furthermore, it is very soluble in paraffinic solvents, and thus it is principally composed of paraffins, naphthenes, and olefins. Fractions 2 and 3 are oils and are principally composed of aromatics, with fraction 2 having a high content of monoaromatics. Fraction 4 is soft resin and is composed principally of higher aromatics. It is a semisolid at room temperature. Fractions 5 and 6 are resins and are solids at room temperature. Fraction 6 is a polar resin and is extremely viscous at $240^{\circ}F$. Fraction 7 is eluted asphaltenes with properties intermediate between those of resins and polar asphaltenes and is very soluble in methylene chloride. Fractions 8–12 are polar asphaltenes and are very friable black

ISBN 0-201-08300-0

powders. Fraction 13 is the noneluted fraction and is probably composed of strongly acidic compounds and inorganic solids such as carbon and catalyst fines.

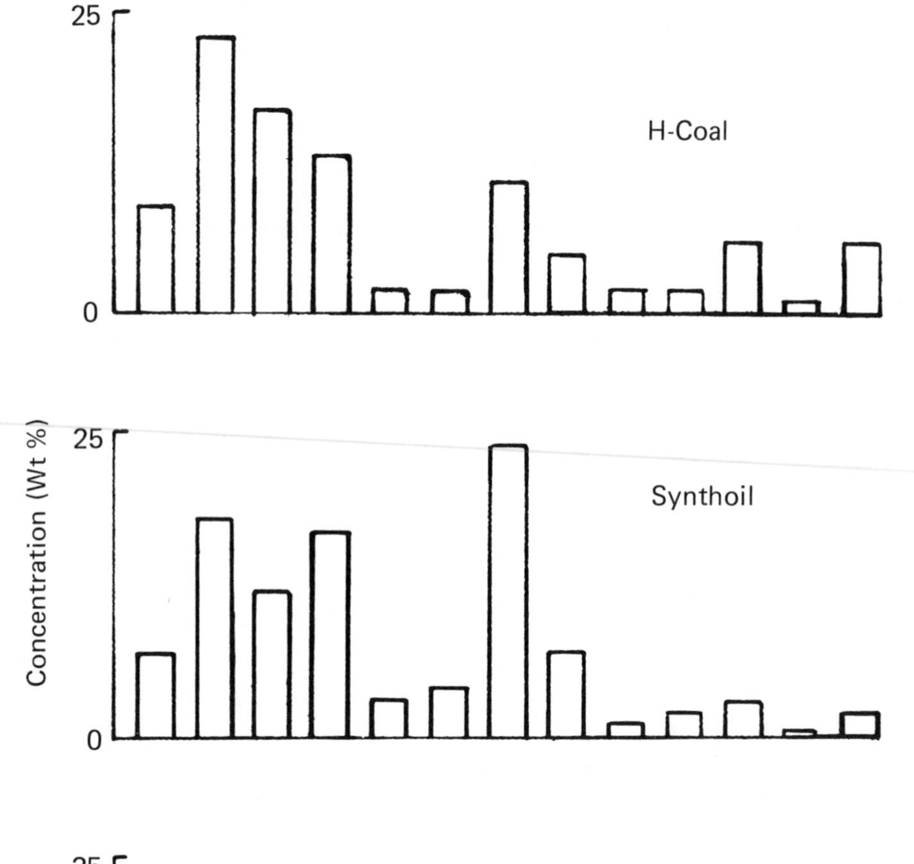

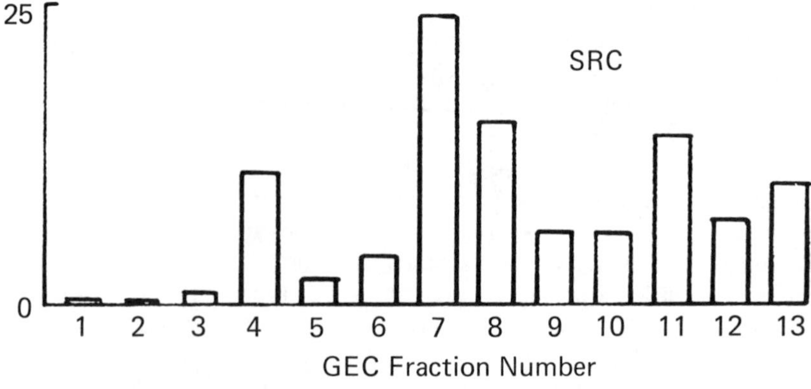

Figure 7.4. GEC fractions of coal liquid. (Callen et al. [78])

ISBN 0-201-08300-0

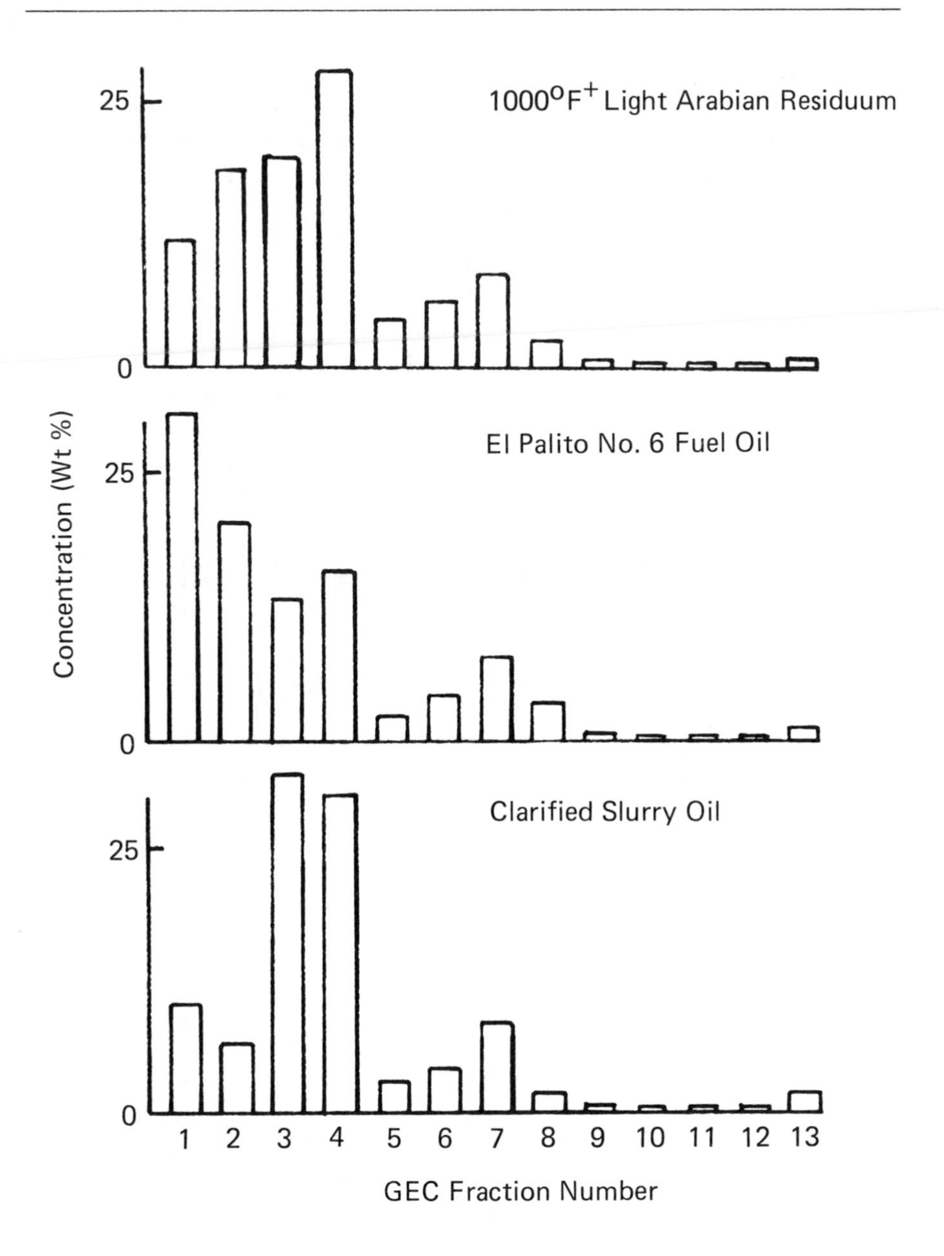

Figure 7.5. GEC fractions of some petroleum liquids. (Callen et al. [78])

ISBN 0-201-08300-0

A comparison of Figures 7.4 and 7.5 is of great interest. Coal liquids contain significantly more asphaltenic fractions than petroleum liquids. For example, 81% of SRC belongs to fractions 7–13, whereas only 12% Arabian residuum belongs to these fractions. Both H-coal and synthoil use catalyst during hydrogenation, and thus their compositions are somewhat similar. Approximately 49% and 36%, respectively, of H-coal and synthoil belong to the low molecular weight fractions 1–3, but only 1% of SRC belongs to fractions 1–3. Thus the effectiveness of catalyst in increasing the degree of hydrogenation is clearly demonstrated in Figure 7.4. For SRC most of the products are in the high molecular asphaltenes range, whereas for H-coal and synthoil most of the asphaltenes are further hydrogenated into lower molecular weight products.

The aromaticity of coal liquid was also investigated by the use of carbon-13 nuclear magnetic resonance (NMR) spectroscopy. The results for various liquids are shown in Figure 7.6 [78], where the aromaticities of the liquids are plotted against the carbon/hydrogen atom ratio. Notice the high aromaticity of the various coal liquids.

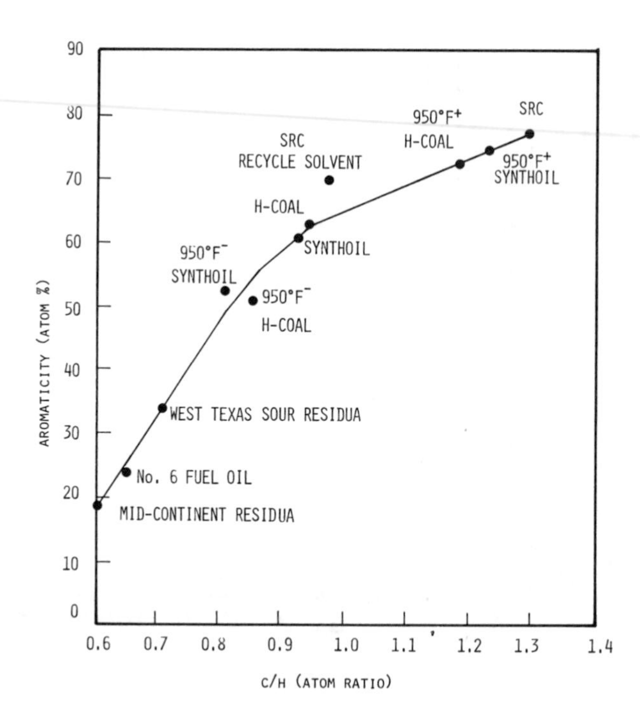

Figure 7.6. Aromaticity versus C/H atom ratio. (Callen et al. [78])

7.8. Asphaltenes and Preasphaltenes

Asphaltenes are the primary end products in dissolution or in SRC. But in catalytic hydrogenation most of the asphaltenes form an intermediate in the conversion process from coal to oil. Thus it is desirable to have a detailed investigation

ISBN 0-201-08300-0

of asphaltenes, not only because they constitute the main material in SRC, but also because of their importance in the mechanism of coal conversion.

Coal is a solid and is insoluble in benzene. Asphaltenes are solids also but, in contrast to coal, are soluble in benzene, anthracene oil, and other solvents. For coals with low organic sulfur, the conversion of coal to asphaltenes is all that is needed to produce a low ash and low sulfur fuel such as SRC. However, for coals with high organic sulfur, further conversion from asphaltenes to low molecular weight products is necessary. This additional conversion must be carried out under more severe conditions with catalyst.

Preasphaltenes and asphaltenes have fairly high molecular weights and, more importantly, viscosities. Furthermore, the viscosity of coal liquids is significantly influenced by the presence of preasphaltenes and asphaltenes. Sternberg et al. [25] have found, at least for the particular case studied, that the presence of asphaltenes and preasphaltenes can increase the viscosity of the centrifuged liquid products from synthoil by a significant amount. For pure oils the viscosity was only 16 centistokes (cSt) at 140°F. This viscosity increased to 907 cSt for a mixture with 42% asphaltenes and 58% oil, at the same temperature. A further increase to 1400 cSt was obtained when 5% of preasphaltenes was added to make a mixture of 55% oil, 40% asphaltenes, and 5% preasphaltenes.

Fu and Batchelder [67] studied the influence of asphaltenes on viscosity in more detail. Their results are reproduced in Figure 6.14. The increase of viscosity due to asphaltenes is apparently fairly large. Since low viscosity is essential for the easy separation of solids or ashes from liquefaction products, this effect of asphaltenes is especially significant.

A comparison of the molecular weights of total liquid products and asphaltenes is of interest. The results obtained by Fu and Batchelder are shown in Table 7.5 [67]. These materials were obtained from a bituminous coal at 450°C and 3000 psi pressure with catalyst and with a coal/solvent ratio of 1:2.3. It should be pointed out that the molecular weights of asphaltenes vary considerably with the type of coal and processing condition. However, coal-derived asphaltenes have a much lower molecular weight than petroleum-derived asphaltenes, whose molecular weight is on the order of 3000.

Sternberg and co-workers [26, 81] suggested that asphaltenes have essentially an acid-base structure. Furthermore, the acidic and basic components can be separated by dissolving the asphaltenes in toluene and passing dry HCl gas through the solution. The basic components precipitate, while the acidic components remain in solution. The following composite structures for the acidic and basic components were suggested:

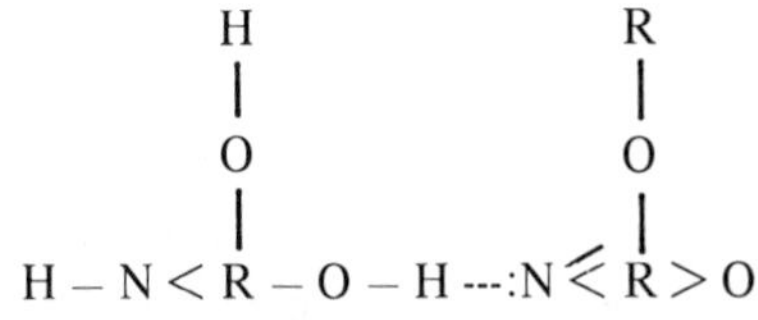

where the left-hand structure is acid, and the right-hand structure base.

ISBN 0-201-08300-0

This hydrogen-bonded structure of asphaltenes is compatible with their solubility characteristics. In nonpolar solvents such as pentane, a large complex is formed because of hydrogen bonding. However, in moderately polar solvents such as benzene, hydrogen bonding does not form, and the separate basic and acidic components become soluble in the solvent.

Farcasiu et al. [66] have investigated the influence of asphaltols on coal hydrogenation. They discovered that up to 80% of the yield was char on pyrolysis of asphaltols in the temperature range used in coal liquefaction; thus asphaltols may well contribute to high coke making during dissolution. Schweighardt et al. [82] provided an interesting summary on asphaltenes and concluded that "asphaltenes and pre-asphaltenes might well be the smallest units into which coal may be divided and yet still retain most of the spectral and some of the chemical properties of the original whole coal."

Table 7.5

Molecular Weights of Coal Liquids and Asphaltenes (Fu and Batchelder [67])

	Molecular Weight			
	syngas (2H:ICO)		H_2	
Time, min	Asphaltene	Total Liquid	Asphaltene	Total Liquid
15	339	218	437	227
30	494	221	544	210
60	395	194	405	194

8. OTHER LIQUEFACTION PROCESSES

8.1. Introduction

In addition to the liquefaction concepts discussed earlier, many new and novel second-generation processes are being developed. Some of the more promising ones are discussed in this section. Other promising techniques which are not discussed are underground liquefaction, catalytic dehydrogenation of coal, and oxidation of coal.

The production of chemicals from coal is another area which will not be discussed in this chapter. Obviously, because of the aromatic nature of coal, one class of chemicals most easily available from coal is the aromatic chemicals such as benzene, toluene, xylenes, naphthalene, and phenol.

ISBN 0-201-08300-0

8.2. New Catalytic Systems

One of the most promising approaches to develop second-generation coal liquefaction processes is the use of new and novel catalytic systems. Intensive research is being carried on to obtain novel catalysts and to improve existing ones. Some of the criteria in the development of new catalysts and the problems involved are summarized in the literature [94, 95]. The technical challenges are the normally encountered problems, namely, controlling selectivity and activity, minimizing activity decline from poisons, sintering, and mechanical degradation, and developing cost-effective catalyst regeneration procedures. High activity can reduce reactor size and pressure in liquefaction, and good selectivity can minimize the consumption of expensive hydrogen. The catalyst performs three distinct functions in coal liquefaction: the cracking of large aromatic molecules, the hydrogenation and hence stabilization of these cracked moledules, and the simultaneous removal of heteroatoms such as sulfur, nitrogen, and oxygen.

Some of the more promising catalyst systems have been summarized by Mills [73]; they are nascent-active hydrogen generated *in situ* with carbon monoxide and steam, organic hydrogen-donor solvents, massive amounts of halide catalysts, complexes of transition metals, akali metals, reductive alkylation, volatile catalysts such as iodine, molecular sieves, high energy excitation such as ultrasonics, and dehydrogenation of coal to form hydrogen. The first two approaches, nascent-active hydrogen and hydrogen-donor solvents, were discussed in preceding sections; they can certainly be improved by further investigation. Some of the other systems are discussed in this section.

8.3. Halide Catalysts

Halide catalysts have been used in two different ways in coal hydrogenation research in recent years. Wiser and co-workers [96, 97] have used 5% $ZnCl_2$ for coal hydrogenation with a very short residence time at a temperature of 500°C or higher. This pyrolysis process appears to offer a promising approach in coal liquefaction.

In the second way of using halide catalysts, which is more directly connected to the topics discussed in this chapter, fairly large amounts of catalyst (catalyst/coal ratio on the order of 1) serve for coal or coal extract effluent hydrogenation. Distinctive features of halide catalysts are the high conversion of asphaltenes to oil and other products, and the high selectivity of these catalysts for producing gasoline.

Kawa et al. [98] have investigated the effectiveness of liquefaction with 10 halide catalysts. Their results, which are reproduced in Table 8.1, were obtained with 50 g of hvA bituminous coal and 1 hr reaction time at 4000 psi and 425°C. High coal conversion was obtained with ZnI_2, $ZnBr_2$, $ZnC\ell_2$, and SnI_2. However, all of the halide catalysts when used in large amount are effective in reducing asphaltene content. Without catalyst the asphaltene obtained is 28%; this is reduced to 14% with the least effective catalyst, SnI_2. When a small amount of catalyst (1%)

ISBN 0-201-08300-0

is used, $ZnC\ell_2$ is greatly inferior to $SnC\ell_2$ catalyst. This confirms earlier statements that $SnC\ell_2$ is a superior coal hydrogenation catalyst when used in the usual amount.

Table 8.1

Hydrogenation Products Using Halide Catalysts (Kawa et al. [98])

		Yield, wt % of maf coal						
Catalyst	Catalyst/ Coal Weight Ratio[a]	Organic Benzene Insolubles	Asphaltene	Heavy Oil	Light Oil	Hydro- carbon Gases	Acid Gases and CO	Net Water
None	–	37	28	13	6	6	3	7
$NiI_2 \cdot 6H_2O$	1.0	18	2	21	38	10	4	12
$NiBr_2$	1.0	20	4	29	33	10	6	5
ZnI_2	1.0	10	1	5	55	17	7	10
I_2	1.0	23	1	8	48	15	3	11
$ZnBr_2$	1.0	10	2	8	56	14	3	13
$ZnCl_2$	1.0	12	2	16	45	14	4	11
$SnCl_2 \cdot 2H_2O$	1.0	18	7	40	29	6	4	1
CdI_2	1.0	17	3	24	37	13	1	12
$FeI_2 \cdot 4H_2O$	1.0	20	9	13	41	10	2	5
S_nI_2	1.0	8	14	42	20	6	5	5
$ZnCl_2$	0.01	33	26	18	8	8	4	7
$SnCl_2 \cdot 2H_2O$	0.01	12	38	34	7	4	5	7
$SnCl_2 \cdot 2H_2O$[b]	0.01	7	7	36	18	26	1	9

[a] Does not include water of hydration.
[b] Thirty minutes at 480°C.

Gorin and co-workers [99–101] have made extensive investigations using molten halide catalysts for the hydrogenation of polynuclear hydrocarbons, coal extract effluents, and coal. They found that by using massive quantities of zinc halide catalyst single-stage conversion of either coal or coal extract effluent to high octane gasoline at relatively mild operating conditions is possible. These catalysts act as Lewis acids, and because of their high activity the residence time can be shortened considerably. This is shown clearly in Figure 8.1 [100], for the case where coal extract effluent is hydrogenated by the use of both $ZnC\ell_2$ and the conventional hydrofining catalyst nickel molybdate under hydrogen pressure. Three hours is required to reach a conversion of 75% with the hydrofining catalyst, as compared with less then 5 min when the molten zinc chloride catalyst is used. The coal extract effluent was obtained by dissolution of a bituminous coal with tetralin with a residence time of 46 min at 380°C, and the unreacted coal and ashes were removed by filtration at 200°C.

ISBN 0-201-08300-0

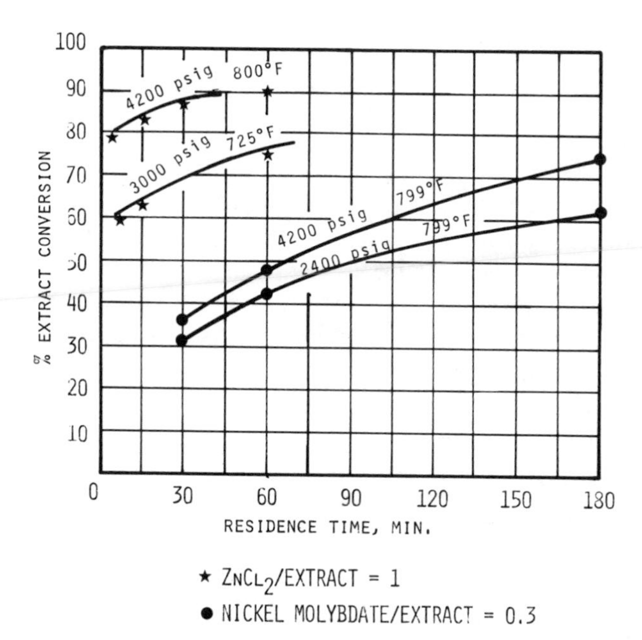

Figure 8.1. Conversion versus residence time. (Zielke et al. [100])

The influence of temperature and pressure on conversion of extract effluent is shown in Figure 8.2; here a residence time of 1 hr is allowed with a $ZnC\ell_2$/ extract weight ratio of 1.

The kinetics of this reaction has also been investigated, and the results can be represented by two or more simultaneous first-order reactions. The reactant which is the extract effluent is assumed to be composed of three species: an unreactive, a fast reacting, and a slow reacting species. The reaction rate equation in a batch unit can be represented by [101]

$$C_e - C = N_1 C_e e^{-k_1 t} + (1 - N_1) C_e e^{-k_2 t} \tag{8.1}$$

where $C_e - C$ is the convertible material remaining at time t, N_1 is the fraction of the convertible material which reacts via the fast reaction, k_1 and k_2 are reaction rate constants for conversion via the fast and slow reactions, respectively, C is the conversion of the reactant at time t based on maf extract, and C_e is the conversion at which the rate of conversion becomes zero.

Zinc chloride was also used to hydrogenate a Pittsburgh seam bituminous coal without a hydrocarbon vehicle under fairly mild conditions of 358°C and 103 atm abs. hydrogen pressure in a 316 stainless steel rocking autoclave. The results are

ISBN 0-201-08300-0

summarized in Table 8.2 [103], where all the data are for a residence time of 1 hr. However, attempts to hydrogenate subbituminous coal under these mild conditions were not successful because of the formation of viscosity mass, and thus tetralin was used as the vehicle solvent. Approximately 87–89% conversions were obtained for both the bituminous and the subbituminous coal. The conversion is defined as gases plus methyl ethyl ketone (MEK) soluble.

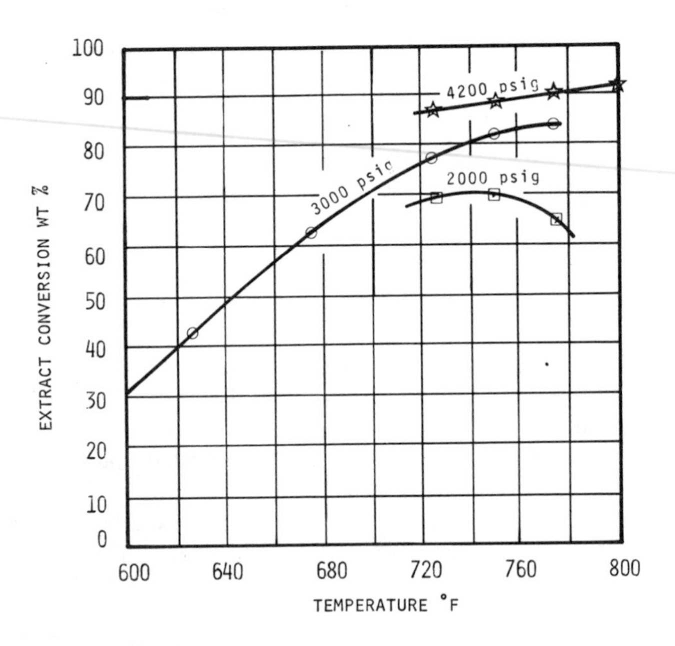

Figure 8.2. Conversion versus temperature and pressure. (Zielke et al. [100])

Under more severe reaction conditions no solvent vehicle is needed even for the subbituminous coal. Within the temperature range of 370–427°C at 205 atm, maximum conversion was obtained at 385°C. Notice the substantially high yield of heavy fuel oil (+400°C MEK soluble fraction) from the subbituminous coal. This yield decreases rapidly as the operating conditions become more severe. Another noticeable feature in Table 8.2 is the high yield obtained for gasoline, which is

ISBN 0-201-08300-0

approximately the C_4 x 200°C fraction. Except for the experiment carried out at 358°C, the C_4 x 400°C distillate fraction is mostly gasoline, with only 1–3% distillates above 200°C.

Table 8.2

Liquefaction Yield with Zinc Chloride Catalyst (Zielke et al. [103])

Variable	Coal				
	subbituminous			bituminous	
Temperature, °C	358	385	413	358	385
H_2 pressure, atm	103	205	205	103	205
$ZnC\ell_2$/mf coal, weight ratio	2.5	2.5	2.5	2.0	2.5
Tetralin/mf coal, weight ratio	0.5	0	0	0	0
Conversion, wt % maf coal	89.5	94.8	91.3	87.2	94.0
Yield, wt % maf coal					
C_1–C_3	1.7	5.4	9.9	5.2	8.7
C_4 x 400°C distillate	27.5	58.7	60.2	56.4	68.6
+400°C MEK soluble	46.9	17.1	9.2	20.3	13.9
+400°C MEK insoluble	10.5	5.2	8.7	12.8	6.0
$CO + CO_2 + H_2O$	17.8	19.9	19.3	8.8	8.6
Sulfur in +400°C MEK soluble, %	0.17	0.04	–	0.17	0.18
Sulfur in mf original coal, %	0.74	0.74	0.74	2.8[a]	4.08
H_2 consumed, % maf coal	5.9	8.7	9.5	6.5	9.1

[a] The bituminous coal used for the 358°C and 103 atm run was precleaned and was about 2% higher in carbon content and 3% lower in ash content than the uncleaned coal.

ISBN 0-201-08300-0

8.4. Regeneration of Zinc Chloride Catalyst

Although $ZnCl_2$ appears to be the most economic and effective catalyst, its practical utilization requires efficient and low cost recovery and regeneration of this relatively high cost material. Corrosion of the catalyst melt can be another practical difficulty. Since $ZnCl_2$ melts at 283°C and boils at 732°C, it is in liquid form in the liquefaction temperature range.

Spent molten $ZnCl_2$ catalyst has been regenerated in a continuous unit by combustion with air in a fluidized bed of inert solids [102, 103]. The combustion process removes the carbon, nitrogen, and sulfur impurities and simultaneously vaporizes the zinc chloride. The inert solid used is silica. The bed is operated at essentially atmospheric pressure with 1800–1900°F temperature and with air as the fluidizing agent. The superficial vapor residence time was on the order of 1 sec. No operability problems such as ash agglomeration or defluidization of the bed were encountered.

During the hydrogenation process the $ZnCl_2$ catalyst becomes contaminated with ZnS, $ZnCl_2 \cdot NH_3$, and $ZnCl_2 \cdot NH_4Cl$ by the reaction of the catalyst with nitrogen and sulfur liberated from coal:

$$ZnCl_2 + H_2S \rightarrow ZnS + 2HCl \tag{8.2}$$

$$ZnCl_2 + xNH_3 \rightarrow ZnCl_2 \cdot xNH_3 \tag{8.3}$$

$$ZnCl_2 \cdot yNH_3 + yHCl \rightarrow ZnCl_2 \cdot yNH_4Cl \tag{8.4}$$

In addition to the above, the catalyst also contains residual carbon and coal ashes.

When pure air was used, about 10% of the catalyst was lost as ZnO solids through the reaction

$$ZnCl_2 + H_2O \rightarrow ZnO + 2HCl \tag{8.5}$$

To avoid this loss, the HCl was recylced with the feed air with a resulting concentration of 5.5% anhydrous HCl. This recycling completely suppressed the above hydrolysis reaction and thus prevented the loss of catalyst through the formation of ZnO [103].

One limitation of this approach is that the coal must have relatively low sodium and potassium contents to avoid large losses of chlorine to these alkali metals. Since lignite usually has high contents of these alkali metals, the zinc chloride system cannot be applied to lignite coals. Loss of chlorine through the formation of calcium chloride and magnesium chloride was not detected.

Iron is another inorganic ash whose reaction with the catalyst must be considered. It appears that some $FeCl_2$ was formed. However, $FeCl_2$ does not appear to affect the catalyst activity.

ISBN 0-201-08300-0

8.5. Transition Metals Complexes [73, 104]

One of the earliest transition metal complex catalysts is cobalt carbonyl, which is active in hydrogenating certain aromatic compounds and coal. Some of these complexes are soluble and hence are homogeneous catalysts. One distinct advantage of this type of catalyst is its high degree of selectivity in terms of causing only certain molecules to react from a mixture and then to form only particular and sometimes unusual products. For example, dicobalt octacarbonyl, $Co_2(CO)_8$, in the presence of carbon monoxide and hydrogen, functions as a selective homogeneous hydrogenation catalyst for polynuclear aromatic hydrocarbons with the following reactions [73]:

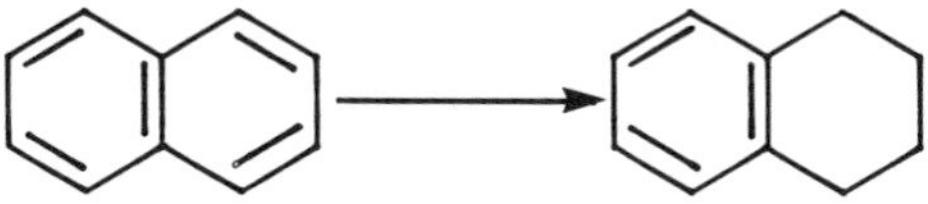

(8.6)

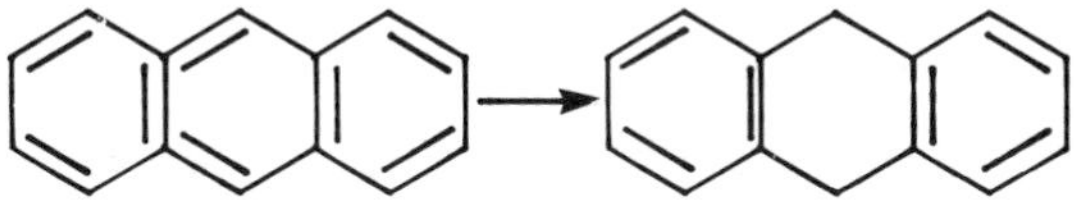

(8.7)

These reactions were carried out with a hydrogen/carbon monoxide mole ratio equal to 1 at 3000 psi pressure and 152–200°C temperature. When coal was treated with carbon monoxide and hydrogen in the presence of dicobalt octacarbonyl at 200°C, both H_2 and CO were added to coal; thus this is one of the most active catalysts for coal hydrogenation. However, this catalyst is inactive with isolated benzene rings. In comparison with the heterogeneous catalyst Co-Mo-Al_2O_3, cobalt carbonyl requires a much lower temperature, as much as 100°C below that for cobalt molybdate catalyst. However, cobalt carbonyl is less active in removing heteroatoms such as sulfur, nitrogen, and oxygen.

8.6. Reductive Alkylation

Since one of the primary purposes of coal liquefaction is the transformation of solid coal into liquid form, an ideal approach would be the use of a minimum amount of expensive hydrogen to produce high solubility. Reductive alkylation appears to be such an approach, which is based on two steps. The first step is the reduction of the aromatic coal, dissolved in a suitable solvent, with alkali metals forming aromatic hydrocarbon anions. The second step is the introduction of alkyl groups into the coal molecule by allowing these anions to react with alkyl halides.

ISBN 0-201-08300-0

For example, naphthalene dissolved in hexamethylphosphoramide (HMPA) reacts with 1 or 2 moles of lithium to form the naphthalene monoanion:

$$Li + C_{10}H_8 \rightarrow [C_{10}H_8]^- + Li^+ \quad (8.8)$$

or the naphthalene dianion:

$$2Li + C_{10}H_8 \rightarrow [C_{10}H_8]^{-2} + 2Li^+ \quad (8.9)$$

depending on the amount of lithium used. Aromatic hydrocarbon anions react easily with alkyl halides to introduce alkyl groups into the aromatic hydrocarbon. For example, anthracene dianion reacts with methyl iodide to convert one of the aromatic nuclei into an alkylated dihydrobenzene:

$$[\text{anthracene}]^{-2} + 2CH_3I \longrightarrow \text{9,10-dihydro-9,10-dimethylanthracene (H, CH}_3\text{; H, CH}_3) + 2I^- \quad (8.10)$$

Since coal is essentially a polyaromatic compound, the above reactions can also be carried out with coal. Sternberg and co-workers [105, 106] have used this approach to add alkyl groups to various bituminous and subbituminous coals. However, this approach was not very successful in adding alkyl groups to anthracite.

The solubility increase by alkylation is tremendous. For example, although only 3% soluble in HMPA, coal became 90% soluble on addition of lithium to a suspension of a lvB coal in HMPA. This tremendous increase in solubility is almost certainly due to the formation of aromatic anions by the transfer of electrons to coal:

$$nLi + \text{coal} \rightarrow nLi^+ + [\text{coal}]^{-n} \quad (8.11)$$

This high solubility of the coal anion offers an opportunity to introduce alkyl groups into the coal molecule.

In the second step this coal is alkylated with ethyl iodide to yield an ethylated coal which was 35% soluble in benzene at room temperature. A second alkylation yielded a poduct which was 85% soluble in benzene. The ethyl groups introduced into the coal after two alkylations correspond to about one methyl group per five carbon atoms.

Other suitable solvents and alkali metals can also be used. Sternberg and co-workers treated various coals with potassium in tetrahydrofuran (THF) in the presence of naphthalene; electrons were added to the coal and coal anions were formed in the same way as that shown in Equation 8.11 except that lithium was

ISBN 0-201-08300-0

replaced by potassium. Alkylation of these anions with ethyl iodide resulted in the addition of 16 and 14 ethyl groups per 100 carbon atoms for bituminous and subbituminous coals, respectively. The alkylated coals were 88% and 45% soluble in benzene. The molecular weights of the benzene-soluble portions of the bituminous and subbituminous coals were 2000 and 700, respectively.

It should be noted that reductive alkylation also serves to add hydrogen, in that the hydrogen/carbon atom ratio in coal after alkylation is increased. Moreover, this addition of hydrogen is much more effective in increasing solubility than direct hydrogenation.

8.7. Alkali Metals

Alkali metals can be used in the hydrogenation of coal in at least three different ways: as hydrogen activation catalysts, as reducing agents together with amines, and as electrolytes in catalytic electrochemical reduction [73].

Alkali metals are active hydrogenation catalysts; the extent of hydrogenation of coal depends on the temperature and the alkali metal used. For example, at 250°C and 1400 psi pressure, phenanthrene is hydrogenated to 9,10-dihydrophenanthrene by sodium and to octahydrophenanthrene by sodium-potassium. On the other hand, with sodium-rubidium, dihydrophenanthrene is obtained at 180°C and octahydrophenanthrene at 200°C. One advantage of the sodium catalyst is that the presence of sulfur does not deactivate the catalysts.

Electrochemical reduction of coal in ethylenediamine in the presence of lithium chloride has been shown to be a useful method for the hydrogenation of coal under very mild conditions [107]. As much as 53 hydrogen atoms can be added per 100 carbon atoms. The reduced coal is 78% soluble in pyridine and 30% soluble in benzene at room temperature. However, the current efficiency is very low, only 10–20% for coal. Much higher efficiency can be obtained for some other compounds. For example, a current efficiency of 80% was obtained for the reduction of benzene ring.

Coal can be reduced by alkali metal-amine systems [108]. For example, chemical reduction of coal by lithium-ethylenediamine proceeds at room temperature and can add as many as 55 hydrogen atoms per 100 atoms of carbon. This reaction appears to proceed in the following manner:

$$R + 2Li + 2RNH_2 \rightarrow RH_2 + 2Li^+ + 2RNH^- \tag{8.12}$$

where R represents the coal. In general, the reduced coal is much more soluble in pyridine. In one instance the solubility increased as much as 35-fold. One reduced vitrain became 97% soluble. Infrared and ultraviolet spectra showed a consistent loss of aromatic rings in the reduced vitrains.

8.8. Supercritical Gases

A novel approach being developed by the British Coal Research Establishment [109] is to use compressed supercritical gases for the extraction or liquefac-

ISBN 0-201-08300-0

tion of coal. The idea is based on the fact that, in the presence of a compressed supercritical gas, the ability of a given solid substance to vaporize can be increased tremendously. For example, the liquids formed when coal is heated to about 400°C cannot be distilled because of the low temperature. Increasing the temperature causes polymerization and is certainly undesirable. However, in the presence of a compressed supercritical gas, these coal liquids are caused to vaporize.

The extract gas is chosen so that its critical temperature is close to the temperature at which the extraction is to be carried out. Toluene, which has a critical temperature of 318°C, was used as the extract gas in the British Coal Research Establishment studies. Organic mixtures such as coal tar or petroleum naphtha fractions were also used.

A stream of supercritical gas, such as toluene, was passed into a heated bed of coal, which is readily penetrated by gas, so that fine grinding is not necessary. Since water can also act as a supercritical solvent, it is not necessary to thoroughly dry the coal. Some of the coal constituents passed into the gas phase, leaving behind an unvolatile char with the mineral matter. The supercritical gas stream with the volatilized coal was passed into another vessel at atmospheric pressure, thus causing the coal extract to precipitate and effecting a complete separation between the extracted coal and the extracted gas. Depending on the rank and volatile content of the coal, up to over 30% of the coal can be extracted. In general, the higher the volatile content of the coal, the higher the yield. Although the yield is less than that obtained by using liquid hydrogen-donor solvents, it is considerably greater than in carbonization processes. Higher yield can be obtained by carrying out the extraction in the presence of reactive species such as hydrogen.

The product obtained is essentially ash-free and solvent-free. There is no mineral or solvent separation problem, and the extract solvent is virtually completely recovered. The product thus obtained has a high hydrogen content and a lower molecular weight. Furthermore, the unreacted solid residue or char can be used easily in fluidized combustion, in gasification, or for the production of hydrogen, since such residues are essentially composed of porous low temperature char which has no agglutinating tendencies and releases little or no tarry matter on heating.

REFERENCES

1. De Marsilly, C., "Solvent Extraction of Coal," *Ann. Chim. Phys.*, **3**, 66 (1862).
2. Kiebler, M. W., "The Action of Solvents on Coal," Chapter 19 of *Chemistry of Coal Utilization,* Vol. 1, H. H. Lowry, Ed., John Wiley & Sons, New York, 1945.
3. Dryden, I. G. C., "Chemical Constitution and Reactions of Coal," Chapter 6 of *Chemistry of Coal Utilization,* Suppl. Vol., H. H. Lowry, Ed., John Wiley & Sons, New York, 1963, pp. 237–252.
4. Dryden, I. G. C., "Behavior of Bituminous Coal towards Solvents," I and II, *Fuel,* **29**, 197 and 221 (1950).

ISBN 0-201-08300-0

5. Van Krevelen, D. W., *Coal: Typology, Chemistry, Physics and Constitution,* Elsevier, New York, 1961.
6. Berthelot, P. E. M., "Method for Hydrogenation of Organic Solids," *Ann. Chim. Phys.,* **20**, 526 (1870).
7. Bergius, F. and Billwiller, J., "Process for Preparing Liquid, and Soluble Organic Compounds from Bituminous Coal," German Patent 301,231, Nov. 26, 1919; Application, Aug. 8, 1913.
8. Wu, W. R. K. and Storch, H. H., "Hydrogenation of Coal and Tar," *U. S. Bur. Mines Bull.* 633, 1968.
9. Donath, E. E., "Hydrogenation of Coal and Tar," Chapter 22 of *Chemistry of Coal Utilization,* Suppl. Vol., H. H. Lowry, Ed., John Wiley & Sons, New York, 1963.
10. Storch, H. H., "Hydrogenation of Coal and Tar," Chapter 28 of *Chemistry of Coal Utilization,* Vol. II, H. H. Lowry, Ed., John Wiley & Sons, New York, 1945.
11. Hawk, C. O. and Hiteshue, R. W., "Hydrogenation of Coal in Batch Autoclave," *U. S. Bur. Mines Bull.* 622, 1965.
12. Pott, A. and Broche, H., "Extracts from Fuels such as Coal, Lignite, or Peat," U.S. Patent 2,123,380, July, 1938.
13. Pichler, H., "Twenty-five Years of Synthesis of Gasoline by Catalytic Conversion of Carbon Monoxide and Hydrogen," *Advances in Catalysis,* Vol. 4, Academic Press, New York, 1952, p. 271.
14. Fischer, F. and Tropsch, H., "Synthesis of Petroleum from Gasification Products of Coal at Normal Pressure," *Chem. Ber.,* **59**, 830, 832, 923 (1926).
15. Storch, H. H., Golumbic, N., and Anderson, R. B., *The Fischer-Tropsch and Related Syntheses,* John Wiley & Sons, New York, 1951.
16. Storch, H H., "Synthesis of Hydrocarbons from Water Gas," Chapter 39 of *Chemistry of Coal Utilization,* Vol. II, H. H. Lowry, Ed., John Wiley & Sons, New York, 1945.
17. Forney, A. J., Haynes, W. P., Eliott, J. J., and Zarochak, M. F., "The Fischer-Tropsch Process: Gasoline from Coal," *Preprints, Am. Chem. Soc., Div. Fuel Chem.,* **20**, No. 3, 171 (August 1975).
18. Rousseau, P. E., "The Robens Coal Science Lecture 1975: The Coal Renaissance, A South African Point of View," *J. Inst. Fuel (London),* pp. 167–175, December 1975.
19. Hoogendoorn, J. C. and Saloman, J. M., "Sasol: World's Largest Oil from Coal Plant" (in four parts), *Bri. Chem. Eng.,* May, June, July, and August 1957.
20. Hoogendoorn, J. C., "Experience with Fischer-Tropsch Synthesis at Sasol," in *Clean Fuels from Coal, Symposium Papers,* Institute of Gas Technology, Chicago, 1973.
21. Fisher, C. H., Sprunk, G. C., Eisner, A., O'Dennell, H. J., Clarke, L., and Storch, H. H., "Hydrogenation and Liquefaction of Coal," Part 2: "Effect of Petrographic Composition and Rank of Coal," *U. S. Bur. Mines Tech. Paper* 642, 1942.
22. Davis, A., Spackman, W., and Given, P. H., "The Influence of the Properties of Coals on Their Conversion into Clean Fuels," *Energy Source,* **3**, 55 (1976).

ISBN 0-201-08300-0

23. Davis, A., Spackman, W., and Given, P. H., "The Influence of the Properties of Coals in Their Conversion into Clean Fuels," paper presented to American Chemical Society, Division of Petroleum Chemistry, Atlantic City, N. J., September 1974.
24. Given, P. H., Cronauer, D. C., Spackman, W., Lovell, H. L., Davis, A., and Biswas, B., "Dependence of Coal Liquefaction Behavior on Coal Characteristics," Parts I and II, *Fuel,* **54**, 34 and 40 (1975).
25. Sternberg, H. W., Raymond, R., and Schweighardt, F. K., "The Nature of Coal Liquefaction Products," paper presented to American Chemical Society, Division of Petroleum Chemistry, Chicago, August 1975.
26. Sternberg, H. W., "A Second Look at the Reductive Alkylation of Coal and at the Nature of Asphaltenes," *Preprints, Am. Chem. Soc., Div. Fuel Chem.,* **21**, No. 7, 1 (August-September 1976).
27. Pott, A. and Broche, H., "The Solution of Coal by Extraction under Pressure," *Fuel,* **13**, 91 (1934).
28. Guin, J. A., Tarrer, A. R., Taylor, Z. L., and Green, S. C., "A Photomicrographic Study of Coal Dissolution," *Preprints, Am. Chem. Soc., Div. Fuel Chem.,* **20**, No. 1, 66 (April 1975).
29. Curran, G. P., Struck, R. T., and Gorin, E., "Mechanism of the Hydrogen-Transfer Process to Coal and Coal Extract," *Ind. Eng. Chem. Process Des. Dev.,* **6**, 166 (1967).
30. Neavel, R. C., "Liquefaction of Coal in Hydrogen-Donor and Nondonor Vehicles," *Fuel,* **55**, 237 (1976).
31. Oele, A. P., Waterman, H. I., Goedkoop, M. L., and Van Krevelen, D. W., "Extractive Disintegration of Bituminous Coal," *Fuel,* **30**, 169 (1951).
32. Van Krevelen, D. W., **Coal**, Elsevier, New York, 1961.
33. Orchin, M. and Storch, H. H., "Solvation and Hydrogenation of Coal," *Ind. Eng. Chem.,* **40**, 1385 (1948).
34. Yen, Y. K., Furlani, D. E., and Weller, S. W., "Batch Autoclave Studies of Catalytic Hydrodesulfurization of Coal," *Ind. Eng. Chem. Prod. Res. Dev.,* **15**, 24 (1976).
35. Kang, C. C., Nongbri, G., and Stewart, N., "The Role of Solvent in the Refined Coal Process," *Preprints, Am. Chem. Soc., Div. Fuel Chem.,* **21**, No. 5, 19 (August-September 1976).
36. Wolk, R. H., Stewart, N. C., and Silver, H. F., "Review of Desulfurization and Denitrogenation in Coal Liquefaction," *Preprints, Am. Chem. Soc., Div. Fuel Chem.,* **20**, No. 2, 116 (April 1975).
37. Gary, J. H., Baldwin, R. M., Bao, C. Y., Bain, R. L., Kirchner, M. S., and Golden, J. O., "Removal of Sulfur from Coal by Treatment with Hydrogen," *Preprints, Am. Chem. Soc., Div. Fuel Chem.,* **19**, No. 1, 167 (April 1974).
38. Wiser, W. H., "A Kinetic Comparison of Coal Pyrolysis and Coal Dissolution," *Fuel,* **47**, 475 (1968).
39. Wiser, W. H., Anderson, L. L., Qader, S. A., and Hill, G. R., "Kinetic Relationship of Coal Hydrogenation, Pyrolysis and Dissolution," *J. Appl. Chem. Biotechnol.,* **21**, 82 (1971).
40. Heredy, L. A. and Fugassi, P., "Phenanthrene Extraction of Bituminous Coal," in *Coal Science, Advances in Chemistry Series* 55, American Chemical Society, Washington, D. C., 1966, p. 448.

ISBN 0-201-08300-0

41. Hill, G. R., Hariri, H., Reed, R. I., and Anderson, L. L., "Kinetics and Mechanism of Solution of High Volatile Coal," in *Coal Science, Advances in Chemistry Series* 55, American Chemical Society, Washington, D. C., 1966, p. 427.
42. Hill, G. R., Anderson, L. L., Wiser, W. H., Qader, S. A., Wood, R. E., and Bodily, D. M., "Project Western Coal: Conversion of Coal into Liquids," Final report, Contract No. 14-01-0001-271, Office of Coal Research, May 1970.
43. Wen, C. Y. and Han, K. W., "Kinetics of Coal Liquefaction," *Preprints, Am. Chem. Soc., Div. Fuel Chem.*, **20**, No. 1, 216 (April 1975).
44. Koltz, R. C., Baldwin, R. M., Bain, R. L., Golden, J. O., and Gary, J. H., "Kinetics of Coal Hydrodesulfurization in a Batch Reactor," *Preprints, Am. Chem. Soc., Div. Fuel Chem.*, **21**, No. 5, 96 (August-September 1976).
45. Guin, J. A., Tarrer, A. R., Pitts, W. S., and Prather, J. W., "Kinetics and Solubility of Hydrogen in Coal Liquefaction Reactions," *Preprints, Am. Chem. Soc., Div. Fuel Chem.*, **21**, No. 5, 170 (August-September 1976).
46. Whitehurst, D. D. and Mitchell, T. O., "Short Contact Time Coal Liquefaction." I: "Techniques and Product Distributions," *Preprints, Am. Chem. Soc., Div. Fuel Chem.*, **21**, No. 5, 127 (August-September 1976).
47. Kawa, W., Friedman, S., Wu, W. R. K., Frank, L. V., and Yavorsky, P. M., "Evaluation of Catalysts for Hydrodesulfurization and Liquefaction of Coal," *Preprints, Am. Chem. Soc., Div. Fuel Chem.*, **19**, No. 1, 192 (April 1974).
48. Yavorsky, P. M., Akhtar, S., and Friedman, S., "Process Developments: Fixed-Bed Catalysis of Coal to Fuel Oil," paper presented at the 65th Annual Meeting American Institute of Chemical Engineers, New York, November 1972.
49. Wright, C. H. and Severson, D. E., "Experimental Evidence for Catalyst Activity of Coal Minerals," *Preprints, Am. Chem. Soc., Div. Fuel Chem.*, **16**, No. 2, 68 (April 1972).
50. Tarrer, A. R., Guin, J. A., Pitts, W. S., Henley, J. P., Prather, J. W., and Styles, G. A., "Effect of Coal Minerals on Reaction Rates During Coal Liquefaction," *Preprints, Am. Chem. Soc., Div. Fuel Chem.*, **21**, No. 5, 59 (August-September 1976).
51. Kang, C. C. and Johanson, E. S., "Deactivation of Co-Mo Catalyst during H-Coal Operations," *Preprints, Am. Chem. Soc., Div. Fuel Chem.*, **21**, No. 5, 32 (August-September 1976).
52. Williams, A., Butler, G. A., and Hammonds, J., "Sintering of Nickel-Alumina Catalyst," *J. Catal.*, **24**, 352 (1972).
53. Kovach, S. M. and Bennett, J., "Coal Liquefaction. I. Catalyst Posions Present in Coal," *Preprints, Am. Chem. Soc., Div. Fuel Chem.*, **20**, No. 1, 143 (April 1975).
54. Stanulonis, J. J., Gates, B. C., and Olson, J. H., "Catalyst Aging in a Process for Liquefaction and Hydrodesulfurization of Coal," *AIChE J.*, **22**, 576 (1976).
55. Akhtar, S., Friedman, S., and Hiteshue, R. W., "Hydrogenation of Coal to Liquids in Fixed Beds of Silica Promoted Cobalt Molybdate Catalyst," *Preprints, Am. Chem. Soc., Div. Fuel Chem.*, **14**, No. 4, Part I, 27 (September 1970).

ISBN 0-201-08300-0

56. Weller, S. W., "Catalysis in the Liquid-Phase Hydrogenation of Coal and Tar," in *Catalysis,* Vol. IV, P. H. Emmett, Ed., Reinhold, New York, 1956.
57. Liebenberg, B. J. and Potgieter, H. G., "The Uncatalyzed Hydrogenation of Coal," *Fuel,* **52**, 130 (1973).
58. Yoshida, R., Maekawa, Y., Ishii, T., and Takeya, G., "Mechanism of High-Pressure Hydrogenolysis of Hokkaido Coals," *Fuel,* **55**, 337 (1976).
59. Potgieter, H. G. J., "Kinetics of Conversion of Tetralin during Hydrogenation of Coal," *Fuel,* **52**, 134 (1973).
60. Angelovich, J. M., Pastor, G. R., and Silver, H. F., "Solvents Used in the Conversion of Coal," *Ind. Eng. Chem. Process Des. Dev.,* **9**, 106 (1970).
61. Pastor, G. R., Angelovich, J. M., and Silver, H. F., "Study of the Interaction of Solvent and Catalyst in Coal Hydrogenation," *Ind. Eng. Chem. Process Des. Dev.,* **9**, 609 (1970).
62. Kiebler, M. W., "Extraction of a Bituminous Coal, Influence of the Nature of Solvents," *Ind. Eng. Chem.,* **32**, 1389 (1940).
63. Blanks, R. F. and Prausnitz, J. M., "Thermodynamics of Polymer Solubility in Polar and Nonpolar Systems," *Ind. Eng. Chem. Fundam.,* **3**, 1 (1964).
64. Hildebrand, J. H. and Scott, R. L., *Solubility of Nonelectrolytes,* Reinhold, New York, 1950.
65. Weller, S., Clark, E. L., and Pelipetz, M. G., "Mechanism of Coal Hydrogenation," *Ind. Eng. Chem.,* **42**, 334 (1950).
66. Farcasiu, M., Mitchell, T. O., and Whitehurst, D. D., "On the Chemical Nature of the Benzene Insolubles of Solvent Refined Coals," *Preprints, Am. Chem. Soc., Div. Fuel Chem.,* **21**, No. 7, 11 (August-September 1976).
67. Fu, Y. C. and Batchelder, R. F., "Catalytic Liquefaction of Coal," *Preprints, Am. Chem. Soc., Div. Fuel Chem.,* **21**, No. 5, 78 (August-September 1976).
68. Mitchell, G. D., Davis, A., and Spackman, W., "A Petrographic Classification of Solid Residues Derived from the Hydrogenation of Bituminous Coal," *Preprints, Am. Chem. Soc., Div. Fuel Chem.,* **21**, No. 5, 189 (August-September 1976).
69. Ruberto, R. G., Cronauer, D. C., Jewell, D. M., and Seshadri, K. S., "Structural Aspects of Subbituminous Coal Deduced from Solvation Studies," Parts I and II, *Fuel,* **56**, 17 and 25 (1977).
70. Akhtar, S., Sharkey, A. G., Jr., Shultz, J. L., and Yavorsky, P. M., "Organic Sulfur Compounds in Coal Hydrogenation Products," *Preprints, Am. Chem. Soc., Div. Fuel Chem.,* **19**, No. 1, 207 (April 1974).
71. Given, P. H., "The Distribution of Hydrogen in Coals and Its Relation to Coal Structure," *Fuel,* **39**, 147 (1960).
72. Wender, I., "Catalytic Synthesis of Chemicals from Coal," *Preprints, Am. Chem. Soc., Div. Fuel Chem.,* **20**, No. 4, 16 (August 1975).
73. Mills, G. A., "Conversion of Coal to Gasoline," *Ind. Eng. Chem.,* **61**, No. 7, 6 (July 1969).
74. Stotler, H. H., "The H-Coal Pilot Plant Program," *Coal Processing Technology,* Vol. 2, American Institute of Chemical Engineers, 1975, p. 15.
75. Mukherjee, D. K. and Chowdhury, P. B., "Catalytic Effect of Mineral Matter Constituents in a North Assam Coal on Hydrogenation," *Fuel,* **55**, 4 (1976).

ISBN 0-201-08300-0

76. Ruberto, R. G., Jewell, D. M., Jensen, R. K., and Cronauer, D. C., "Characterization of Synthetic Liquid Fuels," *Preprints, Am. Chem. Soc., Div. Fuel Chem.,* **19**, No. 2, 258 (April 1974).
77. Dooley, J. E. and Thompson, C. J., "The Analysis of Liquids from Coal Conversion Processes," *Preprints, Am. Chem. Soc., Div. Fuel Chem.,* **21**, No. 5, 243 (August-September 1976).
78. Callen, R. B., Bendoraitis, J. G., Simpon, C. A., and Voltz, S. E., "Upgrading Coal Liquids to Gas Turbine Fuels." I: "Analytical Characterization of Coal Liquids," *Ind. Eng. Chem. Prod. Res. Dev.,* **15**, 222 (1976).
79. Farcasiu, M., "Fractionation and Structural Characterization of Coal Liquids," *Fuel,* **56**, 9 (1977).
80. Haebig, J. E., Davis, B. E., and Dzuna, E. R., "Preliminary Small-Scale Combustion Tests of Coal Liquids," *Preprints, Am. Chem. Soc., Div. Fuel Chem.,* **20**, No. 1, 203 (April 1975).
81. Sternberg, H. W., Raymond, R., and Schweighardt, F. K., "Acid-Base Structure of Coal-Derived Asphaltenes," *Science,* **188**, 49 (April 4, 1975).
82. Schweighardt, F. K., Retcofsky, H. L., and Raymond, R., "Asphaltenes from Coal Liquefaction," *Preprints, Am. Chem. Soc., Div. Fuel Chem.,* **21**, No. 7, 27 (August-September 1976).
83. Appell, H. R. and Wender, I., "The Hydrogenation of Coal with Carbon Monoxide and Water," *Preprints, Am. Chem. Soc., Div. Fuel Chem.,* **12**, No. 3, 220 (September 1968).
84. Appell, H. R., Wender I., and Miller, R. D., "Solubilization of Low Rank Coal with Carbon Monoxide and Water," *Chem. Ind. (London),* p. 1703, Nov. 22, 1969.
85. Appell, H. R., Wender, I., and Miller, R. D., "Dissimilar Behavior of Carbon Monoxide Plus Water and of Hydrogen in Hydrogenation," *Preprints, Am. Chem. Soc., Div. Fuel Chem.,* **13**, No. 4, 39 (September 1969).
86. Appell, H. R., Moroni, E. C., and Miller, R. D., "Costeam Liquefaction of Lignite," *Preprints, Am. Chem. Soc., Div. Fuel Chem.,* **20**, No. 1, 58 (April 1975).
87. Del Bel, E., Friedman, S., Yavorsky, P. M., and Wender, I., "Oil by Liquefaction of Lignite," *Coal Processing Technology,* Vol. 2, American Institute Chemical Engineers, 1975, p. 104.
88. Fu, Y. C. and Illig, E. G., "Catalytic Coal Liquefaction Using Synthesis Gas," *Ind. Eng. Chem. Process Des. Dev.,* **15**, 392 (1976).
89. Office of Coal Research, Development of a Process for Producing an Ashless, Low-Sulfur Fuel from Coal," *R & D Rep.* 53, *Interim Rep.* 11, 1972.
90. Gorin, E., Kulik, C. J., and Lebowitz, H. E., "Deashing of Coal Liquefaction Products via Partial Deasphalting," Parts I and II, *Ind. Eng. Chem. Process Des. Dev.,* **16**, 95 and 102 (1977).
91. Newman, J. O. H., Akhtar, S., and Yavorsky, P. M., "Coagulation and Filtration of Solids from Liquefied Coal of Synthoil Process," *Preprints, Am. Chem. Soc., Div. Fuel Chem.,* **21**, No. 5, 109 (August-September 1976).
92. Katz, S. and Rodgers, R. B., "Reformation of Inorganic Particulates Suspended in Coal Derived Liquids and Improved Separation," *Preprints, Am. Chem. Soc., Div. Fuel Chem.,* **21**, No. 5, 8 (August-September 1976).

ISBN 0-201-08300-0

93. Office of Coal Research, "Pilot-Scale Development of the CSF Process," *R & D Rep.* 39, Vol. IV, Book 3, August 1971.

94. Energy Research and Development Administration, "Scientific Resources Relevant to the Catalytic Problems in the Conversion of Coal," report prepared by Catalytica Associates, Inc., 1976.

95. Levy, R. B. and Cusumano, J. A., "New Materials for Coal Liquefaction," *Preprints, Am. Chem. Soc., Div. Fuel Chem.,* **21**, No. 5, 1 (August-September 1976).

96. Wood, R. E. and Wiser, W. H., "Coal Liquefaction in Coiled Tube Reactors," *Ind. Eng. Chem. Process Des. Dev.,* **15**, 144 (1976).

97. Bodily, D. M., Lee, S. H. D., and Wiser, W. H., "Devolatilization of Coal and Zinc Chloride," *Preprints, Am. Chem. Soc., Div. Fuel Chem.,* **20**, No. 3, 7 (August 1975).

98. Kawa, W., Feldmann, H. F., and Hiteshue, R. W., "Hydrogenation of Asphaltene from Coal Using Halide Catalysts," *Preprints, Am. Chem. Soc., Div. Fuel Chem.,* **14**, No. 4, Part I, 19 (September 1970).

99. Zielke, C. W., Struck, R. T., Evans, J. M., Costanza, C. P., and Gorin, E., "Molten Catalysts for Hydrocracking of Polynuclear Hydrocarbons," *Ind. Eng. Chem. Process Des. Dev.,* **5**, 151, 1966.

100. Zielke, C. W., Struck, R. T., Evans, J. M., Costanza, C. P., and Gorin, E., "Molten Zinc Halide Catalysts for Hydrocracking Coal Extract and Coal," *Ind. Eng. Chem. Process Des. Dev.,* **5**, 158, 1966.

101. Struck, R. T., Clark, W. E., Dudt, P. J., Rosenhoover, W. A., Zielke, C. W., and Gorin, E., "Kinetics of Hydrocracking of Coal Extract with Molten Zinc Chloride Catalysts in Batch and Continuous Systems," *Ind. Eng. Chem. Process Des. Dev.,* **8**, 546, 1969.

102. Zielke, C. W., Struck, R. T., and Gorin, E., "Fluidized Combustion Process for Regeneration of Spent Zinc Chloride Catalysts," *Ind. Eng. Chem. Process Des. Dev.,* **8**, 552 (1969).

103. Zielke, C. W., Rosenhoover, W. R., and Gorin, E., "Direct $ZnCl_2$ Hydrocracking of Subbituminous Coal," *Preprints, Am. Chem. Soc., Div. Fuel Chem.,* **19**, No. 2, 306 (April 1974).

104. Halpern, J., "Catalysts by Coordination Compounds," *Ann. Rev. Phys. Chem.,* **16**, 103 (1965).

105. Sternberg, H. W., Delle Donne, C. L., Pantages, P., Moroni, E. C., and Markby, R. E., "Solubilization of an lvb Coal by Reductive Alkylation," *Fuel,* **50**, 432 (1971).

106. Sternberg, H. W. and Delle Donne, C. L., "Solubilization of Coals by Reductive Alkylation," *Fuel,* **53**, 172 (1974).

107. Sternberg, H. W., Delle Donne, C. L., Markby, R. E., and Wender, I., "Electrochemical Reductions in Ethylenediamine," in *Coal Science,* P. H. Given, Ed., *Advances in Chemistry Series* 55, American Chemical Society, Washington, D. C., p. 516.

108. Reggel, L., Zahn, C. Wender, I., and Raymond, R., "Reduction of Coal by Lithium-Ethylenediamine and Reaction of Model Compounds with Metal-Amine Systems," *U. S. Bur. Mines Bull.* 615, 1965.

109. Whitehead, J. C. and Williams, D. F., "Solvent Extraction of Coal by Supercritical Gases," *Inst. Fuel (London),* p. 182, December 1975.

ISBN 0-201-08300-0

110. Field, J. H., Bienstock, D., Forney, A. J., and Demski, R. J., "Further Studies of the Fischer-Tropsch Synthesis Using Recycle Gas Cooling," *U. S. Bur. Mines Rep. Invest.* 5871, 1961.

111. Field, J. H., Benson, H. G., and Anderson, R. B., "Synthetic Liquid Fuel by Fischer-Tropsch Process," *Chem. Eng. Prog.,* **56**, No. 4, 44 (1960).

112. Forney, A. J., Demski, R. J., Bienstock, D., and Field J. H., "Recent Catalyst Developments in the Hot-Gas-Recycle Process," *U. S. Bur. Mines Rep. Invest.* 6609, 1965.

113. Elliott, J. J., Haynes, W. P., and Forney, A. J., "Gasoline via the Fischer-Tropsch Using the Hot-Gas-Recycle System," *Preprints, Am. Chem. Soc., Div. Fuel Chem.,* **16**, No. 1, 44 (April 1972).

114. O'Hara, J. B., Bela, A., Jentz, N. E., and Khaderi, S. K., "Fischer-Tropsch Plant Design Criteria," paper presented at 68th Annual Meeting, American Institute of Chemical Engineers, Los Angeles, November 1975.

115. Dry, M. E., "Advances in Fischer-Tropsch Chemistry," *Ind. Eng. Chem. Prod. Res. Dev.,* **15**, 282 (1976).

116. Mills, G. A. and Harney, B. M., "Methanol, the New Fuel from Coal," *Chemteck,* p. 26, January 1974.

117. Meisel, S. L., McCullough, J. P., Lechthaler, C. H., and Weisz, P. B., "Gasoline from Methanol in One Step," *Chemteck,* p. 86, February 1976.

118. Weller, S. and Pelipetz, M. G., *Ind. Eng. Chem.,* **43**, 1243 (1951).

119. Weller, S., Pelipetz, M. G., and Friedman, S., *Ind. Eng. Chem.,* **43**, 1572, 1575 (1951).

ISBN 0-201-08300-0

Appendix

ISBN 0-201-08300-0

Table I

SI Base and Supplementary Units[1]

physical quantity	unit	definition
length	meter	the length equal to 1 650 763.73 wavelengths in vacuum of the radiation corresponding to the transition between the levels $2p_{10}$ and $5d_5$ of the krypton-86 atom (adopted by 11th CGPM 1960)
mass	kilogram	the unit of mass; it is equal to the mass of the international prototype of the kilogram (adopted by 1st and 3rd CGPM 1889 and 1901)
time	second	the duration of 9 192 631 770 periods of the radiation corresponding to the transition between the two hyperfine levels of the ground state of the cesium-133 atom (adopted by 13th CGPM 1967)
current	ampere	that constant current which, if maintained in two straight parallel conductors of infinite length, of negligible circular cross section, and placed one meter apart in vacuum, would produce between these conductors a force equal to 2×10^{-7} newton per meter of length (adopted by 9th CGPM 1948)
temperature	kelvin	unit of thermodynamic temperature, the fraction 1/273.16 of the thermodynamic temperature of the triple point of water (adopted by 13th CGPM 1967)
amount of substance	mole	the amount of substance of a system which contains as many elementary entities as there are atoms in 0.012 kilogram of carbon-12 (adopted by 14th CGPM 1971). NOTE: When the mole is used, the elementary entities must be specified and may be atoms, molecules, ions, electrons, other particles, or specified groups of such particles.
uminous intensity	candela	the luminous intensity, in the perpendicular direction, of a surface of 1/600 000 square meter of blackbody at the temperature of freezing platinum under a pressure of 101 325 newtons per square meter (adopted by 13th CGPM 1967)
plane angle	radian	the plane angle between two radii of a circle which cut off on the circumference an arc equal in length to the radius

ISBN 0-201-08300-0

Table I (Continued)

physical quantity	unit	definition
solid angle	steradian	the solid angle which, having its vertex in the center of a sphere, cuts off an area of the surface of the sphere equal to that of a square with sides of length equal to the radius of the sphere

[1]Reproduced with permission from *Handbook for Authors of Papers in American Chemical Society Publications (1978)*. Copyright by American Chemical Society.

Table II

SI Derived Units[a, 1]

quantity	name	symbol	expression in terms of other units	expression in terms of SI base units
	A. SI Derived Units			
area	square meter	m^2		
volume	cubic meter	m^3		
speed, velocity	meter per second	m/s		
acceleration	meter per second squared	m/s^2		
wavenumber	one per meter	m^{-1}		
density, mass	kilogram per cubic meter	kg/m^3		
current density	ampere per square meter	A/m^2		
magnetic field strength	ampere per meter	A/m		
concentration (of amount of substance)	mole per cubic meter[b]	mol/m^3		
specific volume	cubic meter per kilogram	m^3/kg		

ISBN 0-201-08300-0

Table II (Continued)

quantity	name	symbol	expression in terms of other units	expression in terms of SI base units
luminance	candela per square meter	cd/m^2		
		B. SI Derived Units with Special Names		
frequency	hertz	Hz		s^{-1}
force	newton	N		m·kg·s^{-2} [a]
pressure, stress	pascal	Pa	N/m^2	m^{-1}·kg·s^{-2}
energy, work, quantity of heat	joule	J	N·m	m^2·kg·s^{-2}
power, radiant flux	watt	W	J/s	m^2·kg·s^{-3}
quantity of electricity, electric charge	coulomb	C	A·s	s·A
electric potential, potential difference, electromotive force	volt	V	W/A	m^2·kg·s^{-3}·A^{-1}
capacitance	farad	F	C/V	m^{-2}·kg^{-1}·s^4·A^2
electric resistance	ohm	Ω	V/A	m^2·kg·s^{-3}·A^{-2}
conductance	siemens	S	A/V	m^{-2}·kg^{-1}·s^3·A^2
magnetic flux	weber	Wb	V·s	m^2·kg·s^{-2}·A^{-1}
magnetic flux density	tesla	T	Wb/m^2	kg·s^{-2}·A^{-1}
inductance	henry	H	Wb/A	m^2·kg·s^{-2}·A^{-2}
luminous flux	lumen	lm		cd·sr
illuminance	lux	lx	lm/m^2	m^{-2}·cd·sr
activity (radioactive)	becquerel	Bq		s^{-1}
absorbed dose	gray	Gy	J/kg	m^2·s^{-2}
		C. SI Derived Units Expressed by Special Names		
dynamic viscosity	pascal second	Pa·s	m^{-1}·kg·s^{-1}	
moment of force	newton meter	N·m	m^2·kg·s^{-2}	
surface tension	newton per meter	N/m	kg·s^{-2}	
heat flux density, irradiance	watt per square meter	W/m^2	kg·s^{-3}	
heat capacity, entropy	joule per kelvin	J/K	m^2·kg·s^{-2}·K^{-1}	

ISBN 0-201-08300-0

Table II (Continued)

quantity	name	symbol	expression in terms of SI base units
specific heat capacity, specific entropy	joule per kilogram kelvin	J/(kg·K)	$m^2 \cdot s^{-2} \cdot K^{-1}$
specific energy	joule per kilogram	J/kg	$m^2 \cdot s^{-2}$
thermal conductivity	watt per meter kelvin	W/(m·K)	$m \cdot kg \cdot s^{-3} \cdot K^{-1}$
energy density	joule per cubic meter	J/m^3	$m^{-1} \cdot kg \cdot s^{-2}$
electric field strength	volt per meter	V/m	$m \cdot kg \cdot s^{-3} \cdot A^{-1}$
electric charge density	coulomb per cubic meter	C/m^3	$m^{-3} \cdot s \cdot A$
electric flux density	coulomb per square meter	C/m^2	$m^{-2} \cdot s \cdot A$
permittivity	farad per meter	F/m	$m^{-3} \cdot kg^{-1} \cdot s^4 \cdot A^2$
permeability	henry per meter	H/m	$m \cdot kg \cdot s^{-2} \cdot A^{-2}$
molar energy	joule per mole	J/mol	$m^2 \cdot kg \cdot s^{-2} \cdot mol^{-1}$
molar entropy, molar heat capacity	joule per mole kelvin	J/(mol·K)	$m^2 \cdot kg \cdot s^{-2} \cdot K^{-1} \cdot mol^{-1}$

[a] Symbols for compound units are shown with center dots. [b] M will be retained as a special symbol for concentration of amount of substance, mole per cubic decimeter (mole per liter).

[1]Reproduced with permission from *Handbook for Authors of Papers in American Chemical Society Publications (1978)*. Copyright by American Chemical Society.

Table III

Derived Units of the International System Having Special Names[1]

physical quantity	unit	definition
absorbed dose	gray	the absorbed dose when one joule of energy is imparted to one kilogram of matter by ionizing radiation
activity	becquerel	the activity of a radionuclide having one spontaneous nuclear transition per second
electric capacitance	farad	the capacitance of a capacitor between the plates of which there appears a difference of potential of one volt when it is charged by a quantity of electricity equal to one coulomb

ISBN 0-201-08300-0

Table III (Continued)

physical quantity	unit	definition
electric conductance	siemens	the electric conductance of a conductor in which a current of one ampere is produced by an electric potential difference of one volt
electric inductance	henry	the inductance of a closed circuit in which an electromotive force of one volt is produced when the electric current in the circuit varies uniformly at a rate of one ampere per second
electric potential difference, electromotive force	volt	the difference of electric potential between two points of a conductor carrying a constant current of one ampere, when the power dissipated between these points is equal to one watt
electric resistance	ohm	the electric resistance between two points of a conductor when a constant difference of potential of one volt, applied between these two points, produces in this conductor a current of one ampere, this conductor not being the source of, any electromotive force
energy	joule	the work done when the point of application of a force of one newton is displaced a distance of one meter in the direction of the force
force	newton	that force which, when applied to a body having a mass of one kilogram, gives it an acceleration of one meter per second squared
frequency	hertz	the frequency of a periodic phenomenon of which the period is one second
illuminance	lux	the illuminance produced by a luminous flux of one lumen uniformly distributed over a surface of one squared meter
luminous flux	lumen	the luminous flux emitted in a solid angle of one steradian by a point source having a uniform intensity of one candela
magnetic flux	weber	the magnetic flux which, linking a circuit of one turn, produces in it an electromotive force of one volt as it is reduced to zero at a uniform rate in one second

ISBN 0-201-08300-0

Table III (Continued)

physical quantity	unit	definition
magnetic flux density	tesla	the magnetic flux density given by a magnetic flux of one weber per square meter
power	watt	the power which gives rise to the production of energy at the rate of one joule per second
pressure or stress	pascal	the pressure or stress of one newton per square meter
quantity of electricity	coulomb	the quantity of electricity transported in one second by a current of one ampere

[1]Reproduced with permission from *Handbook for Authors of Papers in American Chemical Society Publications (1978)*. Copyright by American Chemical Society.

Table IV

Units in Use with SI[1]

quantity	name	symbol	value in SI unit
		A. General	
time	minute	min	1 min = 60 s
	hour	h	1 h = 60 min = 3600 s
	day	d	1 d = 24 h = 86 400 s
plane angle	degree	°	$1° = (\pi/180)$ rad
	minute	′	$1' = (1/60)° = (\pi/10\,800)$ rad
	second	″	$1'' = (1/60)' = \pi/648\,000)$ rad
volume	liter	L	1 L = 1 $dm^3 = 10^{-3}$ m^3
mass	metric ton	t	1 t = 10^3 kg
temperature	degree Celsius	°C	*a*
		B. Limited	
energy	kilowatt-hour	kWh	1 kWh = 3.6 MJ
area	barn	b	1 b = 10^{-28} m^2
	hectare	ha	1 ha = 1 $hm^2 = 10^4$ m^2
pressure	bar	bar	1 bar = 10^5 Pa
activity	curie	Ci	1 Ci = 3.7×10^{10} Bq
exposure	roentgen	R	1 R = 2.58×10^{-4} C/kg
absorbed dose	rad	rd	1 rd = 0.01 Gy

[a] Temperature intervals in kelvin and degrees Celsius are identical; however, temperature in kelvins equals temperature in degrees Celsius plus 273.15.

[1]Reproduced with permission from *Handbook for Authors of Papers in American Chemical Society Publications (1978)*. Copyright by American Chemical Society.

ISBN 0-201-08300-0

Table V

Non-SI Units That Should Be Discouraged[1]

name	value in SI units
ångström	0.1 nm
kilogram-force	9.80665 N
calorie (thermochemical)	4.184 J
mho	1 S
standard atmosphere	101.325 kPa
technical atmosphere	98.0665 kPa
conventional millimeter of mercury	133.322 Pa
torr	133.322 Pa
grad	$2\pi/400$ rad
metric carat	0.2 g
metric horsepower	735.499 W
micron	1 μm

[1]Reproduced with permission from *Handbook for Authors of Papers in American Chemical Society Publications (1978)*. Copyright by American Chemical Society.

Table VI

Elements[1]

name	symbol	atomic number	atomic weight
Actinium	Ac	89	..
Aluminum	Al	13	26.98154
Americium	Am	95	..
Antimony	Sb	51	121.75
Argon	Ar	18	39.948
Arsenic	As	33	74.9216
Astatine	At	85	..
Barium	Ba	56	137.34
Berkelium	Bk	97	..
Beryllium	Be	4	9.01218
Bismuth	Bi	83	208.9804
Boron	B	5	10.81
Bromine	Br	35	79.904
Cadmium	Cd	48	112.40
Calcium	Ca	20	40.08
Californium	Cf	98	..
Carbon	C	6	12.011
Cerium	Ce	58	140.12
Cesium	Cs	55	132.9054
Chlorine	Cl	17	35.453
Chromium	Cr	24	51.996
Cobalt	Co	27	59.9332
Copper (Cuprum)	Cu	29	63.546
Curium	Cm	96	..
Dysprosium	Dy	66	162.50
Einsteinium	Es	99	..

ISBN 0-201-08300-0

Table VI (Continued)

name	symbol	atomic number	atomic weight
Erbium	Er	68	167.26
Europium	Eu	63	151.96
Fermium	Fm	100	..
Fluorine	F	9	18.99840
Francium	Fr	87	..
Gadolinium	Gd	64	157.25
Gallium	Ga	31	69.72
Germanium	Ge	32	72.59
Gold (Aurum)	Au	79	196.9665
Hafnium	Hf	72	178.49
Helium	He	2	4.00260
Holmium	Ho	67	164.9304
Hydrogen	H	1	1.0079
Indium	In	49	114.82
Iodine	I	53	126.9045
Iridium	Ir	77	192.22
Iron (Ferrum)	Fe	26	55.847
Krypton	Kr	36	83.80
Lanthanum	La	57	138.9055
Lawrencium	Lr	103	..
Lead (Plumbum)	Pb	82	207.2
Lithium	Li	3	6.941
Lutetium	Lu	71	174.97
Magnesium	Mg	12	24.305
Manganese	Mn	25	54.9380
Mendelevium	Md	101	..
Mercury	Hg	80	200.59
Molybdenum	Mo	42	95.94
Neodymium	Nd	60	144.24
Neon	Ne	10	20.179
Neptunium	Np	93	..
Nickel	Ni	28	58.71
Niobium	Nb	41	92.9064
Nitrogen	N	7	14.0067
Nobelium	No	102	..
Osmium	Os	76	190.2
Oxygen	O	8	15.9994
Palladium	Pd	46	106.4
Phosphorus	P	15	30.97376
Platinum	Pt	78	195.09
Plutonium	Pu	94	..
Polonium	Po	84	..
Potassium	K	19	39.098
Praseodymium	Pr	59	140.9077
Promethium	Pm	61	..
Protactinium	Pa	91	..
Radium	Ra	88	..
Radon	Rn	86	..
Rhenium	Re	75	186.2

ISBN 0-201-08300-0

Table VI (Continued)

name	symbol	atomic number	atomic weight
Rhodium	Rh	45	102.9055
Rubidium	Rb	37	85.4678
Ruthenium	Ru	44	101.07
Samarium	Sm	62	150.4
Scandium	Sc	21	44.9559
Selenium	Se	34	78.96
Silicon	Si	14	28.086
Silver (Argentum)	Ag	47	107.868
Sodium	Na	11	22.98977
Strontium	Sr	38	87.62
Sulfur	S	16	32.06
Tantalum	Ta	73	180.9479
Technetium	Tc	43	..
Tellurium	Te	52	127.60
Terbium	Tb	65	158.9254
Thallium	Tl	81	204.37
Thorium	Th	90	232.0381
Thulium	Tm	69	168.9342
Tin (Stannum)	Sn	50	118.69
Titanium	Ti	22	47.90
Tungsten (Wolfram)	W	74	183.85
Uranium	U	92	238.029
Vanadium	V	23	50.9414
Xenon	Xe	54	131.30
Ytterbium	Yb	70	173.04
Yttrium	Y	39	88.9059
Zinc	Zn	30	65.38
Zirconium	Zr	40	91.22

[1] Reproduced with permission from *Handbook for Authors of Papers in American Chemical Society Publications (1978)*. Copyright by American Chemical Society.

ISBN 0-201-08300-0

Table VII

Values of the "Defined" Constants*

Constant	Symbol	Value (exact, by definition)
Unified atomic mass unit	u	1/12 times the mass of an atom of ^{12}C
Mole	mol	The amount of a substance, of specified chemical formula, containing the same number of formula units (molecules, atoms, ions, electrons, or other entities) as there are atoms in 12 g (exactly) of the pure nuclide ^{12}C
Standard acceleration of gravity, in free fall	g	980.665 cm s^{-2}
Normal atmosphere, pressure	atm	1013250 dyn cm^{-2}
Absolute temperature of the triple point of water[a]	T_{tp}	273.16 ^{o}K
Thermochemical calorie	cal	4.184 J
International steam calorie	cal_{IT}	4.1868 J
Inch	in	2.54 cm
Pound, avoirdupois	lb	453.59237 g

[a]The difference between the temperature of the triple point of water and the so-called "ice point" (temperature of equilibrium of solid and liquid water saturated with air at one atmosphere) is accurately known: T (triple point) – T (ice point) = 0.0100 ± 0.0001 ^{o}K.

ISBN 0-201-08300-0

Table VIII

Recommended Values of the "Basic" Constants*,†

Constant	Symbol	Value (with estimated uncertainty)	
Velocity of light *in vacuo*	c	2.997925×10^{10} ±0.000003	cm s^{-1}
Avogadro number	N	6.02252×10^{23} ±0.00028	molecules mol^{-1}
Faraday constant	F	96487.0 ±1.6	coulomb $equiv^{-1}$
		23060.9 ±0.4	cal $volt^{-1}$ $equiv^{-1}$
Planck constant	h	6.6256×10^{-27} ±0.0005	erg s
Pressure-volume product for one mole of gas – 0°C and zero pressure	$(PV)_{0°C}^{P=0}$	2271.06 ±0.12	J mol^{-1}
		22413.6 ±1.2	cm^3 atm mol^{-1}

*The selection of these constants as the five "basic" ones is somewhat arbitrary. A least squares adjustment such as that of Cohen and DuMond actually treats, on an equal basis, both the "basic" constant and those "derived" constants which can be evaluated experimentally. In order to evaluate the accuracy of any constant derived from those on this list it is necessary to use the complete error matrix as more fully explained in the report of Cohen and DuMond.

†Reproduced with permission from *Handbook for Authors of Papers in the Journals of the American Chemical Society (1967)*. Copyright by American Chemical Society.

Table IX

Recommended Values of the "Derived" Constants*

Constant	Symbol and relation	Value (with estimated uncertainty)	
Elementary charge	$e = F/N$	4.80298×10^{-10} ±0.00020	$cm^{3/2}$ $g^{1/2}$ s^{-1} (esu)
Gas constant	$R = (PV)_{0°C}^{P=0}/T_{0°C}$	8.31433 ±0.00044	J deg^{-1} mol^{-1}
		1.98717 ±0.00011	cal deg^{-1} mol^{-1}

ISBN 0-201-08300-0

Table IX (Continued)

Constant	Symbol and relation	Value (with estimated uncertainty)	
Boltzmann constant	$k = R/N$	1.38054×10^{-16} ±0.00009	erg deg^{-1} $molecule^{-1}$
Second radiation constant	$c_2 = hc/k$	1.43879 ±0.00009	cm deg
Einstein constant relating mass and energy	$Y = c^2$	8.987554×10^{13} ±0.000018	J g^{-1}
		2.148076×10^{13} ±0.000004	cal g^{-1}
Constant relating wave-number and energy	$Z = Nhc$	11.96255 ±0.00038	J cm mol^{-1}
		2.85912 ±0.00009	cal cm mol^{-1}

* Reproduced with permission from *Handbook for Authors of Papers in the Journals of the American Chemical Society (1967)*. Copyright by American Chemical Society.

Table X

Conversion Factors[1]

Reproduced from "Selected Values of Properties of Chemical Compounds," Thermodynamics Research Center, Texas A&M University, College Station, Texas.

To convert the numerical value of a property expressed in one of the units in the left-hand column of a table to the numerical value of the same property in one of the units in the top row of the same table, multiply the former value by the factor in the block common to both units. The factors have been carried out to seven significant figures, as derived from the constants discussed on the previous pages and the definitions of the units. However, this does not mean that the factors are always known to that accuracy. Numbers followed by . . . are to be continued indefinitely with repetition of the same pattern of digits. Factors written with fewer than seven significant digits should be taken as exact values. Numbers followed by an asterisk (*) are definitions of the relation between the two units.

[1] Reproduced with permission from *Handbook for Authors of Papers in the Journals of the American Chemical Society (1967)*. Copyright by American Chemical Society.

ISBN 0-201-08300-0

Table X (Continued)

THE MOL UNIT

Units		g-mol (Unified)	g-mol (Chemists')	g-mol (Physicists')	lb-mol (Unified)	lb-mol (Chemists')	lb-mol (Physicists')
1 g-mol (Unified)	=	1	1.000043	1.000318	2.204623×10^{-3}	2.204717×10^{-3}	2.2053237×10^{-3}
1 g-mol (Chemists')	=	0.9999570	1	1.000275	2.204528×10^{-3}	2.204623×10^{-3}	2.205229×10^{-3}
1 g-mol (Physicists')	=	0.9996821	0.999725	1	2.203922×10^{-3}	2.204017×10^{-3}	2.204623×10^{-4}
1 lb-mol (Unified)	=	453.5924	453.6119	453.7366	1	1.000043	1.000318
1 lb-mol (Chemists')	=	453.5729	453.5924	453.7171	0.9999570	1	1.000275
1 lb-mol (Physicists')	=	453.4482	453.4677	453.5924	0.9996821	0.999725	1

UNITS OF LENGTH

Units		cm	m	in.	ft	yd	mile
1 cm	=	1	0.01*	0.3937008	0.03280840	0.01093613	6.213712×10^{-6}
1 m	=	100.	1	39.37008	3.280840	1.093613	6.213712×10^{-4}
1 in.	=	2.54*	0.0254	1	0.08333333...	0.02777777...	1.578283×10^{-5}
1 ft	=	30.48	0.3048	12.*	1	0.3333333...	1.893939×10^{-4}...
1 yd	=	91.44	0.9144	36.	3.*	1	5.681818×10^{-4}...
1 mile	=	1.609344×10^{5}	1.609344×10^{3}	6.336×10^{4}	5280.*	1760.	1

ISBN 0-201-08300-0

ISBN 0-201-08300-0

UNITS OF AREA

Units		cm^2	m^2	in.^{2}	ft^2	yd^2	$mile^2$
1 cm^2	=	1	10^{-4}*	0.1550003	1.076391×10^{-3}	1.195990×10^{-4}	3.861022×10^{-11}
1 m^2	=	10^4	1	1550.003	10.76391	1.195990	3.861022×10^{-7}
1 in.^{2}	=	6.4516*	6.4516×10^{-4}	1	6.944444×10^{-3}...	7.716049×10^{-4}	2.490977×10^{-10}
1 ft^2	=	929.0304	0.09290304	144.*	1	0.1111111...	3.587007×10^{-8}
1 yd^2	=	8361.273	0.8361273	1296.	9.*	1	3.228306×10^{-7}
1 $mile^2$	=	2.589988×10^{10}	2.589988×10^6	4.014490×10^9	2.78784×10^7*	3.0976×10^6	1

UNITS OF VOLUME

Units		cm^3	liter	in.^{3}	ft^3	qt	gal
1 cm^3	=	1	10^{-3}	0.06102374	3.531467×10^{-5}	1.056688×10^{-3}	2.641721×10^{-4}
1 liter	=	1000.*	1	61.02374	0.03531467	1.056688	0.2641721
1 in.^{3}	=	16.38706*	0.01638706	1	5.787037×10^{-4}	0.01731602	4.329004×10^{-3}
1 ft^3	=	28316.85	28.31685	1728.*	1	2.992208	7.480520
1 qt	=	946.353	0.946353	57.75	0.0342014	1	0.25
1 gal (U. S.)	=	3785.412	3.785412	231.*	0.1336806	4.*	1

Table X (Continued)

UNITS OF MASS

Units		g	kg	oz	lb	metric ton	ton
1 g	=	1	10^{-3}	0.03527396	2.204623×10^{-3}	10^{-6}	1.102311×10^{-6}
1 kg	=	1000.	1	35.27396	2.204623	10^{-3}	1.102311×10^{-3}
1 oz (avdp)	=	28.34952	0.02834952	1	0.0625	2.834952×10^{-5}	$5. \times 10^{-4}$
1 lb (avdp)	=	453.5924	0.4535924	16.*	1	4.535924×10^{-4}	0.0005
1 metric ton	=	10^{6}	1000.*	35273.96	2204.623	1	1.102311
1 ton	=	907184.7	907.1847	32000.	2000.*	0.9071847	1

UNITS OF DENSITY

Units		**g cm^{-3}**	**g l.$^{-1}$**	**oz in.$^{-3}$**	**lb in.$^{-3}$**	**lb ft^{-3}**	**lb gal^{-1}**
1 g cm^{-3}	=	1	1000.	0.5780365	0.03612728	62.42795	8.345403
1 g l.$^{-1}$	=	10^{-3}	1	5.780365×10^{-4}	3.612728×10^{-5}	0.06242795	8.345403×10^{-3}
1 oz in.$^{-3}$	=	1.729994	1729.994	1	0.0625	108.	14.4375
1 lb in.$^{-3}$	=	27.67991	27679.91	16.	1	1728.	231.
1 lb ft^{-3}	=	0.01601847	16.01847	9.259259×10^{-3}	5.7870370×10^{-4}	1	0.1336806
1 lb gal^{-1}	=	0.1198264	119.8264	4.749536×10^{-3}	4.3290043×10^{-3}	7.480519	1

ISBN 0-201-08300-0

ISBN 0-201-08300-0

UNITS OF PRESSURE

Units		dyn cm^{-2}	bar	atm	kg (wt) cm^{-2}	mmHg (torr)	in. Hg	lb (wt) in.$^{-2}$
1 dyn cm^{-2}	=	1	10^{-6}	9.869233×10^{-7}	1.019716×10^{-6}	7.500617×10^{-4}	2.952999×10^{-5}	1.450377×10^{-5}
1 bar	=	10^{6}*	1	0.9869233	1.019716	750.0617	29.52999	14.50377
1 atm	=	1013250.*	1.013250	1	1.033227	760.	29.92126	14.69595
1 kg (wt) cm^{-2}	=	980665.	0.980665	0.9678411	1	735.5592	28.95903	14.22334
1 mmHg (torr)	=	1333.224	1.333224×10^{-3}	1.3157895×10^{-3}	1.3595099×10^{-3}	1	0.03937008	0.01933678
1 in. Hg	=	33863.88	0.03386388	0.03342105	0.03453155	25.4	1	0.4911541
1 lb (wt) in.$^{-2}$	=	68947.57	0.06894757	0.06804596	0.07030696	51.71493	2.036021	1

UNITS OF ENERGY*

Units		g mass (energy equiv)	J	int J	cal	cal_{IT}	Btu_{IT}	kW hr	hp hr	ft-lb (wt)	cu ft-lb (wt) in.$^{-2}$	l.-atm
1 g mass (energy equiv)	=	1	8.987554×10^{13}	8.986071×10^{13}	2.148077×10^{13}	2.146640×10^{13}	8.518558×10^{10}	2.496543×10^{7}	3.347919×10^{7}	6.628880×10^{13}	4.603399×10^{11}	8.870026×10^{11}
1 J	=	1.112650×10^{-14}	1	0.999835	0.2390057	0.2388459	9.478172×10^{-4}	$2.777777... \times 10^{-7}$	3.725062	0.7375622	5.121960×10^{-3}	9.869233×10^{-3}
1 int J	=	1.112833×10^{-14}	1.000165	1	0.2390452	0.2388853	9.479735×10^{-4}	2.778236×10^{-7}	3.725676×10^{-7}	0.7376839	5.122805×10^{-3}	9.870862×10^{-3}
1 cal	=	4.655327×10^{-14}	4.184*	4.183310	1	0.9993312	3.965667×10^{-3}	$1.1622222... \times 10^{-6}$	1.558562×10^{-6}	3.085960	2.143028×10^{-2}	0.04129287
1 cal_{IT}	=	4.658442×10^{-14}	4.1868*	4.186109	1.000669	1	3.968321×10^{-3}	1.163000×10^{-6}	1.559609×10^{-6}	3.088025	2.144462×10^{-2}	0.04132050

*The electrical units in these tables are those in terms of which certification of standard cells, standard resistances, etc., is made by the National Bureau of Standards. Unless otherwise indicated, all electrical units are absolute.

Table X (Continued)

UNITS OF ENERGY

1 Btu_{IT}	=	1.173908×10^{-11}	1055.056	1054.882	252.1644	251.9958*	1	2.930711×10^{-4}	3.930148×10^{-4}	778.1693	5.403953	10.41259
1 kW hr	=	4.005539×10^{-8}	3600000.*	3599406.	860420.7	859845.2	3412.142	1	1.341022	2655224.	18439.06	35529.24
1 hp hr	=	2.986930×10^{-8}	2684519.	2684077.	641615.6	641186.5	2544.33	0.7456998	1	1980000.*	13750.	26494.15
1 ft-lb (wt)	=	1.508550×10^{-14}	1.355818	1.355594	0.3240483	0.3238315	1.285067×10^{-3}	3.766161×10^{-7}	$5.050505... \times 10^{-7}$	1	$6.944444... \times 10^{-3}$	0.01338088
1 cu ft-lb (wt) in.$^{-2}$	=	2.172313×10^{-12}	195.2378	195.2056	46.66295	46.63174	0.1850497	5.423272×10^{-5}	$7.272727... \times 10^{-5}$	144.*	1	1.926847
1 l.-atm	=	1.127392×10^{-12}	101.3250	101.3083	24.21726	24.20106	0.09603757	2.814583×10^{-5}	3.774419×10^{-5}	74.73349	0.5189825	1

UNITS OF MOLECULAR ENERGY*

Units		erg molecule^{-1}	J mol^{-1}	int J mol^{-1}	cal mol^{-1}	eV molecule^{-1}	int eV molecule^{-1}	wavenumber (cm^{-1})
1 erg molecule^{-1}	=	1	6.022520×10^{16}	6.021526×10^{16}	1.439417×10^{16}	6.241808×10^{11}	6.239748×10^{11}	5.034474×10^{15}
1 J mol^{-1}	=	1.660435×10^{-17}	1	0.9998350	0.2390057	1.036411×10^{-5}	1.036069×10^{-5}	8.359414×10^{-2}
1 int J mol^{-1}	=	1.660709×10^{-17}	1.000165	1	0.2390452	1.036582×10^{-5}	1.036240×10^{-5}	8.360793×10^{-2}
1 cal mol^{-1}	=	6.947258×10^{-17}	4.1840	4.18331	1	4.336345×10^{-5}	4.334914×10^{-5}	0.3497579
1 eV molecule^{-1}	=	1.602100×10^{-12}	96486.79	96470.87	23060.90	1	0.9996701	8065.730
1 int eV molecule^{-1}	=	1.602393×10^{-12}	96518.63	96502.71	23068.51	1.000330	1	8068.392
1 wavenumber (cm^{-1})	=	1.986305×10^{-16}	11.96256	11.96059	2.859121	1.239813×10^{-4}	1.239404×10^{-4}	1

ISBN 0-201-08300-0

ISBN 0-201-08300-0

UNITS OF SPECIFIC ENERGY*

Units		$J\ g^{-1}$	int $J\ g^{-1}$	cal g^{-1}	$cal_{IT}\ g^{-1}$	$Btu_{IT}\ lb^{-1}$	kW hr lb^{-1}
1 $J\ g^{-1}$	=	1	0.999835	0.2390057	0.2388459	0.4299226	1.259979×10^{-4}
1 int $J\ g^{-1}$	=	1.000165	1	0.2390452	0.2388853	0.4299936	1.260187×10^{-4}
1 cal g^{-1}	=	4.184*	4.18331	1	0.9993312	1.798796	5.271752×10^{-4}
1 $cal_{IT}\ g^{-1}$	=	4.1868*	4.186109	1.000669	1	1.8*	5.275279×10^{-4}
1 $Btu_{IT}\ lb^{-1}$	=	2.326000	2.325616	0.5559273	0.5555555...	1	2.390711×10^{-4}
1 kW hr lb^{-1}	=	7936.641	7935.332	1896.903	1895.643	3414.425	1

UNITS OF SPECIFIC ENERGY PER DEGREE*

Units		$J\ g^{-1}\ °C^{-1}$	int $J\ g^{-1}\ °C^{-1}$	cal $g^{-1}\ °C^{-1}$	$cal_{IT}\ g^{-1}\ °C^{-1}$	$Btu_{IT}\ lb^{-1}\ °F^{-1}$	kW hr $lb^{-1}\ °F^{-1}$
1 $J\ g^{-1}\ °C^{-1}$	=	1	0.999835	0.2390057	0.2388459	0.2388459	6.999883×10^{-5}
1 int $J\ g^{-1}\ °C^{-1}$	=	1.000165	1	0.2390452	0.2388853	0.2388853	7.001037×10^{-5}
1 cal $g^{-1}\ °C^{-1}$	=	4.184*	4.183310	1	0.999312	0.9993312	2.928751×10^{-4}
1 $cal_{IT}\ g^{-1}\ °C^{-1}$	=	4.1868*	4.186109	1.000669	1	1*	2.930711×10^{-4}
1 $Btu_{IT}\ lb^{-1}\ °F^{-1}$	=	4.1868	4.186109	1.000669	1*	1	2.930711×10^{-4}
1 kW hr $lb^{-1}\ °F^{-1}$	=	14285.95	14283.60	3414.425	3412.142	3412.142	1

*The electrical units in these tables are those in terms of which certification of standard cells, standard resistances, etc., is made by the National Bureau of Standards. Unless otherwise indicated, all electrical units are absolute.

Table XI

Molal Heat Capacities of Gases at Zero Pressure[a]

$$C^{\mathrm{o}}_p = a + bT + cT^2 + dT^3 \quad (T = {}^{\mathrm{o}}\mathrm{K})$$

Gas		a	$b \times 10^2$	$c \times 10^5$	$d \times 10^9$	Temperature Range, °K	Error, %	
							Max.	Avg.
Nitrogen	N_2	6.903	−0.03753	0.1930	−0.6861	273 − 1800	0.59	0.34
Oxygen	O_2	6.085	0.3631	−0.1709	0.3133	273 − 1800	1.19	0.28
Air		6.713	0.04697	0.1147	−0.4696	273 − 1800	0.72	0.33
Hydrogen	H_2	6.952	−0.04576	0.09563	−0.2079	273 − 1800	1.01	0.26
Carbon monoxide	CO	6.726	0.04001	0.1283	−0.5307	273 − 1800	0.89	0.37
Carbon dioxide	CO_2	5.316	1.4285	−0.8362	1.784	273 − 1800	0.67	0.22
Water vapor	H_2O	7.700	0.04594	0.2521	−0.8587	273 − 1800	0.53	0.24
Methane	CH_4	4.750	1.200	0.3030	−2.630	273 − 1500	1.33	0.57
Ethane	C_2H_6	1.648	4.124	−1.530	1.740	273 − 1500	0.83	0.28
Propane	C_3H_8	−0.966	7.279	−3.755	7.580	273 − 1500	0.40	0.12
Sulfur	S_2	6.499	0.5298	−0.3888	0.9520	273 − 1800	0.99	0.38
Sulfur dioxide	SO_2	6.157	1.384	−0.9103	2.057	273 − 1800	0.45	0.24
Sulfur trioxide	SO_3	3.918	3.483	−2.675	7.744	273 − 1300	0.29	0.13
Hydrogen sulfide	H_2S	7.070	0.3128	0.1364	−0.7867	273 − 1800	0.74	0.37
Carbon disulfide	CS_2	7.390	1.489	−1.096	2.760	273 − 1800	0.76	0.47
Carbonyl sulfide	COS	6.222	1.536	−1.058	2.560	273 − 1800	0.94	0.49
Ammonia	NH_3	6.5846	0.61251	0.23663	−1.5981	273 − 1500	0.91	0.36

[a]Values from Hougen et al. [42, Chapter 4].

Table XII

Standard Heats of Combustion[a]

Final Products: CO_2(g), H_2O(l), SO_2(g)

Compound	Formula	State	$-\Delta H_c$	
			kcal/g-mole[b]	Btu/scf[c]
Carbon (graphite)	C	s	94.0518	–
Carbon monoxide	CO	g	67.6361	321.67
Hydrogen	H_2	g	68.3174	324.91
Methane	CH_4	g	212.798	1012.04
Ethyne (acetylene)	C_2H_2	g	310.615	1477.24
Ethene (ethylene)	C_2H_4	g	337.234	1603.84
Ethane	C_2H_6	g	372.820	1773.08

ISBN 0-201-08300-0

Table XII (Continued)

Compound	Formula	State	$-\Delta H_c$ kcal/g-mole[b]	$-\Delta H_c$ Btu/scf[c]
Propyne (allylene, methylacetylene)	C_3H_4	g	463.109	2202.48
Propene (propylene)	C_3H_6	g	491.987	2339.82
Propane	C_3H_8	g	530.605	2523.49
1,2-Butadiene	C_4H_6	g	620.71	2952.01
2-Methylpropene (isobutylene, isobutene)	C_4H_8	g	646.134	3072.93
2-Methylpropane (isobutane)	C_4H_{10}	g	686.342	3264.15
n-Butane	C_4H_{10}	g	687.982	3271.95
1-Pentene (amylene)	C_5H_{10}	g	806.85	3837.27
Cyclopentane	C_5H_{10}	l	786.54	–
2,2-Dimethylpropane (neopentane)	C_5H_{12}	g	840.49	3997.26
2-Methylbutane (isopentane)	C_5H_{12}	g	843.24	4010.34
n-Pentane	C_5H_{12}	g	845.16	4019.47
Benzene	C_6H_6	g	789.08	3752.76
Benzene	C_6H_6	l	780.98	–
1-Hexene (hexylene)	C_6H_{12}	g	964.26	4585.89
Cyclohexane	C_6H_{12}	l	936.88	–
n-Hexane	C_6H_{14}	l	995.01	–
Methylbenzene (toluene)	C_7H_8	g	943.58	4487.54
Methylbenzene (toluene)	C_7H_8	l	934.50	–
Cycloheptane	C_7H_{14}	l	1086.9	–
n-Heptane	C_7H_{16}	l	1151.27	–
1,2-Dimethylbenzene (*o*-xylene)	C_8H_{10}	g	1098.54	5224.51
1,2-Dimethylbenzene (*o*-xylene)	C_8H_{10}	l	1088.16	–
1,3-Dimethylbenzene (*m*-xylene)	C_8H_{10}	g	1098.12	5222.51
1,3-Dimethylbenzene (*m*-xylene)	C_8H_{10}	l	1087.92	–
1,4-Dimethylbenzene (*p*-xylene)	C_8H_{10}	g	1098.29	5223.32
1,4-Dimethylbenzene (*p*-xylene)	C_8H_{10}	l	1088.16	–
n-Octane	C_8H_{18}	l	1307.53	–
1,3,5-Trimethylbenzene (mesitylene)	C_9H_{12}	l	1241.19	–
Naphthalene	$C_{10}H_8$	s	1231.6	–

ISBN 0-201-08300-0

Table XII (Continued)

Compound	Formula	State	$-\Delta H_c$ kcal/g-mole[b]	$-\Delta H_c$ Btu/scf[c]
n-Decane	$C_{10}H_{22}$	l	1620.06	–
Diphenyl	$C_{12}H_{10}$	s	1493.5	–
Anthracene	$C_{14}H_{10}$	s	1695	–
Phenanthrene	$C_{14}H_{10}$	s	1693	–
n-Hexadecane	$C_{16}H_{34}$	l	2557.64	–
Hydrogen sulfide	H_2S	g	134.46	639.48
Carbonyl sulfide	COS	g	132.21	628.77
Carbon disulfide	CS_2	g	263.52	1253.26

[a] Source: Hougen et al. [42, Chapter 4].

[b] Reference conditions: 25°C (298.16°K), 1 atm, ideal gaseous state.

[c] Reference conditions: 60°F (15.56°C), 30 in. Hg (1.00267 atm, 762.032 mm Hg); 378.48 scf/lb-mole.

ISBN 0-201-08300-0

Author Index

Author Index

Numbers in parentheses indicate the numbers of the references when these are cited in the text without the name of the author. Numbers set in *italics* designate those page numbers on which the complete literature citations are given.

ISBN 0-201-08300-0

ISBN 0-201-08300-0

ISBN 0-201-08300-0

ISBN 0-201-08300-0

ISBN 0-201-08300-0

ISBN 0-201-08300-0

ISBN 0-201-08300-0

ISBN 0-201-08300-0

ISBN 0-201-08300-0

ISBN 0-201-08300-0

ISBN 0-201-08300-0

Subject Index

Subject Index

ISBN 0-201-08300-0

ISBN 0-201-08300-0

ISBN 0-201-08300-0

ISBN 0-201-08300-0

ISBN 0-201-08300-0

ISBN 0-201-08300-0

About the Contributors

Subhash Dutta is an Assistant Professor of Chemical Engineering at the New Jersey Institute of Technology. He has actively published in the area of chemical kinetics, coal combustion and gasification, and fluidization for the past several years. He is the recipient of a 1970 UNESCO Scholarship from the Government of Japan and a 1975 National Scholarship from the Government of India.

Sabri Ergun holds B.S. and M.S. degrees from Columbia University and the Sc.D. degree from the Technical University of Vienna. He has been a member of the Coal Research Laboratory of Carnegie Institute of Technology; Chief of Special Coal Research and Manager of Solid State Physics at the U.S. Bureau of Mines; Visiting Professor at the University of Karlsruhe; and Principal Consulting Engineer at Bechtel Corporation. Currently he is Program Manager of Coal and Synfuels Projects at Lawrence Berkeley Laboratory of the University of California. He has published more than 200 research papers, served as Associate Editor of *Carbon,* and was the recipient of the Pettinos Award of the American Carbon Society.

Robert H. Essenhigh is Bailey Professor of Energy Conversion and Professor of Mechanical Engineering at the Ohio State University. Dr. Essenhigh has published close to 100 papers and contributed chapters to a number of other volumes. He has done extensive consulting with industry in matters related to fuels and combustion engineering and was awarded the Bone and Wheeler Medal and NIFES Prize of the (London) Institute of Fuel in 1959. A native of England, he was on the Fuel Science faculty at the Pennsylvania State University from 1961 to 1978.

E. Stanley Lee holds M.S. and Ph.D. degrees in Chemical Engineering from North Carolina State University and Princeton University, respectively. He has been a member of the research group of Phillips Petroleum Company and Professor of Chemical and Electrical Engineering at the University of Southern California, Los Angeles. Dr. Lee joined Kansas State University in 1966 and is currently a professor of industrial engineering. In addition to teaching, he is the